AF830857

Corticotropin-Releasing Factor: Basic and Clinical Studies of a Neuropeptide

Editors

Errol B. De Souza
Chief, Neurobiology Laboratory
National Institute on Drug Abuse
Addiction Research Center
Baltimore, Maryland

Charles B. Nemeroff
Professor
Department of Psychiatry
Division of Biological Psychiatry
Duke University Medical Center
Durham, North Carolina

CRC Press
Taylor & Francis Group
Boca Raton London New York

CRC Press is an imprint of the
Taylor & Francis Group, an informa business

First published 1990 by CRC Press
Taylor & Francis Group
6000 Broken Sound Parkway NW, Suite 300
Boca Raton, FL 33487-2742

Reissued 2018 by CRC Press

© 1990 by CRC Press, Inc.
CRC Press is an imprint of Taylor & Francis Group, an Informa business

No claim to original U.S. Government works

This book contains information obtained from authentic and highly regarded sources. Reasonable efforts have been made to publish reliable data and information, but the author and publisher cannot assume responsibility for the validity of all materials or the consequences of their use. The authors and publishers have attempted to trace the copyright holders of all material reproduced in this publication and apologize to copyright holders if permission to publish in this form has not been obtained. If any copyright material has not been acknowledged please write and let us know so we may rectify in any future reprint.

Except as permitted under U.S. Copyright Law, no part of this book may be reprinted, reproduced, transmitted, or utilized in any form by any electronic, mechanical, or other means, now known or hereafter invented, including photocopying, microfilming, and recording, or in any information storage or retrieval system, without written permission from the publishers.

For permission to photocopy or use material electronically from this work, please access www.copyright.com (http://www.copyright.com/) or contact the Copyright Clearance Center, Inc. (CCC), 222 Rosewood Drive, Danvers, MA 01923, 978-750-8400. CCC is a not-for-profit organization that provides licenses and registration for a variety of users. For organizations that have been granted a photocopy license by the CCC, a separate system of payment has been arranged.

Trademark Notice: Product or corporate names may be trademarks or registered trademarks, and are used only for identification and explanation without intent to infringe.

Library of Congress Cataloging-in-Publication Data

Corticotropin-releasing factor.

Includes bibliographies and index.
1. Corticotropin releasing hormone. I. De Souza, Errol B. II. Nemeroff, Charles B. [DNLM: 1. Corticotropin Releasing Hormone. WK 515 C8293]
QP572.C62C67 1990 599'.019'2456 89-9933
ISBN 0-8493-4550-2

A Library of Congress record exists under LC control number: 89009933

Publisher's Note
The publisher has gone to great lengths to ensure the quality of this reprint but points out that some imperfections in the original copies may be apparent.

Disclaimer
The publisher has made every effort to trace copyright holders and welcomes correspondence from those they have been unable to contact.

ISBN 13: 978-1-315-89183-5 (hbk)
ISBN 13: 978-1-351-07093-5 (ebk)

Visit the Taylor & Francis Web site at http://www.taylorandfrancis.com and the
CRC Press Web site at http://www.crcpress.com

FOREWORD

The chemical and biological characterization of corticotropin-releasing factor (CRF) has evolved from two fundamental concepts developed over 30 years ago. The investigations of Scharrer and Scharrer, Harris, and others supported the theory that factors secreted by hypothalamic neurons into the hypothalamo-hypophysial portal vessel system could then reach and modulate the functions of the anterior pituitary gland. Evidence for the existence of a hypothalamic factor controlling ACTH secretion was provided by the *in vivo* studies of Harris, Hume, Fortier, and others and by the *in vitro* experiments of Guillemin and Saffron and Schally. The other key concept, derived from the work of Bernard, Cannon, and Selye, is that when exposed to real or perceived challenge to its internal homeostasis, an organism reacts with an array of hormonal, autonomic, and behavioral responses. Among the principle "stress responses" is the activation of the pituitary-adrenal cortical axis.

Although biologically recognized in the 1950s, the chemical identity of CRF was not elucidated until 1981 when our group reported the primary sequence of ovine CRF, a 41 amino acid straight-chain polypeptide which is a highly potent stimulator of the biosynthesis and secretion of ACTH and other proopiomelanocortin (POMC) derived peptides such as β-endorphin. The development of improved cell culture-based bioassays, the recognition and identification of some of the other substances including vasopressin and epinephrine in hypothalamic extracts capable of releasing ACTH, and the improvement of purification and sequencing techniques contributed to the success of this effort. The following year, the structures of rat CRF isolated by us from rat hypothalamus and human CRF deduced by Numa's group using human genomic clones were found to be identical to one another. More recently, the sequences of bovine, carpine, and porcine CRF were determined.

High affinity CRF binding sites, and cellular actions of CRF have been described for tissues in which the peptide has been localized. CRF activates adenylate cyclase and modifies firing rates in several brain regions in which receptors have been identified by membrane binding and autoradiographic techniques.

In addition to activation of the pituitary-adrenal axis, stressful stimuli evoke many other endocrine, autonomic, and behavioral responses. The intracerebroventricular administration of CRF provokes several stress-like responses including activation of the sympathetic nervous system and inhibition of the parasympathetic nervous system with consequential increases in heart rate, blood pressure, blood sugar, and inhibition of gastrointestinal functions. Furthermore, elevating CRF levels in the brain inhibits the secretion of gonadotropins and growth hormone by modulating the production of gonadotropin-releasing hormone and somatostatin. Finally, significant effects of CRF on behavior have been noted and many are consistent with increased arousal and emotional reactivity to the environment. Recent experiments using CRF antagonists support the hypothesis that endogenous CRF participates in many of the neuroendocrine, autonomic, and behavioral responses to stress. These observations along with evidence for increased production of CRF in certain human affective disorders as revealed by inappropriate pituitary-adrenal activation or elevated CRF levels in the cerebrospinal fluid have raised questions concerning the role of the peptide in conditions such as anorexia nervosa and depression. Future fundamental work on the structure of the CRF receptor(s), on the development of improved CRF receptor antagonists for physiologic and clinical studies, on the functions of brain CRF pathways in behavior and in coordinating the stress response may yield improved pharmacologic tools and provide insight concerning the nature of stress-related diseases.

Wylie Vale
Clayton Foundation Laboratories for Peptide Biology
The Salk Institute for Biological Studies
La Jolla, California

THE EDITORS

Errol B. De Souza, Ph.D., is Chief, Neurobiology Laboratory, Addiction Research Center, National Institute on Drug Abuse, Alcohol, Drug Abuse and Mental Health Administration, Baltimore, Maryland. He also holds a joint appointment as Assistant Professor of Pathology, Johns Hopkins University School of Medicine, Baltimore, Maryland.

Dr. De Souza graduated in 1976 from the University of Toronto with a B.A. degree in Physiology and obtained his Ph.D. degree in 1982 from the Department of Physiology, University of Toronto, Toronto, Canada. Subsequently, in 1983, he spent 2 years in the Department of Neuroscience at Johns Hopkins University School of Medicine where he carried out extensive work imaging receptors using autoradiographic techniques.

Dr. De Souza is a member of the American Association for the Advancement of Science, the Endocrine Society, the International Brain Research Organization, and the Society for Neuroscience.

Dr. De Souza has presented over 40 invited lectures and seminars at national and international meetings and at universities and institutes. He has published approximately 90 research papers and 100 abstracts. His current major research interests have focused on the identification, characterization, and localization of various neurotransmitter receptors in brain and in the periphery. In addition, he has carried out extensive studies examining changes in receptors and receptor-mediated second messengers in neurodegenerative diseases and neuropsychiatric disorders.

Charles B. Nemeroff, M.D., Ph.D., was born in New York City almost 40 years ago and was educated in the New York City Public School System. After graduating from the City College of New York in 1970, he briefly enrolled as a graduate student in evolutionary biology at the American Museum of Natural History. An interest in neuroscience, however, led him to a research assistant's position at the Neurochemistry Laboratory at McLean Hospital in Belmont, Massachusetts. He simultaneously enrolled in graduate school at Northeastern University and received a Master's degree in Biology in 1973. He then relocated to North Carolina and enrolled in the Ph.D. program in Neurobiology at the University of North Carolina at Chapel Hill. His Ph.D. work on thyrotropin-releasing hormone was performed under the direction of a psychiatrist, Arthur J. Prange, Jr. After a year of postdoctoral training in neurochemistry, he enrolled in medical school at the University of North Carolina in Chapel Hill. After residency training in psychiatry both at the University of North Carolina and at Duke University, he joined the faculty of Duke University and is currently Professor of Psychiatry, Pharmacology and Neurobiology and Chief of the Division of Biological Psychiatry at Duke University. He has been certified in psychiatry from the American Board of Psychiatry and Neurology. He has published approximately 260 research reports and reviews, has nine edited books published or in press, and has been an invited speaker on more than 200 occasions. He has received a number of awards for his research including a MERIT Award from the National Institute of Mental Health, the Anna Monika Prize for Research in Endogenous Depression (1987), the Daniel H. Efron Award of the American College of Neuropsychopharmacology (1987), the Curt P. Richter Award from the International Society of Psychoneuroendocrinology (1985), the Jordi Folch-Pi Award from the American Society for Neurochemistry (1987) and the Judith Silver Memorial Young Scientist Award from the National Alliance for the Mentally Ill (1988). He is on the editorial board of ten journals. Finally, he would, on most days, prefer to be at the seashore with his wife, Melissa, and two children, Matthew and Amanda.

CONTRIBUTORS

Greti Aguilera
Chief, Section on Endocrine Physiology
National Institute of Child Health and Human Development
National Institutes of Health
Bethesda, Maryland

George Battaglia
Assistant Professor
Department of Pharmacology
Loyola University of Chicago
Stritch School of Medicine
Maywood, Illinois

Garth Bissette
Assistant Professor and Co-Director
Laboratory of Psychoneuroendocrinology
Duke University Medical Center
Durham, North Carolina

Karen T. Britton
Department of Psychiatry
San Diego Veterans Administration Medical Center
University of California School of Medicine
La Jolla, California

Marvin R. Brown
Professor
Departments of Medicine and Surgery
University of California Medical Center
San Diego, California

Pilar Carvallo
Visiting Fellow
Department of Endocrinology
National Institute of Child Health and Human Development
National Institutes of Health
Bethesda, Maryland

Kevin J. Catt
Chief
Endocrinology and Reproduction Research Branch
National Institutes of Health
Bethesda, Maryland

Errol B. De Souza
Chief
Neurobiology Laboratory
Neuroscience Branch
National Institute on Drug Abuse
Addiction Research Center
Baltimore, Maryland

James O. Douglass
Assistant Staff Scientist
Vollum Institute for Advanced Biomedical Research
Assistant Professor
Departments of Anatomy and Neurology
Oregon Health Sciences University
Portland, Oregon

Cindy L. Ehlers
Assistant Member
Department of Neuropharmacology
Research Institute of Scripps Clinic
La Jolla, California

Laurel A. Fisher
Assistant Professor
Department of Pharmacology
University of Arizona Health Sciences Center
Tucson, Arizona

Mirza Flores
Visiting Fellow
National Institute of Child and Human Development
National Institutes of Health
Bethesda, Maryland

Thackery S. Gray
Associate Professor
Department of Anatomy
Loyola Stritch School of Medicine
Maywood, Illinois

Dimitri E. Grigoriadis
Visiting Staff Fellow
Neuroscience Branch
National Institute on Drug Abuse
Addiction Research Center
Baltimore, Maryland

Mark M. Gunion
Assistant Research Psychologist
Department of Medicine and Brain Research Institute
Los Angeles, California
Sepulveda VA Medical Center
Sepulveda, California

James P. Harwood
Executive Secretary
National Institute of Aging
National Institutes of Health
Bethesda, Maryland

Edward Herbert (Deceased)
Vollum Institute for Advanced Biomedical Research
Oregon Health Sciences University
Portland, Oregon

Florian Holsboer
Professor and Chairman
Department of Psychiatry
University of Freiburg
Freiburg, West Germany

Thomas R. Insel
Senior Staff Scientist
Laboratory of Clinical Science
National Institute of Mental Health
National Institutes of Health
Poolesville, Maryland

Ned H. Kalin
Director of Research
Department of Psychiatry
University of Wisconsin Medical School
Madison, Wisconsin

George F. Koob
Associate Member
Department of Neuropharmacology
Research Institute of Scripps Clinic
La Jolla, California

Allen S. Levine
Deputy ACOS for Research
Department of Research
Veterans Administration Medical Center
Minneapolis, Minnesota

Samuel M. McCann
Professor and Director, Neuropeptide Division
Department of Physiology
University of Texas Health Science Center
Dallas, Texas

Monica Millan
Staff Fellow
National Institute of Child and Human Development
National Institutes of Health
Bethesda, Maryland

John E. Morely
Professor
Department of Medicine
St. Louis University Medical School
St. Louis, Missouri

Charles B. Nemeroff
Professor
Department of Psychiatry
Division of Biological Psychiatry
Duke University Medical Center
Durham, North Carolina

Michael J. Owens
Postdoctoral Fellow
Department of Psychiatry
Duke University Medical Center
Durham, North Carolina

Richard E. Powers
Assistant Professor
Departments of Psychiatry and Pathology
University of Alabama School of Medicine
Birmingham, Alabama

Donald L. Price
Departments of Pathology, Psychiatry, Neurology, and Neuroscience
Johns Hopkins University School of Medicine
Baltimore, Maryland

Catherine Rivier
Associate Research Professor
Clayton Foundation Laboratories for Peptide Biology
The Salk Institute
La Jolla, California

Paul E. Sawchenko
Associate Professor
Laboratory of Neuronal Structure and Function
The Salk Institute
San Diego, California

Audrey F. Seasholtz
Assistant Research Scientist
Mental Health Research Institute
University of Michigan
Ann Arbor, Michigan

George Robert Siggins
Department of Neuropharmacology
Research Institute of Scripps Clinic
La Jolla, California

Mark Smith
Medical Staff Fellow
Clinical Neuroendocrinology Branch
National Institute of Mental Health
National Institutes of Health
Bethesda, Maryland

Robert Stephens
Assistant Professor
The Ohio State University
Department of Physiology
College of Medicine
Columbus, Ohio

Larry W. Swanson
Senior Member
Neural System Laboratory
The Salk Institute
La Jolla, California

Yvette Taché
Professor in Residence
Department of Medicine
VA Medical Center, West Los Angeles
University of California at Los Angeles
Los Angeles, California

Robert C. Thompson
Postdoctoral Fellow
Mental Health Research Institute
University of Michigan
Ann Arbor, Michigan

Wylie Vale
Professor and Head
Clayton Foundation Laboratories for Peptide Biology
The Salk Institute
La Jolla, California

Rita J. Valentino
Assistant Professor
Department of Mental Health Sciences
Hahnemann University
Philadelphia, Pennsylvania

Ulrich von Bardeleben
Department of Psychiatry
University of Freiburg
Freiburg, West Germany

Elizabeth Webster
Graduate Student
Department of Neuroscience
University of Florida School of Medicine
Gainesville, Florida

Peter J. Whitehouse
Director
Alzheimer Center
Department of Neurology
Case Western Reserve University
Cleveland, Ohio

W. Scott Young, III
Senior Staff Fellow
Laboratory of Cell Biology
National Institute of Mental Health
Bethesda, Maryland

TABLE OF CONTENTS

Chapter 1

CLONING AND DISTRIBUTION OF EXPRESSION OF THE RAT CORTICOTROPIN - RELEASING FACTOR (CRF) GENE

Robert C. Thompson, Audrey F. Seasholtz, James O. Douglass, and Edward Herbert

TABLE OF CONTENTS

I. INTRODUCTION

In 1955, Guillemin and Rosenberg[1] and Saffran and Schally[2] demonstrated the presence of hypothalamic factors that stimulated the secretion of adrenocorticotropic hormone (ACTH) by the pituitary gland. In 1981, Vale et al.[3] isolated, sequenced and characterized a 41 amino acid ovine hypothalamic peptide. This peptide has been shown to be a very potent secretagogue for ACTH and appears to be the major physiological corticotropin releasing factor (CRF). Additionally, the CRF peptide has been implicated in mediating behavioral activation,[4] the stimulation of the sympathetic nervous system,[5] and the inhibition of feeding[6] and sexual behaviors[7] through central nervous system pathways.

The CRF peptide has been anatomically localized in the paraventricular nucleus of the hypothalamus.[8-12] These neurons communicate with the anterior pituitary gland by secreting CRF into the portal blood system in response to the appropriate stimulus, for example, stress. Additionally, immunoreactive CRF has been demonstrated in other regions of the brain and in peripheral tissues[8-10,12-19] where it may be involved in mediating other aspects of the mammalian stress response.

Since 1983, the nucleotide sequences encoding the ovine[20] and rat[21] CRF cDNA precursors as well as the human[22] and rat[23] CRF genes have been determined. With the report of the nucleotide sequence of the ovine CRF cDNA,[20] the structure of the CRF protein precursor was revealed. The carboxy-terminus of the precursor contained the CRF peptide sequence preceded by a pair of basic amino acids. Additional pairs of basic amino acids were also found within the prohormone and may represent other proteolytic cleavage sites which produce novel peptides. Later in the same year the isolation and nucleotide sequence of the human CRF gene[22] was reported. Comparisons between the ovine and human sequences revealed many interesting findings. The ovine and human prohormones contained 190 and 196 amino acids, respectively, and both precursors contained the CRF peptide at the carboxy terminus preceded by a pair of basic amino acids and followed by a Gly-Lys. The human and ovine signal peptide regions were highly homologous at both the amino acid (91.6%) and nucleotide (92%) levels. However, the nucleotide sequence analysis revealed that the human CRF peptide contained seven amino acid substitutions from the ovine sequence, all of which represented changes among chemically similar amino acids and were caused by single nucleotide substitutions except for the Ile/Ala replacement at the carboxy-terminus of CRF.

The nucleotide analysis of the human CRF sequences also provided the first description of the structural organization of this gene including the promoter sequences and intron/exon boundaries. The gene was shown to contain two exons interrupted by an 800 bp intron. The cap site marking the 5′ limit of the mRNA sequence was putatively assigned to nucleotide −985 (55 bp upstream of the site corresponding to the 5′ end of the cloned ovine cDNA), because putative TATA and CAAT boxes are found 23 and 58 bp upstream of the putative capping site, respectively.[22] Although no human CRF cDNA sequence was reported to define the intron/exon boundaries, the first exon was described to contain 169 bp of the 5′ untranslated region of the mRNA with the second exon containing 14 bp of the 5′ untranslated region, the protein coding region and the 3′ untranslated region of the mRNA. The 3′ untranslated region of the human CRF mRNA contains two copies of the poly(A) addition signal (AAUAAA), whereas the ovine cDNA has three signals, suggesting that the human CRF mRNA may be polyadenylated at multiple sites as was previously determined for the ovine CRF mRNA.

In 1985, Jingami et al.[21] isolated and sequenced the rat CRF cDNA from a rat hypothalamic cDNA library. The nucleotide sequence corroborated the previous work of Rivier et al.[24] on the amino acid sequence of the rat CRF peptide. These complementary studies demonstrated that the rat and human CRF peptides were identical with regard to size and amino acid composition. The amino acid homology between human and rat CRF was quite high throughout the precursor and this was reflected by the high degree of homology at the nucleotide level within this region.

In the ovine, human, and rat precursors, the putative proteolytic cleavage site to liberate the CRF peptide appears well conserved. However, other putative proteolytic cleavage sites at pairs of basic amino acids are not. The ovine and human proteins share an additional pair of basic amino acids at positions 116—117 and 123—124, respectively, found 30 amino acids in the ovine and 29 amino acids in the human from the amino terminus of the CRF peptide. However, this pair of basic amino acids is not found in the rat protein precursor being changed from an Arg-Arg (in both human and ovine) to a Gln-Arg. In addition, potential posttranslational modification sites exist in some of these precursors. For example, two N-linked glycosylation sites (Asn-X-Ser[Thr]) are found in the rat precursor (at amino acid 71—73 and 81—83), one in the human (at amino acid 80—82) and none in the ovine. These protein precursors also contain a relatively high percentage of proline (9.1% in the rat precursor, 12.8% in the human, and 12.1% in the ovine); this may reflect some interesting structural constraints on the overall configuration of these precursors.

In 1987, the rat CRF gene[23] was isolated and sequenced. Comparisons between the rat and the human genes, the tissue-specific expression, and promoter element analysis of the rat CRF gene will be further elaborated in the remainder of this chapter.

II. CLONING OF THE RAT CRF GENE

The rat CRF gene was isolated employing a two-step strategy. The first step was to isolate a full length CRF cDNA clone and the second step involved the use of this cDNA clone as a hybridization probe to isolate the rat CRF gene from a rat λ genomic library.

Approximately 800,000 λgt-10 cDNA clones were screened in duplicate with a ^{32}P-labeled oligonucleotide (5′-TGGAAGGTGAGATCCAGGGA-3′ synthesized on an Applied Biosystems 300A DNA Synthesizer) complementary to the human CRF gene sequence[22] between bases 478 and 497 found within the CRF peptide encoded region. Hybridization positive clones (in duplicate) were isolated and plaque purified; the cDNA inserts were subcloned into M13 vectors and the nucleotide sequences were determined. The largest rat proCRF cDNA included all but 120 bases of the 5′ untranslated sequence of the previously reported CRF mRNA. A 976 base pair (bp) *Eco*RI insert from this cDNA was subcloned into pUC18 for use as a hybridization probe for genomic Southern analysis and into pSP65 for cRNA synthesis.

A rat lambda genomic library was screened using the 976 bp rat CRF cDNA insert as a hybridization probe, yielding 15 hybridization positives in duplicate. The DNA from these recombinant phages was purified and subjected to Southern blot analysis[25] using probes prepared from the cDNA sequence. Three classes of clones were detected, but all clones contained common *Pst*I, *Sal*I and *Eco*RI restriction fragments. These data suggested that the three classes of clones contained different genomic DNA fragments representing the same CRF gene. To confirm the number of CRF genes present in the rat genome, rat genomic DNA was digested with various restriction endonucleases and analyzed by Southern blot analysis using the 976 bp rat CRF cDNA insert as hybridization probe. This analysis demonstrated the presence of unique *Eco*RI and *Hind*III fragments and appropriately sized *Bam*HI and *Pvu*II fragments supporting the hypothesis that only one CRF gene is present in the rat genome.

III. ANALYSIS OF THE RAT CRF GENE SEQUENCE

The rat and human CRF genes contain the same basic organization having two exons separated by an intervening sequence. The first exon of the rat CRH gene contains approximately 160 bp of the 5′ untranslated region of the mRNA. Exon II contains 15 bp of the 5′ untranslated region of the mRNA, the protein coding region, and the complete 3′ untranslated region of the mRNA. The nucleotide sequences at intron/exon boundaries in the rat CRF gene are well conserved across species and follow the donor and acceptor rules as described by Breathnach et al.[26] Two sets of TATAA and CAAT sequences are found in both the human and rat CRF 5′ flanking regions. In the rat the set found nearest to the putative cap site is positioned 23 bp (TATAA) and 60 bp (CAAT) 5′ to this cap site. The second set is found 191 bp (TATAA) and 268 bp (CAAT) 5′ to the putative cap site. The 3′ region of the rat gene has four poly(A) addition signals (AATAAA).

Nucleotide comparisons between the rat and human sequences indicated that these two genes were highly conserved. Two regions of these genes were particularly homologous at the nucleotide level, the CRF peptide encoded region (94%) and the 5′ flanking DNA region (94%). The homology in the CRF peptide region was not unexpected, but the nucleotide homology in the promoter region was very intriguing. The homology is shown in Figure 1. DNA sequence elements responsible for glucocorticoid regulation,[27-29] tissue specific expression,[30,31] cAMP and phorbol ester regulation,[32-34] and enhancer activity[35] have all been localized to 5′ flanking DNA sequences. These highly conserved 5′ sequences may therefore represent DNA regulatory elements for CRF gene expression.

IV. DISTRIBUTION OF EXPRESSION OF THE RAT CRF GENE

As was mentioned earlier, the CRF peptide has been immunologically detected in many regions of the central nervous system as well as some peripheral tissues.[8-19] Northern blot analysis was performed in order to determine whether this immunoreactivity was synthesized in these regions or transported to these sites from other regions of synthesis.[23] A summary of these data is presented in Table 1. In the rat brain, CRF mRNA is detected in the brainstem (medulla/pons), cerebral cortex, hypothalamus, midbrain, striatum, and hippocampus. CRF mRNA was not detectable in the cerebellum. The rat brain CRF mRNA was found to be approximately 1400 bases in length (data not shown) in all brain areas tested. In the nonbrain tissues analyzed, CRF transcripts were detected in the adrenal gland (whole adrenal), spinal cord, testis, and pituitary gland (whole pituitary). CRF mRNA was undetectable in the liver, thymus, duodenum, and kidney. In general, the presence of CRF mRNA in both brain and nonbrain tissues corresponds to the presence of immunoreactive CRF with the exception of the testis which appears to indicate a novel region of synthesis.

RNA blot analysis of the nonbrain tissue samples indicated that the size of the CRF mRNA was variable. The spinal cord and pituitary transcript sizes were comparable to those found in the brain (1400 bases); however, the adrenal and testicular transcripts were larger by approximately 200 and 700 bases, respectively. Size differences in transcripts produced in different tissues have been reported for a number of other genes including the proopiomelanocortin gene[36,37] and the rat prodynorphin gene.[38] The size differences of the mRNAs may be due to alternate utilization of poly(A) addition signals, alternate splicing, or aberrant initiation of transcription.

V. TRANSCRIPTIONAL REGULATION OF THE RAT CRF GENE

The regulatory mechanisms and factors which act as modulators of CRF mRNA levels have not been well characterized. Jingami et al.[39] reported an increase in hypothalamic CRF mRNA levels 7 d after adrenalectomy to 152% of control levels; this increase could be prevented by dexamethasone administration. Young[40] has used *in situ* hybridization to monitor changes in CRF mRNA levels in the rat hypothalamus under salt loading and dehydration stimuli (see Chapter 2, Section II.B). Secretion studies have suggested that catecholamines,[41]

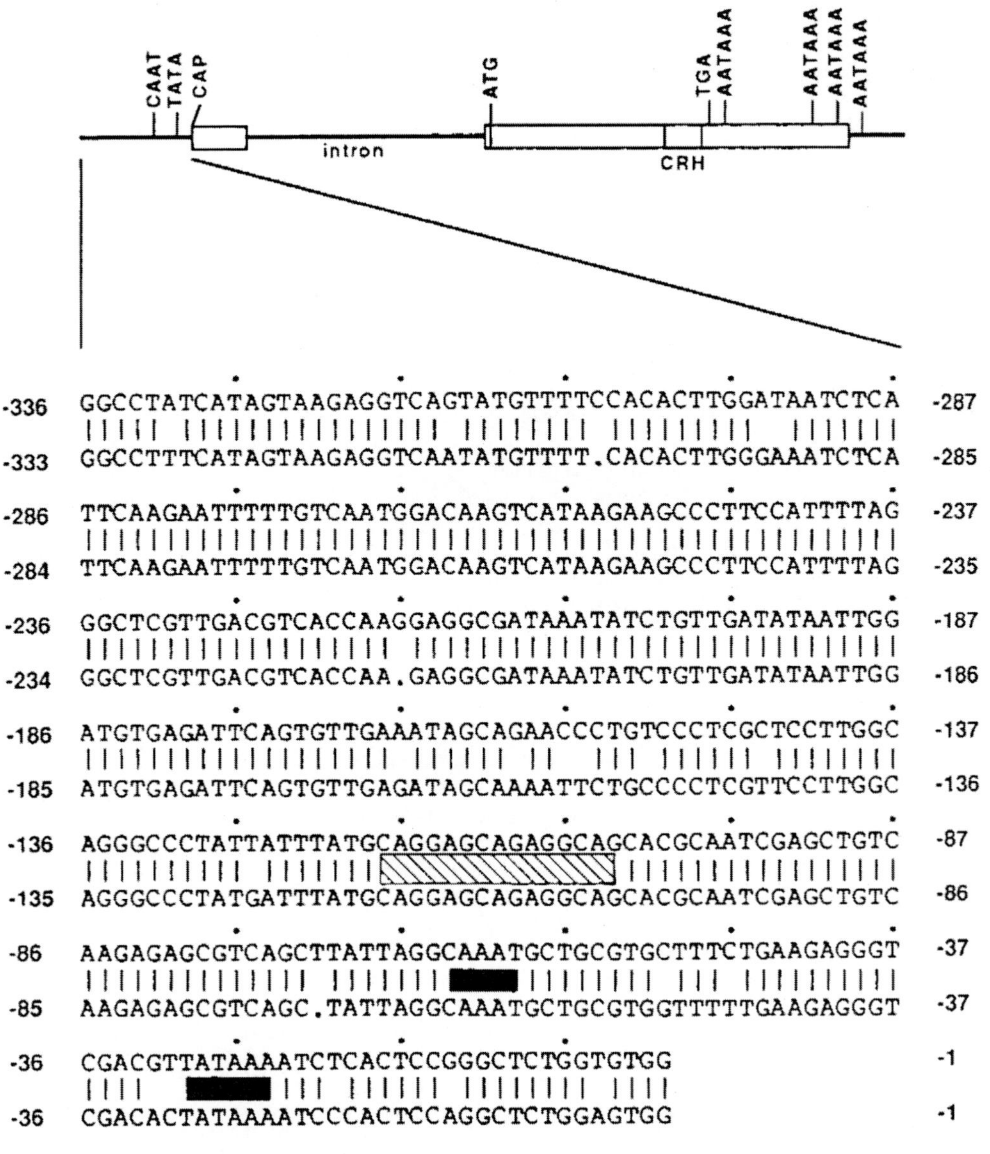

FIGURE 1. Nucleic acid sequence homology between 5′ flanking sequences of rat and human proCRF genes. A schematic representation of the rat CRF gene is shown at the top. The exons are shown as blocks and the intron by a line. The TATA and CAAT sequences, putative cap site, translation initiation ATG, translation terminator TGA, and poly(A) addition signals (AATAAA) are indicated. The location of the CRF peptide is indicated by CRF. The nucleotide sequences of the 5′ flanking DNA from the rat (upper sequence −336 bp) and human (lower sequence −333 bp) CRF genes are shown at the bottom of the figure. The vertical lines denote conserved nucleotides. The TATA and CAAT boxes are shown in black. The location of a putative cAMP "consensus" sequence is indicated by the striped box. Nucleotide −1 corresponds to the first nucleotide 5′ to the putative CRF mRNA cap site.

opioid peptides,[42-45] acetylcholine,[41] serotonin,[42] and angiotensin II[46] alter the secretion of CRF. It is currently not known whether they also alter the transcription of the CRF gene. Finally, the molecular mechanisms responsible for regulation of CRF transcription by any of these compounds are unknown.

New methods in molecular biology have provided us with the tools to examine these questions. The availability of cloned genes and the development of new methods for efficient

TABLE 1
Tissue Specific Expression of CRF mRNA

Brain Regions	
Brainstem (medulla/pons)	+
Cerebellum	−
Cerebral cortex	+
Hippocampus	+
Hypothalamus	+
Midbrain	+
Striatum	+
Nonbrain Regions	
Adrenal gland (whole)	+*
Duodenum	−
Kidney	−
Liver	−
Pituitary gland (whole)	+
Spinal cord	+
Testis	+*
Thymus	−

Note: Northern blot analysis was performed as previously described[23] using poly(A) RNA from various brain regions and peripheral tissues. The results of the Northern analysis are summarized here with + representing the presence of detectable CRF mRNA and − indicating the absence of detectable CRF mRNA. Asterisks denote the presence of CRF mRNA of a different size; CRF transcripts in the brain are approximately 1400 bases in length while adrenal and testis CRF mRNA are approximately 1600 and 2100 bases, respectively.

transfer of DNA into eukaryotic cells have revolutionized the field of gene regulation. Gene transfer methods are now available which allow us to study regulation of expression at a variety of levels (i.e., transcription, processing, posttranslational modifications, and secretion). Such techniques have allowed investigators to determine that transcriptional regulation appears to be controlled by a variety of *cis-* and *trans*-acting regulatory elements. The DNA sequences representing *cis*-acting elements are quite diverse in function and structure, and may affect the general stimulation of transcription (promoters and enhancers), tissue-specific gene expression, or the induction (or repression) of transcription by the action of specific pharmacological compounds. The *trans*-acting regulatory elements are often specific protein factors which interact with the *cis*-acting DNA sequences. ''Activated'' receptors (i.e., glucocorticoid receptor) or cellular phosphoproteins which translocate to the nucleus and interact with specific genomic DNA sequences to modulate transcription of specific genes are examples of *trans*-acting factors.

Gene transfer techniques can be divided into two classes: (1) stable transformants and

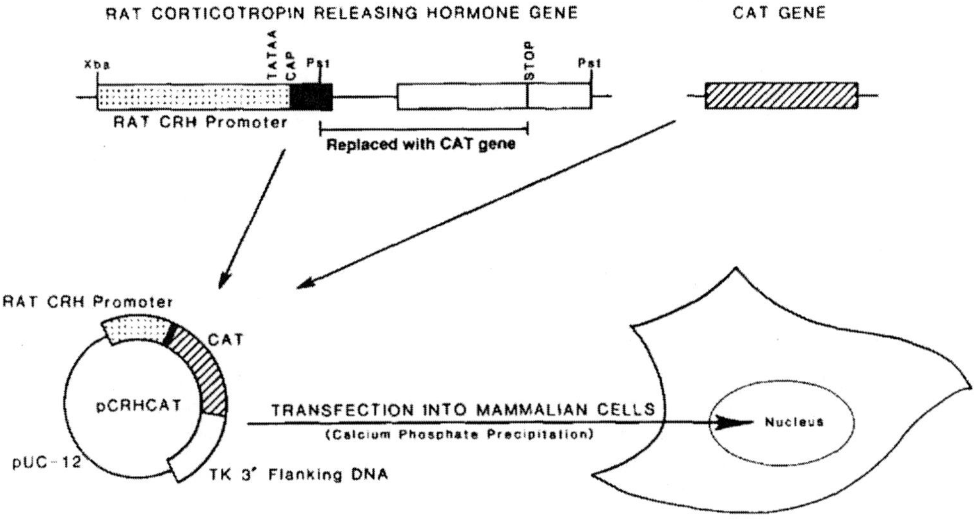

FIGURE 2. Schematic representation of CRHCAT fusion gene construction and introduction into eukaryotic cells. The rat CRF gene is diagrammed on the upper left with the 5′ flanking sequence shown as a stippled box and 5′ untranslated region of the CRF mRNA as a black box. Various lengths of CRF 5′ flanking sequence and 90 bp of 5′ untranslated CRF mRNA are linked to the *E. coli* chloramphenicol acetyltranferase (CAT) gene to create a number of CRHCAT plasmids. The 3′ untranslated sequence, poly(A) addition signals, and 3′ flanking sequence are from the herpes thymidine kinase (tk) gene. (Adapted from Thompson, R. C., Seasholtz, A. F., Douglass, J. O., and Herbert, E., *Ann. N.Y. Acad. Sci.*, 512, 1, 1988.)

(2) transient assay systems. To study stably transfected cell lines, one isolates a clone of eukaryotic cells that has integrated the foreign gene into its chromosomal DNA. At the present time, we are unable to control the region of the genome into which the introduced gene is integrated; thus, different clones carrying the same gene may show significant variability in protein and RNA production depending upon differences in location of the foreign gene in the host chromosome. In transient expression systems, the foreign gene is introduced into eukaryotic cells and its expression is assayed within 24 to 72 h before it can integrate into the host DNA. These transient assay systems can be used to rapidly examine the regulation and expression of cloned genes in a variety of cellular environments. Also, by using site-directed mutagenesis, one can examine the role of specific nucleotide sequences in controlling the expression of the gene.

A number of transient expression systems are presently available for studying transcriptional regulation. We have utilized the fusion gene system in which the control region of the gene of interest is fused to the coding sequence of a gene whose protein product is readily assayed by enzymatic means. The amount of enzyme activity reflects the amount of protein produced under the control of the linked regulatory region. The *E. coli* chloramphenicol acetyltransferase (CAT) reporter function developed by Gorman et al.[47] was selected for these studies. By fusing the 5′ flanking sequence of the rat CRF gene to the coding region of the CAT gene, we can analyze the efficiency of the CRF promoter, test for tissue-specific elements, and map DNA control elements responsible for alterations in transcription via regulatory compounds (i.e., steroid hormones, peptide hormones, and neurotransmitters). A summary of this strategy is shown in Figure 2.

Many extracellular signals (peptide hormones, neurotransmitters) are recognized through specific membrane receptors, and binding of the signal molecule to the receptor initiates a rapid series of intracellular events that eventually translates this external signal into a specific cellular response such as RNA and/or protein synthesis, secretion, cell division, etc. It is now generally accepted that "second messenger" molecules are involved in this signal transduction, at least partially through the action of protein kinases. Activation of adenylate cyclase by a variety of different hormone receptor-mediated mechanisms causes increased levels of the second messenger cAMP. Elevated cAMP concentrations cause increased cAMP-dependent protein kinase activity, and the resultant changes in protein phosphorylation have been postulated to modulate gene transcription. The regulation of CRF transcription by hormones that mediate their effects through activation or inhibition of adenylate cyclase had not previously been examined *in vivo*. However, the rat CRF 5′ flanking sequence contains a DNA element very homologous to a 15 bp cAMP-responsive "consensus sequence" derived from studies on the human proenkephalin,[32] rat preprosomatostatin,[33] and rat phosphoenolpyruvate carboxykinase genes.[34] This putative CRF cAMP "consensus sequence" is shown in Figure 1. The fusion gene system provides us with a rapid method for testing the ability of this cAMP responsive "consensus sequence" to confer regulation by cAMP mediated pathways on the rat CRF gene.

The first fusion gene construct (CRHCAT-1400) contained 1.4kb of rat CRF 5′ flanking sequence and 90 bp of the 5′ untranslated region of the CRF mRNA linked to the CAT reporter gene (see Figure 2). The 3′ untranslated sequence, termination signals, poly(A) addition sites, and 3′ flanking sequence are from the herpes thymidine kinase (tk) gene. This plasmid DNA was transfected into mammalian cells by the calcium phosphate precipitation method, including a glycerol shock after 4 h to aid in DNA uptake. The cells were harvested after 48 h, lysed with Triton® X-100, and assayed for CAT activity as previously described.[47]

The basal levels of CAT activity were examined after transfection of CRHCAT-1400 into a variety of mammalian cells. In all cases, the basal CAT expression was very low, suggesting that the CRF promoter may be relatively weak. The levels of CAT expression were increased dramatically in each cell line in the presence of 8BrcAMP (a cAMP analog and isobutylmethylxanthine (IBMX), a phosphodiesterase inhibitor which prevents the breakdown of intracellular cAMP. Both the RIN (rat pancreas) and CV-1 (monkey kidney) cells showed a 2.5- to 5-fold increase in CAT activity in the presence of 8BrcAMP and IBMX, while PC-12 (rat pheochromocytoma) cells showed a 30-fold induction. Thus, the rat CRF gene is transcriptionally activated via a cAMP-mediated pathway with varied levels of induction in different cell lines. A second CRHCAT construct (CRHCAT-500) containing 500 bp of CRF 5′ flanking DNA and 90 bp of 5′ untranslated CRF mRNA fused to the CAT gene showed similar levels of induction with 8BrcAMP and IBMX in all cell lines tested.

A third fusion gene (CRHCAT-131) was constructed which contained only 131 bp of rat CRF 5′ flanking DNA (and 90 bp of 5′ untranslated CRF mRNA) fused to the CAT gene. When this plasmid was transfected into PC-12 cells, the basal level of CAT expression was similar to cells transfected with CRHCAT-500, but treatment of the transfected cells with 8BrcAMP and IBMX no longer caused an induction in CAT expression (Figure 3). Similar results were obtained in CV-1 cells transfected with CRHCAT-131. This finding was most interesting, since the CRF DNA sequence homologous to the 15 bp cAMP "consensus" sequence is located at position −104 to −118 bp upstream of the putative CRF mRNA cap site (as shown in Figure 1). Thus, even though this CRHCAT construct contains a sequence similar to the putative cAMP regulatory element, it no longer shows cAMP induction. This observation may be due to the fact that upstream sequences are required for this cAMP consensus sequence element to be functional, since deletions just upstream of the 15 bp cAMP consensus sequence in the human proenkephalin gene also lack observable cAMP induction following transfection.

In addition to its induction by cAMP, the CRHCAT-1400 plasmid has also demonstrated induction by TPA (a phorbol ester) in the presence of IBMX in CV-1 cells, a result similar to that found with the human proenkephalin gene. A preliminary experiment in AtT-20 cells

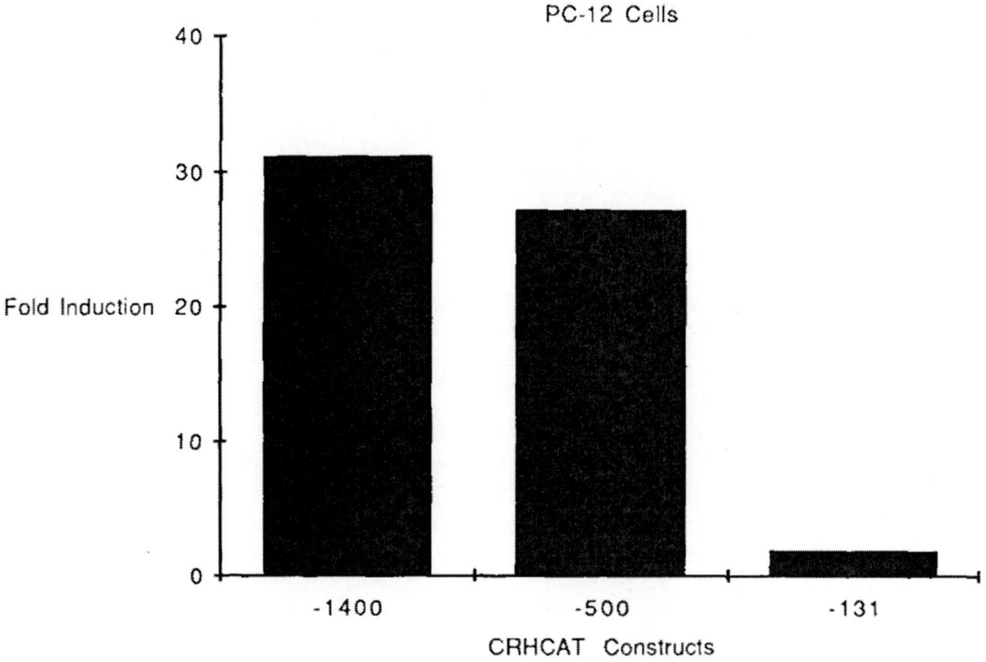

FIGURE 3. Regulation of CRHCAT Expression by cAMP. PC-12 cells were transfected with the three CRHCAT constructs, CRHCAT-1400, CRHCAT-500, and CRHCAT-131. Regulator treatments were for 6 h beginning at 42 h after transfection. Control plates received fresh media while regulated plates received fresh media containing 1 m*M* 8BrcAMP and 0.5 m*M* IBMX. At the end of 6 h cells were harvested and lysed, and CAT assays were performed on the cell extracts as previously described.[32,47] The fold induction in CAT activity by 8BrcAMP and IBMX was determined by dividing CAT activity in the presence of 8BrcAMP and IBMX by the CAT activity in the absence of regulators for each construct.

has also demonstrated not only induction of CAT expression by cAMP, but also suggests that this induction is antagonized by glucocorticoids (dexamethasone). A number of additional CRHCAT constructs have been created containing varying lengths of 5′ flanking sequences in order to localize and define these regulatory elements in a variety of cell lines.

Both genetic and biochemical analyses are necessary to elucidate the molecular mechanisms involved in transcriptional regulation. A complete mutational dissection of a promoter and its flanking sequence is required to map all the *cis*-acting domains in the DNA sequence surrounding a gene. Once a detailed map of *cis*-acting domains in the CRF 5′ flanking sequence is available, careful biochemical analysis of cellular extracts will permit the identification of protein factors that interact with these domains in a sequence specific fashion to affect CRF transcription.

VI. SUMMARY

In this chapter we have described the isolation and characterization of the rat corticotropin releasing factor gene. Nucleotide sequence comparisons with the human CRF gene have demonstrated several interesting regions of homology and suggest that the gene was highly conserved through evolution. Additionally we have described the tissue-specific expression of the rat CRF gene. The regional distribution of expression parallels previously documented immunocytochemical demonstrations and supports the hypothesis that CRF peptides have multiple roles in different tissues.

Additionally we have described initial studies using gene transfer techniques to localize regulatory elements in the 5′ flanking region of the rat CRF gene. We are presently utilizing fusion genes (CRHCAT plasmids) to localize a cAMP-responsive DNA element in the rat CRF gene, and compare it to the previously reported cAMP-responsive "consensus sequence". Similarly, we also plan to utilize the CRHCAT constructs to examine regulation of the rat CRF gene by phorbol esters, glucocorticoids, and other hormone-mediated regulatory pathways. Through these gene transfer studies we hope to gain a better understanding of the role of conserved sequences in the 5′ flanking DNA for transcriptional control of the rat (and human) CRF genes.

ADDENDUM

The cAMP responsive DNA element in the rat CRF gene has recently been localized to a 59 bp region located between 238 and 180 bp 5′ to the putative CRF mRNA cap site (49). The CRHCAT-131 plasmid construct does not contain this sequence, explaining its lack of induction of CAT activity by cAMP analogs. This 59 bp sequence (-238 to -180 bp) can confer cAMP responsiveness on a heterologous promoter in an orientation independent fashion and also has homology to the cAMP responsive element core sequence (TGACGTCA) defined by comparison of cAMP responsive DNA sequences from a number of eukaryotic genes.

REFERENCES

1. **Guillemin, R. and Rosenberg, B.,** Humoral hypothalamic control of anterior pituitary: study with combined tissue cultures, *Endocrinology*, 57, 599, 1955.
2. **Saffran, M. and Schally, A. V.,** The release of corticotropin by anterior pituitary tissue in vitro, *Can. J. Biochem. Physiol.*, 33, 408, 1955.
3. **Vale, W., Spiess, J., Rivier, C., and Rivier, J.,** Characterization of a 41-residue ovine hypothalamic peptide that stimulates secretion of corticotropin and beta-endorphin, *Science*, 213, 1394, 1981.
4. **Sutton, R. E., Koob, G. F., LeMoal, M., Rivier, J., and Vale, W.,** Corticotropin releasing factor produces behavioural activation in rats, *Nature*, 297, 331, 1982.
5. **Brown, M. R., Fisher, L. A., Spiess, J., Rivier, C., Rivier, J., and Vale, W.,** Corticotropin-releasing factor: actions on the sympathetic nervous system and metabolism, *Endocrinology*, 111, 928, 1982.
6. **Britton, D. R., Koob, G. F., Rivier, J., and Vale, W.,** Intraventricular corticotropin-releasing factor enhances behavioral effects of novelty, *Life Sci.*, 31, 363, 1982.
7. **Sirinathsinghji, D. J. S., Rees, L. H., Rivier, J., and Vale, W.,** Corticotropin-releasing factor is a potent inhibitor of sexual receptivity in the female rat, *Nature*, 305, 232, 1983.
8. **Olschowka, J. A., O'Donohue, T. L., Mueller, G. P., and Jacobowitz, D. M.,** Hypothalamic and extrahypothalamic distribution of CRF-like immunoreactive neurons in the rat brain, *Neuroendocrinology*, 35, 305, 1982.
9. **Merchenthaler, I., Vigh, S., Petrusz, P., and Schally, A. V.,** Immunocytochemical localization of corticotropin-releasing factor (CRF) in the rat brain, *Am. J. Anat.*, 165, 385, 1982.
10. **Swanson, L. W., Sawchenko, P. E., Rivier, J., and Vale, W. W.,** Organization of ovine corticotropin-releasing factor immunoreactive cells and fibers in the rat brain: an immunohistochemical study, *Neuroendocrinology*, 36, 165, 1983.
11. **Bloom, F. E., Battenberg, E. L. F., Rivier, J., and Vale, W.,** Corticotropin releasing factor (CRF) immunoreactive neurons and fibers in rat hypothalamus, *Regul. Pept.*, 4, 43, 1982.
12. **Bugnon, C., Fellman, D., Gouget, A., and Cardot, J.,** Ontogeny of the corticoliberin neuroglandular systems in the rat brain, *Nature*, 298, 159, 1982.
13. **Petrusz, P., Merchenthaler, I., Maderdrut, J. L., and Hertz, P. U.,** Central and peripheral distribution of corticotropin-releasing factor, *Fed. Proc.*, 44, 229, 1985.
14. **Merchenthaler, I., Hynes, M. A., Vigh, S., Schally, A. V., and Petrusz, P.,** Immunocytochemical localization of corticotropin releasing factor (CRF) in the rat spinal cord, *Brain Res.*, 275, 373, 1983.
15. **Bruhn, T. O., Engeland, W. C., Anthony, E. L. P., Gann, D. S., and Jackson, I. M. D.,** Corticotropin releasing factor in the dog adrenal medulla is secreted in response to hemorrhage, *Endocrinology*, 120, 25, 1987.

16. **Petrusz, P., Merchenthaler, I., Maderdrut, J. L., Vigh, S., and Schally, A. V.**, Corticotropin releasing factor (CRF)-like immunoreactivity in the vertebrate endocrine pancreas, *Proc. Natl. Acad. Sci. U.S.A.*, 80, 1721, 1983.
17. **Suda, T., Tomori, N., Tozawa, F., Mouri, T., Demura, H., and Shizume, K.**, Distribution and characterization of immunoreactive corticotropin-releasing factor in human tissues, *J. Clin. Endocrinol. Metab.*, 59, 861, 1984.
18. **Morel, G., Hemming, F., Tonon, M-C., Vaudry, H., Dubois, M. P., Coy, D., and Dubois, P. M.**, Ultrastructural evidence for corticotropin-releasing factor (CRF)-like immunoreactivity in the rat pituitary gland, *Cell Biol.*, 44, 89, 1982.
19. **Nieuwenhuyzen Kruseman, A. C., Linton, E. A., Ackland, J., Besser, G. M., and Lowry, P. J.**, Heterogeneous immunocytochemical reactivities of oCRF-41-like material in the human hypothalamus, pituitary and gastrointestinal tract, *Neuroendocrinology*, 38, 212, 1984.
20. **Furutani, Y., Morimoto, Y., Shibahara, S., Noda, M., Takahashi, H., Hirose, T., Asai, M., Inayama, S., Hayashida, H., Miyata, T., and Numa, S.**, Cloning and sequence analysis of cDNA for ovine corticotropin-releasing factor precursor, *Nature*, 301, 537, 1983.
21. **Jingami, H., Mizuno, N., Takahashi, H., Shibahara, S., Furutani, Y., Imura, H., and Numa, S.**, Cloning and sequence analysis of cDNA for rat corticotropin-releasing factor precursor, *FEBS Lett.*, 191, 63, 1985.
22. **Shibahara, S., Morimoto, Y., Furutani, Y., Notake, M., Takahashi, H., Shimizu, S., Horikawa, S., and Numa, S.**, Isolation and sequence of the human corticotropin-releasing factor precursor gene, *EMBO J.*, 2, 775, 1983.
23. **Thompson, R. C., Seasholtz, A. F., and Herbert, E.**, Rat corticotropin-releasing hormone gene: sequence and tissue-specific expression, *Mol. Endocrinol.*, 1, 363, 1987.
24. **Rivier, J., Spiess, J., and Vale, W.**, Characterization of rat hypothalamic corticotropin-releasing factor, *Proc. Natl. Acad. Sci. U.S.A.*, 80, 4851, 1983.
25. **Southern, E. M.**, Detection of specific sequences among DNA fragments separated by gel electrophoresis, *J. Mol. Biol.*, 98, 503, 1975.
26. **Breathnach, R., Benoist, C., O'Hare, K., Gannon, F., and Chambon, P.**, Ovalbumin gene: evidence for a leader sequence in mRNA and DNA sequences at the exon-intron boundaries, *Proc. Natl. Acad. Sci. U.S.A.*, 75, 4853, 1978.
27. **Hynes, N., Van Ooyen, A. J. J., Kennedy, N., Herrlich, P., Ponta, H., and Groner, B.**, Subfragment of the large terminal repeat causes glucocorticoid responsive expression of mouse mammary tumor virus and of an adjacent gene, *Proc. Natl. Acad. Sci. U.S.A.*, 80, 3637, 1983.
28. **Chandler, V. L., Maler, B. A., and Yamamoto, K. R.**, DNA sequences bound specifically by glucocorticoid receptor in vitro render a heterologous promoter hormone responsive in vivo, *Cell*, 33, 489, 1983.
29. **Karin, M., Haslinger, A., Holtgreve, H., Richards, R. I., Krauter, P., Westphal, H. M., and Beato, M.**, Characterization of DNA sequences through which cadmium and glucocorticoid hormones induce human metallothionein-II gene, *Nature*, 308, 513, 1984.
30. **Walker, M. D., Edlund, T., Boulet, A., and Rutter, W. J.**, Cell-specific expression controlled by the 5′-flanking region of insulin and chymotrypsin genes, *Nature*, 306, 557, 1983.
31. **Ciliberto, G., Dente, L., and Cortese, R.**, Cell-specific expression of a transfected human anti-trypsin gene, *Cell*, 41, 531, 1985.
32. **Comb, M. Birnberg, N., Seasholtz, A., Herbert, E., and Goodman, H.**, A cyclic AMP- and phorbol ester-inducible DNA element, *Nature*, 323, 353, 1986.
33. **Montiminy, M. R., Sevarino, K. A., Wagner, J. A., Mandel, G., and Goodman, R. H.**, Identification of a cyclic AMP-responsive element within the rat somatostatin gene, *Proc. Natl. Acad. Sci. U.S.A.*, 83, 6682, 1986.
34. **Short, J. M., Wynshaw-Boris, A., Short, H. P., and Hanson, R. W.**, Characterization of the phosphoenolpyruvate carboxykinase (GTP) promoter-regulatory region, *J. Biol. Chem.*, 261, 9721, 1986.
35. **Gluzman, Y. and Shenk, T., Eds.**, *Enhancers and Eukaryotic Gene Expression*, Cold Spring Harbor Laboratory, Cold Spring Harbor, N.Y., 1983.
36. **Civelli, O., Birnberg, N., and Herbert, E.**, Detection and quantitation of pro-opiomelanocortin mRNA in pituitary and brain tissues from different species, *J. Biol. Chem.*, 257, 6783, 1982.
37. **Jingami, H., Nakanishi, S., Imura, H., and Numa, S.**, Tissue distribution of messenger RNAs coding for opioid peptide precursors and related RNA, *Eur. J. Biochem.*, 142, 441, 1984.
38. **Civelli, O., Douglass, J., Goldstein, A., and Herbert, E.**, Sequence and expression of the rat prodynorphin gene, *Proc. Natl. Acad. Sci. U.S.A.*, 82, 4291, 1985.
39. **Jingami, H., Matsukura, S., Numa, S., and Imura, H.**, Effects of adrenalectomy and dexamethasone administration on the level of preprocorticotropin-releasing factor messenger ribonucleic acid (mRNA) in the hypothalamus and adrenocorticotropin/β-lipotropin precursor mRNA in the pituitary in rats, *Endocrinology*, 117, 1314, 1985.

40. **Young, W. S.,** Corticotropin-releasing factor mRNA in the hypothalamus is affected differently by drinking saline and by dehydration, *FEBS Lett.*, 208, 158, 1986.
41. **Hillhouse, E. W., Burden, J., and Jones, M. T.,** The effect of various putative neurotransmitters on the release of corticotropin releasing hormone from the hypothalamus of the rat *in vitro*. I. The effects of acetylcholine and noradrenaline, *Neuroendocrinology*, 17, 1, 1975.
42. **Owens, M. J., Smith, M. A., and Nemeroff, C. B.,** Neurotransmitter regulation of release of hypothalamic corticotropin releasing factor (CRF), *Neurosci. Abstr.*, 12, 1494, 1986.
43. **Buckingham, J. C. and Cooper, T. A.,** Differences in hypothalamo-pituitary-adrenocortical activity in the rat after acute and prolonged treatment with morphine, *Neuroendocrinology*, 38, 411, 1984.
44. **Yajima, F., Suda, T., Tomori, N., Sumitomo, T., Nakagami, Y., Ushiyama, T., Demuro, H., and Shizume, K.,** Effects of opioid peptides on immunoreactive corticotropin-releasing factor release from the rat hypothalamus in vitro, *Life Sci.*, 39, 181, 1986.
45. **Buckingham, J.,** Stimulation and inhibition of corticotropin releasing factor secretion by beta-endorphin, *Neuroendocrinology*, 42, 148, 1986.
46. **Suda, T., Yajima, F., Tomori, N., Demura, H., and Shizume, K.,** In vitro study of immunoreactive corticotropin releasing factor release from the rat hypothalamus, *Life Sci.*, 37, 1499, 1985.
47. **Gorman, C. M., Moffat, L. F., and Howard, B. H.,** Recombinant genomes which express chloramphenicol acetyltransferase in mammalian cells, *Mol. Cell. Biol.*, 2, 1044, 1982.
48. **Thompson, R. C., Seasholtz, A. F., Douglass, J. O., and Herbert, E.,** The rat corticotropin releasing hormone gene, *Ann. N.Y. Acad. Sci.*, 512, 1, 1988.
49. **Seasholtz, A. F., Thompson, R. C., and Douglass, I. O.,** Identification of a cyclic adenosine monophosphate-responsive element in the rat corticotropin-releasing hormone gene, *Mol. Endocrinol.*, 2, 1311, 1988.

Chapter 2

DISTRIBUTION AND REGULATION OF CORTICOTROPIN-RELEASING FACTOR mRNA IN BRAIN USING *IN SITU* HYBRIDIZATION HISTOCHEMISTRY

W. Scott Young, III

TABLE OF CONTENTS

I. INTRODUCTION

The amino acid sequence of corticotropin-releasing factor (CRF) was established in 1981.[1] A wealth of information has accumulated since on the distribution, pharmacology, and physiology of CRF in the central nervous system and pituitary. As befits its name, CRF has been primarily studied in relationship to the hypothalamic-hypophysial-adrenal axis. Numerous studies (reviewed in this volume) have revealed the presence of CRF in the hypothalamus, transport to and release into the portal blood circulation that flows to the anterior pituitary, and subsequent stimulation of adrenocorticotropin (ACTH) release for control of adrenal function.[2] CRF has also been implicated in various human diseases. In the dementia of the Alzheimer's type, CRF is reduced in the neocortex whereas levels of its receptor are increased.[3,4] In major depression, CRF is increased in the cerebrospinal fluid and adrenocorticotropin and cortisol levels are increased in blood.[5,6] We have pursued our interest in the biosynthesis of CRF by localizing and quantifying the levels of its mRNA in brain tissues of various species, including man, and after various physiological, pharmacological, and surgical manipulations. These studies were made possible by the sequencing of the rat[7] and human[8] complementary DNAs (cDNAs) coding for the CRF precursor. They relied on *in situ* hybridization histochemistry (ISHH) using a synthetic ^{35}S-labeled oligodeoxyribonucleotide probe complementary to the CRF mRNA[9] in a region well conserved between species.[7,8] In the following sections I will describe our current approach to ISHH in detail, present and discuss some of our findings, and suggest areas for future exploration.

II. *IN SITU* HYBRIDIZATION HISTOCHEMISTRY FOR CRF mRNA

Currently, the most widely used application of ISHH is for the localization of mRNA species in tissue sections. This approach relies on the ability to fix mRNA within the tissue sections while maintaining its ability to hybridize with labeled (usually, radioactively) complementary polynucleotide probes.[10] We find the best preservation of tissue and hybridization when sections from fresh-frozen tissue are thaw-mounted and post-fixed with formaldehyde on slides. The sections are next incubated in 0.25% acetic anyhydride and a series of ethanols and chloroform to reduce nonspecific binding of the probe. Subsequently, each section is incubated with 0.5 to 2 $\times$ 10^6 dpm of probe in a hybridization buffer containing 0.66 *M* sodium salts and 50% formamide as well as 10% dextran sulfate, 100 m*M* dithiothreitol, DNA, tRNA, bovine serum albumin, ficoll, and polyvinylpyrrolidone to further reduce nonspecific binding. After the 20 to 24 h incubations at 37°C, the sections are washed in 0.33 *M* sodium salts/50% formamide at 40°C, dried, and apposed to X-ray film (Kodak, X-Omat) or NTB3 (Kodak) nuclear emulsion (by dipping or coated coverslips[11]). Quantitation is obtained by simultaneously exposing calibrated ^{35}S-impregnated brain-paste standards to create a standard curve of radioactivity vs. optical density (film) or light reflectance (nuclear emulsion examined under darkfield microscopy).[9] These readings are obtained and analyzed with a Loats Image Analysis System, RAAS 1000 (Westminster, MD).

The 48 base oligonucleotide probe is directed against bases 496—543 of the rat cDNA (amino acids 22—37 of CRF proper).[7] It has a 94% homology with the corresponding human sequence.[8] The probe was made on an Applied Biosystems DNA synthesizer, purified on an 8% polyacrylamide/8 *M* urea preparative sequencing gel, and labeled on the 3′ end using terminal deoxynucleotidyl transferase (Boerhinger-Mannheim) and ^{35}S-deoxyadenosine 5′-(α-thio)triphosphate (>1000Ci/mmol, New England Nuclear) to specific activities of >3000Ci/mmol.

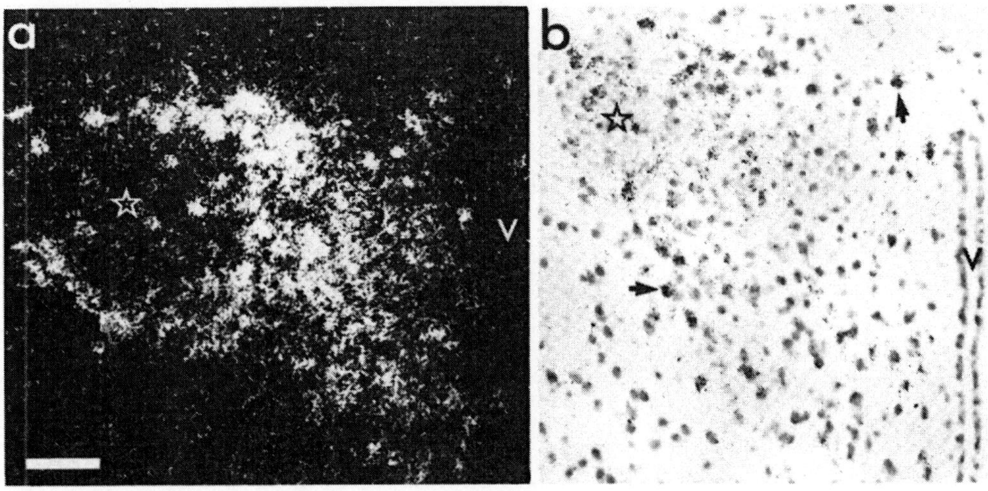

FIGURE 1. CRF mRNA in paraventricular neurons of the rat (a) (white grain clusters in darkfield photomicrograph) and mouse (b) hypothalami demonstrated by *in situ* hybridization histochemistry. In the rat, labeled cells are located around the vasopressin magnocellular core (star) and in the parvocellular region (densest labeling) between the ventricle (v) and the magnocellular core. Labeled cells (arrows point to two examples) in the mouse are more dispersed. Bar equals 100 µm for a and 60 µm for b.

III. CRF mRNA LOCALIZATION AND QUANTITATION

A. DISTRIBUTION IN THE HYPOTHALAMUS OF RAT AND OTHER SPECIES

Our initial studies concentrated on sections containing the hypothalamus from the rat.[9,12,13] Immunohistochemical studies had revealed 900 to 2000 prominent CRF cells in the parvocellular divisions of the two hypothalamic paraventricular nuclei (PVN).[14,15] Our observation of CRF mRNA in this same region (Figure 1a) confirmed that CRF is made in those neurons that have been shown to project to the median eminence[16,17] and release their products into the portal circulation. We also identified CRF mRNA in the PVN of the mouse (Figure 1b), rabbit, oppossum, and human (Figure 2).

The supraoptic nucleus (SON) of the rat contains CRF immunoreactivity in oxytocin-containing magnocellular neurons.[18,19] In agreement with these observations, we found prominent labeling with the CRF probe of magnocellular neurons in both the rat SON and PVN.[12,13] Although we have not used simultaneous immunohistochemistry to determine whether the vasopressin or oxytocin magnocellular neurons contain the CRF transcripts, the patterns of distribution, especially in the PVN (Figure 1a), favor colocalization with oxytocin.[20,21] Scattered cells in the lateral hypothalmus, central nucleus of the amygdala, and neocortex of the rat at the level of the PVN and SON also contained CRF mRNA.

B. EFFECTS OF ADRENALECTOMY AND STRESS ON PVN AND SON CRF mRNA LEVELS

1. Effects of Adrenalectomy

As noted above, CRF that is synthesized in the parvocellular neurons of the PVN is transported to the median eminence where it is released into the portal circulation. The CRF is then carried to the anterior pituitary where it stimulates ACTH synthesis and release. ACTH, in turn, stimulates glucocorticoid synthesis and release by the adrenal gland.[2] Therefore, it was expected that the animal would attempt to compensate for the glucocorticoid deficiency of adrenalectomy by increasing CRF synthesis. This was verified by radioimmunoassay and immunohistochemistry of CRF in portal plasma and hypothalami, respectively (see References 22 and 23 for further references), and by Northern analysis of CRF

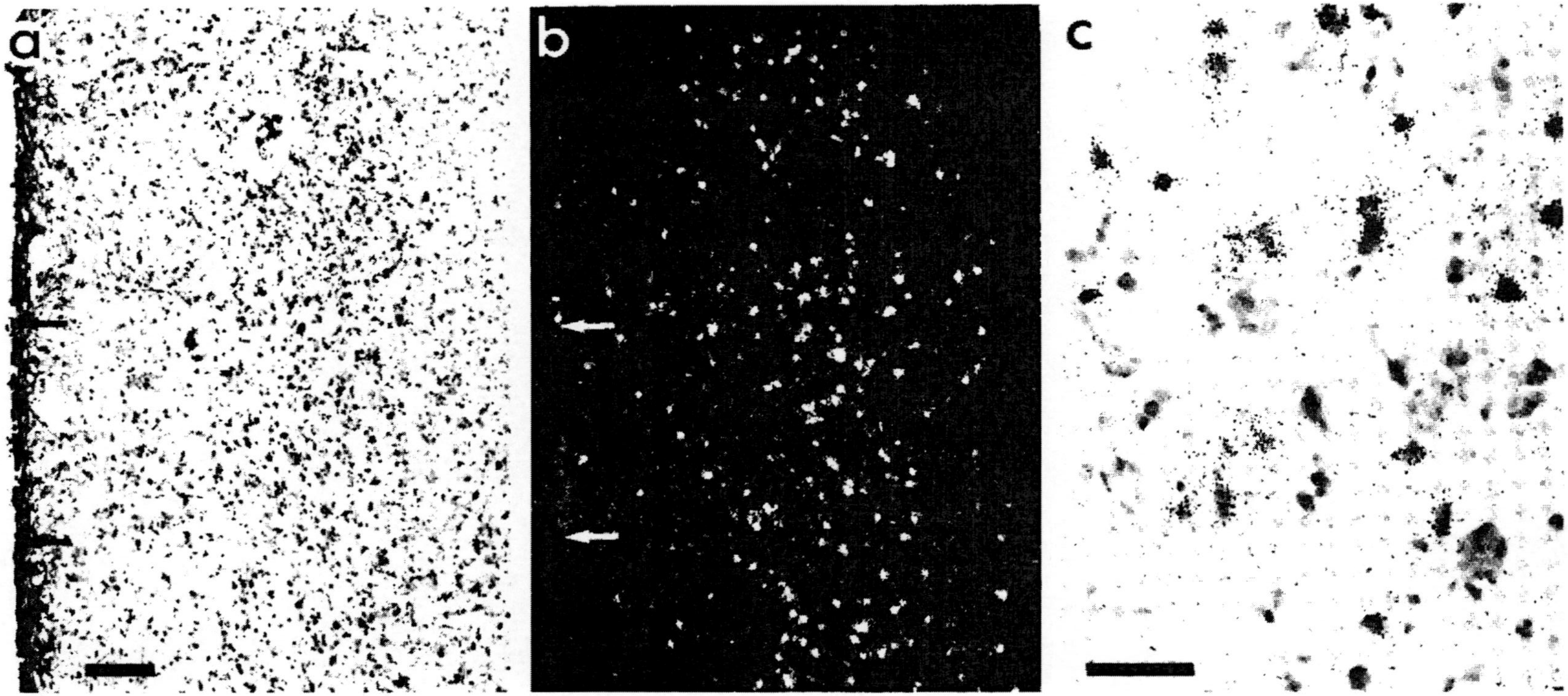

FIGURE 2. Neurons of the human paraventricular nucleus contain CRF mRNA. Low power photomicrographs show Nissl staining (a) and ISHH (labeled cells in b are shown as white grain clusters). Panel c shows a higher power view of some of the labeled cells. The tissue was obtained from a 74-year-old man without neurological disease 6 h after death (BRC #292) and frozen and stored at −80°C until processed for ISHH. Arrows delineate the third ventricular surface. Bars equal 200 μm in a and b and 50 μm in c.

mRNA levels in the rat hypothalamus.[24] We examined this paradigm using ISHH and found an increase in levels of CRF mRNA per volume of tissue as well as the volume expressing it in the parvocellular division of the PVN.[9] At the level of the PVN that we examined, this yielded a 220% increase. These cells were also shown to increase their levels of vasopressin[25-27] and vasopressin mRNA[28,29] after adrenalectomy, presumably to augment the effects of CRF on the corticotrophs.[22] Adrenalectomy had little or no effect on magnocellular CRF, vasopressin, or oxytocin mRNA levels.[9,29]

ISHH was used to examine the site of glucocorticoid feedback to the PVN.[30] When pellets of the synthetic glucocorticoid, dexamethasone, were implanted near the PVN, levels of CRF mRNA were reduced to background levels, even after adrenalectomy.[30] Unilateral implants had unilateral effects while implants of cholesterine near the PVN or of dexamethasone in the CRF-rich amygdala had no effect. Similar results were noted on the levels of CRF and vasopressin immunostaining.[31] A recent study reported the presence of glucocorticoid receptor immunoreactivity in parvocellular neurons of the PVN that contain CRF.[32]

2. Effects of Stress

The hypothalamic-hypophysial-adrenal axis is able to respond rapidly to various stresses that confront an animal.[2] In order to assess the role of the PVN, we stressed rats by precipitating withdrawal after chronic morphine administration or after acute intraperitoneal injections of hypertonic saline. Not surprisingly, we saw increased levels of CRF mRNA in the parvocellular PVN as early as 2 h after the onset of the stress followed by increased CRF immunostaining by 6 h.[44] Interestingly, we also observed increased levels of enkephalin mRNA in the same region.[33] The potential effects of increased release of enkephalin and CRF into the portal system during stress should be a rich area for study.

C. EFFECTS OF HYPEROSMOTIC STIMULI ON PVN AND SON CRF mRNA LEVELS

The function of CRF synthesized by the oxytocin magnocellular neurons[18,19] is unclear. Since these neurons respond to hyperosmotic stimuli by increasing oxytocin biosynthesis and release,[34-36] we studied the effect of drinking 2% saline on CRF mRNA levels in the PVN and SON.[12] This treatment lead to elevated levels of the CRF mRNA in magnocellular neurons in the PVN and SON that are likely to be oxytocinergic based on their anatomical distribution.[20,21] We were surprised to find a simultaneous reduction in the levels of CRF mRNA in the parvocellular subdivision of the PVN. The mechanism by which this inhibition occurred is unclear, but may involve intranuclear or afferent synaptic activity or, perhaps, increased glucocorticoid levels produced by a salt-loading stress. Consistent with the latter possibility is the reversal by adrenalectomy of the parvocellular CRF mRNA reduction.[12] Interestingly, another hyperosmotic stimulus, 3 d of dehydration, did not increase message levels in the magnocellular oxytocin neurons but, instead, decreased the levels in some animals.[12]

D. CRF IN THE INFERIOR OLIVARY NUCLEUS

The other CRF system that we have investigated is the olivocerebellar climbing fiber system. Immunohistochemical studies had demonstrated that inferior olivary neurons contain CRF and that they can be labeled by retrograde transport of dye from the cerebellum.[37,38] We confirmed that these neurons actually synthesize CRF by demonstrating that they contain CRF mRNA.[39] In subsequent light and electron microscopic immunohistochemical studies, we traced the complete CRF pathway of the climbing fibers from the inferior olive through the inferior cerebellar peduncle to the synapses on the Purkinje cell dendrites.[40]

These findings suggested that a lack of CRF might cause some of the symptoms seen in the olivopontocerebellar atrophies.[41] We examined the inferior olives of humans and observed that most, if not all, the inferior olivary neurons contain CRF mRNA.[39] Immunohistochemical examination of the inferior olive and cerebellum from humans showed CRF immunoreactivity in the neurons and climbing fibers, respectively.[42] We attempted to create a model of the olivopontocerebellar atrophies by eliminating the inferior olivary neurons by injecting rats with the neurotoxin 3-acetylpyridine, along with harmaline and niacinamide.[43] After treatment, the rats contained no CRF mRNA-containing inferior olivary neurons and they displayed a coarse tremor and a splayed posturing of their feet, but we detected no change in cerebellar CRF receptors as measured by homogenate or autoradiographic assays 1 and 3 weeks after the lesions were made (Warden and Young, unpublished observations). We were also unable to correct the neurological symptoms by intraventricular injections of CRF (Warden, et al., unpublished observations); however, it is likely that the inferior olive synthesizes a variety of transmitters which would need to be replaced. Whether or not affected patients would benefit from CRF administration deserves further study.

IV. CONCLUSIONS

These initial studies demonstrate some of the potential uses of ISHH in elucidating the role of CRF in the central nervous system. Future investigators may wish to map the distribution of CRF mRNA-containing neurons in tissues from the developing and adult rat or other species' tissues as the sequence we chose for our probe is apparently highly conserved: we have been able to detect CRF mRNA in neurons of the rat, mouse, rabbit, oppossum, baboon, and human. Exciting insights are likely to come from studies of the levels of CRF mRNA after various physiological and pharmacological manipulations and in human diseases.

ACKNOWLEDGMENTS

I would like to express my appreciation for helpful discussions and assistance from M. J. Brownstein, É. Mezey, R. E. Siegel, S. L. Lightman, D. L. Price, M. K. Warden, and E. Shepard. The human tissue was kindly provided by Drs. D. L. Price and R. Struble, Brain Resource Center, Neuropathology Laboratory, Department of Pathology, The Johns Hopkins University Medical School, Baltimore, MD.

REFERENCES

1. **Vale, W., Spiess, J., Rivier, C., and Rivier, J.,** Characterization of a 41-residue ovine hypothalamic peptide that stimulates secretion of corticotropin and β-endorphin, *Science*, 213, 1394, 1981.
2. **Gillies, G. E. and Lowry, P. J.,** Adrenal function, in *Neuroendocrinology*, Lightman, S. L. and Everitt, B. J., Eds., Blackwell Scientific, London, 1986, 360.
3. **Bissette, G., Reynolds, G. P., Kilts, C. D., Widerlöv, E., and Nemeroff, C. B.,** Corticotropin-releasing factor-like immunoreactivity in senile dementia of the Alzheimers type: reduced cortical and striatal concentrations, *JAMA*, 254, 3067, 1985.
4. **De Souza, E. B., Whitehouse, P. J., Kuhar, M. J., Price, D. L., and Vale, W. W.,** Reciprocal changes in corticotropin-releasing factor (CRF)-like immunoreactivity and CRF receptors in cerebral cortex of Alzheimer's disease, *Nature*, 319, 593, 1986.
5. **Nemeroff, C. B., Widerlöv, E., Bissette, G., Walléus, H., Karlsson, I., Eklund, K., Kilts, C. D., Loosen, P. T. and Vale, W.,** Elevated concentrations of CSF corticotropin-releasing factor-like immunoreactivity in depressed patients, *Science*, 226, 1342, 1984.

6. **Gold, P. W., Loriaux, L., Roy, A., Kling, M. A., Calabrese, J. R., Kellner, C. H., Nieman, L. K., Post, R. M., Pickar, D., Gallucci, W., Avgerinos, P., Paul, S., Oldfield, E. H., Cutler, G. B., Jr., and Chrousos, G. P.**, Responses to corticotropin-releasing hormone in the hypercortisolism of depression and Cushing's disease. Pathophysiologic and diagnostic implications, *N. Engl. J. Med.*, 314, 1329, 1986.
7. **Jingami, H., Mizuno, N., Takahashi, H., Shibahara, S., Furutani, Y., Imura, H., and Numa, S.**, Cloning and sequence analysis of cDNA for rat corticotropin-releasing factor precursor, *FEBS Lett.*, 192, 63, 1985.
8. **Shibahara, S., Morimoto, Y., Furutani, Y., Notake, M., Takahashi, H., Shimizu, S., Horikawa, S., and Numa, S.**, Isolation and sequence analysis of the human corticotropin-releasing factor precursor gene, *EMBO J.*, 2, 775, 1983.
9. **Young, W. S., III, Mezey, É., and Siegel, R. R.**, Quantitative *in situ* hybridization histochemistry reveals increased levels of corticotropin-releasing factor mRNA after adrenalectomy in rats, *Neurosci. Lett.*, 20, 198, 1986.
10. **Uhl, G. R.**, *In Situ Hybridization in Brain*, Plenum Press, New York, 1986.
11. **Young, W. S., III and Kuhar, M. J.**, A new method for receptor autoradiography: ^{3}H-opioid receptors in rat brain, *Brain Res.*, 179, 255, 1979.
12. **Young, W. S., III**, Corticotropin-releasing factor mRNA in the hypothalamus is affected differently by drinking saline and by dehydration, *FEBS Lett.*, 208, 158, 1986.
13. **Lightman, S. L. and Young, W. S., III**, Vasopressin, oxytocin, dynorphin, enkephalin and corticotropin-releasing factor mRNA stimulation in the rat, *J. Physiol. (London)*, 394, 23, 1987.
14. **Antoni, F. A., Palkovits, M., Makara, G. B., Linton, E. A., Lowry, P. J., and Kiss, J. Z.**, Immunoreactive corticotropin-releasing hormone in the hypothalamo-infundibular tract, *Neuroendocrinology*, 36, 415, 1983.
15. **Swanson, L. W., Sawchenko, P. E., Rivier, J., and Vale, W. W.**, Organization of ovine corticotropin-releasing factor immunoreactive cells and fibers in the rat brain: an immunohistochemical study, *Neuroendocrinology*, 36, 165, 1983.
16. **Wiegand, S. J. and Price, J. L.**, Cells of origin of the afferent fibers to the median eminence in the rat, *J. Comp. Neurol.*, 192, 1, 1980.
17. **Lechan, R. M., Nestler, J. L., and Jacobson, S. J.**, The tuberoinfundibular system of the rat as demonstrated by immunohistochemical localization of retrogradely transported wheat germ agglutinin (WGA) from the median eminence, *Brain Res.*, 245, 1, 1982.
18. **Burlet, A., Tonon, M. -C., Tankosic, P., Coy, D., and Vaudry, H.**, Comparative immunocytochemical localization of corticotropin releasing factor (CRF-41) and neurohypophyseal peptides in the brain of Brattleboro and Long-Evans rats, *Neuroendocrinology*, 37, 64, 1983.
19. **Sawchenko, P. E., Swanson, L. W., and Vale, W. W.**, Corticotropin-releasing factor: co-expression within distinct subsets of oxytocin-, vasopressin-, and neurotensin-immunoreactive neurons in the hypothalamus of the male rat, *J. Neurosci.*, 4, 1118, 1984.
20. **Swaab, D. F., Nijveldt, F., and Pool, C. W.**, Distribution of oxytocin and vasopressin cells in the rat supraoptic and paraventricular nucleus, *J. Endocrinol.*, 67, 461, 1975.
21. **Vandesande, F. and Dierickx, K.**, Identification of the vasopressin producing and of the oxytocin producing neurons in the hypothalamic magnocellular system of the rat, *Cell Tissue Res.*, 164, 153, 1975.
22. **Antoni, F. A.**, Hypothalamic control of adrenocorticotropin secretion, advances since the discovery of 41-residue corticotropin-releasing factor, *Endocr. Rev.*, 7, 351, 1986.
23. **Plotsky, P. M. and Sawchenko, P. E.**, Hypophysial-portal plasma levels, median eminence content, and immunohistochemical staining of corticotropin-releasing factor, arginine vasopressin, and oxytocin after pharmacological adrenalectomy, *Endocrinology*, 120, 1361, 1987.
24. **Jingami, H., Matsukura, S., Numa, S. and Imura, H.**, Effects of adrenalectomy and dexamethasone administration on the level of preprocorticotropin-releasing factor messenger ribonucleic acid (mRNA) in the hypothalamus and adrenocorticotropin/β-lipotropin precursor mRNA in the pituitary in rats, *Endocrinology*, 117, 1314, 1985.
25. **Tramu, G., Croix, C., and Pillez, A.**, Ability of the CRF immunoreactive neurons of the paraventricular nucleus to produce a vasopressin-like material, *Neuroendocrinology*, 37, 467, 1983.
26. **Kiss, J. Z., Mezey, É., and Skirboll, L.**, Corticotropin-releasing factor-immunoreactive neurons of the paraventricular nucleus become vasopressin positive after adrenalectomy, *Proc. Natl. Acad. Sci. U.S.A.*, 81, 1854, 1984.
27. **Sawchenko, P. E., Swanson, L. W., and Vale, W. W.**, Co-expression of corticotropin-releasing factor and vasopressin immunoreactivity in parvocellular neurosecretory neurons of the adrenalectomized rat, *Proc. Natl. Acad. Sci. U.S.A.*, 81, 1883, 1984.
28. **Wolfson, B., Manning, R. W., Davies, L. G., Arentzen, R., and Baldino, F., Jr.**, Co-localization of corticotropin-releasing factor and vasopressin mRNA in neurones after adrenalectomy, *Nature*, 315, 59, 1985.

29. **Young, W. S., III, Mezey, É., and Siegel, R. E.,** Vasopressin and oxytocin mRNAs in adrenalectomized and Brattleboro rats: analysis by quantitative in situ hybridization histochemistry, *Mol. Brain Res.*, 1, 231, 1986.
30. **Kovács, K. J. and Mezey, É.,** Dexamethasone inhibits corticotropin releasing factor gene expression in the rat paraventricular nucleus, *Neuroendocrinology*, 46, 365, 1987.
31. **Kovács, K., Kiss, J. Z., and Makara, G. B.,** Glucocorticoid implants around the hypothalamic paraventricular nucleus prevent the increase of corticotropin-releasing factor and arginine vasopressin immunostaining induced by adrenalectomy, *Neuroendocrinology*, 44, 229, 1986.
32. **Cintra, A., Fuxe, K., Härfstrand, A., Agnati, L. F., Wikström, A.-C., Okret, S., Vale, W., and Gustafsson, J.-Å.,** Presence of glucocorticoid receptor immunoreactivity in corticotrophin releasing factor and growth hormone releasing factor immunoreactive neurons of the rat di- and telencephalon, *Neurosci. Lett.*, 77, 25, 1987.
33. **Lightman, S. L. and Young, W. S., III,** Changes in hypothalamic preproenkephalin A mRNA following stress and opiate withdrawal, *Nature*, 328, 643, 1987.
34. **Cheng, S. W. T. and North, W. G.,** Responsiveness of oxytocin-producing neurons to acute salt-loading in rats: comparisons with vasopressin-producing neurons, *Neuroendocrinology*, 42, 174, 1986.
35. **Majzoub, J. A., Rich, A., vanBoom, J., and Habener, J. F.,** Vasopressin and oxytocin mRNA regulation in the rat assessed by hybridization with synthetic oligonucleotides, *J. Biol. Chem.*, 258, 14061, 1983.
36. **Burbach, J. P. H., VanTol, H. H. M., Bakkus, M. H. C., Schmale, H., and Ivell, R.,** Quantitation of vasopressin mRNA and oxytocin mRNA in hypothalamic nuclei by solution hybridization assays, *J. Neurochem.*, 47, 1814, 1986.
37. **Schipper, J., Werkman, T. R., and Tilders, F. J. H.,** Quantitative immunocytochemistry of corticotropin-releasing factor (CRF). Studies on nonbiological models and on hypothalamic tissues of rats after hypophysectomy, adrenalectomy, and dexamethasone treatment, *Brain Res.*, 293, 111, 1984.
38. **Cummings, S., Elde, R., and Sharp, B.,** CRF-immunoreactive neurons within the inferior olive project to the flocculus and dorsal and ventral paraflocculi, *Soc. Neurosci. Abstr.*, 11, 883, 1985.
39. **Young, W. S., III, Walker, L. C., Powers, R. E., De Souza, E. B., and Price, D. L.,** Corticotropin-releasing factor mRNA is expressed in the inferior olives of rodents and primates, *Mol. Brain Res.*, 1, 189, 1986.
40. **Palkovits, M., Léránth, C., Görcs, T., and Young, W. S., III,** Corticotropin-releasing factor in the olivocerebellar tract of rats: demonstration by light- and electron-microscopic immunohistochemistry and in situ hybridization histochemistry, *Proc. Natl. Acad. Sci. U.S.A.*, 84, 3911, 1987.
41. **Koeppen, A. H. and Barron, K. D.,** The neuropathology of olivopontocerebellar atrophy, in *The Olivopontocerebellar Atrophies*, Duvoisin, R. C. and Plaitakis, A., Eds., Raven Press, New York, 1984, 13.
42. **Powers, R. E., De Souza, E. B., Walker, L. C., Price, D. L., Vale, W. W., Kuhar, M. J., and Young, W. S., III,** Corticotropin-releasing factor as transmitter in the human olivocerebellar pathway, *Brain Res.*, 415, 347, 1987.
43. **Llinás, R., Walton, K., Hillman, D. E., and Sotelo,** Inferior olive: its role in motor learning, *Science*, 190, 1230, 1975.
44. **Lightman S. L. and Young, W. S., III,** Response of hypothalamic corticotropin releasing factor mRNA and pituitary proopiomeclanocortin mRNA to stress, opiates and opiate withdrawl, *J. Physiol. (London)*, 403, 511,1988.

Chapter 3

CENTRAL NERVOUS SYSTEM CRF IN STRESS: RADIOIMMUNOASSAY STUDIES

Garth Bissette

TABLE OF CONTENTS

I. INTRODUCTION

Corticotropin-releasing factor (CRF) was isolated at the end of a 40 year search for the hypothalamic releasing factor responsible for the release of corticotropin (ACTH) from the anterior pituitary which, in turn, regulates the release of corticosteroids from the adrenal gland. The involvement of the hypothalamic-pituitary-adrenal (HPA) axis in the stress response was first discovered by Selye,[1] when he noticed the hypertrophied adrenal glands of animals exposed to chronic stress. The secretion of ACTH and corticosterone in response to a variety of experimental stressors has been copiously documented by many laboratories around the world. A thorough discussion of these data is beyond the scope of this chapter, which will focus instead on the evidence implicating regional brain CRF concentration changes in the response to stress.

Several lines of evidence adumbrate a crucial role for central nervous system CRF in the physiological, behavioral, and endocrine effects of stress. Several research groups have shown that the peripheral administration of CRF antisera will block the ACTH response to ether stress,[2-5] confirming CRF release as the major regulator of ACTH secretion in response to stress. The intraventricular administration of CRF has been found to elicit peripheral endocrine changes similar to those seen in response to stress.[6]

There is now a growing literature documenting CRF-induced behaviors that are similar to behaviors seen in response to stress.[7] (Also see Chapter 17 for review.) These CRF-induced behaviors can be blocked by concomitant administration of the CRF antagonist, α-helical CRF_{9-41}.[8-11] These effects of CRF do not occur through CRF release of ACTH, as blockade of ACTH secretion by the synthetic glucocorticoid dexamethasone does not attenuate CRF-induced stress behaviors.[12] These data suggest that both hypothalamic and extra-hypothalamic CRF systems mediate many of the behavioral, physiological, and endocrine responses to stressful stimuli. Thus it is reasonable to expect the concentration of CRF to be altered in brain regions containing this peptide that are activated by stressful stimuli.

II. RADIOIMMUNOASSAY OF CRF

At present, the most sensitive way to quantify CRF is by use of radioimmunoassay techniques. This necessitates raising antisera directed toward some portion of the 41 amino acid sequence of the CRF molecule. A variety of methods have been used to render CRF antigenic, but the most successful have been conjugations of CRF to larger molecules such as hemocyanin, albumin, or thyroglobulin. A variety of reactions have been used to perform this conjugation (carbodiimide, glutaraldehyde, etc.) and may be chosen to preferentially select for a C- or N-terminal directed antisera. The conjugated peptide-protein complex is injected into an animal species with robust immunoresponsivity and with a body size appropriate for the volume of antiserum needed. Animals are bled at 2 to 3 week intervals, serum is checked for antibody titer, and booster injections of conjugate are administered as needed to preserve a useful titer. Antisera are then characterized for specificity of the immunoreactive site using fragments of the CRF molecule or peptides with homologous amino acid sequences. There should be no immunoreactivity toward the protein that CRF has been conjugated with. Dilution curves of synthetic and endogenous CRF should be parallel in binding to CRF antisera. A radioactive tracer is prepared by one of several methods (Bolton-Hunter, chloramine T, iodogen, etc.) using ^{125}I and a CRF molecule with an N-terminal tyrosine moiety. The radioactive fraction containing the most immunoreactivity is determined by binding to several concentrations of antiserum. Bound and free radioactivity can be separated by several methods (charcoal, talc, etc.) but maximum sensitivity is usually achieved by the use of a second, species-specific (e.g., goat anti-rabbit) antiserum and manual vacuum aspiration of supernatant. By adjusting the amount of primary and secondary antiserum, the amount of radioactive tracer and the time of reaction, the maximum assay sensitivity can be titrated.

III. BRAIN DISSECTION AND EXTRACTION

Brain dissection techniques are chosen for the ability to isolate areas of interest from other regions that might obscure the changes expected if allowed to remain. As such, the anatomic resolution demanded by the experiment will dictate the dissection method of choice. The various methods of dissection are thoroughly examined in a volume entitled *Brain Microdissection Techniques*, edited by Cuello, 1983.[13] The microdissection of the various organizational brain nuclei has been made possible by the use of the micropunch technique pioneered by Palkovits (see Reference 14 for a detailed review of this procedure). This method has the advantage of being able to sample small brain regions containing as little as 20 μg of protein with great reproducibility. A detailed, working knowledge of neuroanatomy is required by the investigator performing this procedure. Rat brains are removed and frozen, without distortion, on a block of dry ice covered with aluminum foil. Brains are then vertically mounted on a microtome specimen stage and frozen into place. The rostral pole of the brain faces the stage if lower brain stem regions are of predominate interest, otherwise, the brain stem faces the stage and coronal sections (300 μm) are cut starting at the olfactory bulb. Sections are placed on glass slides and these are placed on a freezing stage or on a dry ice-filled dish for dissection. Using a dissecting microscope to aid in identification and orientation of the region to be micropunched, the investigator takes a cannula with an outside diameter smaller than the region to be sampled and presses it into the brain section at the location of the nucleus of interest. Because most brain nuclei have a rostral-caudal orientation, two to four successive slides may be sampled to include the full extent of the nucleus. The sample is then extruded into the vessel in which it will be homogenized. Most neuropeptides are extracted in acid (1 *N* hydrochloric or 2 *N* acetic acid) which denatures peptidases. We have found that the larger peptides such as CRF are more efficiently extracted in acid solutions containing at least 40 volumes by weight of solvent per weight of tissue (unpublished observation). The percent recovery of sample must be calculated using spiked samples containing known amounts of synthetic peptide compared to a sample containing only endogenous peptide. Protein estimation is performed on the extracted tissue pellets after centrifugation and the neuropeptide concentration of the supernatant fraction is calculated as weight (nanogram or picogram) or molar (picomole or femtomole) amounts of peptide per milligram of protein. This important protein assay must be carefully performed by reliable laboratory personnel because the typical micropunch only contains from 20 to 50 μg of protein and thus does not readily allow duplicate measurements. Likewise, because of the small amount of tissue, individual assay samples will have low total neuropeptide content, necessitating maximal assay sensitivity. Table 1 lists the CRF concentration determined by radioimmunoassay in 36 micropunch dissected brain regions of normal control rats.

While there have been many immunohistochemical studies of CRF distribution in the rat and various other species (for review see Chapter 4), there have been relatively few quantitative radioimmunoassay studies and only two other micropunch dissection studies describing CRF concentrations in rat brain. The proportional distribution of CRF in our micropunch study agrees well with the macrodissected regional brain areas assayed by Fischman and Moldow.[16]

The Skofitsch and Jacobowitz[17] study compared four different CRF antisera in 50 micropunched brain nuclei and regions. No detectable CRF was found in a total of 19 of the 50 brain regions including most rat telencephalic structures with two separate antisera raised against rat/human CRF. They were, however, able to detect CRF immunoreactivity in these

TABLE 1
CRF Concentration of Various Rat Brain Nuclei or Regions

Brain nucleus or region	CRF concentration (pg/mg protein ± SEM)[a]
Cerebral cortex	
Medial prefrontal cortex	22 ± 1
Pyriform cortex	28 ± 2
Cingulate cortex	Not detectable
Entorhinal cortex	27 ± 3
Limbic system	
Olfactory tubercles	94 ± 7
Nucleus tractus diagonalis	64 ± 4
Lateral septal nucleus	68 ± 7
Medial septal nucleus	86 ± 9
Nucleus accumbens	94 ± 12
Bed nucleus of the stria terminalis	136 ± 14
Dorsal hippocampus	31 ± 2
Ventral hippocampus	34 ± 2
Lateral habenula	72 ± 8
Substantia innominata	98 ± 5
Cortical amygdaloid nucleus	148 ± 21
Basal amygdaloid nucleus	85 ± 3
Central amygdaloid nucleus	192 ± 17
Medial amygdaloid nucleus	110 ± 10
Lateral amygdaloid nucleus	77 ± 5
Hypothalamus	
Median eminence/arcuate nucleus	5054 ± 674
Paraventricular nucleus	498 ± 44
Periventricular nucleus	316 ± 15
Medial preoptic nucleus	142 ± 11
Lateral preoptic nucleus	86 ± 6
Anterior hypothalamic nucleus	93 ± 7
Ventromedial nucleus	97 ± 15
Dorsomedial nucleus	114 ± 7
Midbrain	
Ventral tegmental area	134 ± 11
Zona reticularis of the substantia nigra	94 ± 7
Zona compacta of the substantia nigra	190 ± 13
Pars lateralis of the substantia nigra	129 ± 7
Periaqueductal gray	144 ± 18
Pons	
Medial raphe nucleus	241 ± 18
Dorsal raphe nucleus	236 ± 28
Locus coeruleus	148 ± 9
Medulla	
Dorsal vagal complex	193 ± 15

[a] Values represent means of ten individual animals. All regions were measured in a single radioimmunoassay using antiserum oC-33 raised against ovine CRF and generously provided by W. Vale, Salk Institute, La Jolla, CA. This antiserum recognizes the 33—41 region of the CRF molecule and, using fresh ^{125}I-rat-human Tyr-oCRF as a tracer, has a sensitivity of 0.625 pg/tube with 50% displacement of trace (IC_{50}) at 30 pg. See Reference 15 for details.

regions using two additional antisera raised against ovine CRF and all four antisera detected CRF-like immunoreactivity in the remaining 31 brain regions. Their study highlights the different absolute values obtained with different antisera and suggests that there may be regional differences in different immunoreactive forms of CRF. Their proportional differences among the 25 regions in common with our study are in good agreement.

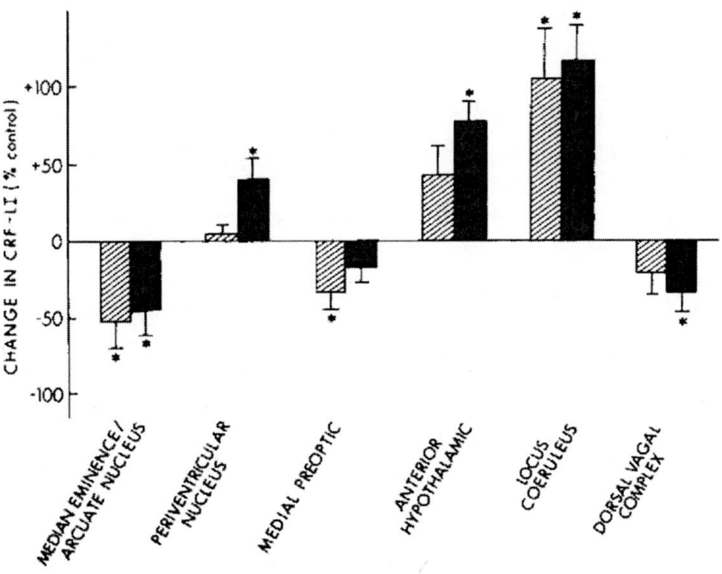

FIGURE 1. Alterations in the concentration of CRF-like immunoreactivity in brain regions from rats exposed to acute and chronic stress. The hatched bars represent rats exposed to an acute stress and the solid bars represent rats exposed to chronic (14 d) stress. Values represent percent CRF-like immunoreactivity concentration change in these brain regions relative to control concentrations. Asterisks denote statistical significance equal to or greater than 0.05 as determined by ANOVA and Student Newman-Keuls test. See Reference 15 for details.

Palkovits et al.[18] surveyed 70 micropunched rat brain nuclei for content of CRF immunoreactivity and, due to relative insensitivity of the rat/human CRF antisera used (300 pg/mg protein) was unable to detect CRF immunoreactivity in 9 of the 27 regions in common with our more sensitive (10 pg/mg protein) study. There was still good proportional agreement between the common remaining detectable regions examined by Palkovits et al.[18] and ourselves. Thus there is reasonable agreement about the rank order of brain regions containing CRF-like immunoreactivity detectable by radioimmunoassay and these data agree well with the combined results of the various immunohistochemical studies.

IV. CRF ALTERATIONS IN STRESS

The focus of several recent studies has been the examination of brain regions where CRF concentrations may be altered by stress. In the most comprehensive of these studies[15] the effects of acute (3 h immobilization at 4°C) and chronic stress on CRF immunoreactivity in 36 rat brain regions compared to nonstressed controls (n = 10 per group). Chronic stress was a 14-d regimen of daily exposure to one of several different stressors (ether stress, tail pinch, immobilization, cold swim, etc.) and alternating single and group housing with strange cagemates. This paradigm prevents development of tolerance to a single chronic stressor. The effects of acute and chronic stress on the adrenal weight and the plasma concentrations of corticosterone and ACTH were similar to previous findings. Adrenal weight was increased 20% and plasma corticosterone was doubled in the chronic group; while plasma ACTH (↑ 50%) and corticosterone (↑ 300%) were increased in the acute stress group compared to the controls. Figure 1 lists the percent of control CRF concentrations in six brain regions from a total of 36 examined that demonstrate statistically significant changes in CRF-like immunoreactivity with acute and chronic stress. Both acute and chronic stress significantly lowered the concentration of CRF in the arcuate/median eminence and both stress paradigms significantly increased CRF in the locus coeruleus. Of the remaining four regions represented in Figure 1 only one of the two stress paradigms (chronic stress) was successful in significantly changing CRF concentrations in the periventricular nucleus. The same trend in CRF concentration changes, although not significant, is clearly seen in the same regions (medial preoptic nucleus, anterior hypothalamic nucleus, dorsal vagal complex) where statistical significance is achieved with the other stress paradigm. None of the 30 other regions listed in Table 1 had significant alterations from control values with either the acute or chronic stress paradigm.

In a further examination of stress-induced changes in neuropeptides, Deutch et al.[19] examined the concentration of CRF, somatostatin, and neurotensin in six brain regions that were expertly microdissected without micropunches (free-hand) from rats exposed to mild (0.2 mA) uncontrollable foot-shock. Neurotensin was increased (20%) in the ventral tegmental area with no effect on either CRF or somatostatin. None of these peptides was changed in the substantia nigra (pars lateralis), retrorubral field, prefrontal cortex, striatum, or the nucleus accumbens. These results agree with our observed lack of CRF alterations in these regions after acute or chronic stress.

Moldow et al.[20] have used restraint stress to initiate alterations in CRF concentrations of rat hypothalamus in animals with or without cyclohexamide pretreatment to block *de novo* protein synthesis. Hypothalamic CRF concentrations fell throughout the first 30 min of restraint stress in both cyclohexamide treated and untreated stressed rats. Between 60 and 90 min of restraint stress in rats without cyclohexamide, hypothalamic CRF concentration increased followed by further decreases through 2 h. Restrained rats receiving cyclohexamide did not show this concentration increase which presumably reflects synthesis of CRF or the enzymes that cleave CRF from the prohormone.

The consensus of these data at present is that acute and chronic stress cause decreases in the CRF concentration of the median eminence/arcuate nucleus, presumably due to release of CRF. Besides release of CRF, other mechanisms that could result in decreased concentrations of CRF include increased degradation or decreased synthesis; these mechanisms cannot be distinguished by radioimmunoassay alone. The hypothalamic CRF system is extremely well connected to cutaneous and visceral nociceptive pathways that impinge on paraventricular CRF neurons, whose axons, in turn, project to the arcuate/median eminence (see Reference 21 for review of sensory afferents to CRF-containing neurons in the hypothalamus). If this decrease in the hypothalamus is due to release of CRF, chronic stress should induce synthesis of CRF and the data of Moldow et al.,[20] with cyclohexamide treatment, suggests that this is, indeed, occurring.

One might also expect to see down-regulation of pituitary CRF receptors and decreased release of ACTH by corticotrophs during adaptation to chronic stress. Young and Akil[22] have demonstrated that acute (1-d footshock, 30 min at 1 mA every 5 s for 1 s duration), but not chronic (14 d of same daily regimen as for acute) stress, decreased ACTH and β-endorphin release by isolated rat pituitaries exposed to CRF or vasopressin *in vitro*. High concentrations of CRF augmented the release of ACTH and β-endorphin in animals exposed to chronic stress and these animals had higher pituitary contents of ACTH and β-endorphin than the acute stress group. The failure of pituitary receptors to down-regulate after chronic stress could be explained by adaptation of the animals to the repeated presentation of the same stressor. Thus, the hypothalamic CRF system responds to stress exposure by release of CRF from axon terminals in the median eminence, whose cell bodies reside predominantly in the paraventricular nucleus of the hypothalamus. Chronic stimulation of this pathway results in increased CRF synthesis and down-regulation of CRF receptors on pituitary cells.

Among the extrahypothalamic brain regions showing altered CRF concentrations after acute and chronic stress, the locus coeruleus is perhaps of most interest. This region has been long known to be responsive to noxious stimuli and to mediate arousal and attention. Sawchenko and Swanson[23] have demonstrated the extensive and intimate association of CRF-containing nerve terminals with locus coeruleus noradrenergic neurons. These investigators[24] have also mapped the projection of locus coeruleus axons to the paraventricular nucleus of the hypothalamus. Alonso et al.[25] have recently shown this input from the locus coeruleus stimulates synthesis and release of CRF from paraventricular neurons. Valentino et al.[26] have elegantly demonstrated that CRF iontophoresis onto locus coeruleus neurons results in depolarization of 9 neurons out of 14 tested. Several other lines of evidence from endocrine and anatomic studies have indicated a close association of the noradrenergic system with CRF in the response to stress. Many of these data are reviewed by Valentino (see Chapter 15).

The CRF component of the HPA axis has recently been implicated in human psychiatric illnesses such as affective and anxiety disorders (see review, Chapter 23). Our laboratory (Owens et al., 1988, submitted for publication) has recently reported the involvement of the locus coeruleus, hypothalamus and cortex in the CRF neuronal response to acute treatment with the triazalobenzodiazepine anxiolytic drugs, adinazolam and alprazolam. Acute treatment of rats with alprazolam (1 mg/kg) or adinazolam (10 mg/kg) 1 h before sacrifice, dissection, and subsequent radioimmunoassay for CRF revealed alterations in CRF concentration of these brain regions compared to imipramine (10 mg/kg) treated or vehicle controls. The CRF concentration changes in the locus coeruleus (50% decrease) and hypothalamus (50% increase) were exactly opposite in direction to those we have observed in the acute and chronic stress studies. Thus it would appear that anxiolytic drugs may have effects on CRF systems that directly oppose the effects of stress on these same systems.

A growing body of evidence from a variety of disciplines now implicates CRF as being a major neurotransmitter in the physiological, endocrine, and behavioral response to stress. The further exploration of this fascinating interaction promises to yield tangible benefits in the pharmacological treatment of diverse human psychiatric illnesses such as major depression and anxiety disorders as well as furthering our knowledge of the brain regions and neurochemicals involved in the integration of the biological responses to stress.

REFERENCES

1. **Selye, H.,** A syndrome produced by diverse noxious agents, *Nature,* 138, 32, 1936.
2. **Rivier, C. and Vale, W.,** Modulation of stress-induced ACTH release by corticotropin-releasing factor, catecholamines and vasopressin, *Nature,* 305, 325, 1983.
3. **Linton, E. A., Tilders, F. J. H., Hodgkinson, S., Berkenbosch, F., Vermes, I., and Lowry, P. J.,** Stress-induced secretion of adrenocorticotropin in rat is inhibited by administration of antisera to ovine corticotropin-releasing factor and vasopressin, *Endocrinology,* 116, 966, 1985.
4. **Nakane, T., Audhya, T., Kanie, N., and Hollander, C. S.,** Evidence for a role of endogenous corticotropin-releasing factor in cold, ether, immobilization and traumatic stress, *Proc. Natl. Acad. Sci. U.S.A.,* 82, 1247, 1985.
5. **Ono, N., Samson, W. K., McDonald, J. K., Lumpkin, M. D., de Castro, J. C. B., and McCann, S. M.,** Effects of intravenous and intraventricular injection of antisera directed against corticotropin-releasing factor on the secretion of anterior pituitary hormones, *Proc. Natl. Acad. Sci. U.S.A.,* 82, 7787, 1985.
6. **Lenz, H. J., Raedler, A., Greten, H., and Brown, M. R.,** CRF initiates biological actions within the brain that are observed in response to stress, *Am. J. Physiol.,* 252, R34, 1987.
7. **Koob, G. F.,** Stress, corticotropin-releasing factor and behavior, in *Perspectives on Behavioral Medicine,* Vol. 2, Academic Press, New York, 1985, 39.
8. **Britton, K. T., Lee, G., Vale, W., Rivier, J., and Koob, G. F.,** Corticotropin releasing factor (CRF) receptor antagonist blocks activating and "anxiogenic" actions of CRF in the rat, *Brain Res.,* 369, 303, 1986.
9. **Krahn, D. D., Gosnell, B. A., Grace, M., and Levine, A. S.,** CRF antagonist partially reverses CRF- and stress-induced effects on feeding, *Brain Res. Bull.,* 17, 285, 1986.

10. **Tazi, A., Dantzer, R., LeMoal, M., Rivier, J., Vale, W., and Koob, G. F.,** Corticotropin releasing factor antagonist blocks stress-induced fighting in rats, *Regul. Peptides,* 18, 37, 1987.
11. **Berridge, C. W. and Dunn, A. J.,** A corticotropin-releasing factor antagonist reverses the stress-induced changes of exploratory behavior in mice, *Horm. Behav.,* 21, 393, 1987.
12. **Britton, K. T., Lee, G., Dana, R., Risch, S. C., and Koob, G. F.,** Activating and "anxiogenic" effects of corticotropin-releasing factor are not inhibited by blockade of the pituitary-adrenal system with dexamethasone, *Life Sci.,* 39, 1281, 1986.
13. **Cuello, A. C., Ed.,** *Brain Microdissection Techniques,* John Wiley & Sons, New York, 1983, 1.
14. **Palkovits, M.,** Microdissection of individual brain nuclei and areas, in *Neuromethods,* Boulton, A. A. and Baker, G. B., Eds., Humana Press, Clifton, NJ, 1985, 1.
15. **Chappell, P. B., Smith, M. A., Kilts, C. D., Bissette, G., Ritchie, J., Anderson, C., and Nemeroff, C. B.,** Alterations in corticotropin-releasing factor-like immunoreactivity in discrete rat brain regions after acute and chronic stress, *J. Neurosci.,* 6, 2908, 1986.
16. **Fischman, A. J. and Moldow, R. L,** Extrahypothalamic distribution of CRF-like immunoreactivity in the rat brain, *Peptides,* 1, 149, 1982.
17. **Skofitsch, G. and Jacobowitz, D. M.,** Distribution of corticotropin-releasing factor-like immunoreactivity in the rat brain by immunohistochemistry and radioimmunoassay: comparison and characterization of ovine and rat/human CRF antisera, *Peptides,* 6, 319, 1985.
18. **Palkovits, M., Brownstein, M. J., and Vale, W.,** Distribution of corticotropin-releasing factor in rat brain, *Fed. Proc.,* 44, 215, 1985.
19. **Deutch, A. Y., Bean, A. J., Bissette, G., Nemeroff, C. B., Robbins, R. J., and Roth, R. H.,** Stress-induced alterations in neurotensin, somatostatin and corticotropin-releasing factor in mesotelencephalic dopamine system regions, *Brain Res.,* 417, 350, 1987.
20. **Moldow, R. L., Kastin, A. J., Graf, M., and Fischman, A. J.,** Stress mediated changes in hypothalamic corticotropin-releasing factor-like immunoreactivity, *Life Sci.,* 40, 413, 1987.
21. **Palkovits, M.,** Organization of the stress response at the anatomical level, in *Progress in Brain Research,* Vol. 72, de Kloet, E. R., Wiegant, V. M., and de Wied, D., Eds., Elsevier, Amsterdam, 1987, 47.
22. **Young, E. A. and Akil, H.,** Corticotropin-releasing factor stimulation of adrenocorticotropin and β-endorphin release: effects of acute and chronic stress, *Endocrinology,* 117, 23, 1985.
23. **Swanson, L. W., Sawchenko, P. E., Rivier, J., and Vale, W. W.,** Organization of ovine corticotropin-releasing factor immunoreactive cells and fibers in the rat brain: an immunohistochemical study, *Neuroendocrinology,* 36, 165, 1983.
24. **Sawchenko, P. E. and Swanson, L. W.,** The organization of noradrenergic pathways from the brainstem to the paraventricular and supra optic nuclei in the rat brain, *Brain Res. Rev.,* 4, 285, 1982.
25. **Alonso, G., Szafarczyk, A., Balmfrezol, M., and Assenmacher, I.,** Immunocytochemical evidence for stimulatory control by the ventral noradrenergic bundle of parvocellular neurons of the paraventricular nucleus secreting corticotropin-releasing hormone and vasopressin in rats. *Brain Res.,* 397, 297, 1986.
26. **Valentino, R. J., Foote, S. L., and Aston-Jones, G.,** Corticotropin-releasing factor activates noradrenergic neurons of the locus coeruleus, *Brain Res.,* 270, 363, 1983.

Chapter 4

ORGANIZATION OF CRF IMMUNOREACTIVE CELLS AND FIBERS IN THE RAT BRAIN: IMMUNOHISTOCHEMICAL STUDIES

P. E. Sawchenko and L. W. Swanson

TABLE OF CONTENTS

I. INTRODUCTION

Close on the heels of the initial description[125] of the isolation and characterization of ovine corticotropin-releasing factor (oCRF) came a predictable flurry of immunohistochemical studies describing aspects of its distribution in the central nervous system of the rat and a number of other species.[6,10,12-14,17,21,38,40,41,62-64,73,76,105,111,121] The subsequent identification of the rat/human sequence[88] and prediction of the prohormonal structures from cloned cDNA sequences[25,37] have allowed the development of immunologic and molecular probes that provide more detailed and penetrating characterization of the central CRF system in the rat model. There now exists consensual agreement as to the primary localization of the CRF neuron population responsible for initiating the stress cascade, and a growing body of evidence providing further characterization of peptide dynamics in this cell type. It is also clear that CRF is among the more ubiquitously distributed of the ''hypophysiotropic'' neuropeptides, and the organization of extrahypophysial CRF-immunoreactive (CRFir) systems, and the extent to which they might subserve functions complementary to, or integrated with, stress-related neuroendocrine mechanisms, remain topics in need of further scrutiny.

The purpose of this chapter is to provide an update on the central distribution of CRFir from a systems-level perspective. We will not itemize and weigh the evidence for each reported localization; each of these will need to be evaluated individually as the need or interest arises. Instead, we will summarize what is known about the anatomic organization of CRF in particular systems and pathways. Neuroendocrine, and particularly hypophysiotropic, CRFir projections have commanded the greatest attention, and will constitute the principal focus here. We will, in addition, summarize recent morphologic studies that have attempted to establish and characterize the participation of CRF in other anatomically defined systems.

II. NEUROENDOCRINE SYSTEMS

A. PARVOCELLULAR NEUROSECRETORY SYSTEM

1. Localization of Hypophysiotropic Neurons

Evidence gleaned from immunohistochemical and hybridization histochemical studies in normal and steroid manipulated rats,[22,43,76,93,104,121,124,133,134,137] from combined ablation-immunohistochemical studies,[6,61] and from ablation studies coupled with the assay of circulating levels of ACTH or corticosterone,[35,58] have all converged to identify the paraventricular nucleus of the hypothalamus (PVH)[117] as the principal source for the delivery of CRF to the hypophysial portal vasculature for the stimulatory control of ACTH secretion. Estimates of the total size of the CRFir neuron population of the PVH are on the order of 2000 cells per side,[68,121] and while these are distributed in each of the subdivisions of the nucleus that have been identified on the basis of cytoarchitectonic and connectional criteria, they are concentrated in one, the dorsal aspect of the medial parvocellular part (or PVHmpd; see Figure 1), as defined initially by Swanson and Kuypers,[116] and subsequently refined.[120] This region of the PVH has been shown to project prominently to the median eminence,[48,131] and there is little doubt that a majority of CRFir neurons in this region contribute substantially to the hypophysiotropic projection. It must be noted, however, that combined retrograde transport-immunohistochemical technology has not yet been applied to delineate fully the origins of CRF-containing projections to the median eminence. Until it is, questions will remain as to whether some, most, or all CRF-stained neurons in PVHmpd project to the median eminence, and as to the extent to which CRF-stained neurons in other subdivisions of the PVH, and cell groups outside the PVH, might also contribute. Based on combined lesion-immunohistochemical data, CRFir projections to the median eminence from the medial preoptic and posterior hypothalamic areas, and from the dorsomedial nucleus of the hypo-

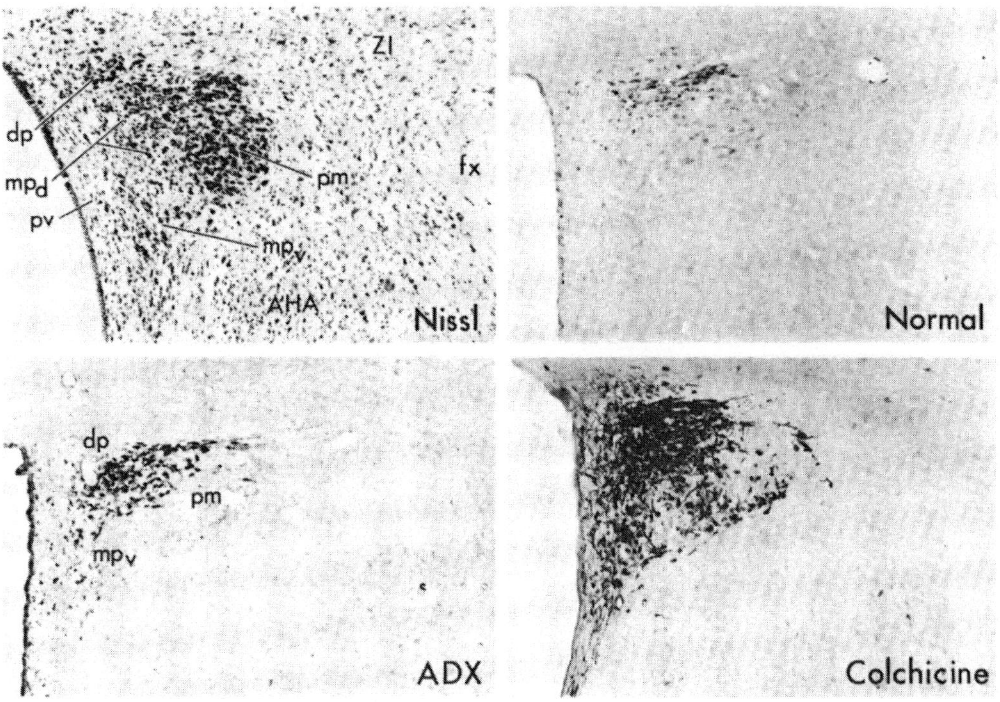

FIGURE 1. Comparison of CRF immunostaining in the PVH of untreated (normal), adrenalectomized (ADX), and colchicine-treated rats; a section through a comparable level stained for Nissl material is shown for reference. In the untreated rat, a relatively small number of cells, centered in PVHmpd, are stained weakly for CRFir. After ADX, the number and staining intensity of cells displaying CRFir is augmented, but stained cells remain concentrated in PVHmpd. Colchicine treatment enhances immunostaining for CRF not only in PVHmpd, but in autonomic-related (dp, mpv), other parvocellular neurosecretory (pv), and magnocellular neurosecretory (pm) regions as well. All micrographs × 60.

thalamus have been suggested,[61] and remain to be evaluated critically. It is worthy of emphasis that even within the PVH, CRFir cells clearly do not constitute a homogeneous population with respect to either their efferent projections or the regulatory influences that may play upon them (see below). In addition to cells that give rise to the hypophysiotropic projection, which in all likelihood predominate numerically, subpopulations of CRFir neurons, occupying cytoarchitectonically distinct regions of the nucleus, have been shown to project to the posterior pituitary,[15,75,103] or to autonomic centers in the brainstem and spinal cord.[94]

At the ultrastructural level, CRF-stained neurons in the parvocellular division of the PVH have been described that display morphologic specializations typical of neurosecretory neurons. These include 70 to 120 nm neurosecretory vesicles, well-developed Golgi-GERL complexes, and nuclear invaginations.[50,53] CRF has been localized within terminals in the external lamina of the median eminence, where staining is associated with secretory vesicles.[50,53,77] In untreated animals, stained terminals have been localized in contact with the perivascular space, but are frequently segregated from it by interposed tanycyte processes. In response to adrenalectomy (ADX), an acknowledged stimulus for enhanced production and release of CRF (see below), the parvocellular CRF neuron displays morphologic alterations consistent with elevated secretory activity. Increased secretory vesicle concentration, dilatation of rough endoplasmic reticulum cisternae, and the appearance of regions of tight somatic membrane apposition between CRF-stained perikarya and other (labeled and unlabeled) neurons have been described in the ADX rat, as has increased incidence of labeled

TABLE 1
Colocalization of Neuroactive Peptides Within the Parvocellular CRF Neuron

Peptide	Extent of colocalization	Minimal conditions	Ref.	See also
Angiotensin II	Extensive	Colch + ADX	52, 119	—
Cholecystokinin	Extensive	Colch + ADX	69, 70	93
Enkephalin	Extensive	Colch	31, 32, 93	30, 51
Neurotensin	Slight	Colch	103	—
PHI	Slight	Colch	32	7, 31
Vasopressin	Extensive	ADX	43, 104, 124	133—135
VIP	Slight	Colch	32	66, 67

terminals abutting the limiting membrane of the pericapillary space in the median eminence, presumably as a result of withdrawal of tanycyte processes.[53]

The course of CRFir fibers to the median eminence has been studied in normal immunohistochemical material[13,41,62,73,121] and using combined ablation-immunohistochemical approaches.[6,61] In summary, the hypophysiotropic projection appears to exit the PVH in a primarily lateral direction, sweep through the perifornical region (some fibers passing over the fornix, a majority under), and turn caudomedially to gather in the caudal retrochiasmatic area and travel over the optic tracts to the portal plexus. A smaller contingent of fibers may descend through the periventricular zone.[61,62] Within the median eminence, CRF-stained fibers are distributed across the breadth of the external lamina, and in well-stained preparations can also be seen decorating capillary loops that extend into the internal zone. Lesion studies have established that the projection as a whole is predominantly ipsilateral,[6,61] (see also Reference 2).

The median eminence may not constitute the sole target of the hypophysiotropic pathway. Ultrastructural studies of the distribution of CRF immunoreactivity within the PVH have provided evidence for interaction of CRF-stained terminals with magnocellular and parvocellular elements, including some CRF-stained somata[54] (see also Reference 108). This raised the possibility of the existence of an "ultra-short loop feedback" regulation imposed by axon collaterals of CRF-containing neurosecretory neurons. It remains to be established whether such terminals arise from neurosecretory neurons specifically, or even from within the PVH. Support for the existence of collateral projections of CRF-stained neuroendocrine neurons has been provided using an approach that combines intracellular filling with retrograde transport and immunohistochemical techniques.[84] This study described short axon collaterals of identified CRF-stained hypophysiotropic neurons lying primarily within the cell-sparse region immediately beyond the morphological borders of the PVH. The cell types that may be contacted by such collaterals remain to be identified.

2. Colocalization

With the possible exception of magnocellular neurosecretory neurons, nowhere else in the brain does the expression of multiple neuroactive agents appear more pervasive than in the parvocellular neurosecretory CRF neuron. Current evidence indicates that at least seven additional peptides may be expressed in variously sized subsets of CRFir neurons in the parvocellular division of the PVH (see Table 1).

a. Enkephalin

A substantial majority of CRFir neurons seen in the PVH of colchicine treated rats have been shown to stain with antisera to met-enkephalin.[32,93] Reports of the distribution of various proenkephalin-derived peptides and recent studies using *in situ* hybridization methods to localize proenkephalin mRNA,[51] provide a compatible view, though the particular moieties that may predominate within parvocellular neurons remain to be identified. Met-enkephalin- and CRF-immunoreactivities have been demonstrated within at least some of the same secretory granules in terminals in the median eminence.[30] Prodynorphin-derived peptides have been localized broadly within the PVH, including the medial parvocellular part, but the extent to which these may be coexpressed with CRF in parvocellular neurosecretory neurons remains to be demonstrated.

b. Peptide Histidine-Isoleucine (PHI)

PHI immunoreactivity was reported initially to be present in the great majority of parvocellular CRFir neurons of the colchicine-treated rat.[32] This view was challenged on the basis of evidence suggesting that the amidated C-terminal isoleucine residue common to both agents could provide a basis for cross-reactivity.[7] A careful reanalysis[31] of the issue confirmed this possibility, but nonetheless showed a small number of CRF-stained cells also expressing PHI immunoreactivity (demonstrable with N-terminally directed antisera).

c. Vasoactive Intestinal Polypeptide (VIP)

VIP and PHI display a considerable sequence homology, and are now known to be derived from a common precursor. Initial studies described relatively small numbers of VIPir neurons in the parvocellular division of the PVH,[66,67] but failed to provide evidence for any correspondence with CRFir neurons. Most recently, VIP has been demonstrated in a subset of CRF-stained neurons in the PVH, the size of which is comparable to that stained for PHI.[31]

d. Vasopressin

In the adrenalectomized (ADX) or hypophysectomized rat, vasopressin immunoreactivity is demonstrable in a majority of CRF-stained cells in the PVHmpd.[43,104,124] In untreated or colchicine-treated animals, vasopressin immunoreactivity is seen in only a handful of CRF-stained cells (Figure 2). This steroid-dependent enhancement of vasopressin immunoreactivity in the PVHmpd is accompanied by increases in its mRNA,[136,138] an observation that provides independent support for the authenticity of the vasopressin immunoreactivity at this locus. The two peptides have been colocalized within neurosecretory vesicles in terminals in the median eminence of nonmanipulated animals, and the extent of such colocalization is enhanced following ADX.[133] These data indicate that vasopressin and its congeners are normally expressed at low levels in at least a subset of CRFir neurons. Whitnall and colleagues have provided evidence to support the notion that under normal conditions, vasopressin-deficient and vasopressin-expressing subpopulations of CRFir neurons can be distinguished, and that ADX renders parvocellular CRF neurons more uniformly vasopressin positive.[133-135]

e. Neurotensin

Like CRF-stained cells, neurotensin-ir neurons are concentrated in the PVHmpd, though the distributions of the two cell types show only a partial overlap. Consistent with this, neurotensin staining has been demonstrated in a small subset of CRF-stained cells in colchicine-treated animals.[103]

f. Angiotensin II

Angiotensin II immunoreactivity has been demonstrated in a sizeable subset of CRF-stained cells in PVHmpd, many of which also display vasopressin immunoreactivity.[52] Both ADX and colchicine treatment were required to demonstrate extensive angiotensin II immunostaining in parvocellular neurons.

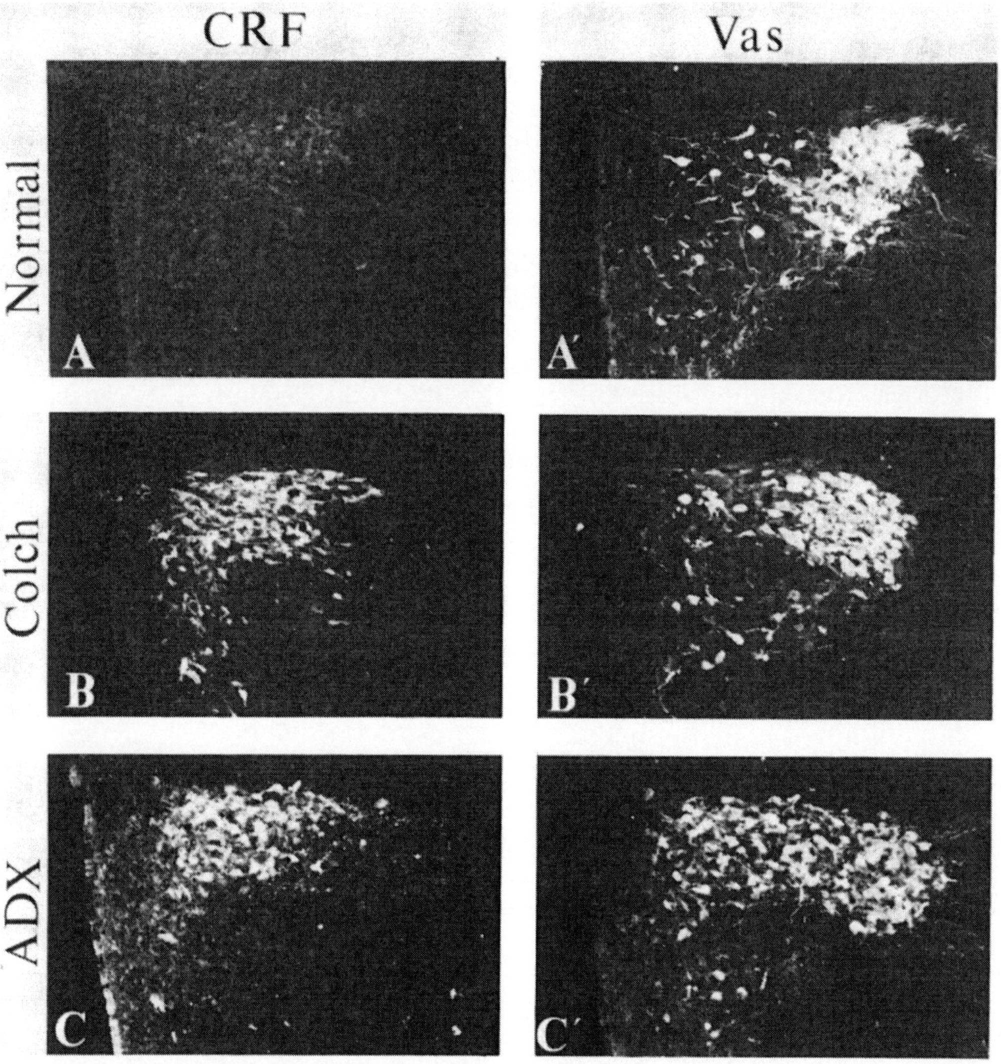

FIGURE 2. Distribution of CRF- and vasopressin-immunoreactive neurons in the PVH in untreated (normal), colchicine-treated (Colch) and adrenalectomized (ADX) rats. Each pair of fluorescence micrographs shows a single section through a common level of the PVH stained sequentially with antisera against CRF (left) and vasopressin (right). In the untreated animal, few CRF-positive cells are apparent, while in the same section many neurons centered in the compact magnocellular division of the nucleus are stained with antivasopressin. Colchicine treatment enhances CRF staining in a discrete subset of neurons in the parvocellular division of the PVH (i.e., in PVHmpd), while the distribution of vasopressin-stained neurons is similar to that seen in the normal rat. ADX enhances CRF immunostaining cells in the parvocellular division of the PVH; vasopressin-immunoreactive neurons are again seen in the magnocellular division of the nucleus and now are also seen in the same region (in fact in many of the same neurons) of the parvocellular division in which CRF-stained neurons are found. All micrographs ×75. (From Sawchenko, P. E., Swanson, L. W., and Vale, W. W., *Proc. Natl. Acad. Sci. U.S.A.*, 81, 1883, 1984. With permission.)

g. Cholecystokinin

Cholecystokinin-8 immunoreactivity has been demonstrated within parvocellular CRF- (and vasopressin-) stained cells in ADX rats treated with colchicine.[69,70] Evidence for a direct stimulatory effect of cholecystokinin on ACTH secretion, and an additive interaction with vasopressin, but not CRF, in this context has been provided.[70]

The presence of multiple neuroactive peptides in the parvocellular neurosecretory "CRF neuron" is, at first blush, perplexing in view of the vital and closely regulated function this cell group subserves. Although we are far from being able to provide a comprehensive account of the significance of this phenomenon, the fact that some of these coexisting peptides themselves exhibit corticotropin-releasing activity, and that their expression appears to follow lawfully fluctuations in circulating corticosteroid titers, has provided some initial insight. Additional clues may be gleaned from evidence to indicate that some of these peptides may be active in hypophysiotropic functions apart from those related directly to the control of corticotropin secretion.

3. Regulation

The manner in which the expression of CRF and related corticotropin-releasing peptides may be regulated within the parvocellular neurosecretory system has been the subject of a number of recent reviews[96,97,101,119,120] and will be summarized here only briefly.

Several of the early descriptions of the distribution of CRF immunoreactivity in the CNS noted that following ADX, staining was enhanced in the PVH and median eminence;[12,62,121] these reports differed as to whether, and to what extent, other cell groups in which CRF staining had been reported in colchicine-treated rats might similarly be affected. Although the generality of the effect of steroid withdrawal on CRF expression in extrahypothalamic regions remains at issue, the basic observation at the level of the PVH provided support, and a morphological focus, for the well-known negative feedback regulation by adrenal steroids of corticotropin-releasing activity of the hypothalamus that had been gleaned from earlier bioassay data.[5,42]

The effect of ADX on the parvocellular neurosecretory system is not limited to influences on CRF expression. Several groups have reported that vasopressin immunoreactivity is also enhanced in a majority of these very neurons,[43,104,124] this in addition to its "normal" localization in magnocellular neurosecretory somata. It now seems clear, despite the fact that vasopressin immunoreactivity is not readily demonstrable in parvocellular somata of untreated or colchicine-treated animals, that some level of expression of vasopressin and/or of vasopressin-related peptides occurs under normal conditions and is enhanced following steroid withdrawal.[133-135] This basic finding suggested the existence of a form of plasticity which is relevant to function, and suggested a substrate which could help explain the well-documented synergistic action of the two peptides in promoting corticotropin secretion.[27,87,126]

Subsequent work has shown that the ADX-induced enhancement of CRF and vasopressin immunoreactivity to be specific to peptides implicated as having direct stimulatory actions on ACTH secretion. This includes not only CRF and AVP, but also angiotensin II[52] and cholecystokinin as well.[70] Staining for at least two additional peptides known to be expressed in subsets of the parvocellular CRF neuron, met-enkephalin and neurotensin, is not perceptibly altered in ADX rats[93] (Figure 3). Moreover, within the PVH, ADX fails to produce a perceptible enhancement of CRF immunoreactivity in magnocellular neurosecretory cell groups[104] or in cells that give rise to descending, autonomic-related, projections,[94] in both of which staining for the peptide is readily demonstrated in colchicine-treated animals (Figure 1). Thus, evidence exists that the ADX-induced enhancement of staining for corticotropin-releasing peptides in the PVH shows at least some measure of specificity to particular peptides and cell types.

The enhancement in CRF- and vasopressin immunoreactivity in the system is accompanied by increases in their respective mRNAs[136-138] and by increases in the secretion of both peptides into hypophyseal portal plasma.[82] Systemic steroid replacement studies in ADX rats have shown the effect to be primarily glucocorticoid mediated,[22,93] with staining for both CRF and vasopressin varying in tandem to various replacement regimens[93] (Figures 4 and 5). The results of intracerebral replacement studies suggest the effect to be exerted, at least in part, at the level of receptors in or very near the PVH itself,[44,45,92] a view supported

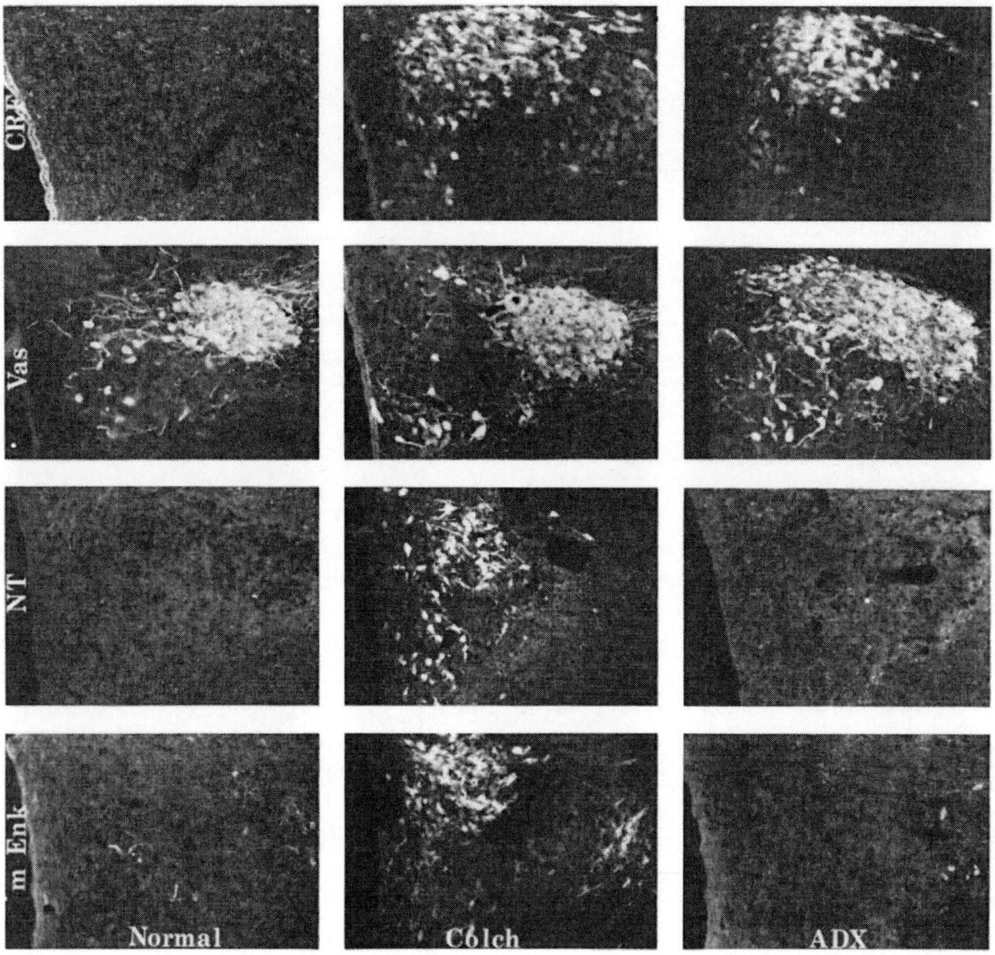

FIGURE 3. Comparison of the distribution of CRF, vasopressin (Vas), met-enkephalin (m-Enk), and neurotensin (NT) immunoreactivity at a similar level of the PVH in untreated (normal), colchicine-treated (Colch), and ADX rats. In the untreated rat, only magnocellular Vas-ir cells are prominent. Colchicine treatment reveals the presence of CRF-, m-Enk-, and NT-ir cells in the parvocellular division of the PVH. Colocalization studies have indicated that m-Enk- and NT-ir coexist with CRF in subsets of parvocellular neurons. ADX enhances CRF- and Vas-ir in parvocellular neurons, while staining for m-Enk and NT is not visibly affected. The effect of ADX appears to be specific to CRF and vasopressin, which alone among the peptides in question are known to stimulate ACTH secretion directly. All micrographs ×60. (From Sawchenko, P. E., *J. Neurosci.*, 7, 1093, 1987. With permission.)

by the localization of glucocorticoid receptor immunoreactivity within CRF-stained neurons in the parvocellular division.[1,55] Evidence exists to implicate other cell groups as playing a role in the feedback effects of corticosteroids on the central limb of this system, raising the possibility of remote, presumably transsynaptic, mediation.

Prominently implicated as modulators, or even mediators, of negative feedback effects of corticosteroids are catecholaminergic inputs from the brainstem. It has been suggested repeatedly that an intact catecholamine system is necessary for steroid feedback effects to be exerted on the CRF neuron,[23,39,112] and each of the aminergic cell groups that contribute to the innervation of the PVH[19,59,100,102,118,122] have been shown to display glucocorticoid receptor-IR.[29] In addition, various pharmacologic[68,113] and surgical[3] manipulations of ascending catecholaminergic pathways have been found to affect markedly staining for CRF and/or AVP in the PVH. Experiments in which unilateral transection of ascending aminergic

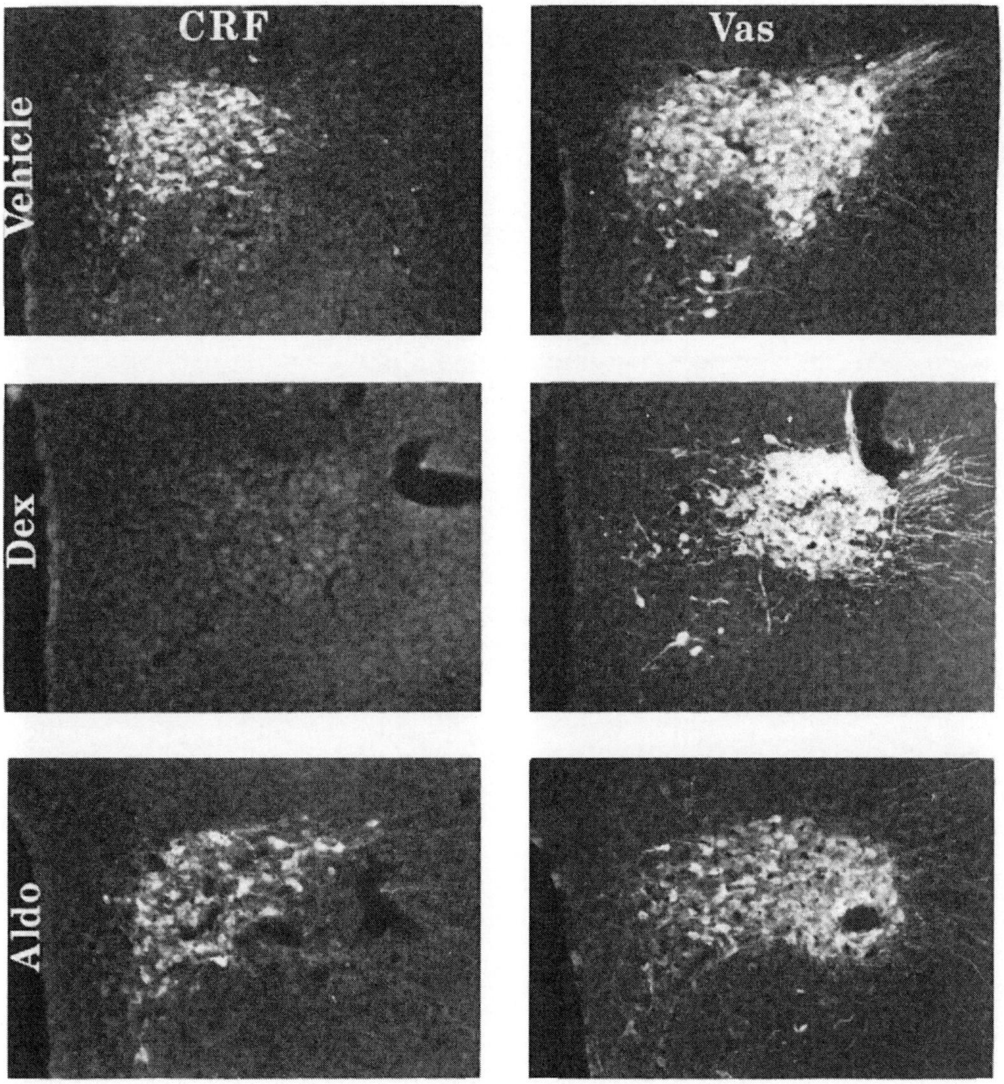

FIGURE 4. Adjacent sections through the PVH stained for CRF and vasopressin (Vas) immunoreactivity from ADX rats treated systemically with either vehicle (control), dexamethasone (Dex, a synthetic glucocorticoid), or a similar dose of aldosterone (Aldo, a relatively pure mineralocorticoid). At these dosage levels, replacement with Dex virtually abolished the ADX enhancement of both CRF- and Vas-ir in the parvocellular division of the PVH (Vas-ir magnocellular neurons remain prominently stained). In contrast, replacement with Aldo had virtually no impact on the response to ADX. The potency of adrenal steroids in antagonizing the ADX-dependent enhancement of both CRF and Vas staining appeared to be a direct function of their glucocorticoid activity. All micrographs ×75. (From Sawchenko, P. E., *J. Neurosci.*, 7, 1093, 1987. With permission.)

afferents was carried out in otherwise untreated ADX and ADX-dexamethasone treated rats[95] have suggested that while the ADX effect on CRF and vasopressin staining appears blunted on the lesioned side, some level of enhancement remains obvious. Moreover, the mitigating effect of steroid administration on the ADX-induced enhancement in staining for both peptides remained unaffected. While the mechanisms by which aminergic afferents influence peptide dynamics in the CRF neuron remain obscure, they appear to be largely independent of those through which corticosteroid effects are exerted.

Collectively, these data have begun to define the secretory *capabilities* of the CRF

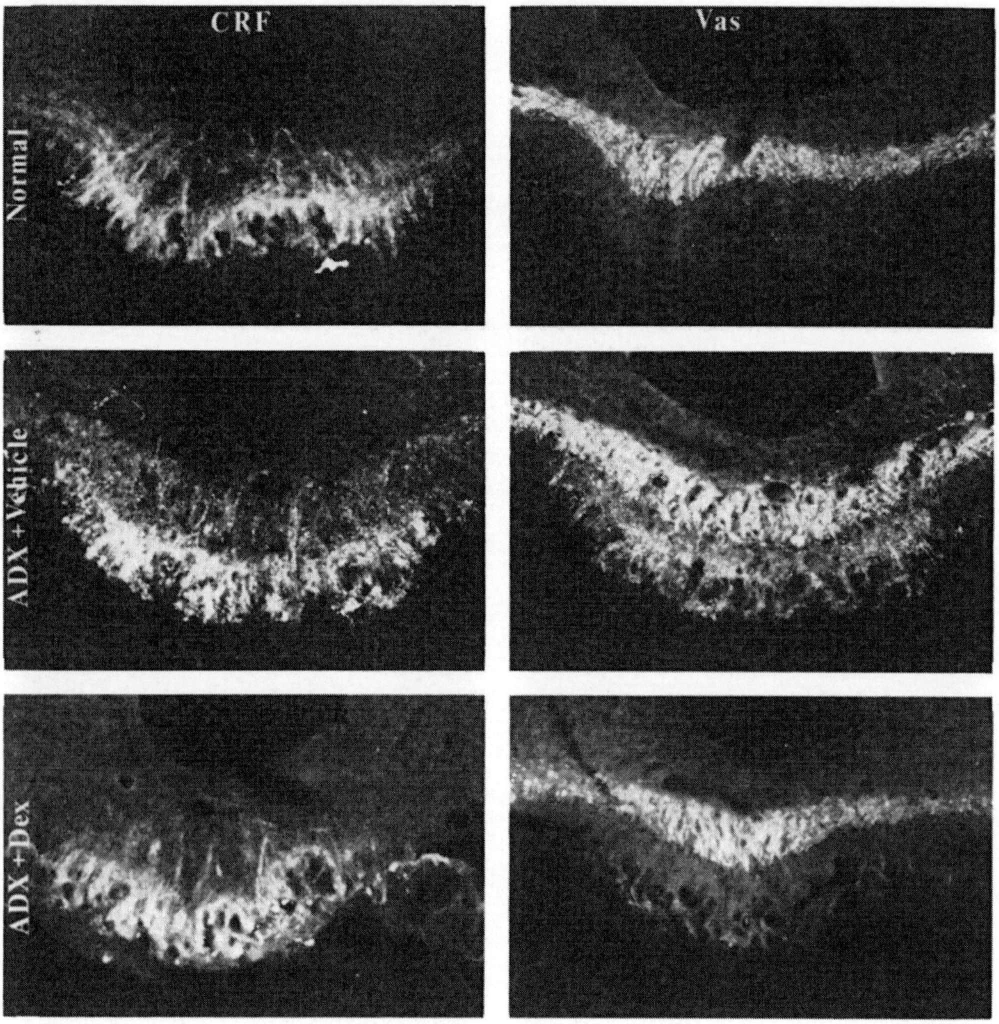

FIGURE 5. Adjacent sections through the median eminence of untreated (normal), vehicle-treated ADX, and dexamethasone-treated ADX rats, stained for CRF or vasopressin. In the normal rat, CRF-ir fibers are concentrated in the external lamina of the median eminence, while the preterminal axons of vasopressin-ir neurons course through the internal lamina; only very few vasopressin-stained fibers in the external zone are evident. ADX results in an enhancement in the staining intensity of CRF-ir fibers and in the appearance of a far more prominent vasopressinergic plexus in the external lamina. Treatment with dexamethasone through the duration of the 7-d period after ADX resulted in staining patterns that resemble those of the intact rat. All micrographs ×95. (From Sawchenko, P. E., *J. Neurosci.*, 7, 1093, 1987. With permission.)

neuron particularly as regards its classical role as the initiator of the stress response. Questions remain as to the relevance of these capabilities to the normal physiology of the animal, and as to the existence and extent of the involvement of peptides colocalized within the CRF neuron in different functional arenas. Recent studies have begun to address the former of these issues. Silverman and colleagues have reported an increase in the number of cells in the PVH displaying CRF- and oxytocin-ir following exposure to a chronic behavioral stress paradigm.[109] In a similar vein, rapid increases in CRF and proenkephalin mRNAs in the PVHmpd have been reported following stresses associated with a hypertonic saline challenge or opiate withdrawal.[51] Much needs to be done to identify the mechanisms and pathways by which the multiple secretory capabilities of the "CRF neuron" may be modified under defined naturalistic conditions.

In view of the organization of the hypothalamo-adenohypophyseal system, where multiple hypophysiotropic agents are secreted into a common soup (portal plasma), the diversity of functional roles served by any given secreted peptide is limited only by the number of target cell classes that contain matching receptors. It has recently been reported that AVP[56] and angiotensin II[106] are capable of acting directly on the anterior lobe, at physiological concentrations, to stimulate the secretion of thyroid-stimulating hormone and prolactin, respectively. Neurotensin has been reported to stimulate TSH secretion at both CNS and pituitary levels.[107,128] Met-enkephalin has long been thought to act within the CNS to regulate prolactin secretion,[20,60] a view buttressed by the observation that hypothalamic preproenkephalin mRNA is elevated in lactating animals.[132] The fact that each of these peptides can be demonstrated to coexist within the parvocellular "CRF neuron", which has access to portal plasma, raises the possibility that this "stress-related" cell group may indeed be subject to multiple regulatory influences, and, indirectly, that it may be functionally active in theaters other than that which subserves the stress response.

B. MAGNOCELLULAR NEUROSECRETORY SYSTEM

In the initial description of CRF immunoreactivity in the central nervous system, Bloom and colleagues[10] noted the presence of stained fibers in the posterior pituitary. This has since been confirmed repeatedly, and the bulk of the CRF-like material in the neurointermediate lobe has been shown to be chromatographically indistinguishable from synthetic rat CRF.[36] Ultrastructural localization of CRFir terminals in the posterior lobe showed these to be associated principally with vascular elements,[49,50] often directly apposing or ending in close proximity to the pericapillary space. Intriguingly, CRF-stained terminals in the posterior lobe are reported to contain small electron-lucent vesicles, and dense-cored secretory granules whose size (80 to 120 μm) is more characteristic of the parvocellular than magnocellular systems.[49] Subsequent mapping studies[17,63,121] described CRF-stained cells in the magnocellular division of the PVH and in the supraoptic nucleus (SO), which in the rat consists overwhelmingly, and perhaps exclusively, of magnocellular neurosecretory neurons. The distribution of CRF-stained elements in the magnocellular regions suggested at least a preferential association with oxytocinergic neurons,[121] and the weight of the evidence from subsequent colocalization studies has supported this view.

Descriptions of CRF-oxytocin colocalization in the rat and other species have been provided,[15,75,98,103] (Figure 6). Though estimates of the proportion of oxytocin cells in which CRF immunoreactivity may be demonstrated vary from 10 to 50%, it seems clear that CRF is typically only demonstrable in a subset of oxytocinergic neurons in the PVH of the colchicine-treated rat, and that those expressing both peptides show some degree of topographic organization. They are most frequently encountered rostrally, in the anterior magnocellular part of the PVH; this corresponds to the anterior commissural nucleus of Peterson,[78] and is acknowledged as an overwhelmingly oxytocinergic cell cluster.[34,85,99] The extent of colocalization tapers caudally such that at the level of the maximal development of the parvocellular neurosecretory CRF-stained cell group (i.e., in PVHmpd), relatively few examples of cells stained for both peptides are typically seen. Such topographic considerations may help to explain a failure to demonstrate CRF-oxytocin colocalization in the PVH;[123] it is not clear in this study that the more rostral aspects of the nucleus were sampled.

The situation as regards the supraoptic nucleus appears roughly analogous in that CRF immunoreactivity is typically seen only in a small subset of neurons.[15,103] In the principal part of the nucleus they are concentrated in the dorsal (oxytocin-rich) aspects, and have been colocalized within oxytocinergic somata. Kawata and colleagues[41] have described a considerably more widespread distribution of CRF-stained cells in the SO of a number of species including the rat, raising the possibility that the capacity to produce the peptide may be more widespread than is currently appreciated.

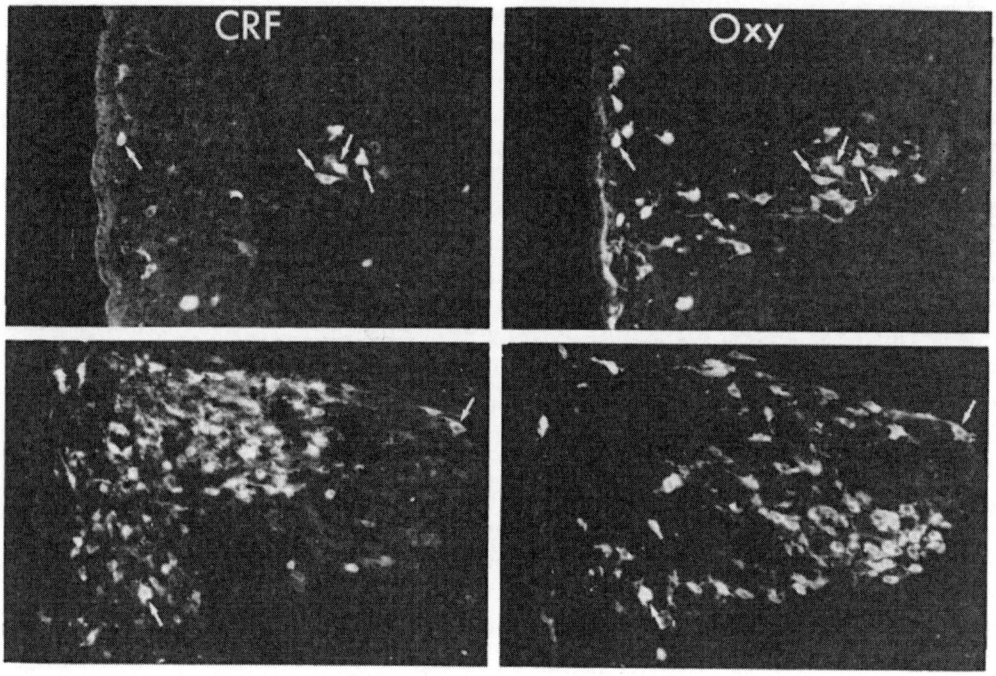

FIGURE 6. CRF-oxytocin (Oxy) colocalization in the PVH of a colchicine-treated female rat. Fluorescence photomicrographs of sections through the rostral (top) and caudal (bottom) magnocellular regions of the PVH stained concurrently for rat CRF and oxytocin. A sizeable subset of oxytocin-stained neurons in the rostral PVH contain CRF immunoreactivity (arrows show a few prominent examples), as does a smaller complement in the caudal region. A large cluster of cells in the parvocellular division at the more caudal level are immunoreactive only for CRF. All micrographs ×95.

Direct evidence to support an association of CRFir with vasopressinergic elements in the magnocellular system is available,[89] though other more or less direct examinations of this possibility have failed to support it. This study employed adrenalectomized, colchicine-treated rats and, in light of the effects of ADX on vasopressin expression outlined above, it is not clear whether magno- or parvocellular elements were sampled. In summary, the weight of the available evidence indicates that CRF immunoreactivity is contained at least primarily within magnocellular oxytocin neurons, with the anterior and medial parts of the magnocellular division of the PVH comprising the principal foci. This is consistent with the results of lesions studies suggesting that the PVH comprises the principal source of CRFir projections to the posterior lobe.[49]

Some data pertinent to the regulation of CRFir in the magnocellular system is available. In the course of studies on the effect of ADX on CRF- and vasopressin-ir in the parvocellular neurosecretory system, it was noted that CRFir in the magnocellular division was not noticeably enhanced following ADX.[104] By contrast, it has been reported that the neurointermediate lobe content of CRFir shows a dexamethasone-suppressible elevation in long term ADX rats,[36] and plasma CRF levels show similar effects.[115] It is therefore possible that the failure to observe enhanced CRF immunostaining in magnocellular perikarya following ADX may reflect a quantitative difference in the magnitude of the response of magnocellular and parvocellular systems to steroid withdrawal, rather than to qualitatively different manners of regulation.

Gonadal steroids also appear to be active in the modulation of CRF dynamics in the magnocellular system. Oxytocin immunoreactivity in the anterior and medial magnocellular parts of the PVH (where CRF immunostaining of magnocellular elements appears most pronounced) has been shown in quantitative immunohistochemical studies to be more responsive to manipulations such as water deprivation and estrogen treatment than that in the remainder of the system.[86] CRFir in the posterior lobe has been found[98] to be enhanced in ovariectomized, vehicle-treated rats relative to those that received estradiol replacement; oxytocin terminals showed the opposite tendencies. Thus, both major classes of steroids may be active in regulating CRF expression in the magnocellular system, and differential effects on coexisting principles may be exerted. The relative potencies of these influences remain to be evaluated.

The functional significance of the presence of CRF in the magnocellular system is not clear. Like most other peptides that coexist within magnocellular neurosecretory somata, it is expressed in the posterior lobe at levels 3 to 4 orders of magnitude lower than those of the nonapeptides,[11] making it unlikely that CRF of neurohypophyseal origin serves a hormonal function (CRF is detectable at low levels in rat plasma,[115] though its origins have not been established). By analogy with the more thoroughly characterized roles of coexisting opiate peptides in the magnocellular system,[9] it seems most likely that the peptide will be shown to act locally at the level of the posterior lobe to modify oxytocin and/or vasopressin secretion. Evidence for a stimulatory effect on vasopressin release from the isolated neurointermediate lobe has been provided;[4] the physiologic relevance of this remains to be demonstrated.

It should be pointed out in this context that oxytocin has been demonstrated in high concentrations in hypophyseal portal plasma,[26,80-82] and that strong evidence is available to suggest that it may serve as one of the several modulators of the effect of CRF on corticotropes. The route(s) by which oxytocin may be delivered to the portal vasculature are unknown, and retrograde blood flow from the posterior lobe represents one possibility. In light of indications that hormone may be released from axons of magnocellular neurosecretory neurons passing through the internal lamina of the median eminence,[33] the possibility must also be entertained that oxytocin (and CRF?) of magnocellular origin may reach the portal vasculature. In contrast to the situation with vasopressin, there presently exist no viable alternatives to magnocellular neurons as potential central source(s) for the delivery of oxytocin to the portal circulation.

III. CENTRAL PATHWAYS

There is now a great deal of immunohistochemical evidence to suggest that CRF is contained within a broad range of central neural systems.[17,38,63,111,121] In general, however, most of these pathways are related in one way or another to the medial forebrain bundle system, including limbic parts of the telencephalon, the hypothalamus, midline parts of the thalamus, and autonomic-related cell groups in the brainstem and spinal cord, with the notable exception of broadly distributed CRF in cortical interneurons (See Figures 7 and 8).

Most of this work has been carried out in normal material, where it is often not possible to determine with certainty the origin of particular pathways and terminal fields, or the sites to which CRFir neuronal cell bodies send their axons and terminals. However, over the last several years some experimental work has been directed at dissecting the organization of these pathways, and it is the results of this more recent work that will now be reviewed briefly.

A. PROJECTIONS FROM THE AMYGDALA

It now seems likely that at least a few CRFir neurons are found in almost all parts of the amygdala,[90] although the densest cluster lies in the central nucleus (Figure 7B). Lesion and combined retrograde tracer/immuno studies have shown that such neurons in the central nucleus project through, and most probably end within, lateral parts of the bed nucleus of

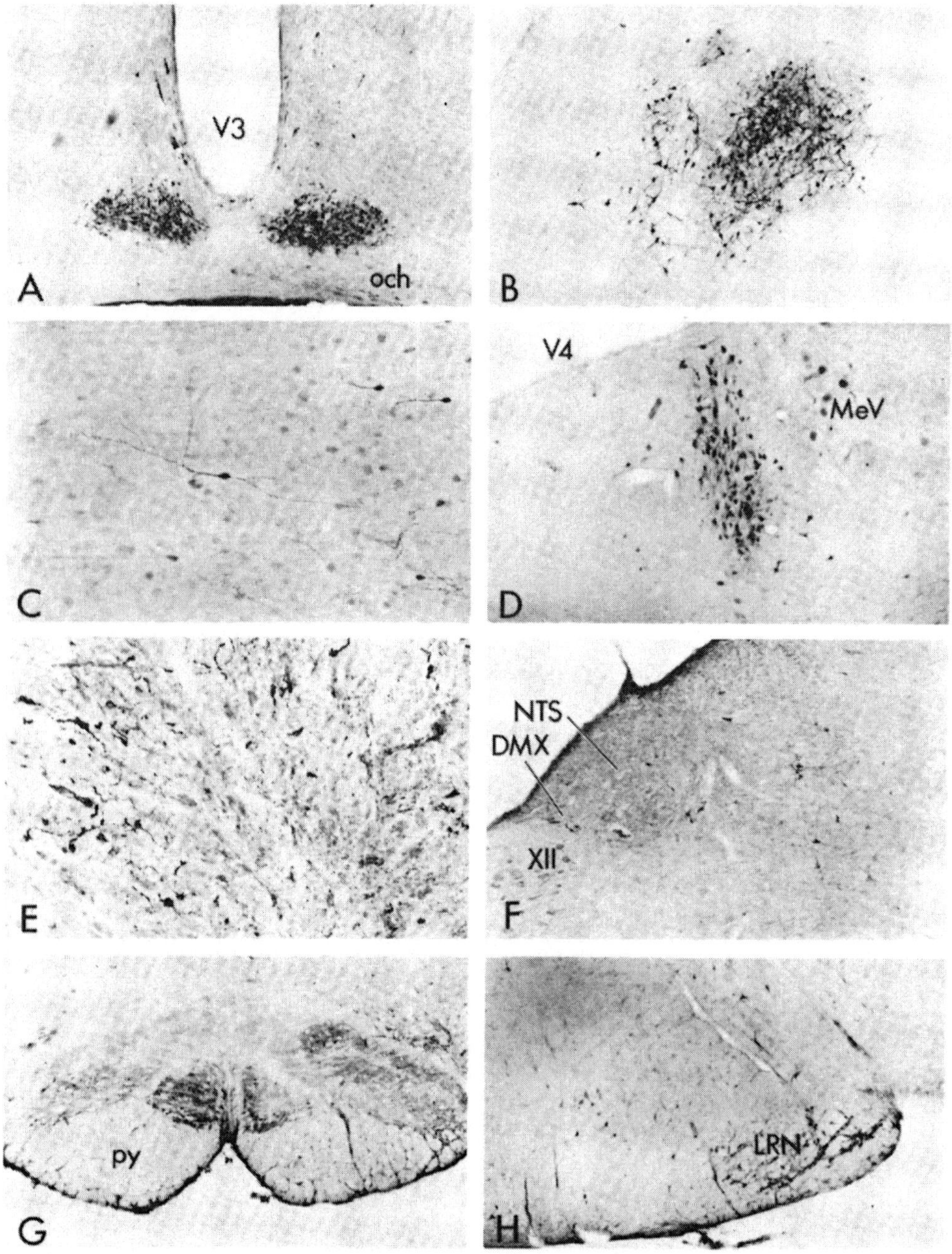

FIGURE 7. Some nonneuroendocrine CRFir cell groups, seen in colchicine-treated rats. (A) suprachiasmatic nuclei; (B) central nucleus of the amygdala; (C) perirhinal cortex (layers II and III); (D) laterodorsal tegmental (Barrington's) nucleus; (E) superior olive; (F) nucleus of the solitary tract; (G) inferior olive; (H) lateral reticular nucleus. Magnifications: Panels G and H (×50) A, B, D, and F (×80) C and E (×200). Abbreviations: DMX, dorsal motor nucleus of the vagus; LRN, lateral reticular nucleus; MeV, mesencephalic nucleus of the trigeminal nerve; NTS, nucleus of the solitary tract; och, optic chiasm; py, pyramidal tract; V3(4), third (fourth) ventricle; XII, hypoglossal nucleus.

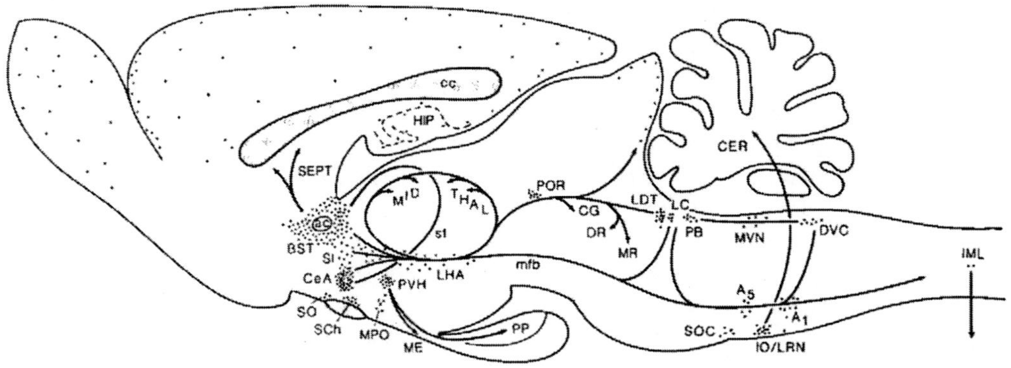

FIGURE 8. Summary of major CRFir cell groups (black dots) and fiber systems illustrated schematically in a sagittal view of the rat brain. Many of the immunoreactive cells and fibers appear to be associated with systems that regulate the output of the pituitary and the autonomic nervous system, though cortical interneurons and precerebellar relays are also CRFir. Most of the longer central fibers course either ventrally through the medial forebrain bundle and its caudal extension in the reticular formation, or dorsally through a periventricular system in the thalamus and brainstem central gray. The direction of fibers in many of these systems is unclear because they appear to interconnect regions that contain CRF-stained cell bodies. Abbreviations: A1(r), catecholamine cell groups; ac, anterior commissure; BST, bed nucleus of the stria terminalis; CeA, central nucleus of the amygdala; cc, corpus callosum; DR, dorsal raphe nucleus; DVC, dorsal vagal complex; HIP, hippocampal formation; IML, intermediolateral column; IO, inferior olive; LC, locus coeruleus; LDT, laterodorsal tegmental nucleus; LHA, lateral hypothalamic area; LRN, lateral reticular nucleus; ME, median eminence; MID THAL, midline thalamic nuclei; mfb, medial forebrain bundle; MPO, medial preoptic area; MR, median raphe nucleus; MVN, medial vestibular nucleus; PB, parabrachial nucleus; POR, perioculomotor region; PP, posterior pituitary; SCh, suprachiasmatic nucleus; SEPT, septal region; SI, substantia innominata; st, stria terminalis. (Modified from Swanson, L. W., Sawchenko, P. E., Rivier, J., and Vale, W. W., *Neuroendocrinology*, 36, 165, 1983.)

the stria terminalis,[90] the lateral hypothalamic area and midbrain reticular formation,[90] medial and lateral parts of the parabrachial nucleus,[72,132] the mesencephalic nucleus of the trigeminal,[90] and the dorsal vagal complex.[28,127] It would appear that the central nucleus projection to the parabrachial nucleus is substantial, since over half of the CRFir neurons in lateral parts of the nucleus can be retrogradely labeled after fast blue injections in the parabrachial nucleus,[72] whereas a considerably smaller proportion of CRFir neurons centered in lateral parts of the nucleus project to the dorsal vagal complex.[28,127] It has also been reported that a small proportion of the CRFir neurons in the intercollated nuclei of the amygdala project to the parabrachial nucleus as well.[71] For the sake of completeness, it should also be mentioned that a few CRFir neurons in lateral parts of the bed nucleus of the stria terminalis also project to the dorsal vagal complex.[28]

B. HYPOTHALAMIC PROJECTIONS

The hypothalamus contains several nonhypophysiotropic CRFir cell groups and some progress has been made in clarifying their projections. Thus, double-labeling studies have shown that at least some such neurons in the lateral hypothalamic area (and in adjacent parts of the zona incerta) send fibers to the inferior colliculus (which also contains CRFir cell bodies[91] and perhaps some CRFir terminals from neurons in the superior olive, Figure 7E) and dorsal vagal complex,[28] although they undoubtedly have much more widespread projections than this. It would appear that many CRFir neurons in this part of the hypothalamus cross-react with antisera to α-melanocyte stimulating hormone as well,[21] and perhaps melanin-concentrating hormone.[24] Because these peptides share common dipeptide sequences, the authenticity of the peptide recognized by antirat CRF sera remains to be established.[94] It may also be noted that combined retrograde/immuno studies indicate that many CRFir neurons in adjacent parts of the substantia innominata appear to send fibers at least as far caudally as the pedunculopontine nucleus, but not to the cerebral cortex (Swerdlow and Swanson, unpublished observations).

It is now clear that the ventral (visuoceptive) part of the suprachiasmatic nucleus contains a substantial number of CRFir neurons[24,73] (Figure 7A), although it has been suggested that they may primarily give rise to local axonal plexuses within the nucleus itself.[130]

On the other hand, a rather substantial population of CRFir neurons has been reported in various parts of the mammillary body of the rat, and there is some evidence to suggest that they may give rise to fibers in the mammillothalamic tract.[46] If so, they presumably end in the anterior thalamic group, although substantial CRFir terminal fields have not been reported in this region. On the other hand, it now seems clear that midline parts of the thalamus contain a relatively small population of CRFir cell bodies and fibers.[65]

C. BRAINSTEM PROJECTIONS

Undoubtedly the most interesting recent development in this field is the demonstration of a population of CRFir neurons in the inferior olive of rats, cats, baboons, and humans that give rise to climbing fibers to the cerebellum[17,74,83,139] (Figure 7G). It would appear that such neurons are found in all subdivisions of the inferior olive in the rat.[74] Interestingly, what appear to be CRFir mossy fibers in the granular layer of the cat and sheep cerebellum have also been described recently, although their source was not identified.[18] However, it is possible that they arise, in part at least, from the lateral reticular nucleus (Figure 7H).

At least two CRFir cell groups in the brainstem appear to send descending fibers to the spinal cord. One population is centered in the laterodorsal tegmental (or Barrington's) nucleus and sends fibers to the intermediolateral column at sacral levels of the cord.[129] This pathway may play a role in the parasympathetic regulation of micturition. The second population of cells is centered in the region of the classically defined Edinger-Westphal nucleus (however, see Reference 114), and is retrogradely labeled after HRP injections into cervical levels of the spinal cord in the cat.[16] The precise distribution of these fibers in the cord is not known. Cells in the region of the Edinger-Westphal nucleus that are stained with antisera to substance P[79] or cholecystokinin[57] also project to the spinal cord, although it is not yet clear whether any or all of these peptides are colocalized within individual cells.

D. THE SPINAL CORD

Two aspects of CRF immunostaining in the spinal cord are of particular interest. The first concerns reports of dense CRFir in the central terminal fields of small dorsal root (and trigeminal) ganglion cells.[105,110,111] However, it now seems clear, based on an extensive study of the problem using a gelatin model system,[8] that some antisera to ovine CRF cross-react with substance P, which is found in high concentrations in many such ganglion cells. Other antisera to ovine CRF (e.g., Reference 121), and antisera to rat CRF (e.g., Reference 8) do not appear to stain sensory ganglion cells and their central terminal fields. And second, a report[64] of CRFir neurons in lower parts of the thoracolumbar intermediolateral column (which are presumably preganglionic sympathetic neurons) of the rat has been confirmed and extended in the cat.[47] Here, CRFir neurons are found at all levels of the intermediolateral column from T_2 to L_3, although the vast majority are concentrated between segments T_2 to T_7 and between segments L_2 and L_3.

IV. CONCLUSIONS

Clearly, a good deal has been learned about the organization of central CRFir systems in their relatively brief history. By way of summary, we shall simply list some outstanding issues considered deserving of experimental attention.

A. PARVOCELLULAR NEUROSECRETORY SYSTEM

Work summarized above has made it apparent that the parvocellular "CRF neuron" is not merely or simply a precorticotrope. It may well, in fact, prove inappropriate to refer to this population generically. Subpopulations may ultimately be defined on the basis of unique complements of coexisting peptides and/or their responsiveness to regulatory factors. Among the most pressing needs at present are (1) thorough analysis of the origins of CRFir projections to the median eminence; (2) profiles of coexisting peptide expression in animals exposed to naturalistic situations and stressors; (3) an evaluation of the role of afferent inputs (e.g., aminergic, limbic) to the CRF neuron in regulating peptide dynamics, either as targets for corticosteroid feedback, or as independent modulators — the level at which any such effects may be exerted (transcriptional, translational, posttranslational) also remain to be clarified; and (4) an evaluation of nonsteroidal factors that may regulate the expression of peptides that coexist within subsets of the parvocellular CRF neuron, some of which play no obvious role in modifying corticotropin secretion. Might the "CRF neuron" under certain conditions function prominently to modify secretions of somatotropes, thyrotropes, or lactotropes, as well as that of corticotropes?

B. MAGNOCELLULAR NEUROSECRETORY SYSTEM

For CRF, as well as the many other peptides thought to coexist within magnocellular perikarya, the major question that remains is one of function, specifically, what is it doing there? The levels at which coexisting peptides are expressed in the magnocellular system are suggestive of local (autocrine or paracrine), rather than classical endocrine, roles. The cellular resolution afforded by morphologic (immunohistochemical and hybridization histochemical) approaches will no doubt contribute to solving this riddle, and assessments of the response(s) of coexisting principles to putative regulatory agents may provide initial clues as to functional associations.

A corollary to this, and of paramount importance in understanding the central limb of the stress response, is a determination of the extent to which magnocellular neurons may comprise a source for the delivery of CRF and other secretagogues for ACTH to the portal vasculature. This, too, is an issue that will require converging anatomical and functional methodologies to resolve.

C. NONNEUROSECRETORY SYSTEMS

The preponderance of localizations in central neuroendocrine- and autonomic-related systems described in initial surveys of CRFir in the rat brain provided a basis for suggesting that the central CRF system might function in a unitary manner to orchestrate endocrine, autonomic, and behavioral responses to stress. Indeed, physiologic and behavioral evidence (described elsewhere in this volume) has been marshalled to support this view. It must be pointed out, however, that some of the more recently established CRFir systems, such as in precerebellar and auditory relays, as well as in neocortical interneurons, make it appear unlikely that such a generalization applies in any universal sense. That components of the CRFir system involved in different modalities of stress-related responses may function in concert remains a viable possibility, though any preferential, or situation-specific, role for CRFir mechanisms in stress-induced autonomic or behavioral arousal remains to be established.

REFERENCES

1. **Agnati, L. F., Fuxe, K., Yu, Z.-Y., Harfstrand, A., Okret, S., Wilkstrom, A.-C., Goldstein, M., Zoli, M., Vale, W., and Gustafsson, J.-A.,** Morphometrical analysis of the distribution of corticotropin-releasing factor, glucocorticoid receptor and phenylethanolamine-N-methyltransferase immunoreactive structures in the paraventricular hypothalamic nucleus of the rat, *Neurosci. Lett.*, 54, 147, 1985.
2. **Alonso, G. and Assenmacher, I.,** Radioautographic studies on the neurohypophyseal projections of the supraoptic and paraventricular nuclei in the rat, *Cell Tissue Res.*, 219, 525, 1981.
3. **Alonso, G., Szafarczyk, A., Balmefrezol, M., and Assenmacher, I.,** Immunocytochemical evidence for stimulatory control by the ventral noradrenergic bundle of the paraventricular nucleus secreting corticotropin releasing factor and vasopressin in rats, *Brain Res.*, 397, 297, 1986.
4. **Alzein, M., Jeandel, L., Lutz-Bucher, B., and Koch, B.,** Evidence that CRF stimulates vasopressin secretion from isolated neurointermediate pituitary, *Neuroendocrinol. Lett.*, 6, 151, 1984.
5. **Antoni, F. A.,** Hypothalamic control of adrenocorticotropin secretion: advances since the discovery of 41-residue corticotropin-releasing factor, *Endocr. Rev.*, 7, 351, 1986.
6. **Antoni, F. A., Palkovits, M., Makara, G., Linton, E. A., Lowry, P. J., and Kiss, J. Z.,** Immunoreactive corticotropin-releasing hormone in the hypothalamoinfundibular tract, *Neuroendocrinology*, 36, 415, 1983.
7. **Berkenbosch, F., Linton, E. A., and Tilders, F. J. H.,** Colocalization of PHI- and CRF-immunoreactivity in neurons of the rat hypothalamus: a surprising artefact, *Neuroendocrinology*, 44, 338, 1986.
8. **Berkenbosch, F., Schipper, J., and Tilders, F. J. H.,** Corticotropin-releasing factor immunostaining in the rat spinal cord and medulla oblongata: an unexpected form of cross-reactivity with substance P, *Brain Res.*, 399, 87, 1986.
9. **Bicknell, R. J.,** Endogenous opioid peptides and hypothalamic neuroendocrine neurons, *J. Endocrinol.*, 107, 437, 1985.
10. **Bloom, F. E., Battenberg, E. L. F., Rivier, J., and Vale, W.,** Corticotropin-releasing factor: immunoreactive neurons and fibers in the rat hypothalamus, *Regul. Pep.*, 4, 43, 1982.
11. **Brownstein, M. J. and Mezey, E.,** Multiple chemical messengers in hypothalamic magnocellular neurons, in *Progress in Brain Research,* Vol. 68, Hökfelt, T., Fuxe, K., and Pernow, B., Eds., Elsevier, Amsterdam, 1986, 161.
12. **Bugnon, C., Fellmann, D., and Gouget, A.,** Changes in corticoliberin and vasopressin-like immunoreactivities in the zona externa of the median eminence in adrenalectomized rats. Immunocytochemical study, *Neurosci. Lett.*, 37, 43, 1983.
13. **Bugnon, C., Fellmann, D., Gouget, A., and Cardot, J.,** Corticoliberin in rat brain: immunocytochemical localization of a novel neuroglandular system, *Neurosci. Lett.*, 30, 25, 1982.
14. **Bugnon, C., Hadjiyiassemis, M., Fellmann, D., and Cardot, J.,** Reserpine-induced depletion of corticoliberin (CRF)-like immunoreactivity in the zona externa of the rat median eminence, *Brain Res.*, 275, 198, 1983.
15. **Burlet, A., Tonon, M.-C., Tankosic, P., Coy, D., and Vaudry, H.,** Comparative immunocytochemical localization of corticotropin releasing factor (CRF-41) and neurohypophyseal peptides in the brain of Brattleboro and Long-Evans rats, *Neuroendocrinology*, 37, 64, 1983.
16. **Chung, R. Y., Mason, P., Strassman, A., and Maciewicz, R.,** Edinger-Westphal nucleus: cells that project to spinal cord contain corticotropin-releasing factor, *Neurosci. Lett.*, 83, 13, 1987.
17. **Cummings, S., Elde, R., Ells, J., and Lindall, A.,** Corticotropin releasing factor immunoreactivity is widely distributed within the central nervous system of the rat: an immunohistochemical study, *J. Neurosci.*, 3, 1355, 1983.
18. **Cummings, S., Sharp, B., and Elde, R.,** Corticotropin-releasing factor in cerebellar afferent systems: a combined immunohistochemistry and retrograde transport study, *J. Neurosci.*, 8, 543, 1988.
19. **Cunningham, E. T., Jr. and Sawchenko, P. E.,** Anatomical specificity of noradrenergic inputs to the paraventricular and supraoptic nuclei of the rat hypothalamus, *J. Comp. Neurol.*, in press.
20. **Cusan, L., Dupont, A., Klendzik, G. S., Labrie, F., Coy, D. H., and Schally, A. V.,** Potent prolactin and growth hormone releasing activities of more analogues of met-enkephalin, *Nature,* 268, 544, 1977.
21. **Daikoku, S., Okamura, Y., Kawano, H., Tsuruo, Y., Maegawa, M., and Shibasaki, T.,** CRF-containing neurons of the rat hypothalamus, *Cell Tissue Res.*, 240, 575, 1985.
22. **Davis, L. G., Arentzen, R., Reid, J. M., Manning, R. W., Wolfson, B., Lawrence, K. L., and Baldino, F.,** Glucocorticoid sensitivity of vasopressin mRNA levels in the paraventricular nucleus of the rat, *Proc. Natl. Acad. Sci. U.S.A.*, 83, 1145, 1986.
23. **Feldman, S., Siegel, R. A., Wiedenfeld, J., Conforti, N., and Melamed, E.,** Role of medial forebrain bundle catecholaminergic fibers in the modulation of glucocorticoid negative feedback effects, *Brain Res.*, 260, 297, 1983.
24. **Fellman, D., Bugnon, C., and Risold, P. Y.,** Unrelated peptide immunoreactivities coexist in neurons of the rat lateral dorsal hypothalamus: human growth hormone-releasing factor$_{1\text{-}37}$-, salmon melanin-concentrating hormone- and α-melanotropin-like substances, *Neurosci. Lett.*, 74, 275, 1987.

25. **Furutani, Y., Morimoto, Y., Shibahara, S., Noda, M., Takahashi, H., Hirose, T., Asai, M., Inayama, S., Hayashida, H., Miyata, T., and Numa, S.,** Cloning and sequence analysis of cDNA for ovine corticotropin-releasing factor precursor, *Nature,* 301, 537, 1983.
26. **Gibbs, D. M.,** Measurement of hypothalamic corticotropin-releasing factors in hypophyseal portal blood, *Fed. Proc.,* 44, 203, 1985.
27. **Gillies, G. E., Linton, E. A., and Lowry, P. J.,** Corticotropin releasing activity of the new CRF is potentiated several times by vasopressin, *Nature,* 299, 355, 1982.
28. **Gray, T. S. and Magnuson, D. J.,** Neuropeptide neuronal efferents from the bed nucleus of the stria terminalis and central amygdaloid nucleus to the dorsal vagal complex in the rat, *J. Comp. Neurol.,* 262, 365, 1987.
29. **Harfstrand, A., Fuxe, K., Cintra, A., Agnati, L. F., Zinin, I., Wikstrom, A.-C., Okret, S., Yu, Z.-Y., Goldstein, M., Steinbusch, H., Verhofstad, A., and Gustafsson, J.-A.,** Glucocorticoid receptor immunoreactivity in monoaminergic neurons of rat brain, *Proc. Natl. Acad. Sci. U.S.A.,* 83, 9779, 1986.
30. **Hisano, S., Daikoku, S., Yanihora, N., and Shibasaki, T.,** Intragranular colocalization of CRF and met-enk-8 in nerve terminals in the rat median eminence, *Brain Res.,* 370, 321, 1986.
31. **Hökfelt, T., Fahrenkrug, J., Ju, G., Ceccatelli, S., Tsuruo, Y., Meister, B., Mutt, V., Rundgren, M., Brodin, E., Terenius, L., Hulting, A.-L., Werner, S., Björklund, H., and Vale, W.,** Analysis of peptide histidine-isoleucine/vasoactive intestinal polypeptide-immunoreactive neurons in the central nervous system with special reference to their relation to corticotropin releasing factor- and enkephalin-like immunoreactivities in the paraventricular hypothalamic nucleus, *Neuroscience,* 23, 827, 1987.
32. **Hökfelt, T., Fahrenkrug, J., Tatemoto, K., Mutt, V., Werner, S., Hultings, A.-L., Terenius, L., and Chang, K. J.,** The PHI (PHI-27)/corticotropin-releasing factor/enkephalin immunoreactive hypothalamic neuron: possible morphological basis for integrated control of prolactin, corticotropin, and growth hormone secretion, *Proc. Natl. Acad. Sci. U.S.A.,* 80, 895, 1983.
33. **Holmes, M. C., Antoni, F. A., Aguilera, G., and Catt, K. J.,** Magnocellular axons in passage through the median eminence release vasopressin, *Nature,* 319, 326, 1986.
34. **Hou-Yu, A., Lamme, A. T., Zimmerman, E. A., and Silverman, A.-J.,** Comparative distribution of vasopressin and oxytocin neurons in the rat brain using a double-label procedure, *Neuroendocrinology,* 44, 235, 1986.
35. **Ixart, G., Alonso, G., Szarfarczyk, A., Malaval, F., Nougier-Soule, J., and Assenmacher, I.,** Adrenocorticotropic regulations after bilateral lesions of the paraventricular or supraoptic nuclei and in Brattelboro rats, *Neuroendocrinology,* 35, 270, 1982.
36. **Jeandel, L., Van Dorsselar, A., Lutz-Bucher, B., and Koch, B.,** Characterization and modulation of corticotropin-releasing factor in the neurointermediate pituitary gland, *Neuroendocrinology,* 45, 146, 1987.
37. **Jingami, H., Mizuno, N., Takahashi, H., Shibahara, S., Furutani, Y., Imura, H., and Numa, S.,** Cloning and sequence analysis of cDNA for rat corticotropin-releasing factor, *FEBS Lett.,* 191, 63, 1985.
38. **Joseph, S. A. and Knigge, K. M.,** Corticotropin releasing factor: immunocytochemical localization in rat brain, *Neurosci. Lett.,* 35, 135, 1983.
39. **Kaneko, M. and Hiroshige, T.,** Site of fast, rate-sensitive feedback inhibition of adrenocorticotropin secretion during stress, *Am. J. Physiol.,* 234, R46, 1978.
40. **Kawata, M., Hashimoto, K., Takahara, J., and Sano, Y.,** Immunohistochemical identification of the corticotropin releasing factor (CRF)-containing nerve fibers in the pig hypophysis, with special reference to the relationship between CRF and posterior lobe hormones, *Arch. Histol. Jpn.,* 46, 183, 1983.
41. **Kawata, M., Hashimoto, K., Takahara, J., and Sano, Y.,** Immunohistochemical identification of neurons containing corticotropin-releasing factor in the rat hypothalamus, *Cell Tissue Res.,* 230, 239, 1983.
42. **Keller-Wood, M. E. and Dallman, M. F.,** Corticosteroid inhibition of ACTH secretion, *Endocr. Rev.,* 5, 1, 1984.
43. **Kiss, J. Z., Mezey, E., and Skirboll, L.,** Corticotropin-releasing factor-immunoreactive neurons of the paraventricular nucleus become vasopressin positive after adrenalectomy, *Proc. Natl. Acad. Sci. U.S.A.,* 81, 1854, 1984.
44. **Kóvács, K., Kiss, J. Z., and Makara, G.,** Glucocorticoid implants around the hypothalamic paraventricular nucleus prevent the increase of corticotropin-releasing factor and arginine vasopressin immunostaining induced by adrenalectomy, *Neuroendocrinology,* 44, 229, 1986.
45. **Kovács, K. and Mezey, E.,** Dexamethasone inhibits corticotropin releasing factor gene expression in the paraventricular nucleus of the rat, *Neuroendocrinology,* 46, 365, 1987.
46. **Kovács, M., Lengvári, I., Liposits, Z., Vigh, S., and Flerkó, B.,** Corticotropin-releasing factor (CRF)-immunoreactive neurons in the mammillary body of the rat, *Cell Tissue Res.,* 240, 455, 1985.
47. **Krukoff, T. L.,** Segmental distribution of corticotropin-releasing factor-like and vasoactive intestinal peptide-like immunoreactivities in presumptive sympathetic preganglionic neurons of the cat, *Brain Res.,* 382, 153, 1986.

48. **Lechan, R. M., Nestler, J. M., Jacobson, S., and Reichlin, S.,** The hypothalamic tuberoinfundibular system of the rat as demonstrated by horseradish peroxidase (HRP) microiontophoresis, *Brain Res.*, 111, 55, 1980.
49. **Lenguari, I., Liposits, Z., Vigh, S., Schally, A. V., and Flerko, B.,** The origin and characteristics of corticotropin-releasing factor (CRF)-immunoreactive nerve fibers in the posterior pituitary of the rat, *Cell Tissue Res.*, 240, 467, 1985.
50. **Léranth, C., Antoni, F. A., and Palkovits, M.,** Ultrastructural demonstration of ovine CRF-like immunoreactivity (oCRF-LI) in the rat hypothalamus: processes of magnocellular neurons establish membrane specializations with parvocellular neurons containing oCRF-LI, *Regul. Pept.*, 6, 179, 1983.
51. **Lightman, S. L. and Young, W. S., III,** Changes in hypothalamic preproenkephalin mRNA following stress and opiate withdrawal, *Nature*, 328, 643, 1987.
52. **Lind, R. W., Swanson, L. W., and Sawchenko, P. E.,** Anatomical evidence that neural circuits related to the subfornical organ contain angiotensin II, *Brain Res. Bull.*, 15, 79, 1985.
53. **Liposits, Zs. and Paull, W. K.,** Ultrastructural alterations of the hypothalamo-infundibular corticotropin releasing factor (CRF)-immunoreactive neuronal system in long term adrenalectomized rats, *Peptides*, 6, 1021, 1985.
54. **Liposits, Zs., Paull, W. K., Setalo, G., and Vigh, S.,** Evidence for local corticotropin-releasing factor (CRF)-immunoreactive neuronal circuits in the paraventricular nucleus of the rat hypothalamus, *Histochemistry*, 83, 5, 1985.
55. **Liposits, Zs., Uht, R. M., Harrison, R. W., Gibbs, F. P., Paull, W. K., and Bohn, M. C.,** Ultrastructural localization of glucocorticoid receptor (GR) in hypothalamic paraventricular neurons synthesizing corticotropin-releasing factor (CRF), *Histochemistry*, 87, 407, 1987.
56. **Lumpkin, M. D., Samson, W. K., and McCann, S. M.,** Arginine vasopressin as a thyrotropin-releasing hormone, *Science*, 235, 1070, 1987.
57. **Maciewicz, R., Phipps, B. S., Grenier, J., and Poletti, C. E.,** Edinger-Westphal nucleus: cholecystokinin immunocytochemistry and projections to spinal cord and trigeminal nucleus in the cat, *Brain Res.*, 299, 139, 1984.
58. **Makara, G., Stark, E., Karteszi, M., Palkovits, M., and Rappay, G.,** Effects of paraventricular lesions on stimulated ACTH release and CRF in stalk-median eminence of the rat, *Am. J. Physiol.*, 240, E441, 1981.
59. **McKellar, S. and Loewy, A. D.,** Organization of some brain stem afferents to the paraventricular nucleus of the hypothalamus in the rat, *Brain Res.*, 217, 351, 1981.
60. **Meites, J., Bruni, J. F., Van Vugt, D. A., and Smith, A. F.,** Relation of endogenous opioid peptides and morphine to neuroendocrine functions, *Life Sci.*, 24, 1325, 1979.
61. **Merchenthaler, I., Hynes, M. A., Vigh, S., Schally, A. V., and Petrusz, P.,** Corticotropin releasing factor (CRF): origin and course of afferent pathways to the median eminence (ME) of the rat hypothalamus, *Neuroendocrinology*, 39, 296, 1984.
62. **Merchenthaler, I., Vigh, S., Petrusz, P., and Schally, A. V.,** The paraventriculo-infundibular corticotropin releasing factor (CRF) pathway as revealed by immunocytochemistry in long-term hypophysectomized or adrenalectomized rats, *Regul. Pept.*, 5, 295, 1983.
63. **Merchenthaler, I., Vigh, S., Petrusz, P., and Schally, A. V.,** Immunocytochemical localization of corticotropin-releasing factor (CRF) in the rat brain, *Am. J. Anat.*, 165, 385, 1982.
64. **Merchenthaler, I., Hynes, M. A., Vigh, S., Schally, A. V., and Petrusz, P.,** Immunocytochemical localization of corticotropin releasing factor (CRF) in the rat spinal cord, *Brain Res.*, 275, 373, 1983.
65. **Merchenthaler, I., Vigh, S., Schally, A. V., Stumpf, W. E., and Arimura, A.,** Immunocytochemical localization of corticotropin releasing factor (CRF)-like immunoreactivity in the thalamus of the rat, *Brain Res.*, 323, 119, 1984.
66. **Mezey, E.,** Vasoactive intestinal polypeptide immunopositive neurons in the paraventricular nucleus of homozygous Brattleboro rats, *Neuroendocrinology*, 42, 88, 1986.
67. **Mezey, E. and Kiss, J. Z.,** Vasoactive intestinal peptide containing neurons in the paraventricular nucleus may participate in regulating prolactin secretion, *Proc. Natl. Acad. Sci. U.S.A.*, 82, 245, 1985.
68. **Mezey, E., Kiss, J. Z., Skirboll, L. R., Goldstein, M., and Axelrod, J.,** Increase of corticotropin releasing factor staining in rat paraventricular nucleus neurons by depletion of hypothalamic adrenaline, *Nature*, 310, 140, 1984.
69. **Mezey, E., Reseine, T. D., Skirboll, L., Beinfeld, M., and Kiss, J. Z.,** Cholecystokinin in the medial parvocellular subdivision of the paraventricular nucleus, *Ann. N.Y. Acad. Sci.*, 448, 152, 1985.
70. **Mezey, E., Reisine, T. D., Skirboll, L., Beinfeld, M., and Kiss, J. Z.,** Role of cholecystokinin in corticotropin release: coexistence with vasopressin and corticotropin-releasing factor in cells of the rat hypothalamic paraventricular nucleus, *Proc. Natl. Acad. Sci. U.S.A.*, 83, 3510, 1986.
71. **Moga, M. M. and Gray, T. S.,** Peptidergic efferents from the intercalated nuclei of the amygdala to the parabrachial nucleus in the rat, *Neurosci. Lett.*, 61, 13, 1985a.

72. **Moga, M. M. and Gray, T. S.,** Evidence for corticotropin-releasing factor, neurotensin, and somatostatin in the neural pathway from the central nucleus of the amygdala to the parabrachial nucleus, *J. Comp. Neurol.*, 241, 275, 1985b.
73. **Olschowka, J. A., O'Donohue, T. L., Mueller, G. P., and Jacobowitz, D. M.,** The distribution of corticotropin releasing factor-like immunoreactive neurons in rat brain, *Peptides*, 3, 995, 1982.
74. **Palkovits, M., Léránth, C., Görcs, T., and Young, W. S., III,** Corticotropin-releasing factor in the olivocerebellar tract of rats: demonstration by light- and electron-microscopic immunohistochemistry and *in situ* hybridization histochemistry, *Proc. Natl. Acad. Sci. U.S.A.*, 84, 3911, 1987.
75. **Papadopoulos, G. C., Karamanlidis, A. N., Michaloudi, H., Dinopoulos, A., Antonopoulos, J., and Parnavelas, J. G.,** The coexistence of oxytocin and corticotropin-releasing factor in the hypothalamus: an immunocytochemical study in the rat, sheep and hedgehog, *Neurosci. Lett.*, 62, 213, 1985.
76. **Paull, W. K. and Gibbs, F. P.,** The corticotropin releasing factor (CRF) neurosecretory system in intact, adrenalectomized, and adrenalectomized-dexamethasone treated rats, *Histochemistry*, 78, 303, 1983.
77. **Pelletier, G., Desy, L., Cote, J., Lefevre, G., Vaudry, H., and Labrie, F.,** Immunoelectron microscopic localization of corticotropin-releasing factor in the rat hypothalamus, *Neuroendocrinology*, 35, 402, 1982.
78. **Peterson, R. P.,** Magnocellular neurosecretory centers in the rat hypothalamus, *J. Comp. Neurol.*, 128, 181, 1966.
79. **Phipps, B. S., Maciewicz, R., Sandrew, B. B., Poletti, C. E., and Foote, W. E.,** Edinger-Westphal neurons that project to spinal cord contain substance P, *Neurosci. Lett.*, 36, 125, 1983.
80. **Plotsky, P. M.,** Hypophyseotropic regulation of adenohypophyseal adrenocorticotropin secretion, *Fed. Proc.*, 44, 207, 1985.
81. **Plotsky, P. M., Bruhn, T. O., and Vale, W.,** Hypophysiotropic regulation of adrenocorticotropin secretion in response to insulin-induced hypoglycemia, *Endocrinology*, 117, 323, 1985.
82. **Plotsky, P. M. and Sawchenko, P. E.,** Hypophysial-portal plasma levels, median eminence content and immunohistochemical staining of corticotropin-releasing factor, arginine vasopressin and oxytocin following pharmacological adrenalectomy, *Endocrinology*, 120, 1361, 1987.
83. **Powers, R. E., De Souza, E. B., Walker, L. C., Price, D. L., Vale, W. W., and Young, W. S., III,** Corticotropin-releasing factor as a transmitter in the human olivocerebellar pathway, *Brain Res.*, 415, 347, 1987.
84. **Rho, J.-H. and Swanson, L. W.,** Neuroendocrine CRF motoneurons: intrahypothalamic axon terminals shown with a new retrograde-Lucifer-immuno method, *Brain Res.*, 436, 143, 1987.
85. **Rhodes, C. H., Morrell, J. I., and Pfaff, D. W.,** Immunohistochemical analysis of magnocellular elements in rat hypothalamus. Distribution and numbers of cells containing neurophysin, oxytocin, and vasopressin, *J. Comp. Neurol.*, 198, 45, 1981.
86. **Rhodes, C. H., Morrell, J. I., and Pfaff, D. W.,** Changes in oxytocin content in magnocellular neurons of the rat hypothalamus following water deprivation or estrogen treatment, *Cell Tissue Res.*, 126, 47, 1981.
87. **Rivier, C. and Vale, W.,** Effects of corticotropin-releasing factor, neurohypophyseal peptides, and catecholamines on pituitary function, *Fed. Proc.*, 44, 189, 1985.
88. **Rivier, J., Speiss, J., and Vale, W.,** Characterization of rat hypothalamic corticotropin-releasing factor, *Proc. Natl. Acad. Sci. U.S.A.*, 80, 4851, 1983.
89. **Roth, K. A., Weber, E., and Barchas, J. D.,** Immunoreactive corticotropin releasing factor (CRF) and vasopressin are colocalized in a subpopulation of the immunoreactive vasopressin cells in the paraventricular nucleus of the hypothalamus, *Life Sci.*, 31, 1857, 1983.
90. **Sakanaka, M., Shibasaki, T., and Lederis, K.,** Distribution and efferent projections of corticotropin-releasing factor-like immunoreactivity in the rat amygdaloid complex, *Brain Res.*, 382, 213, 1986.
91. **Sakanaka, M., Shibasaki, T., and Lederis, K.,** Corticotropin-releasing factor-containing afferents to the inferior colliculus of the rat brain, *Brain Res.*, 414, 68, 1987.
92. **Sawchenko, P. E.,** Evidence for a local site of action for glucocorticoids in inhibiting CRF and vasopressin in the paraventricular nucleus, *Brain Res.*, 403, 213, 1987.
93. **Sawchenko, P. E.,** Adrenalectomy-induced enhancement of CRF- and vasopressin-immunoreactivity in parvocellular neurosecretory neurons: anatomic, peptide, and steroid specificity, *J. Neurosci.*, 7, 1093, 1987.
94. **Sawchenko, P. E.,** Evidence for differential regulation of corticotropin-releasing factor and vasopressin immunoreactivities in parvocellular neurosecretory and autonomic-related projections of the paraventricular nucleus, *Brain Res.*, 437, 253, 1987.
95. **Sawchenko, P. E.,** The effects of catecholamine-depleting medullary knife cuts on CRF- and vasopressin-immunoreactivity in the hypothalamus of normal and steroid-manipulated rats, *Neuroendocrinology*, 48, 459, 1988.
96. **Sawchenko, P. E.,** Neuropeptides, the paraventricular nucleus, and the integration of hypothalamic neuroendocrine and autonomic function, in *Neuropeptides and Stress*, Tache, Y., Morley, J. E., and Brown, M. R., Eds., Springer-Verlag, New York, 1989, 73.

97. **Sawchenko, P. E.,** Functional neuroanatomy of stress-related circuitry in the rat brain, in *Neuronal Control of Bodily Function*, Weiner, H., Florin, I., Hellhammer, D., and Murison, R., Eds., Hans Huber, in press.
98. **Sawchenko, P. E. and Levin, M. C.,** Immunohistochemical evidence for modulation by estrogen of neuropeptide coexpression in the magnocellular neurosecretory system of the female rat, *J. Comp. Neurol.*, submitted.
99. **Sawchenko, P. E. and Swanson, L. W.,** Immunohistochemical identification of neurons in the paraventricular nucleus of the hypothalamus that project to the medulla or to the spinal cord in the rat, *J. Comp. Neurol.*, 205, 1982.
100. **Sawchenko, P. E. and Swanson, L. W.,** The organization of noradrenergic pathways from the brainstem to the paraventricular and supraoptic nuclei in the rat, *Brain Res. Rev.*, 4, 275, 1982.
101. **Sawchenko, P. E. and Swanson, L. W.,** Localization, co-localization and plasticity of corticotropin-releasing factor immunoreactivity in the rat brain, *Fed. Proc.*, 44, 221, 1985.
102. **Sawchenko, P. E., Swanson, L. W., Grzanna, R., Howe, P. R. C., Bloom, S. R., and Polak, J. M.,** Colocalization of neuropeptide Y immunoreactivity in brainstem catecholaminergic neurons that project to the paraventricular and supraoptic nuclei in the rat, *J. Comp. Neurol.*, 241, 138, 1985.
103. **Sawchenko, P. E., Swanson, L. W., and Vale, W. W.,** Corticotropin-releasing factor: co-expression within distinct subsets of oxytocin-, vasopressin-, and neurotensin-immunoreactive neurons in the hypothalamus of the male rat, *J. Neurosci.*, 4, 1118, 1984.
104. **Sawchenko, P. E., Swanson, L. W., and Vale, W. W.,** Co-expression of corticotropin-releasing factor and vasopressin immunoreactivity in parvocellular neurosecretory neurons of the adrenalectomized rat, *Proc. Natl. Acad. Sci. U.S.A.*, 81, 1883, 1984.
105. **Schipper, J., Steinbusch, H. W. M., Vermes, I., and Tilders, F. J. H.,** Mapping of CRF-immunoreactive nerve fibers in the medulla oblongata and spinal cord of the rat, *Brain Res.*, 267, 145, 1983.
106. **Schramme, C. and Denef, C.,** Stimulation of spontaneous and dopamine-inhibited prolactin release from anterior pituitary reaggregate cell cultures by angiotensin peptides, *Life Sci.*, 34, 1651, 1984.
107. **Sheppard, M. C. and Shennan, K. I. J.,** The effect of thyroid hormones *in vitro* and *in vivo* on hypothalamic neurotensin release and content, *Endocrinology*, 112, 1996, 1983.
108. **Shioda, S., Nakai, Y., Kitazawa, S., and Sunayama, H.,** Immunocytochemical observations of corticotropin-releasing factor-containing neurons in the rat hypothalamus with special reference to neuronal communication, *Acta Anat.*, 124, 58, 1985.
109. **Silverman, A. J., Hou-Yu, A., and Kelly, D. D.,** Permanence of changes in corticotropin-releasing hormone (CRF) in rat paraventricular neurons (PVN) in response to chronic behavioral stress, *Soc. Neurosci. Abstr.*, 12, 783, 1986.
110. **Skofitsch, G., Hamill, G. S., and Jacobowitz, D. M.,** Capsaicin depletes corticotropin releasing factor-like immunoreactive neurons in the rat spinal cord and medulla oblongata, *Neuroendocrinology*, 38, 514, 1984.
111. **Skofitsch, G. and Jacobowitz, D. M.,** Distribution of corticotropin releasing factor-like immunoreactivity in the rat brain by immunohistochemistry and radioimmunoassay: comparison and characterization of ovine and rat/human CRF antisera, *Peptides*, 6, 319, 1985.
112. **Smythe, G. A., Bradshaw, J. E., and Vining, R. F.,** Hypothalamic monoamine control of stress-induced adrenocorticotropin release in the rat, *Endocrinology*, 113, 1062, 1983.
113. **Suda, T., Tomori, N., Yajima, F., Sumimoto, T., Nakagami, Y., Ushiyama, T., Demura, H., and Shizume, K.,** Time course study on the effect of reserpine on hypothalamic immunoreactive CRF levels in rats, *Brain Res.*, 405, 247, 1987.
114. **Sugimoto, T., Itoh, K., and Mizuno, N.,** Localization of neurons giving rise to the oculo-motor parasympathetic outflow: a HRP study in cat, *Neurosci. Lett.*, 7, 301, 1978.
115. **Sumimoto, T., Suda, T., Tomori, N., Yajima, F., Nakagami, Y., Ushiyama, T., Demura, H., and Shizume, K.,** Immunoreactive corticotropin-releasing factor in rat plasma, *Endocrinology*, 120, 1391, 1987.
116. **Swanson, L. W. and Kuypers, H. G. J. M.,** The paraventricular nucleus of the hypothalamus: cytoarchitectonic subdivisions and the organization of projections to the pituitary, dorsal vagal complex, and spinal cord as demonstrated by retrograde fluorescence double-labeling methods, *J. Comp. Neurol.*, 194, 555, 1980.
117. **Swanson, L. W. and Sawchenko, P. E.,** Hypothalamic integration: organization of the paraventricular and supraoptic nuclei, *Annu. Rev. Neurosci.*, 6, 275, 1983.
118. **Swanson, L. W., Sawchenko, P. E., Berod, A., Hartman, B. K., Helle, K. B., and Van Orden, D. E.,** An immunohistochemical study of the organization of catecholaminergic cells and terminals in the paraventricular and supraoptic nuclei of the hypothalamus, *J. Comp. Neurol.*, 196, 271, 1981.
119. **Swanson, L. W., Sawchenko, P. E., and Lind, R. W.,** Regulation of multiple peptides in CRF parvicellular neurosecretory neurons: implications for the stress response, in *Progress in Brain Research*, Vol. 68, Hökfelt, T., Fuxe, K., and Pernow, B., Eds., Elsevier, Amsterdam, 1986, 169.

120. **Swanson, L. W., Sawchenko, P. E., Lind, R. W., and Rho, J.-H.,** The CRH motoneuron: differential peptide regulation in neurons with possible synaptic, paracrine, and endocrine outputs, *Ann. N.Y. Acad. Sci.*, 512, 12, 1987.
121. **Swanson, L. W., Sawchenko, P. E., Rivier, J., and Vale, W. W.,** Organization of ovine corticotropin-releasing factor immunoreactive cells and fibers in the rat brain: an immunohistochemical study, *Neuroendocrinology*, 36, 165, 1983.
122. **Szafarczyk, A., Alonso, G., Ixart, G., Malaval, F., and Assenmacher, I.,** Diurnal-stimulated and stress-induced ACTH release is mediated by ventral noradrenergic bundle, *Am. J. Physiol.*, 249, E219, 1985.
123. **Taniguchi, Y.,** Immunohistochemical evidence against the coexistence of a corticotropin-releasing factor and oxytocin or vasopressin in the rat paraventricular nucleus, *Arch. Histol. Jpn.*, 47, 475, 1984.
124. **Tramu, G., Croix, C., and Pillez, A.,** Ability of the CRF immunoreactive neurons of the paraventricular nucleus to produce a vasopressin-like material, *Neuroendocrinology*, 37, 467, 1983.
125. **Vale, W., Speiss, J., Rivier, C., and Rivier, J.,** Characterization of a 41-residue ovine hypothalamic peptide that stimulates secretion of corticotropin and β-endorphin, *Science*, 213, 1394, 1981.
126. **Vale, W., Vaughan, J., Smith, M., Yamamoto, G., Rivier, J., and Rivier, C.,** Effects of synthetic ovine corticotropin-releasing factor, glucocorticoids, neurohypophyseal peptides and other substances on cultured corticotropic cells, *Endocrinology*, 113, 1121, 1983.
127. **Veening, J. G., Swanson, L. W., and Sawchenko, P. E.,** The organization of projections from the central nucleus of the amygdala to brainstem sites involved in central autonomic regulation: a combined retrograde transport-immunohistochemical study, *Brain Res.*, 303, 337, 1984.
128. **Vijayan, E. and McCann, S. M.,** Effects of substance P and neurotensin on growth hormone and thyrotropin release *in vivo* and *in vitro*, *Life Sci.*, 26, 321, 1980.
129. **Vincent, S. R. and Satoh, K.,** Corticotropin-releasing factor (CRF) immunoreactivity in the dorsolateral pontine tegmentum: further studies on the micturition reflex system, *Brain Res.*, 308, 387, 1984.
130. **Watts, A. G. and Swanson, L. W.,** Efferent projections of the suprachiasmatic nucleus. II. Studies using retrograde transport of fluorescent dyes and simultaneous peptide immunohistochemistry in the rat, *J. Comp. Neurol.*, 258, 230, 1987.
131. **Weigand, S. J. and Price, J. L.,** The cells of origin of afferent fibers to the median eminence in the rat, *J. Comp. Neurol.*, 192, 1, 1980.
132. **White, J. D. and McKelvy, J. F.,** Enkephalin biosynthesis and processing during lactation, *Neuroendocrinology*, 43, 377, 1986.
133. **Whitnall, M., Mezey, E., and Gainer, H.,** Co-localization of corticotropin-releasing factor and vasopressin in median eminence neurosecretory vesicles, *Nature*, 317, 248, 1985.
134. **Whitnall, M. H., Key, S., and Gainer, H.,** Vasopressin-containing and vasopressin-deficient subpopulations of corticotropin-releasing factor axons are differentially affected by adrenalectomy, *Endocrinology*, 120, 2180, 1987.
135. **Whitnall, M. H., Smyth, D., and Gainer, H.,** Vasopressin coexists in half of the corticotropin-releasing factor axons in the external zone of the median eminence in normal rats, *Neuroendocrinology*, 45, 420, 1987.
136. **Wolfson, B., Manning, R. W., Davis, L. G., Arentzen, R., and Baldino, F., Jr.,** Co-localization of corticotropin releasing factor and vasopressin mRNA in neurones after adrenalectomy, *Nature*, 315, 59, 1985.
137. **Young, W. S., III, Mezey, E., and Siegel, R. E.,** Quantitative in situ hybridization histochemistry reveals increased levels of corticotropin-releasing factor mRNA after adrenalectomy in rats, *Neurosci. Lett.*, 70, 198, 1986.
138. **Young, W. S., III, Mezey, E., and Siegel, R. E.,** Vasopressin and oxytocin mRNAs in adrenalectomized and Brattleboro rats: analysis by quantitative in situ hybridization histochemistry, *Mol. Brain Res.*, 1, 231, 1986.
139. **Young, W. S., III, Walker, L. C., Powers, R. E., De Souza, E. B., and Price, D. L.,** Corticotropin-releasing factor mRNA is expressed in the inferior olives of rodents and primates, *Mol. Brain Res.*, 1, 189, 1986.

Chapter 5

THE ORGANIZATION AND POSSIBLE FUNCTION OF AMYGDALOID CORTICOTROPIN-RELEASING FACTOR PATHWAYS

Thackery S. Gray

TABLE OF CONTENTS

I. INTRODUCTION

The amygdala is a part of a limbic system circuitry that mediates behavioral, autonomic, and neuroendocrine adjustments in response to environmental demands. More recent studies have emphasized the role of the amygdala in responses to stressful or defense mobilizing stimuli.[1-4] The amygdala, especially the central amygdaloid nucleus (Ce), contains large numbers of corticotropin-releasing factor (CRF) containing cell bodies and terminals. Since administration of CRF within the central nervous system (CNS) initiates a constellation of visceral and behavioral changes resembling the stress or defense response, it is reasonable to believe that amygdaloid pathways may be important for mediation of the some of the central actions of CRF. The purpose of this chapter is (1) to briefly review the studies that implicate the Ce and associated regions in mediation of autonomic and neuroendocrine responses to fear- or stress-producing stimuli; and (2) to review the possible pathways that could mediate amygdaloid autonomic and neuroendocrine responses.

II. AMYGDALA'S ROLE IN AUTONOMIC RESPONSES

A. FUNCTIONAL STUDIES

Electrical stimulation of the central amygdaloid nucleus in several animal species including rats, cats, and rabbits produces a collection of autonomic responses that could be characterized as stress- or defense-like.[1,5-11] Increases in heart rate, blood pressure, and respiratory responsiveness have been observed in awake, unanesthetized rats and cats. At lower current levels these increases occur in the absence of obvious behavioral changes in animals. At higher current levels increased arousal and overt defense-like responses are more evident. The increases in heart rate and blood pressure are associated with reduced sensitivity or altering of the baroreceptor reflex.[12] Electrical stimulation of the amygdala in awake humans also elicits increases in heart rate and blood pressure as well as feelings of fear and/or anxiety.[13,14] Unlike hypothalamic stimulation-induced pressor responses, in order to obtain a pressor response from amygdaloid stimulation it is critical that the subject is awake. For example, electrical stimulation-induced increases in cardiovascular responsiveness are attenuated or are decreased in animals that are anesthetized or asleep.[7,8,15] Central amygdaloid stimulation also increases gastric motility and acid secretion in rats.[16] Activation of the Ce also alters phrenic nerve activity.[17] Increases in heart rate and/or blood pressure can also be chemically induced by injections of calcitonin gene-related peptide, thyrotropin releasing factor, angiotensin II, bombesin, and somatostatin into the Ce of awake rats.[18,19]

The Ce is probably not intimately involved in homeostatic regulation of autonomic functions. Rather, the Ce participates in altering basal autonomic activity in response to arousal-inducing, stressful, or threatening stimuli within the environment. Ablation of the Ce results in behavioral changes that are consistent with a reduction in responsivity to fear-provoking stimuli (for review see Reference 20). Auditory stimuli that signal immobilization treatment or electric shock in rats are correlated with changes in multiple cell activity within the Ce of rats.[21-23] Bilateral lesions of the Ce attenuates learned heart rate responses to shock in rabbits.[24] In the rabbit injection of the beta-adrenergic blocker, *dl*-propranolol, within the Ce impairs learning of conditioned heart rate responses.[25] Injections of the alpha-adrenergic antagonist, *l*-isoproterenol, into the Ce does not impair learned heart rate responses. Lesions of the Ce in rats blocks learned heart rate responses to a tone associated with shock.[26] In this paradigm amygdaloid lesions do not block heart rate responses to the shock (i.e., unconditioned stimulus) alone. Cryogenic blockade of the Ce reduces learned blood pressure and respiratory responses in the cat,[27] but not learned heart rate responses. In pigs, lesions of the amygdala block stress-induced sympathetic activity and increased plasma noradrenalin levels.[28] Bilateral lesions of the Ce attenuates stimulus-induced exaggerated increases in

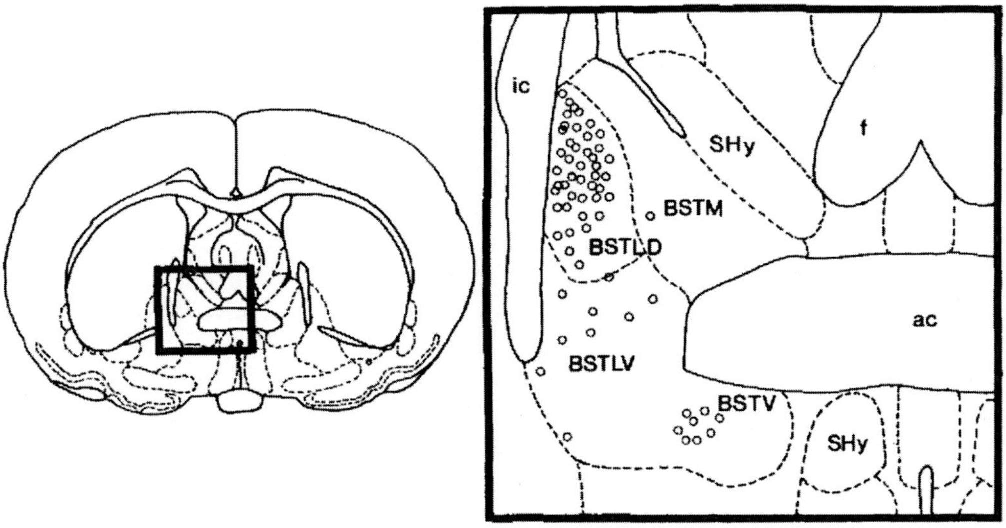

FIGURE 1. Line drawing illustrating the location of corticotropin-releasing factor immunoreactive cell bodies (open circles) in the lateral bed nucleus of the stria terminalis of the rat brain (coronal plane).

cardiovascular responses in spontaneously hypertensive rats.[29,30] Stimulation of alpha-2 adrenoreceptors in the Ce prevents increased renal sympathetic nerve activity and antinatriuresis caused by noise stress in spontaneously hypertensive rats.[31] Lesions of the amygdala, but not the red nucleus or cerebellum, block fear conditioned startle responses.[32] Finally, lesions of the Ce result in increases in punished responding in the conflict test.[33]

Recent studies have suggested that in addition to the neuroendocrine function of CRF, it has important physiological and behavioral actions within the CNS. As previously indicated, intracerebroventricular (ICV) injection of CRF in animals produces centrally mediated autonomic and behavioral changes that are characteristic of the stress or defense reaction, similar to changes observed in awake amygdaloid-stimulated animals. Administration of CRF into the lateral ventricle of the rat or dog results in behavioral activation, increases in heart rate, blood pressure, and plasma and brain tissue levels of catecholamines.[34-37] The effects of ICV administration of CRF are also opposite of those observed following amygdaloid lesions (i.e., increases in fear related behaviors as opposed to decreases). Central administration of CRF produces behavioral changes and "anxiogenic-like" effects consistent with increased emotionality.[36-39] For example, ICV CRF injections potentiate the acoustic startle response.[39] And, ICV CRF injections produce a suppression of responding in the conflict test and exploratory behavior.[38]

B. AMYGDALOID AUTONOMIC PATHWAYS

The brain pathways through which CRF acts to alter autonomic nervous system activity are as yet unknown. Single injections of CRF into specific subregions of the brain do not seem to produce the full effect typically observed after a single ICV injection of CRF.[40] Alterations of CRF immunoreactivity after chronic or acute stress in the rat have been observed within several regions containing CRF terminals.[41] These findings suggest the likely possibility that CRF acts at multiple sites within the CNS. A large number of CRF immunoreactive neurons have been localized within parts of the rat forebrain which are interconnected with autonomic regions of the brainstem and spinal cord.[42] Prominent among the forebrain regions that contain CRF-immunoreactive neurons are the Ce and lateral part of the bed nucleus of the stria terminalis (BSTL) as well as the paraventricular nucleus (PVN) and lateral hypothalamus (LH) (see Figures 1 to 4).

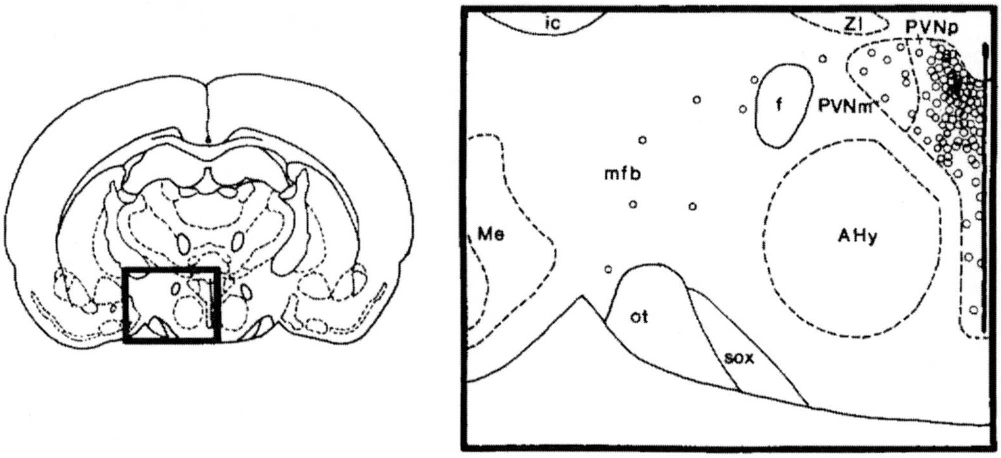

FIGURE 2. Line drawing illustrating the location of corticotropin-releasing factor immunoreactive cell bodies (open circles) in the paraventricular nucleus region of the hypothalamus of the rat brain (coronal section).

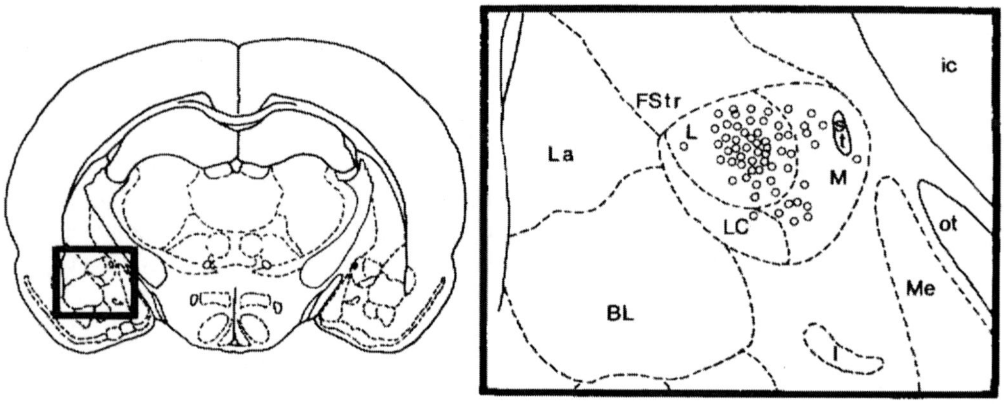

FIGURE 3. Line drawing illustrating the location of corticotropin-releasing factor immunoreactive cell bodies within the central nucleus of the rat amygdala (coronal section).

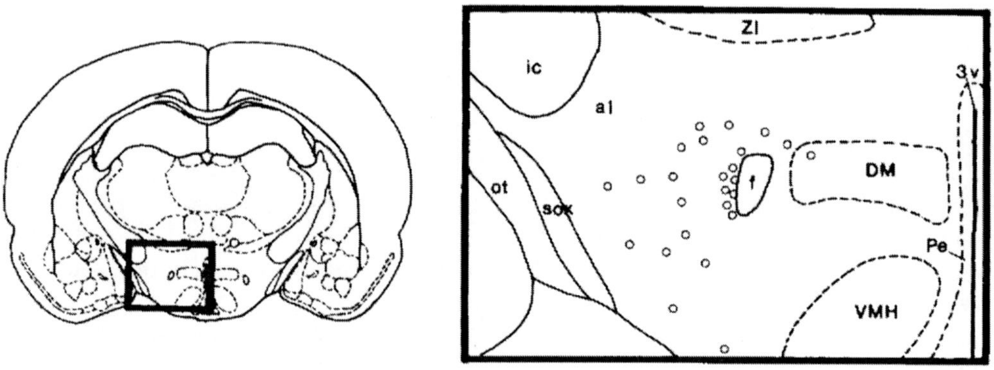

FIGURE 4. Line drawing illustrating the location of corticotropin-releasing factor immunoreactive cell bodies in region of the lateral hypothalamus of the rat (coronal section).

A recent series of studies in our laboratory investigated the possible forebrain sources of CRF and other peptide-containing terminals within the several "autonomic" regions of the brainstem of the rat.[48-53] Hopkins[43] was the first to demonstrate that the amygdala (mainly the Ce) projects to brainstem regions involved in autonomic and behavioral components of the defense response in the cat, rat, and monkey. Subsequently, similar projections have been demonstrated in the rabbit.[44] Brainstem regions that are directly innervated by the Ce include *substantia nigra*-ventral tegmental region, midbrain central gray, parabrachial nucleus, locus coeruleus, A5 catecholaminergic region, reticular formation, raphe nuclei, dorsal vagal complex (i.e., nucleus tractus solitarii and dorsal vagal nucleus), and ventrolateral medulla.[43-47]

We investigated the possible contribution of CRF-immunoreactive forebrain neurons to pathways that terminate within the central gray, parabrachial region, and the dorsal vagal complex. The method we used is a modification of the combined retrograde tracing and immunohistochemical procedure.[54] The retrograde tracer Fluoro-Gold was injected into the central gray, parabrachial nuclei, or the dorsal vagal complex. The brain tissue was also processed for visualization of CRF-like immunoreactivity using the avidin-biotin Texas Red immunofluorescence procedure.[51] Figure 5 illustrates the representative injection sites within the brainstem that resulted in maximal retrograde labeling within regions of the forebrain containing CRF-immunoreactive cells. Examination of the material reveals three populations of cells: (1) retrogradely (Fluoro-Gold) labeled cells; (2) CRF-immunoreactive (Texas Red fluorescent) cells; and (3) cells that contain both the retrograde tracer and immunoreactivity. Figures 6 to 9 illustrate examples of these three populations of cells after injections of Fluoro-Gold into the brainstem and subsequent immunohistochemical processing of the amygdala and hypothalamus. The distribution of retrograde labeling within forebrain following injections of tracer into the parabrachial region, central gray, and dorsal vagal complex has been previously reported in detail.[43,45-47,49,50,52] Also, details of the distribution of CRF immunoreactive neurons within the forebrain have also been published.[41] Since the present results do not differ significantly from previous reports with respect to the distribution of CRF-immunoreactive or retrogradely labeled cells, the following discussion will focus on the distribution of cells containing both CRF-like immunoreactivity and retrograde labeling. CRF-immunoreactive retrogradely labeled neuronal cell bodies were located within the Ce, BSTL, PVN, and the posterior LH after injections of tracer into each brainstem region (see Figures 6 to 9). A few cells also were observed in the intercalated nuclei of the amygdala after injections of tracer into the parabrachial nucleus. The actual number of CRF-immunoreactive retrogradely labeled neurons varied as a function of injection site and region examined within the forebrain. In order to assess this, we counted the total number of CRF-immunoreactive neurons in Ce, bed nucleus of the stria terminalis, LH, and paraventricular hypothalamic nucleus in our best cases. There were approximately 1750 CRF-immunoreactive cells in the Ce as compared to 820 in the BSTL, 600 in the LH, and 2000 in the PVN. We then counted the total number of CRF cells that contained retrograde tracer in these regions and calculated the percentage of CRF cells that contained retrograde tracer. Table 1 presents the percentage of CRF immunoreactive cells within the Ce, BSTL, LH, and PVN that were retrogradely labeled after injections of Fluoro-Gold tracer into the dorsal vagal complex, parabrachial nucleus and midbrain central gray. For the most part, the Ce and the BSTL contained the largest percentage of retrogradely labeled CRF-immunoreactive neurons after injections of Fluoro-Gold into the three sites within the brainstem. The LH provided a substantial input to the central gray and minor input to the parabrachial nucleus and dorsal vagal complex. In spite of the large number of CRF-immunoreactive neurons within the PVN only a small percentage of these cells project to the brainstem. This finding is not surprising since most CRF cells are located in regions of the PVN that project to the median eminence.[42] The parabrachial nucleus and the midbrain central gray receive the bulk of input

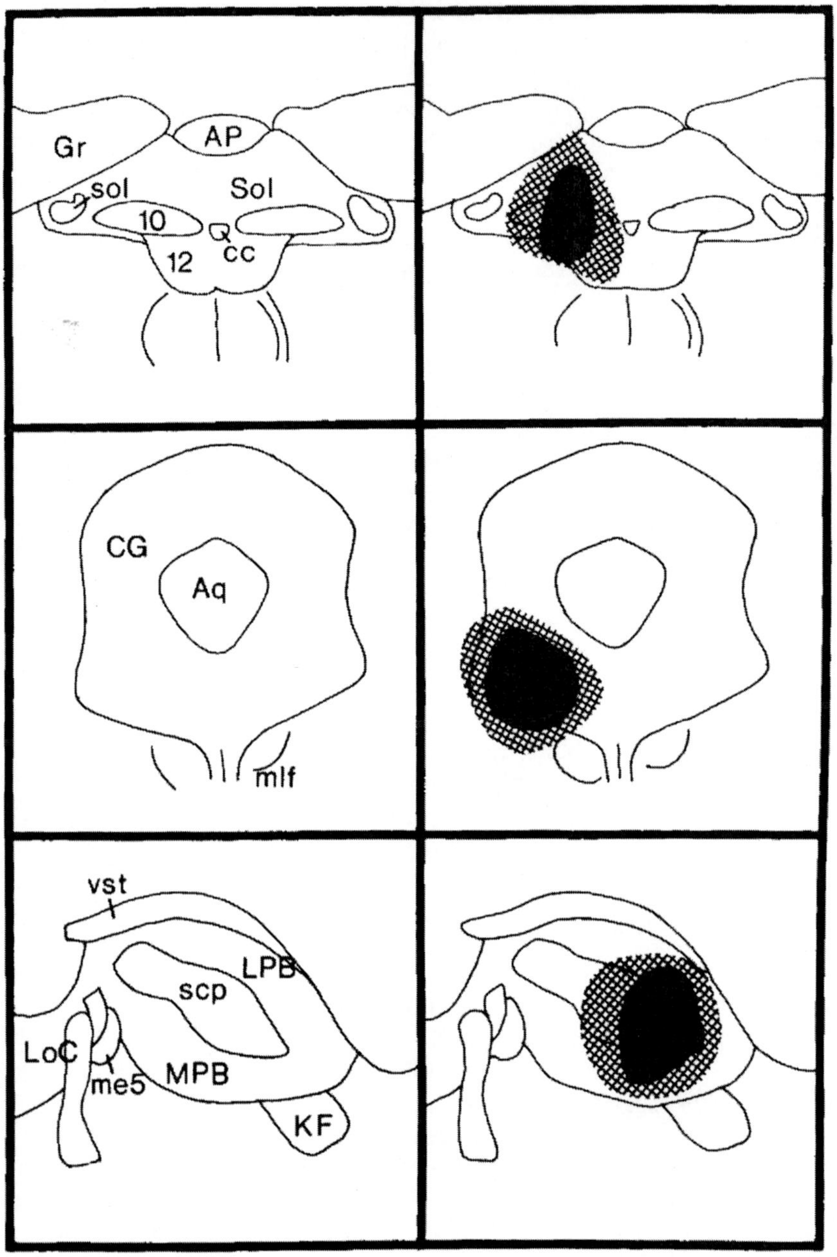

FIGURE 5. Line drawing illustrating the location and spread of fluorescent tracer within the dorsal vagal complex (top two panels), midbrain central gray (middle panels), and the parabrachial regions (bottom panels).

from CRF forebrain neurons compared to the dorsal vagal complex. Overall the results indicate that the Ce and BSTL contain the vast majority of the forebrain CRF-immunoreactive neurons that innervate these three brainstem regions. Veening et al.[55] had previously demonstrated a CRF projection from the Ce to the dorsal vagal complex. And, more recently, Sakanaka et al.[56] have confirmed the Ce's projections to parabrachial nuclei. They also reported that CRF cells in the Ce innervate the mesencephalic trigeminal nucleus and the mesencephalic reticular formation, and possibly the lateral hypothalamus and bed nucleus

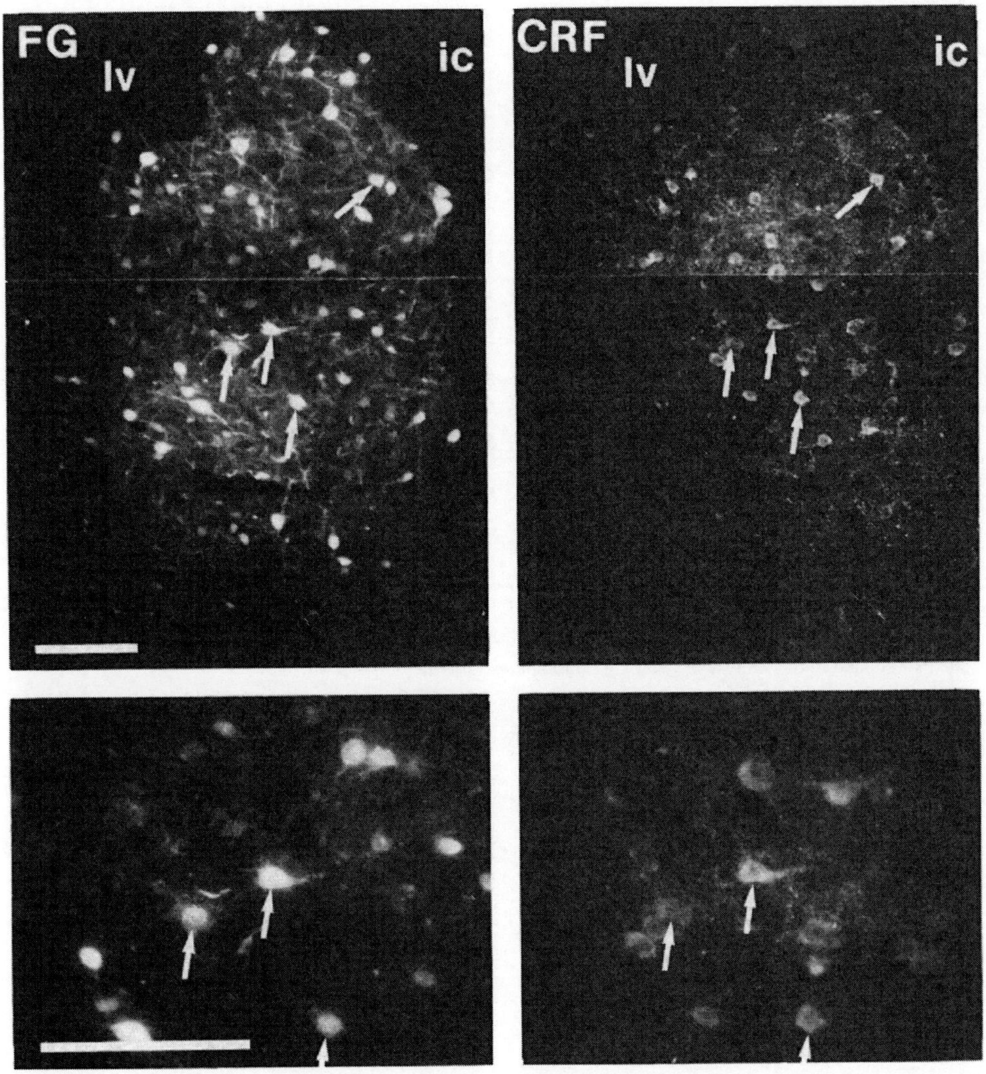

FIGURE 6. Fluorescent photomicrographs of a coronal section through the lateral bed nucleus of the stria terminalis in a rat brain illustrating the location of Fluoro-Gold (FG, left panels) labeled cells after an injection of retrograde tracer into the parabrachial nucleus and corticotropin immunoreactive cells (CRF, right panels). Arrows indicate position of cells that contain both Fluoro-Gold and CRF-immunoreactivity.

of the stria terminalis.[56] In addition, they found that the CRF cells in corticomedial nucleus innervate the ventromedial hypothalamus. Thus, the above studies demonstrate widespread amygdaloid projections to regions of both the hypothalamus and brainstem that are thought to integrate autonomic functions.

The Ce and BSTL are amygdaloid structures that have extensive reciprocal interconnections with other central autonomic nuclei,[2-4,47,57-59] and many of these interconnected pathways contain CRF as well as other peptides (for review see Reference 59). Thus, these findings attach a special importance to the amygdala as a source of descending CRF pathways. Release of CRF from amygdaloid neurons at multiple sites within the brainstem may be important for the expression of behavioral and/or autonomic components of defense or fear-related behaviors. Further studies are required to see if amygdaloid CRF-containing neurons innervate other nuclei within this central autonomic circuitry (e.g., locus coeruleus, ventro-

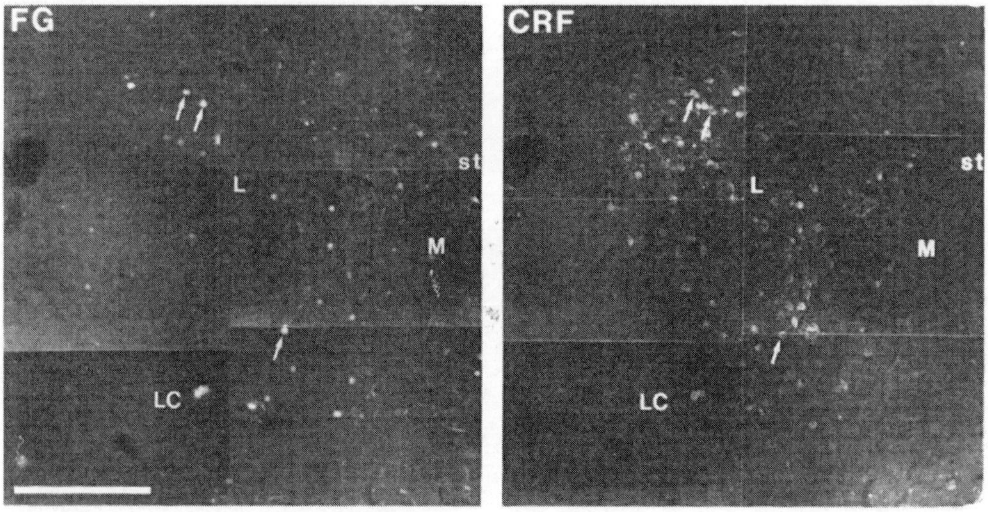

FIGURE 7. Fluorescent photomicrographs of a coronal section through the central nucleus of the amygdala in a rat brain illustrating the location of Fluoro-Gold (FG, left panels) labeled cells after an injection of retrograde tracer into the central gray and corticotropin immunoreactive cells (CRF, right panels). Arrows indicate the position of cells that contain both Fluoro-Gold and CRF immunoreactivity.

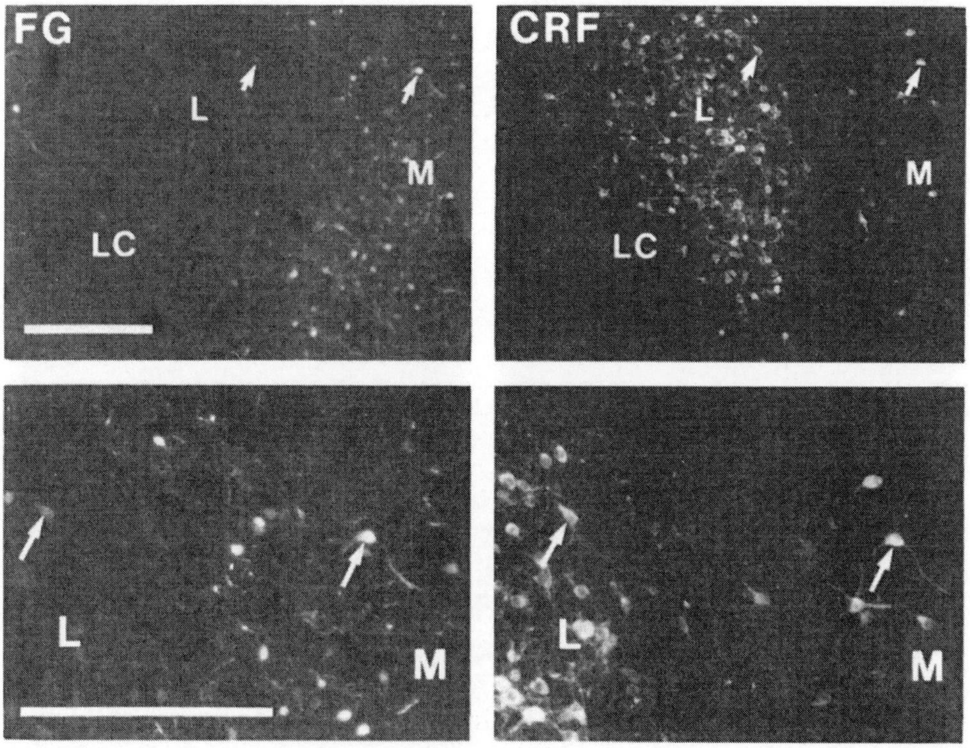

FIGURE 8. Fluorescent photomicrographs of a coronal section through the central nucleus of the amygdala in a rat brain illustrating the location of Fluoro-Gold (FG, left panels) labeled cells after an injection of retrograde tracer into the dorsal vagal complex and corticotropin immunoreactive cells (CRF, right panels). Arrows indicate the position of cells that contain both Fluoro-Gold and CRF immunoreactivity.

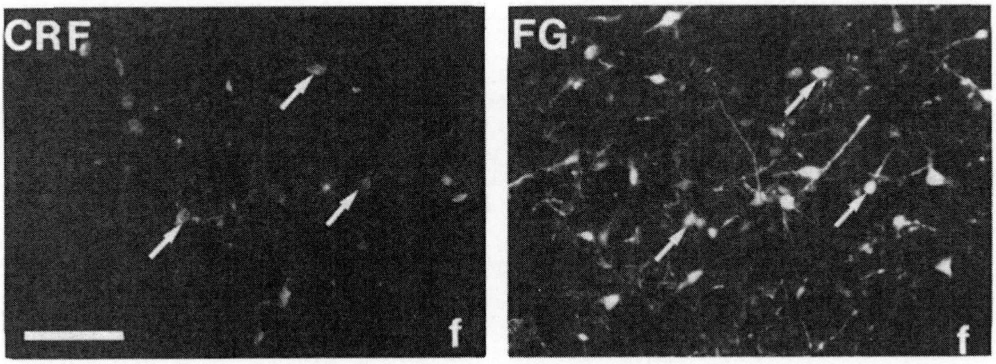

FIGURE 9. Fluorescent photomicrographs of a coronal section through the lateral hypothalamus in a rat brain illustrating the location of Fluoro-Gold (FG, right panel) labeled cells after an injection of retrograde tracer into the central gray and corticotropin immunoreactive cells (CRF, left panel). Arrows indicate the position of cells that contain both Fluoro-Gold and CRF immunoreactivity.

TABLE 1
Percent of Retrogradely Labeled Corticotropin-Releasing Factor Immunoreactive (CRF-IR) Cells in Various Forebrain Areas After Injections of Fluorescent Tracer into the Parabrachial Nucleus (PBN), Midbrain Central Gray (CG), or Dorsal Vagal Complex (DVC)

		Injection sites			# CRF cells per area
		PBN	CG	DVC	
Location of Retrogradely CRF-IR Cells	Ce	66%	34%	15%	+1750
	BSTL	28%	24%	4%	+820
	LH	7%	27%	7%	+600
	PVN	1%	1%	2%	+2000

lateral medulla). Figure 10 summarizes the organization of descending CRF projections from the forebrain to the brainstem autonomic regions that have been examined to date.

III. AMYGDALA'S ROLE IN NEUROENDOCRINE RESPONSES

A. FUNCTIONAL STUDIES

The amygdala participates in mediation of neuroendocrine as well as autonomic responses to stress. Stimulation of the amygdala in monkeys,[60] cats,[61] and rats[62] increases plasma corticosterone levels. Destruction of the amygdala inhibits adrenocortical responses to stress-inducing somatosensory and olfactory stimuli.[63] Destruction of the amygdala also attenuates compensatory hypersecretion of adrenocorticotrophic hormone (ACTH) resulting from adrenalectomy.[64] This suggests that the amygdala is also part of the circuitry involved with

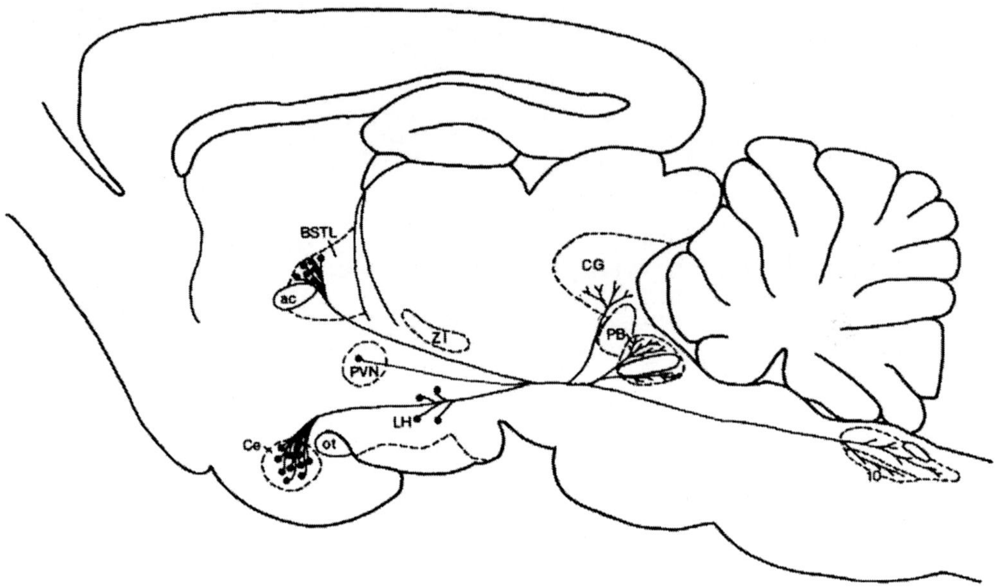

FIGURE 10. Line drawing schematically illustrating the distribution of forebrain corticotropin-releasing factor cells and their projections to the brainstem.

glucocorticoid feedback. Recent studies have implicated a specific subdivision of the amygdala, the central nucleus, in mediation of ACTH secretion in response to immobilization stress. Lesions of the central amygdaloid nucleus dramatically reduce ACTH responses to immobilization stress.[65,66] However, ablation of the Ce does not alter normal basal secretion of ACTH. Localized stimulation of the Ce in urethane-anesthetized rats produces increases and decreases in plasma ACTH levels, although decreases in ACTH are more frequently observed. Immobilization stress is associated with increased noradrenergic and decreased dopaminergic activity within the Ce.[66] Thus, the central amygdaloid nucleus is necessary for the full expression of the ACTH responses to certain types of stress. Lesions of the cortical and medial nuclei or basolateral nucleus of the amygdala do not alter corticosterone or prolactin responses to stress.[67]

B. AMYGDALOID NEUROENDOCRINE PATHWAYS

The anatomic pathways that are involved in amygdaloid-mediated neuroendocrine responses to stress are not known. Presumably these pathways would involve direct or indirect synaptic connections with the pituitary activating regions of the PVN of the hypothalamus. A few retrogradely labeled cells were observed within the Ce following injections of the retrograde tracer, horseradish peroxidase, into the PVN.[68] Amygdaloid projections could also be relayed through the bed nucleus of the stria terminalis which heavily innervates the PVN.[69] So far, anterograde tracing studies have not conclusively demonstrated a direct pathway from the Ce to the PVN.

Recently, we injected the anterograde tracer, *Phaseolus vulgaris* leucoagglutinin lectin into the Ce for the purpose of more precisely defining the organization of the pathways between the amygdala and brainstem.[70] This lectin can be identified by standard immunohistochemical techniques and is probably the most sensitive anterograde tracing technique available.[71] This technique offers the unique advantage of permitting identification of both the cells that originate a pathway as well as a precise localization of their axon terminals. In the course of these preliminary studies, we have obtained data demonstrating that the Ce directly innervates the PVN of the hypothalamus. Injections of the lectin into the medial Ce

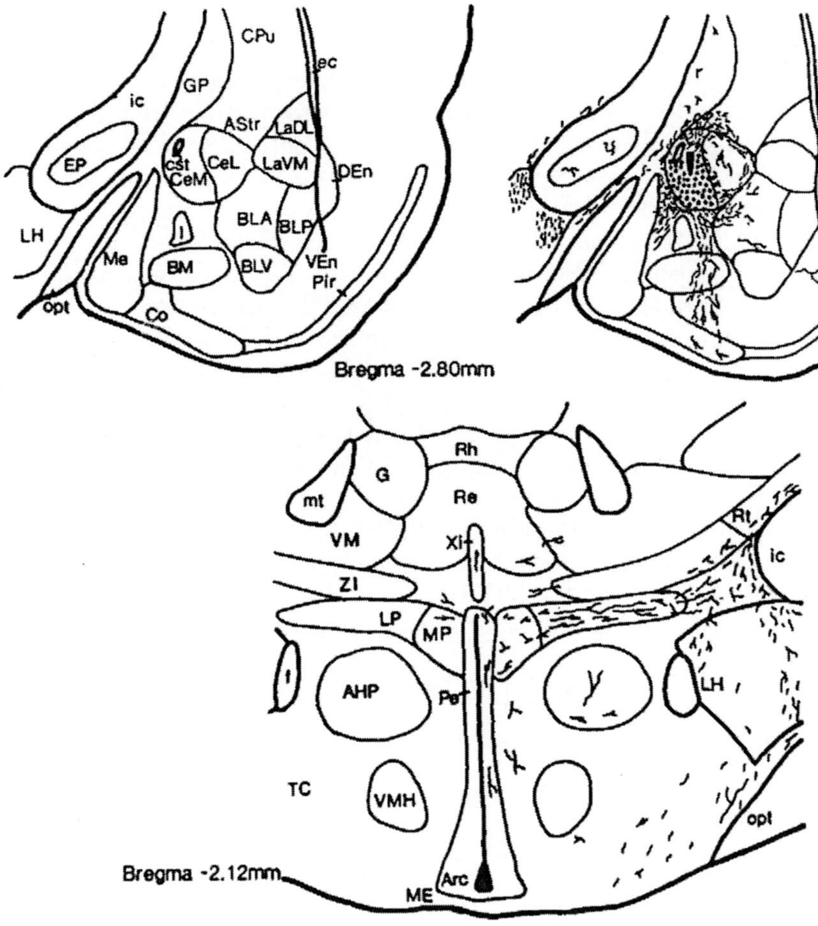

FIGURE 11. Line drawing illustrating the location of a *Phaseolus vulgaris* leucoagluttinin lectin injection within the medial part of the central amygdaloid nucleus (top right) and the resulting axonal labeling within the hypothalamus (below).

resulted in anterograde labeling of axons and terminal-like processes mostly within the medial and lateral parvocellular subdivisions of the PVN at posterior levels (Figure 11). Sparse labeling was also observed in most other subdivisions of the PVN including the magnocellular parts. Lectin injections centered within the lateral Ce failed to result in significant labeling within the PVN, although other regions such as the parabrachial nucleus in the pons contained dense axonal and presumed-terminal labeling.

These findings suggest an anatomical substrate for direct amygdaloid activation of the ACTH releasing cells of the hypothalamus. Further studies are needed in order to determine whether the Ce innervates vasopressin, oxytocin, and/or CRF containing cells. The Ce is also of particular interest because, as previously noted, it also contains large numbers of CRF containing cell bodies. However, most CRF neurons are located within the lateral Ce, although they are also scattered within the medial Ce.[72] So although it is possible that CRF containing amygdaloid neurons directly innervate CRF producing neurons within the PVN of the hypothalamus, it is more probable that the important transmitter(s) involved in this pathway remains to be determined.

IV. CONCLUSIONS

This chapter has summarized the amygdaloid pathways to the hypothalamus and brainstem that are involved in the expression of behavioral, neuroendocrine, and autonomic responses to stressful or defense mobilizing stimuli. The functional studies reviewed in this chapter clearly point to the amygdala as essential for full expression of defense, arousal, and other emotive responses. This is especially apparent in studies which demonstrate marked attenuation of stimulus-induced somatomotor and visceromotor behaviors and neuroendocrine responses following bilateral amygdalectomy. The amygdala then is important for the expression of adaptive emotional and behavioral responses that are essential for normal physical and mental health. Thus, it is not surprising that the amygdala has been implicated as a participant in several stress-related pathologies including hypertension, cardiac arrhythmias, and gastric ulcers.[16,29,30,73] However, the amygdala is only one interactive site and must be studied in relationship to regions of the hypothalamus and brainstem that are similarly involved. A number of studies now point to CRF as a possible neurotransmitter/neuromodulator within amygdaloid pathways. However, it is also clear that many other putative transmitters are also contained within these pathways including other neuropeptides and the more established transmitters such as the catecholamines, serotonin, and GABA. Studies that examine how CRF interacts with other neuroactive substances are necessary to provide a better understanding of the role of CRF in amygdaloid functions.

ABBREVIATIONS

AHP — anterior hypothalamic area
al — ansa lenticularis
AP — area postrema
AQ — cerebral aqueduct
Arc — arcuate hypothalamic nucleus
AStr — fundus striati
BLA — anterior part of basolateral amygdaloid nucleus
BLP — posterior part of basolateral amygdaloid nucleus
BLV — ventral part of basolateral amygdaloid nucleus
BM — basomedial amygdaloid nucleus
BSTL — lateral bed nucleus of the stria terminalis
  BSTLD = dorsal subdivision
  BSTLV = ventral subdivision
BSTM — medial bed nucleus of the stria terminalis
BSTV — ventral bed nucleus of the stria terminalis
cc — central canal
Ce — central amygdaloid nucleus
  L = lateral subdivision
  LC = lateral capsular subdivision
  M = medial subdivision
CG — central gray
Co — cortical amygdaloid nucleus
CPu — caudate-putamen
CRF — corticotropin-releasing factor
cst — commissural part of the stria terminalis
DEn — dorsal part of endopiriform nucleus
DM — dorsomedial hypothalamic nucleus
DVC — dorsal vagal complex

ec — external capsule
EP — entopeduncular nucleus
f — fornix
FStr — fundus striati
G — gelatinosus nucleus of the thalamus
GP — globus pallidus
Gr — nucleus gracilis
I — intercalated nucleus of the amygdala
ic — internal capsule
KF — Kolliker-Fuse nucleus
LaDL — dorsolateral part of lateral amygdaloid nucleus
LaVM — ventromedial part of lateral amygdaloid nucleus
LH — lateral hypothalamus
LoC — locus coeruleus
lv — lateral ventricle
ME — median eminence
Me — medial amygdaloid nucleus
me5 — mesencephalic tract of the trigeminal
mlf — medial longitudinal fasciculus
mfb — medial forebrain bundle
mt — mammilothalamic tract
ot, opt — optic tract
PBN — parabrachial nucleus
LPB = lateral subdivision
MPB = medial subdivision
Pe — periventricular hypothalamic nucleus
Pir — pirirform cortex
PVN — paraventricular nucleus
PVNm = magnocellular subdivision
PVNp = parvocellular subdivision
MP = medial parvocellular subdivision
LP = lateral parvocellular subdivision
Re — nucleus reuniens
Rh — rhomboid nucleus
Rt — reticular thalamic nucleus
scp — superior cerebellar peduncle
SHy — septohypothalamic nucleus
Sol — nucleus tractus solitarii
sol — solitary tract
sox — supraoptic decussation
TC — tuber cinerium
VEn — ventral part of endopiriform nucleus
VMH — ventromedial hypothalamic nucleus
VM — ventromedial thalamic nucleus
vst — ventral spinocerebellar tract
Xi — xiphoid thalamic nucleus
ZI — zona incerta
3v — third ventricle
10 — dorsal motor nucleus of the vagus
12 — hypoglossal nucleus

ACKNOWLEDGMENTS

Preparation of this chapter was supported by NIH Grant NS 20041 and the POTTS Foundation Loyola University Stritch School of Medicine. The author would like to thank Debra J. Magnuson for her critical comments and technical assistance.

REFERENCES

1. **Kaada, B. R.,** Stimulation and regional ablation of the amygdaloid complex reference to functional representations, in *Neurobiology of the Amygdala,* Elftheriou, B. E., Ed., Plenum Press, New York, 1972, 205.
2. **Loewy, A. D. and McKellar, S.,** The neuroanatomic basis of central cardiovascular control, *Fed. Proc.,* 39, 2495, 1980.
3. **Caleresu, F. R., Ciriello, J., Caverson, M. M., Cechetto, D. F., and Krukoff, T. L.,** Functional neuroanatomy of central pathways controlling the circulation, in *Hypertension and the Brain,* Guthrie, G. P., Jr., and Kotchen, T. A., Eds., Futurea, New York, 1984, 3.
4. **Smith, O. A. and DeVito, J. L.,** Central integration for the control of autonomic responses associated with emotion, *Annu. Rev. Neurosci.,* 7, 43, 1984.
5. **Hilton, S. B. and Zbrozyna, A. W.,** Amygdaloid region for defense reaction and its efferent pathway to the brain stem, *J. Physiol. London,* 165, 106, 1963.
6. **Heinemann, H., Stock, G., and Schaefer, H.,** Temporal correlation of responses in blood pressure and motor reaction under electrical stimulation of limbic structures in unanaesthetized, unrestrained cats, *Pflugers Arch.,* 343, 27, 1973.
7. **Mogenson, G. J. and Calaresu, F. R.,** Cardiovascular responses to electrical stimulation of the amygdala in the rat, *Exp. Neurol.,* 39, 166, 1973.
8. **Stock, G., Schlor, K.-H., Heidt, H., and Buss, J.,** Psychomotor behaviour and cardiovascular patterns during stimulation of the amygdala, *Pfluger's Arch. Ges. Physiol.,* 376, 177, 1978.
9. **Kapp, B. S., Gallagher, M., Underwood, M. D., McNall, C. L., and Whitehorn, D.,** Cardiovascular responses elicited by electrical stimulation of the amygdala central nucleus in the rabbit, *Brain Res.,* 234, 251, 1982.
10. **Galeno, T. M. and Brody, M. J.,** Hemodynamic responses to amygdaloid stimulation in spontaneously hypertensive rats, *Am. J. Physiol.,* 245, 281, 1983.
11. **Harper, R. M., Frysinger, R. C., Trelease, R. B., and Marks, J. D.,** State dependent alteration of respiratory cycle timing by stimulation of the central nucleus of the amygdala, *Brain Res.,* 306, 1, 1984.
12. **Schlor, K.-H., Stumpf, H., and Stock, G.,** Baroreceptor reflex during arousal induced by electrical stimulation of the amygdala or by natural stimuli, *J. Autonom. Nerv. Syst.,* 10, 157, 1984.
13. **Chapman, W. P., Schroeder, H. R., Guyer, G., Brazier, M. A. B., Fager, C., Poppen, H. L., Solomon, H. C., and Yakoloev, P. I.,** Physiological evidence concerning the importance of the amygdaloid nuclear region in the integration of circulating function and emotion in man, *Science,* 129, 949, 1954.
14. **Gloor, P., Olivier, A., and Quesney, L. F.,** The role of the amygdala in the expression of psychic phenomena in temporal lobe seizures, in *The Amygdaloid Complex,* Ben-Ari, Y., Ed., Elsevier/North-Holland, New York, 1981, 489.
15. **Frysinger, R. C., Marks, J. D., Trelease, R. B., Schechtman, and Harper, R. M.,** Sleep states attenuate the pressor response to central amygdala stimulation, *Exp. Neurol.,* 83, 604, 1984.
16. **Henke, P. G.,** The telencephalic limbic system and experimental gastric pathology: a review, *Neurosci. Biobehav. Rev.,* 6, 381, 1982.
17. **Bonvallet, M. and Bobo, E. G.,** Changes in phrenic activity and heart rate by localized stimulation of the amygdala and adjacent structures, *Electroencephalogr. Clin. Neurophysiol.,* 32, 1, 1972.
18. **Nguyen, K. Q., Sills, M. A., and Jacobowitz, D. M.,** Cardiovascular effects produced by microinjection of calcitonin gene-related peptide into the rat central amygdaloid nucleus, *Peptides,* 7, 337, 1986.
19. **Brown, M. R. and Gray, T. S.,** Peptide injections into the amygdala of conscious rats: effects on blood pressure, heart rate and plasma catecholamines, *Regul. Pept.,* submitted.
20. **Ursin, H., Jellestad, F., and Gabrera, I. G.,** The amygdala, exploration and fear, in *The Amygdaloid Complex,* Ben-Ari, Y., Ed., Elsevier/North Holland, New York, 1981, 317.
21. **Applegate, C. D., Frysinger, R. C., Kapp, B. S., and Gallagher, M.,** Multiple unit activity recorded from the amygdala central nucleus during Pavlovian heart rate conditioning in the rabbit, *Brain Res.,* 238, 457, 1982.

22. **Henke, P. G.,** Unit-activity in the central amygdalar nucleus of rats in response to immobilization stress, *Brain Res. Bull.*, 10, 833, 1983.
23. **Pascoe, J. P. and Kapp, B. S.,** Electrophysiological characteristics of amygdaloid central nucleus neurons during Pavlovian fear conditioning in the rabbit, *Behav. Brain Res.*, 16, 117, 1985.
24. **Kapp, B. S., Frysinger, R. C., Gallagher, M., and Haselton, J.,** Amygdala central nucleus lesions: effects on heart rate conditioning in the rabbit, *Physiol. Behav.*, 23, 1109, 1984.
25. **Gallagher, M., Kapp, B. S., Frysinger, R. C., and Rapp, P. R.,** Beta-adrenergic manipulation in amygdala central nucleus alters rabbit heart rate conditioning, *Pharmacol. Biochem. Behav.*, 12, 419, 1980.
26. **Iwata, J., Ledoux, J. E., Meeley, M. P., Arneric, S., and Reis, D. J.,** Intrinsic neurons in the amygdaloid field projected to by the medial geniculate body mediate emotional responses conditioned to acoustic stimuli, *Brain Res.*, 383, 195, 1986.
27. **Zhang, J. X., Harper, R. M., and Ni, H.,** Cryogenic blockade of the central nucleus attenuates aversively conditioned blood pressure and respiratory responses, *Brain Res.*, 386, 136, 1984.
28. **Johansson, G., Olsson, K., Haggendal, J., Jonsson, L., and Thoren-Tolling, K.,** Effect of amygdalectomy upon stress-induced myocardial necroses and blood levels of catecholamines in pigs, *Acta Physiol. Scand.*, 113, 553, 1981.
29. **Folkow, B., Hallback-Nordlander, M., Martner, J., and Nordbork, C.,** Influence of amygdala lesions on cardiovascular response to alerting stimuli, on behaviour and on blood pressure development in spontaneously hypertensive rats, *Acta Physiol. Scand.*, 116, 133, 1982.
30. **Galeno, T. M., Van Hoesen, G. W., and Brody, M. J.,** Central amygdaloid nucleus lesions attenuates exaggerated hemodynamic responses to noise stress in the spontaneously hypertensive rat, *Brain Res.*, 291, 249, 1984.
31. **Koepke, J. P., Jones, S., and DiBona, G. F.,** Alpha-2-adrenoreceptors in amygdala control renal sympathetic nerve activity and renal function in conscious spontaneously hypertensive rats, *Brain Res.*, 404, 80, 1987.
32. **Mondlock, J. M. and David, M.,** Lesions of the amygdala, but not of the cerebellum or red nucleus, block conditioned fear as measured with potentiated startle paradigm, *Behav. Neurosci.*, 100, 11, 1986.
33. **Shibata, K., Kataoka, Y., Yamashita, K., and Ueki, S.,** An important role of the central amygdaloid nucleus and mammillary body in the mediation of conflict behavior in rats, *Brain Res.*, 372, 159, 1986.
34. **Brown, M. R., Fisher, L. A., Spiess, J., Rivier, C., Rivier, J., and Vale, W.,** Corticotropin-releasing factor: actions on sympathetic nervous system and metabolism, *Endocrinology*, 111, 928, 1982.
35. **Lenz, H. J., Raedler, A., Greten, H., and Brown, M. R.,** CRF initiates biological actions with the brain that are observed in response to stress, *Am. J. Physiol.*, 363, R34, 1987.
36. **Britton, D. R., Koob, G. F., Rivier, J., and Vale, W.,** Intraventricular corticotropin-releasing factor enhances behavioral effects of novelty, *Life Sci.*, 31, 363, 1982.
37. **Sutton, R. E., Koob, G. F., LeMoal, M., Rivier, J., and Vale, W.,** Corticotropin releasing factor produces behavioral activation in rats, *Nature (London)*, 297, 331, 1982.
38. **Britton, K. T., Lee, G., Dana, R., Risch, S. C., and Koob, G. F.,** Activating and "anxiogenic" effects of corticotropin releasing factor are not inhibited by blockade of the pituitary-adrenal system with dexamethasone, *Life Sci.*, 39, 1281, 1986.
39. **Swerdlow, N. R., Geyer, M. A., Vale, W. W., and Koob, G. F.,** Corticotropin-releasing factor potentiates acoustic startle in rats, *Nature*, 88, 147, 1986.
40. **Brown, M. R.,** Corticotropin releasing factor: central nervous system sites of action, *Brain Res.*, 399, 10, 1986.
41. **Chappell, P. B., Smith, M. A., Kilts, C. D., Bissette, G., Ritchie, J., Anderson, C., and Nemeroff, C. B.,** Alterations in corticotropin-releasing factor-like immunoreactivity in discrete rat brain regions after acute and chronic stress, *J. Neurosci.*, 6, 2908, 1986.
42. **Swanson, L. W., Sawchenko, P. E., Rivier, J., and Vale, W. W.,** Organization of ovine corticotropin-releasing factor immunoreactive cells and fibers in the rat brain: an immunohistochemical study, *Neuroendocrinology*, 36, 165, 1983.
43. **Hopkins, D.,** Amygdalotegmental projections in the rat, cat and rhesus monkey, *Neurosci. Lett.*, 1, 263, 1975.
44. **Schwaber, J. S., Kapp, B. S., Higgins, G. A., and Rapp, P. R.,** Amygdaloid and basal forebrain direct connections with nucleus of the solitary tract and the dorsal motor nucleus, *J. Neurosci.*, 2, 1424, 1982.
45. **Hopkins, D. A. and Holstege, G.,** Amygdaloid projections to the mesencephalon, pons and medulla oblongata in the cat, *Exp. Brain Res.*, 32, 529, 1978.
46. **Holstege, G., Meiners, L., and Tan, K.,** Projections of the bed nucleus of the stria terminalis to the mesencephalon, pons, and medulla oblongata in the cat, *Exp. Brain Res.*, 58, 379, 1985.
47. **Van Der Kooy, D., Koda, L. Y., McGinty, J. F., et al.,** The organization of projections from the cortex, amygdala, and hypothalamus to the nucleus of the solitary tract in rat, *J. Comp. Neurol.*, 224, 1, 1984.
48. **Moga, M. M. and Gray, T. S.,** Evidence for corticotropin-releasing factor, neurotensin and somatostatin from the central nucleus of the amygdala to the parabrachial nucleus, *J. Comp. Neurol.*, 241, 275, 1985.

49. **Moga, M. M. and Gray, T. S.,** Peptidergic efferents from the intercalated nuclei of the amygdala to the parabrachial nucleus in the rat, *Neurosci. Lett.*, 61, 13, 1985.
50. **Moga, M. M. and Gray, T. S.,** Hypothalamic peptidergic efferents to the parabrachial nucleus in the rat, *Soc. Neurosci. Abstr.*, 11, 1985, 681.
51. **Moga, M. M. and Gray, T. S.,** Neuropeptide neuronal efferents from the bed nucleus of the stria terminalis to the parabrachial nucleus, *Anat. Rec.*, 211, 131A, 1985.
52. **Gray, T. S. and Magnuson, D. J.,** Neuropeptide neuronal efferents from the bed nucleus of the stria terminalis and central amygdaloid nucleus to the dorsal vagal complex in the rat, *J. Comp. Neurol.*, 262, 365, 1987.
53. **Gray, T. S.,** Neuropeptide neuronal efferents from the amygdala to the central gray in the rat, *Anat. Rec.*, 214, 1986, 44A.
54. **Sawchenko, P. E. and Swanson, L. W.,** A method for tracing biochemically defined pathways in the central nervous system using combined fluorescence retrograde transport and immunohistochemical techniques, *Brain Res.*, 210, 31, 1981.
55. **Veening, J. G., Swanson, L. W., and Sawchenko, P. E.,** The organization of projections from the central nucleus of the amygdala to the brainstem sites involved in autonomic regulation: a combined retrograde transport-immunohistochemical study, *Brain Res.*, 303, 337, 1984.
56. **Sakanaka, M., Shibasaki, T., and Lederis, K.,** Distribution and efferent projections of corticotropin-releasing factor-like immunoreactivity in the rat amygdaloid complex, *Brain Res.*, 382, 213, 1986.
57. **Saper, C. B.,** Anatomical substrates for the hypothalamic control of the autonomic nervous system, in *Integrative Functions of the Autonomic Nervous System*, Brooks, C. M., Koizumi, K., and Sato, A., Eds., Elsevier/North Holland, New York, 1979, 334.
58. **De Olmos, J., Alheid, G. F., and Beltramino, C. A.,** Amygdala, in *The Rat Nervous System*, Paxinos, G., Ed., Academic Press, Sydney, 1985, 223.
59. **Gray, T. S.,** Autonomic neuropeptide connections of the amygdala, in *Hans Selye Symposium: Neuropeptides and Stress*, Tache, Y., Morley, J. E., and Brown, M. R., Eds., Springer-Verlag, New York, in press.
60. **Mason, J. W.,** Plasma 17-hydroxycorticosteroid levels during electrical stimulation of the amygdaloid complex in conscious monkeys, *Am. J. Physiol.*, 196, 44, 1959.
61. **Setekleiv, J., Skaug, O. E., and Kaada, B. R.,** Increase of plasma 17-hydroxy-corticosteroids by cerebral cortical and amygdaloid stimulation in the cat, *J. Endocrinol.*, 22, 119, 1961.
62. **Redgate, E. S. and Fahringer, E. E.,** A comparison of the pituitary-adrenal activity elicited by electrical stimulation of preoptic, amygdaloid and hypothalamic sites in the rat brain, *Neuroendocrinology*, 12, 334, 1973.
63. **Feldman, S. and Conforti, N.,** Amygdalectomy inhibits adrenocortical responses to somatosensory and olfactory stimulation, *Neuroendocrinology*, 32, 330, 1981.
64. **Allen, J. P. and Allen, C. F.,** Amygdalar participation in tonic ACTH secretion in the rat, *Neuroendocrinology*, 19, 115, 1975.
65. **Beaulieu, S., Di Paolo, T., and Barden, N.,** Control of ACTH secretion by the central nucleus of the amygdala: implication of the serotonergic system and its relevance to the glucocorticoid delayed negative feedback mechanism, *Neuroendocrinology*, 44, 247, 1986.
66. **Beaulieu, S., Di Paolo, T., Cote, J., and Barden, N.,** Participation of the central amygdaloid nucleus in the response of adrenocorticotropin to immobilization stress: opposing roles of the noradrenergic and dopaminergic systems, *Neuroendocrinology*, 45, 37, 1987.
67. **Seggie, J.,** Differential responsivity of corticosterone and prolactin to stress following lesions of the septum or amygdala: implications for psychoneuroendocrinology, *Prog. Neuro-Psychopharmacol. Biol. Psych.*, 11, 315, 1987.
68. **Tribollet, E. and Dreifuss, J. J.,** Localization of neurones projecting to the hypothalamic paraventricular nucleus area of the rat: a horseradish peroxidase study, *Neuroscience*, 6, 1315, 1981.
69. **Swanson, L. W. and Cowan, W. M.,** The connections of the septal region in the rat, *J. Comp. Neurol.*, 186, 621, 1975.
70. **Gray, T. S., Moga, M. M., and Magnuson, D. J.,** Efferent projections of the central nucleus of the amygdala: a reexamination using the PHA-L anterograde tracing method, *Soc. Neurosci. Abstr.*, 12, 1173, 1986.
71. **Gerfen, C. and Sawchenko, P. E.,** An anterograde neuroanatomical tracing method that shows the details of neurons, their axons and terminals: immunohistochemical localization of an axonally transported plant lectin, *Phaseolus vulgaris* leucoagglutinin (PHA-L), *Brain Res.*, 290, 219, 1984.
72. **Cassell, M. D., Gray, T. S., and Kiss, J. Z.,** Neuronal architecture in the rat central nucleus of the amygdala: a cytological, hodological, and immunocytochemical study, *J. Comp. Neurol.*, 246, 478, 1986.
73. **Margraph, C. G., Kapp, B. S., and Khazam, C. D.,** Fear induced cardiac arrhythmias in digitalis predisposed rabbit: amygdaloid central nucleus contribution, *Soc. Neurosci. Abstr.*, 11, 1270, 1985.

Chapter 6

CORTICOTROPIN-RELEASING FACTOR (CRF) RECEPTORS IN THE RAT CENTRAL NERVOUS SYSTEM: AUTORADIOGRAPHIC LOCALIZATION STUDIES

Errol B. De Souza and Thomas R. Insel

TABLE OF CONTENTS

I. INTRODUCTION

The anatomical distribution of corticotropin-releasing factor (CRF) in the central nervous system (CNS) and the wide spectrum of autonomic, behavioral, and electrophysiological effects following CRF administration has led to the suggestion that CRF may act as a stress neurotransmitter in brain. A precise localization of CRF receptors in the CNS is necessary in order to fully understand the neuronal circuits through which CRF produces its varied effects in brain. This chapter provides a summary of the data from autoradiographic studies examining in detail the distribution of CRF receptors in rat CNS.

II. METHODS

A. CHOICE OF RADIOLIGAND

The iodine-125-labeled analogs of CRF that have been used to label CRF receptors on slide-mounted sections of brain include ^{125}I-Tyr$^{\circ}$-ovine-CRF, Nle^{21}, ^{125}I-Tyr^{32}-ovine CRF, and ^{125}I-Tyr$^{\circ}$-rat/human CRF. While all of the iodine-125-labeled analogs of CRF provide good probes to study CRF receptors, the ovine CRF (oCRF) radiolabels have lower levels of nonspecific binding and higher levels of total binding. At present, we routinely use commercial preparations of ^{125}I-Tyr$^{\circ}$-ovine CRF (New England Nuclear Corp., Boston, MA; specific activity 2200 Ci/mmol).

B. TISSUE PREPARATION

Male Sprague-Dawley (Madison, WI) rats weighing 200 to 300 g were anesthetized with pentobarbital and perfused intracardially with 500 to 700 ml of a mixture of equal parts of phosphate-buffered saline and 0.32 *M* sucrose (pH 7.4). The brains were then rapidly removed, embedded in homogenized brain paste, and frozen in powdered dry ice. The tissues were sectioned (8 or 10 μm) using a microtome (Harris) at -16°C, thaw-mounted onto chrome alum/gelatin subbed microscope slides, and stored at -20°C until used. In biochemical studies involving characterization of the receptor, we used slide-mounted coronal sections of rat forebrain through the diencephalon (two 10-μm sections per slide). Sections from at least 15 animals were subsequently used to generate autoradiograms.

C. RECEPTOR LABELING IN SLIDE-MOUNTED BRAIN SECTIONS

Slide-mounted tissue sections were brought to room temperature and incubated with 0.1 to 0.2 n*M* Nle^{21}, ^{125}I-Tyr^{32}-oCRF or ^{125}I-Tyr$^{\circ}$-ovine-CRF in 50 m*M* Tris-HCl (pH 7.4) containing 10 m*M* $MgCl_2$, 2 m*M* EGTA, 0.1% BSA, aprotinin (100 KIU/ml), and 0.1 m*M* bacitracin (Sigma) at room temperature. Blanks were incubated in the same medium with the addition of 1 μ*M* Nle^{21}, Tyr^{32}-oCRF, oCRF, or rat/human CRF. Before exposure to iodine-125-labeled CRF, all tissue sections were incubated at room temperature for two 15-min periods in 50 m*M* Tris-HCl (pH 7.4) containing 5 m*M* $MgCl_2$ and 2 m*M* EGTA in order to displace endogenous CRF from its receptor. After incubation with the radioligand, tissue sections were washed in Dulbecco's phosphate-buffered saline containing 1% BSA (Fraction V, Sigma) or 0.01% Triton X-100, dipped in deionized water, and dried rapidly under a stream of cold, dry air. In preliminary nonautoradiographic studies to determine the time courses of dissociation and association kinetics and pharmacology of Nle^{21}, ^{125}I-Tyr^{32}-oCRF binding to slide-mounted tissue sections, the tissue sections were dried and the tissue was scraped off the glass microscope slide and the tissue-associated radioactivity was assayed in a gamma counter. In subsequent autoradiographic studies, the tissue sections were incubated with Nle^{21}, ^{125}I-Tyr^{32}oCRF or ^{125}I-Tyr$^{\circ}$-ovine-CRF for 60 to 120 min at room temperature and washed for two 5-min periods in Dulbecco's phosphate-buffered saline at 4°C.

D. AUTORADIOGRAPHY

The dried, labeled slide-mounted sections were apposed to either ^{3}H-Ultrofilm (LKB Products) or Kodak NTB-3 emulsion-coated glass, and after 7 to 14 d of exposure at 4°C, the autoradiograms were developed and the tissue was stained with toluidine blue. Following the staining procedures, the sections were dried and the coverslips were reapposed and set with Permount.

E. DATA ANALYSIS

In autoradiograms prepared with NTB-3 emulsion, grain counts were made by eye using a ×100 oil immersion objective on a Zeiss microscope equipped with a grid-containing eyepiece. In autoradiograms prepared with tritium-sensitive film, optical density readings, construction of standard curves, and rapid quantification of the data were carried out using a Loats PC-based computerized image analysis system (Amersham). Autoradiograms of tissue sections and iodinated standards were generated with ^{3}H-Ultrofilm, and the film optical density was related to the molar concentration of radioactivity by using the standard curve generated concomitantly with the autoradiograms.

III. BIOCHEMICAL CHARACTERIZATION OF CRF RECEPTORS ON SLIDE-MOUNTED BRAIN SECTIONS: KINETICS AND PHARMACOLOGY

A series of kinetic and pharmacologic studies were carried out to define the incubation parameters and assess the properties of CRF receptors on slide-mounted brain sections. First, the rate of dissociation of Nle^{21}, ^{125}I-Tyr^{32}-oCRF binding was determined as a function of temperature. Slide-mounted sections of rat brain through the diencephalon were incubated with 0.2 n*M* Nle^{21}, ^{125}I-Tyr^{32}-oCRF for 60 min at 22°C and then washed for various times in Dulbecco's phosphate-buffered saline containing 1% BSA at either 4°C or at 22°C. Nonspecific binding was determined in the presence of 1 μ*M* oCRF. The dissociation rates at 4°C and at 22°C were 0.046 and 0.066 min^{-1}, respectively.

Next, the time course of association of Nle^{21}, ^{125}I-Tyr^{32}-oCRF with receptors at 22°C in slide-mounted sections of rat forebrain was examined. Slide-mounted tissue sections were incubated with 0.2 n*M* Nle^{21}, ^{125}I-Tyr^{32}-oCRF for various times at 22°, rinsed for two 5-min periods at 4°C, and assayed for radioactivity. The specific binding increased with time during the first 60 min of incubation, after which there was no further significant increase. Based on these observations, all subsequent autoradiographic studies utilized a 60 to 120 min incubation at 22°C followed by two 5-min washes at 4°C.

To further define the characteristics of Nle^{21}, ^{125}I-Tyr^{32}-oCRF binding in rat brain under the equilibrium conditions defined above, we incubated serial slide-mounted tissue sections with increasing concentrations of CRF-related and unrelated peptides. The binding of Nle^{21}, ^{125}I-Tyr^{32}-oCRF in rat striatum was saturable and, on Scatchard analysis, revealed a high-affinity component with an apparent K_D of 1.8 n*M* and a low-affinity binding site with a K_D of around 100 n*M*. In competition studies, 1 μ*M* rat/human CRF and ovine CRF inhibited 105 and 81%, respectively, of the specific Nle^{21}, ^{125}I-Tyr^{32}-oCRF binding in rat striatum. Two fragments of oCRF, CRF(1-39) and CRF(1-22), which are weak in stimulating POMC-derived peptide secretion from the pituitary gland, displaced about 30%, and the unrelated peptide arginine vasopressin did not significantly inhibit the binding.

The receptor binding studies carried out in slide-mounted brain sections using conditions identical to those used in autoradiographic experiments demonstrate the presence of high affinity and pharmacologically specific binding sites for Nle^{21}, ^{125}I-Tyr^{32}-oCRF in rat striatum. The kinetic characteristics and peptide specificity of the Nle^{21}, ^{125}I-Tyr^{32}-oCRF binding site in rat striatum were comparable to those for the CRF receptor in homogenates of rat anterior pituitary,[1-6] neocortex,[7] and olfactory bulb[7] and in slide-mounted sections of bovine[2,3] and human anterior pituitary[8] and rat intermediate pituitary.[1] These data substantiate earlier suggestions[9-11] that some structural requirements for CRF activity are shared by brain and pituitary receptors.

IV. REGIONAL DISTRIBUTION AND RELEVANCE OF CRF BINDING SITES IN RAT CNS

The autoradiographic distribution of CRF receptors in rat CNS is shown in Figures 1 to 5 and quantified in Table 1. Overall, nearly all of the binding was to gray matter areas with nonsignificant levels present in white matter tracts. Details of the quantities and pattern of distribution of CRF receptors in rat CNS have been described elsewhere.[12] In this section of the chapter we will discuss the distribution of CRF receptors in the context of their correlation with immunocytochemical distribution of CRF-like immunoreactivity (CRF-IR) and the possible anatomical loci mediating the pharmacological actions of CRF in rat brain.

A. OLFACTORY SYSTEM

The distribution of CRF receptors in the rat olfactory system is striking. Within the olfactory bulb (Figure 2), CRF binding sites are discretely localized in the glomerular layer where the primary olfactory afferent axons terminate and in the adjacent external plexiform layer and in the mitral cell body layer where the olfactory output neurons originate. Low to negligible concentrations of receptors were found in the remainder of the bulb. In the pyriform cortex, a high density of binding sites was present only in lamina I with much lower levels found in the remaining laminae. A high grain density was evident in the external plexiform layer of the olfactory tubercle, where as low density binding was observed in the pyramidal and polymorphic layers of the tubercle. No CRF binding sites were present in the lateral olfactory tract. There is evidence in the literature suggesting that sexual, reproductive, social, and feeding behaviors of most mammalian species are mediated by olfactory substances and pheromonal agents acting through the olfactory system.[13] CRF influences both feeding behavior[9,14-16] and sexual activity.[17,18] Although the hypothalamus is the primary anatomical locus regulating sexual activity and feeding behavior, hypothalamic neuronal activity can be modulated by olfactory impulses relayed through the circuit involving the olfactory bulb, pyriform cortex, amygdala, and subiculum (see Reference 12); moderate to high densities of CRF receptors are present throughout the circuits.

B. CEREBRAL CORTEX

CRF receptors were found throughout the neocortex. Within the neocortex, high grain densities were present in laminae I and IV with moderate levels observed in laminae II, III, V, and VI (Figure 3). Although this laminar distribution of CRF binding sites was maintained throughout the neocortex, on the average, more binding sites were found in somatosensory, striate, and entorhinal cortex than in motor and cingulate cortex. CRF cell bodies in the neocortex are concentrated in laminae II and III, with projections to laminae I and IV,[19,20] areas rich in CRF receptors. While the role of CRF receptors in the cerebral cortex remains undefined, there is clear evidence suggesting that CRF binding sites localized in the neocortex are functional:

1. Of all the brain regions examined, CRF receptor-stimulated adenylate cyclase activity is highest in the cerebral cortex.[21,22]
2. Iontophoretic application of CRF excited spontaneously occurring action potentials in the cerebral cortex.[23]
3. CRF stimulates somatostatin secretion *in vitro* in primary cultures of cerebral cortex.[10]

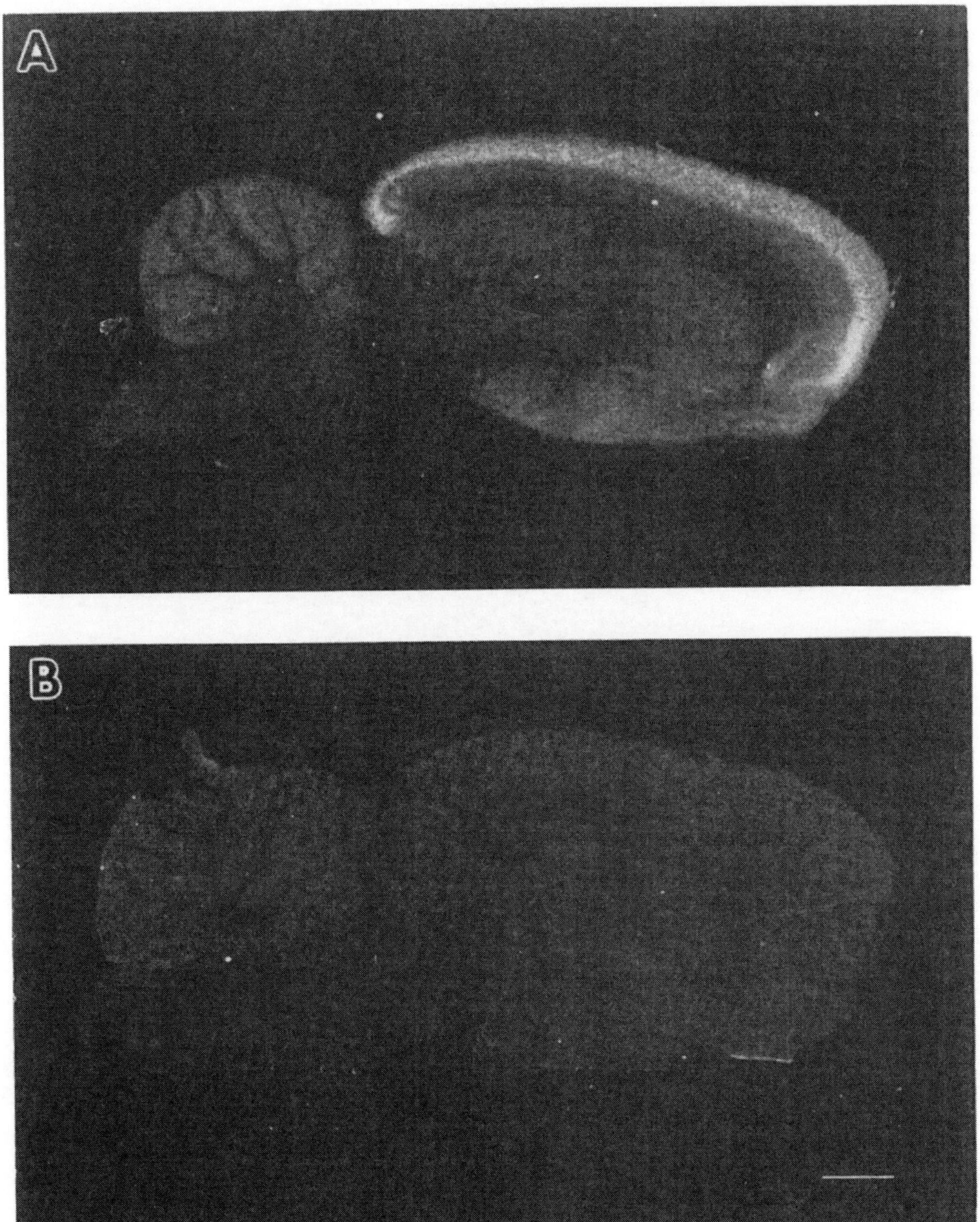

FIGURE 1. Autoradiographic distribution of CRF receptor-binding sites in rat brain, labeled with ^{125}I-Tyr^{0}-ovine-CRF. (A) and (B) are darkfield photomicrographs (^{3}H-Ultrofilm), showing the distribution of autoradiographic grains in sagittal sections of rat brain. In darkfield illumination the autoradiographic grains, i.e., binding sites, appear as white spots and the tissue is not visible. Thus, the brightest areas have the highest concentration of binding sites. (A) shows the "total" binding and (B) shows the absence of specific receptor binding when 1 μM unlabeled rat/human CRF is included in the incubation buffer (blank). Bar = 2 mm.

4. CRF receptors in the cerebral cortex are modulated under conditions in which CRF is altered; in Alzheimer's disease, there is a decrease in CRF-IR in different regions of the cerebral cortex and a reciprocal up-regulation of CRF receptors in the same cortical areas.[24] In contrast, in depression, the concentration of CRF receptors is decreased in the frontal cortex.[25]

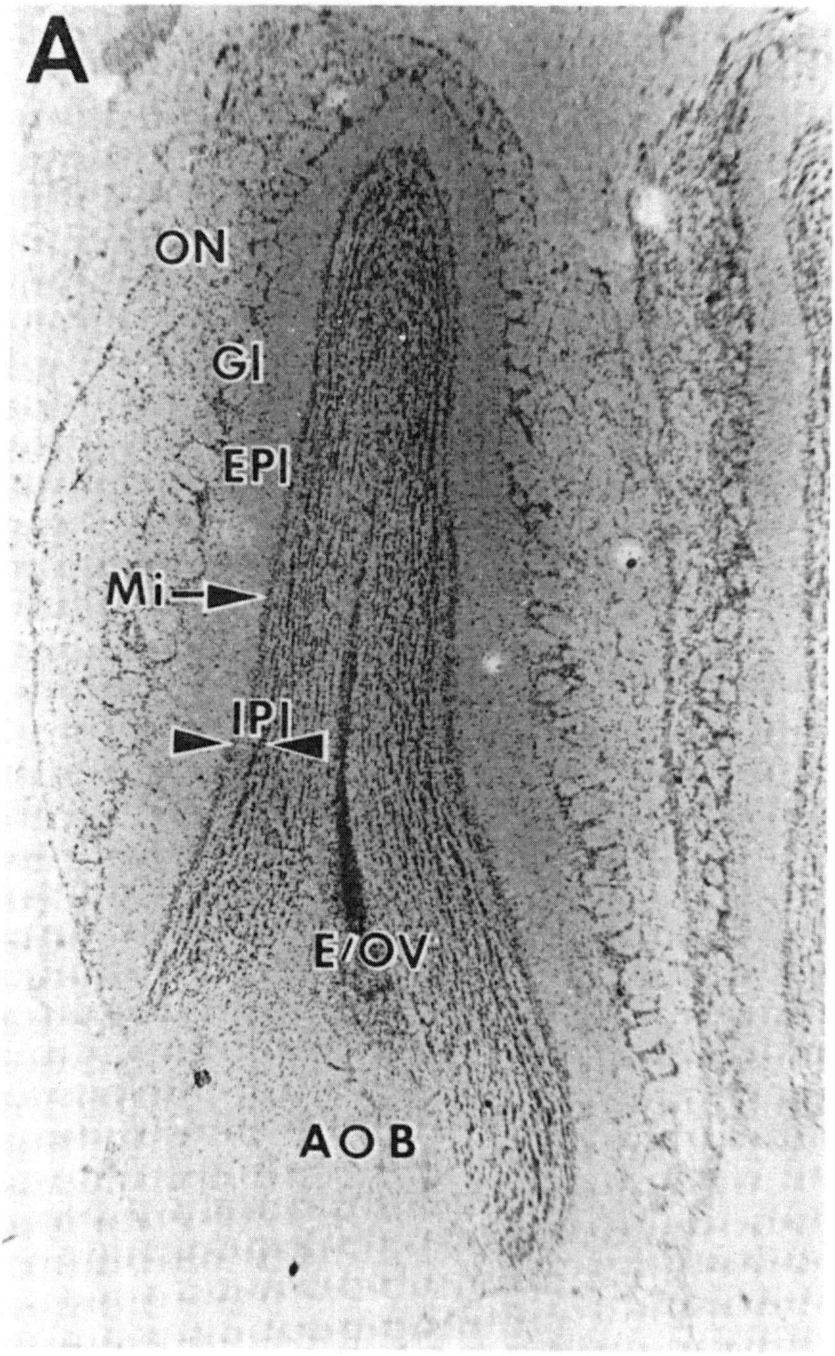

FIGURE 2. Autoradiographic localization of CRF receptors in rat olfactory bulb. (A) Brightfield photomicrograph of a toluidine blue-stained tissue showing the histology of the olfactory bulb sectioned in the horizontal plane. (B) Darkfield photomicrograph showing the autoradiographic grain distribution on emulsion-coated coverslips over the same area shown in (A). In darkfield illumination the autoradiographic silver grains appear as white spots and the tissue is not visible. In (B), note the high concentration of grains in the external plexiform layer and glomerular layer with a somewhat lower concentration in the internal plexiform layer (arrowheads in A). Bar = 200 μm. (From De Souza, E. B., Insel, T. R., Perrin, M. H., Rivier, J., Vale, W. W., and Kuhar, M. J., *J. Neurosci.*, 5, 3189, 1985. With permission.)

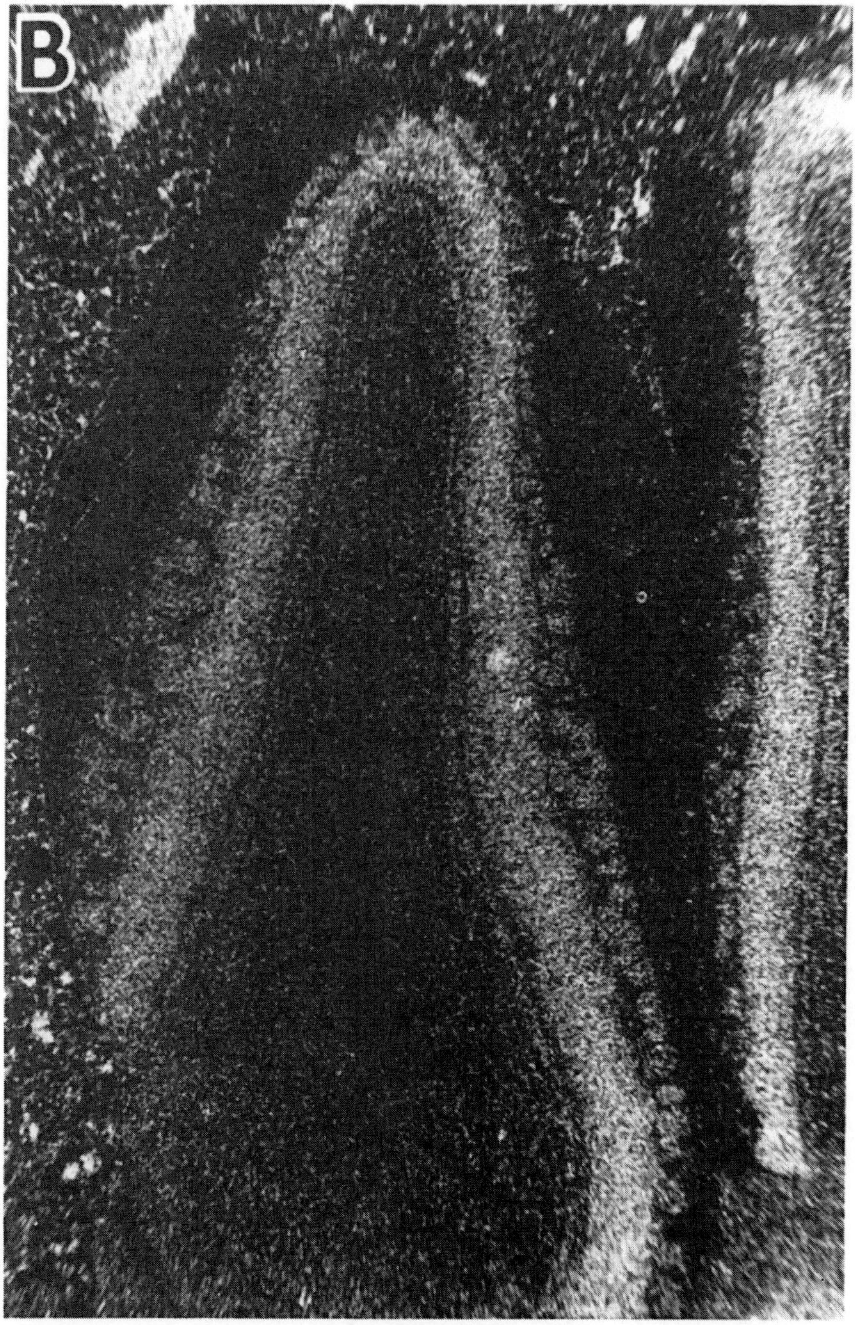

FIGURE 2B.

5. CRF receptors can be selectively up-regulated in rat cerebral cortex following chronic treatment with atropine.[26] Data demonstrating changes in CRF receptors in clinical disorders such as Alzheimer's disease and depression suggest a role for CRF in the cerebral cortex in modulating cognitive function.
6. In behavioral studies, CRF injected into the rat frontal cortex did not initiate, but significantly inhibited the effects of carbachol, a muscarinic cholinergic receptor antagonist, on a stereotyped motor "boxing" behavior.[27]

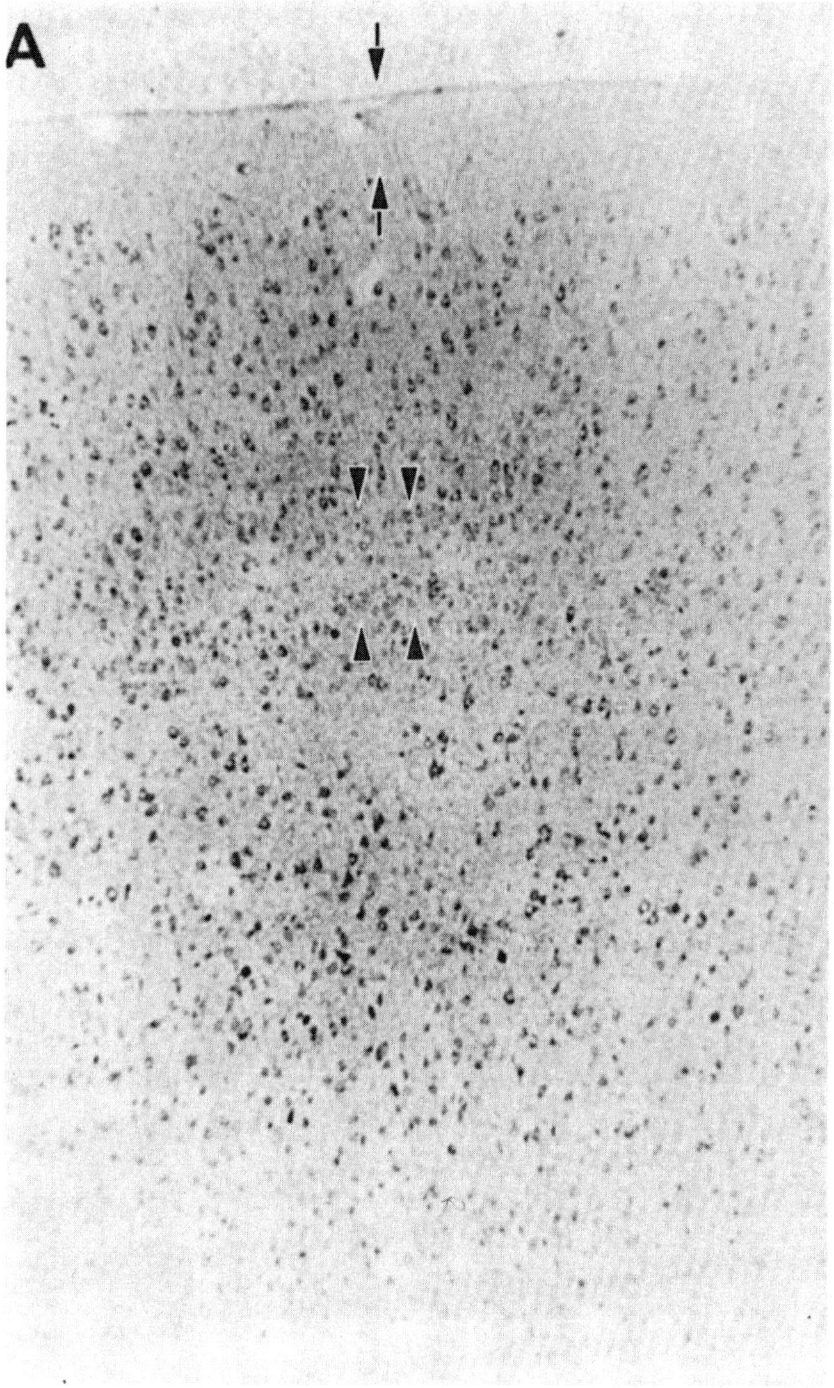

FIGURE 3. Autoradiographic localization of CRF receptors in the cerebral cortex (parietal region). (A) Brightfield photomicrograph of a toluidine blue-stained section (coronal plane) showing the six layers (1 to 6) of the neocortex. (B) Darkfield photomicrograph showing the autoradiographic grain distribution on emulsion-coated coverslips over the same area shown in (A). In (B) note the high concentration of grains in lamina 1 (between arrows) and lamina 4 (between arrowheads) of the neocortex. Bar = 25 μm. (From De Souza, E. B., Insel, T. R., Perrin, M. H., Rivier, J., Vale, W. W., and Kuhar, M. J., *J. Neurosci.*, 5, 3189, 1985. With permission.)

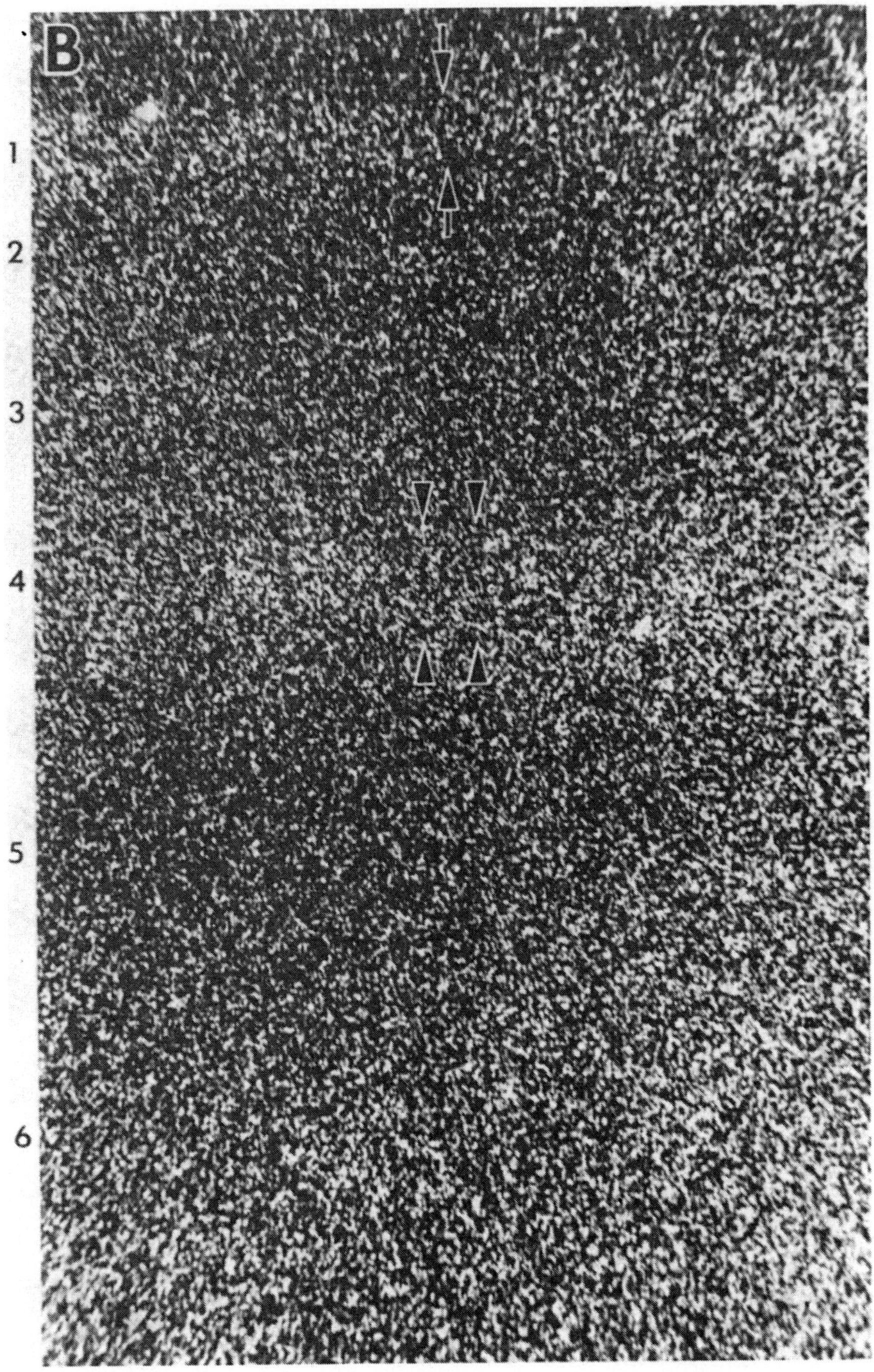

FIGURE 3B.

Additional studies are necessary to further define the role of CRF receptors in the cerebral cortex.

C. LIMBIC SYSTEM

Moderately high levels of CRF receptors were associated with the nucleus accumbens, with slightly higher densities observed over the medial than the lateral parts of this nucleus.

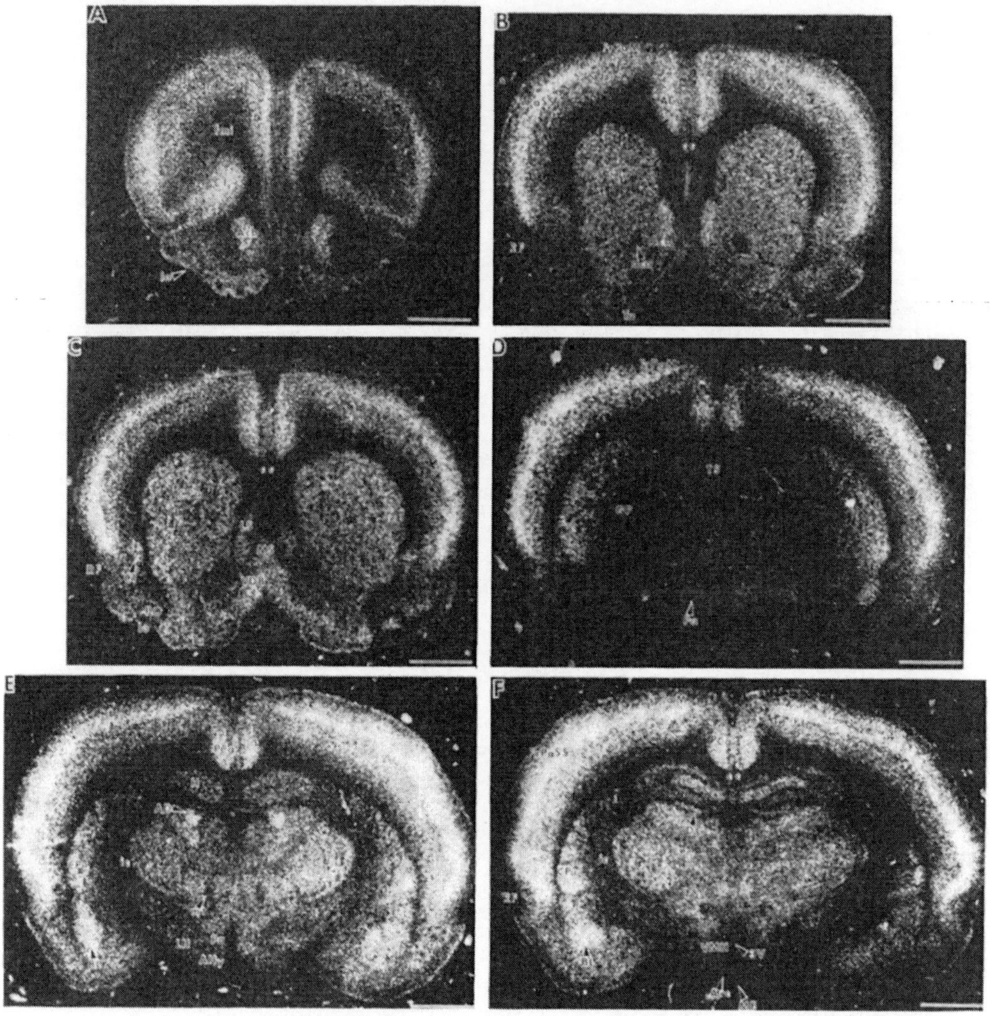

FIGURE 4. Autoradiographic mapping of the distribution of CRF receptor-binding sites in coronal sections of rat brain, labeled with Nle^{21}, ^{125}I-Tyr^{0}-oCRF. (A) to (L) are darkfield photomicrographs (tritium-sensitive Ultrofilm) showing the distribution of autoradiographic grains in coronal sections of rat forebrain (A to F) and brain stem (G to L). The autoradiograms pictured here are from one animal and are organized rostrocaudally. The anatomy of the CRF binding sites was confirmed in sections where autoradiograms were generated by apposition of emulsion-coated coverslips that were permanently affixed to the slides. A concentration of 0.2 n*M* Nle^{21}, ^{125}I-Tyr^{0}-oCRF was used to label these sections. Nonspecific binding, determined in the presence of 1 μ*M* unlabeled Nle^{21}, ^{125}I-Tyr^{0}-oCRF was uniform and comparable to that shown in Figure 1B. Bar = 2 mm. (From De Souza, E. B., Insel, T. R., Perrin, M. H., Rivier, J., Vale, W. W., and Kuhar, M. J., *J. Neurosci.*, 5, 3189, 1985. With permission.)

Low to moderate grain densities were observed over both the vertical and horizontal limbs of the nucleus and tract of the diagonal band. Low levels of CRF binding were present in the areas of medial and lateral septal nuclei and triangular septal nucleus. In the amygdala, low concentrations of receptors were observed in the medial and central nuclei, whereas a moderate concentration of CRF receptors was present in the basolateral nucleus. Moderate levels of CRF receptors were present in the bed nucleus of the stria terminalis. Low to moderate concentrations of receptors were found in the hippocampus. Within the hippocampus, the highest concentrations of binding sites were present in the subiculum, the molecular

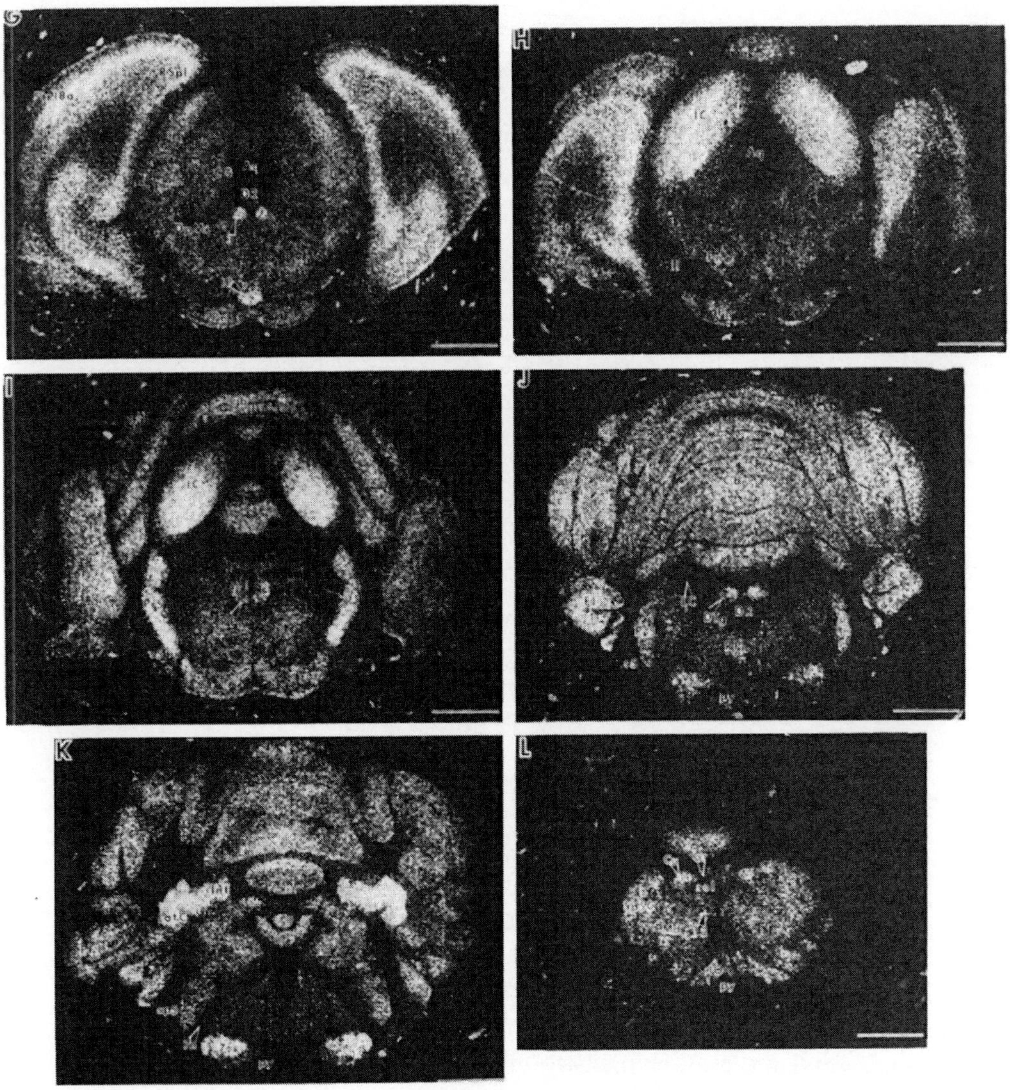

FIGURE 4(G—L)

layer of the dentate gyrus, and the CA1 region. Lower concentrations of binding sites were found in the remaining areas of the hippocampus.

CRF receptors in the limbic system may be responsible for mediating a variety of the effects of CRF that have been observed in rodents following intracerebroventricular administration of CRF. The limbic system plays an important role in regulating behavioral responses, especially those involved with emotion, autonomic responses, and endocrine function (see Chapter 5). Centrally administrated CRF produces a broad spectrum of autonomic and behavioral effects, including increases in plasma levels of catecholamines[28] and glucose,[28] increases in cardiovascular parameters including heart rate and mean arterial pressure,[29] and a variety of behavioral changes associated with generalized arousal, such as increased locomotor activity in familiar surroundings,[9,30-32] and increased "emotionality" in a novel environment.[9,30-32] Iontophoretic application of CRF has been reported to inhibit spontaneously occurring action potentials in the lateral septal area.[23] Hippocampal slice preparations have been employed to demonstrate the electrophysiological effects of CRF *in vitro,*[33] and

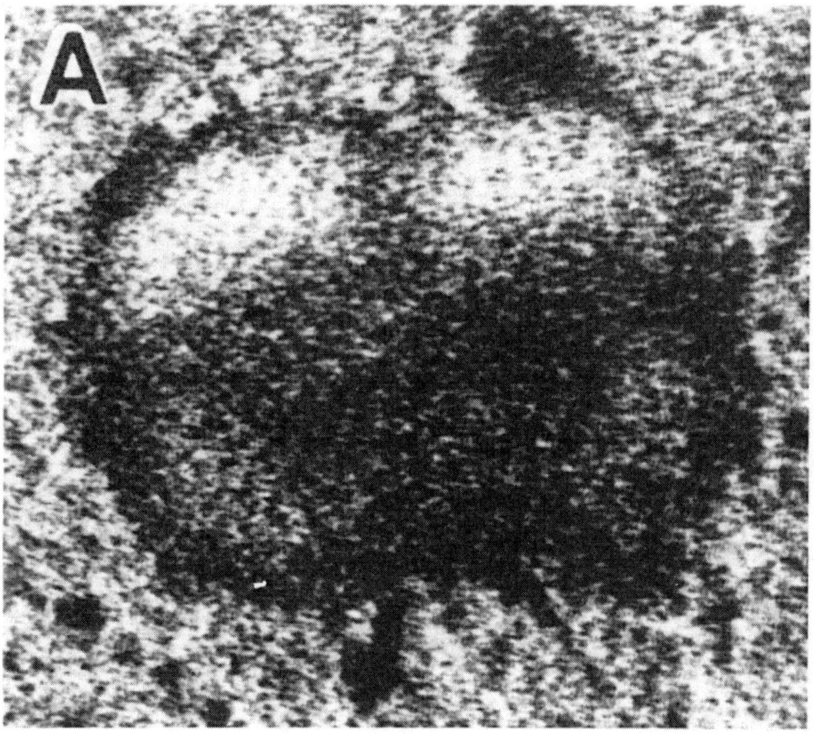

FIGURE 5. Distribution of CRF receptors labeled with ^{125}I-Tyr^{0}-ovine-CRF in rat spinal cord. Darkfield photomicrographs (^{3}H-Ultrofilm) showing the autoradiographic grain distribution. (A) shows the "total" binding and (B) shows the absence of specific receptor binding when 1 μM unlabeled rat/human CRF is included in the incubation buffer (blank). In (A), note the high concentration of grains in the superficial layers (laminae I and II; substantia gelatinosa) of the dorsal horn with progressively lower concentrations present in the motoneurons of the ventral horn and in the intermediate and central zones of the spinal cord, respectively. Bar = 50 μm.

intracerebroventricular administration of high doses of CRF have been shown to elicit epileptiform seizures that appear, on the basis of associated electroencephalographic patterns, to originate in the amygdala.[34,35]

D. THALAMUS-HYPOTHALAMUS

There was a fairly uniform, low to moderate level of CRF binding sites throughout the thalamus with a slightly higher concentration of grains observed in the lateral compared to the medial thalamus. In the dorsal thalamic complex, a higher concentration of receptors was associated with the anterodorsal nucleus. The habenular nuclei of the epithalamus contained a moderate level of CRF binding sites. While the function of CRF receptors in the thalamus is unclear, an inhibition of firing has been demonstrated following iontophoretic application of CRF in the thalamus.[23]

In general, the hypothalamus contained a low density of CRF receptors. Two exceptions included the paraventricular nucleus and the mammillary peduncle where moderate levels of receptors were present. Low to moderate concentrations of CRF binding sites were found in the external layer of the median eminence.

The hypothalamus is a primary brain region involved in the regulation of pituitary function. High concentrations of CRF-IR are found in cell bodies in the paraventricular nucleus of the hypothalamus and in the median eminence where the neurons originating in

FIGURE 5B.

the paraventricular nucleus terminate.[19,20] The hypothalamus may represent a primary site for integrating the endocrine, autonomic, and behavioral effects of CRF. Microinfusions of CRF into the hypothalamus suppress sexual behavior.[17] In addition, intracerebroventricular injections of CRF have been shown to have antipyretic effects,[36] to inhibit the release of growth hormone,[37,38] luteinizing hormone,[37,39] vasopressin, and oxytocin,[40] and to have deleterious effects on reproductive functions.[39] CRF can also stimulate the secretion of somatostatin,[10] β-endorphin,[41] and dynorphin[41] and inhibit the release of gonadotropin-releasing hormone[42] in hypothalamus *in vitro*. A role has been proposed for CRF in the control of food intake;[9,14-16] the medial and lateral hypothalamus represent important brain areas that are involved in regulating food intake.[43]

E. BRAIN STEM

In the midbrain, a moderate level of CRF receptors were seen in the inferior colliculus. Within the inferior colliculus, the ventrolateral aspects had higher grain densities than did the dorsomedial portion. Low to moderate concentrations of CRF receptors were found in the superior colliculus, the trochlear nuclei and the interpeduncular nucleus. Low levels of CRF receptors were present in the periaqueductal gray and dorsal raphe nucleus and very low concentrations were seen in the substantia nigra.

In the pons, the cranial nerve nuclei, including the facial, trigeminal, cochlear, and vestibular nuclei, had very high concentrations of CRF binding sites. The medial portions of the vestibular nucleus had much higher concentrations of CRF binding sites, whereas the lateral and superior aspects of this nucleus had moderate to low concentrations of autoradiographic grains. High grain densities were found in the lateral cervical nucleus and at the point of entry of the eighth nerve. All portions of the nucleus of the spinal tract of the trigeminal nerve had comparable high levels of CRF receptors. Moderate numbers of grains were associated with hypoglossal, lateral reticular, paragigantocellular reticular, parabrachial, and dorsal and ventral tegmental nuclei. The pontine nuclei themselves and the raphe

TABLE 1
Distribution of Corticotropin-Releasing Factor (CRF) Receptors in the Rat Central Nervous System

Region	Relative density of CRF binding sites
Frontoparietal cortex (somatosensory)	
Lamina I and II	3+
Lamina III and IV	4+
Lamina V and VI	3+
Frontoparietal cortex (motor)	
Lamina I and II	2+
Lamina III and IV	3+
Lamina V and VI	3−
Striate cortex (area 18a)	
Lamina I and II	2+
Lamina III and IV	4−
Lamina V and VI	3−
Cingulate cortex (all laminae)	3−
Entorhinal cortex (all laminae)	3+
Olfactory tubercle	3
Basal ganglia	
Caudate-putamen	3−
Globus pallidus	1
Claustrum	4−
Septal area	
Nucleus accumbens	3
Nucleus of the diagonal band (Broca)	2
Lateral septal nucleus	1+
Medial septal nucleus	1+
Triangular septal nucleus	1
Bed nucleus of the stria terminalis	1+
Amygdala	
Basolateral nucleus	3−
Central nucleus	2+
Medial nucleus	2+
Thalamus	
Lateral thalamus	2
Medial thalamus	1+
Anterodorsal thalamic nucleus	2+
Habenula (medial plus lateral)	2
Hypothalamus	
Lateral hypothalamus	1+
Medial hypothalamus	1+
Anterior hypothalamus	2−
Paraventricular nucleus	2
Arcuate nucleus	0
Mammillary peduncle	3−
Median eminence	3
Hippocampus (whole)	1+
Dentate gyrus	1+
Subiculum	2−
Midbrain	
Inferior colliculus	3−
Dorsal raphe nucleus	1−
Central gray	2−
Medulla-pons	
Inferior olive	3+
Superior olive	3+
Lateral reticular nucleus	3

TABLE 1 (continued)
Distribution of Corticotropin-Releasing Factor (CRF) Receptors in the Rat Central Nervous System

Region	Relative density of CRF binding sites
Paragigantocellular reticular nucleus	3−
Nucleus of the spinal tract of the trigeminal nerve	4+
Cuneate nucleus	4−
Gracile nucleus	4−
Nucleus of the solitary tract	2+
Lateral vestibular nucleus	2
Medial vestibular nucleus	3+
Facial nucleus	5+
Lateral cervical nucleus	4+
Vestibulocochlear nucleus	5
Cochlear nucleus	4−
Dorsal tegmental nucleus	3−
Ventral tegmental nucleus	3
Parabrachial nucleus	3
Pontine nucleus	2
Hypoglossal nucleus	3
Locus coeruleus	2
Cerebellum (all lobules)	4+
Interpositus nucleus	3+
Medial nucleus	4+
Paraflocculus	5
Spinal cord (cervical)	
Dorsal horn (rexed 1-2); substantia gelatinosa	4
Dorsal horn (rexed 4-6)	2−
Ventral horn	2+
White matters tracts	
Corpus callosum	0
Forces minor corpus callosum	0
Fornix	0
Anterior commisure	0
Optic chiasm	0

Note: The data are based on observations from two or three animals. The anatomical terminology is derived from G. Paxinos and C. Watson, *The Rat Brain in Stereotaxic Coordinates*, Academic Press, New York, 1982. CRF binding sites were visualized using 0.7 n*M* Nle,[21], ^{125}I-Tyr^{32}-oCRF. At this concentration of ligand, approximately 10% of the total number of binding sites were occupied. Autoradiograms of rat brain, rat spinal cord, and iodinated standards (all 8 μm) were generated with tritium-sensitive Ultrofilm (LKB). An analysis of Nle,[21] ^{125}I-Tyr^{32}-oCRF binding in the different regions of rat CNS was performed by computerized image analysis densitometry, and the film optical density was related to the molar concentration of radioactivity by use of the standard curve generated and analyzed concomitantly with the autoradiograms. The relative density of CRF receptors corresponds to the following range of specific Nle,[21] ^{125}I-Tyr^{32}-oCRF values: 0, no specific binding; 1, 0-20 fmol/mg protein; 2, 20 to 40 fmol/mg protein; 3, 40 to 80 fmol/mg protein; 4, 80 to 100 fmol/mg protein; 5, 100 to 150 fmol/mg protein; + and − correspond to receptor densities above and below the mean of the range, respectively.

pontis nucleus contained moderate to low numbers of grains, whereas the pontine reticular nuclei had a very low density of CRF receptors. Moderate levels of receptors were found in the locus coeruleus.

In the medulla, the highest receptor densities were found in the cuneate and gracilis nuclei. Moderate grain densities were also found in the inferior olive. The concentration of

CRF receptors was variable (average of low to moderate) in the nucleus of the solitary tract, the solitary tract, and throughout the extent of the pyramids.

The localization of CRF receptors in certain regions of the brain stem provide an anatomical basis for some of the observed effects of CRF on autonomic function. CRF may produce some of its autonomic effects by acting on receptors in the parabrachial nucleus, medullary reticular formation, nucleus of the solitary tract, or other brain stem areas known to participate in autonomic regulation. In addition, some of the cardiovascular effects of CRF may also be mediated through the locus coeruleus; CRF has been shown to directly activate noradrenergic neurons of the locus coeruleus,[11] an area with a moderate concentration of CRF binding sites. The locus coeruleus represents an important brain region involved in stress responses and recent data have demonstrated that the levels of CRF are selectively increased in this nucleus following both acute and chronic stress.[44] Direct microapplication of CRF to parabrachial neurons has also been demonstrated to increase discharge rate of these cells.[11]

F. CEREBELLUM

In the cerebellum, CRF receptors were present in all parts of the cerebellar cortex with highest concentrations in the granular layer. The medial nucleus and the interpositus nucleus had high receptor levels. In the rat cerebellum, moderate,[45] low,[19] and undetectable[20] densities of CRF-IR fibers have been reported, while very high concentrations of CRF binding sites are present in this brain region. Possible reasons for the discrepancy in the distribution of CRF-IR between the different studies will be discussed below (see Section V). More recent data provide compelling evidence that CRF is a major neurotransmitter in the olivocerebellar pathway in a variety of species including rat, cat, monkey, and human.[46-49] Using *in situ* hybridization with a synthetic oligonucleotide probe for CRF mRNA and immunocytochemistry with antibodies recognizing CRF, recent studies support the hypothesis that CRF is present in nerve cells of the inferior olivary complex and that CRF-IR fibers are associated with cell bodies and dendrites of Purkinje cells.[47-49] Thus, CRF may be a peptidergic transmitter in inferior olivary complex climbing fiber system and CRF receptors in the cerebellum may provide cellular targets for these neurons.

G. SPINAL CORD

Within the spinal cord, CRF receptors are present in highest concentrations in the superficial layers (laminae I and II) of the dorsal horn with progressively lower concentrations present in the motoneurons of the ventral horn and in the intermediate and central zones of the spinal cord, respectively. The density and distribution of CRF receptors in the rat spinal cord corresponds well with the localization of CRF-IR nerve fibers. For example, large numbers of CRF-IR fibers are present in the substantia gelatinosa of the spinal cord[50-52] where the highest concentrations of CRF binding sites are localized.[53] In addition, treatment of newborn rats with capsaicin causes a disappearance of CRF-immunoreactive fibers within laminae I and II of the dorsal horn of the spinal cord[52] suggesting that CRF is localized primarily to C-fibers within the spinal cord. The presence of high concentrations of CRF binding sites in the dorsal horn of the spinal cord suggest a possible role for this neuropeptide in sensory mechanisms related to processing of noxious stimuli. The functional importance of CRF receptors in the spinal cord is demonstrated by evidence from recent electrophysiological studies demonstrating effects of CRF to depolarize the ventral root in a concentration-related manner.[54]

CRF-immunoreactive fibers and cell bodies are also present in laminae V, VII, and X and in the intermediolateral sympathetic column.[50,51] The CRF receptors that are in close association with the CRF fibers innervating the intermediolateral cell column may, in part, be responsible for mediating the effects of CRF in the CNS to stimulate sympathetic outflow, which in turn results in increases in adrenal epinephrine secretion, norepinephrine secretion, and increases in heart rate and blood pressure.[28,29] Furthermore, CRF receptors within the spinal cord may be important in regulating parasympathetic outflow; intrathecal injections of CRF have been reported to have inhibitory effects on gastric acid secretion.[55]

Moderate concentrations of CRF receptors are present on motoneurons in the ventral horn of the spinal cord. Data from electrophysiological studies clearly demonstrate direct actions of CRF on receptors in motoneurons to produce a slow, long-lasting depolarization response.[54] Consistent with the role of a stress neurotransmitter, the data suggest that CRF may act functionally to modulate the excitability of motoneurons.

V. CORRELATION OF CRF RECEPTORS WITH CRF-LIKE IMMUNOREACTIVITY

The localization of CRF receptors in the rat CNS is generally consistent with the relative distribution of CRF-immunoreactive terminals.[12] For example, CRF receptors are present in low concentrations in most of the hypothalamus with higher densities of binding sites present in the external layer of the median eminence, where the most intense concentration of CRF fibers is also found.[19,20,45] Also, CRF cell bodies in the neocortex are concentrated in laminae II and III, with projections to laminae I and IV,[19,20,45] areas rich in CRF receptors. Additional areas of good correspondence between the distribution of CRF-IR nerve terminals and CRF binding sites include the olfactory tubercle, caudate-putamen, nucleus of the diagonal band of Broca, medial, and lateral septal nuclei, the cranial nerve nuclei in the brain stem, cerebellum, and spinal cord. There are areas in the rat CNS in which the distributions of CRF-immunoreactive nerve fibers and CRF receptors do not correlate as well. Some of these areas include the olfactory bulb and amygdala. In the amygdala, CRF cell bodies and nerve terminals appear primarily confined to the central nucleus,[19,20,45] whereas the highest concentration of CRF receptors is present in the basolateral nucleus.

Similar mismatches between the level of receptor and neurotransmitter have been noted in previous studies, and the "mismatch problem" has been the topic of a recent report.[56] There should not be a match between receptor and neurotransmitter at the light microscopic level in whole sections. The neurotransmitter is contained throughout a neuron, whereas the receptor is contained primarily in the postsynaptic neuron which has a different spatial distribution. Perhaps only when both neurons and interneurons are contained within the same region would a match be expected. Technical limitations in both autoradiographic and immunocytochemical procedures may prevent a complete and appropriate identification of receptor and neurotransmitter molecules, respectively. Subtypes of the CRF receptors may exist which only bind under an appropriate set of conditions such as ligand concentration, salts, and guanine nucleotides. Also, agonist ligands similar to the radioiodinated CRF analog used in the present study primarily label high affinity states of the receptor which may not be found in constant proportions in all regions of the CNS. Immunocytochemistry, in contrast, is dependent on the specificity of the antisera used. Most of the reports (see above) on immunocytochemical distribution of CRF in rat brain have utilized primary antibodies directed against ovine CRF. Different distributions of CRF immunoreactivity in the rat CNS have been reported using a variety of antibodies directed against rat and ovine CNS.[57] Future studies utilizing more specific antisera in combination with better radioligands may minimize the discrepancies between the distribution of CRF receptors and CRF immunoreactivity.

Low to moderate concentrations of CRF binding sites are present in several regions of the rat CNS known to contain primary groups of CRF-immunoreactive cell bodies such as the paraventricular nucleus of the hypothalamus, the central nucleus of the amygdala, the bed nucleus of the stria terminalis, and the inferior olive. These CRF binding sites may represent presynaptic autoreceptors or receptors for internuclear communication which may play important roles in regulating peptide release originating from these perikarya.

VI. SUMMARY AND CONCLUSIONS

We have used the radioiodinated analogs of CRF to map CRF binding sites in the rat CNS using *in vitro* autoradiographic techniques. The areas of distribution of CRF binding sites are correlated well with the immunocytochemical distribution of CRF pathways and pharmacologic sites of action of CRF. These data strongly support a physiological role for endogenous CRF in regulating and integrating CNS activity and suggest the importance of this neuropeptide in regulating endocrine and visceral functions and behavior, especially in response to stressful stimuli. The exact mechanisms by which CRF can modulate these complex functions remain unknown. Studies to characterize CRF receptors and CRF-containing pathways in brain provide a means for better understanding the various functions of this neuropeptide in different areas of the CNS.

ABBREVIATIONS

3 — principal oculomotor nucleus
3V — third ventricle
7 — facial nucleus
7n — facial nerve or root of facial nerve
12 — hypoglossal nucleus
aca — anterior commissure, anterior part
Acb — nucleus accumbens
ACg — anterior cingulate cortex
AD — anterodorsal thalamic nucleus
AHy — anterior hypothalamic area
AOB — accessory olfactory bulb
Aq — cerebral aqueduct (Sylvius)
Arc — arcuate hypothalamic nucleus
BL — basolateral amygdaloid nucleus
Cb — cerebellum
cc — corpus callosum
Ce — central amygdaloid nucleus
CG — central gray
Cl — claustrum
Co — cochlear nucleus
CPu — caudate-putamen
Cu — cuneate nucleus
DG — dentate gyrus
DpMe — deep mesencephalic nucleus
DR — dorsal raphe nucleus
DTg — dorsal tegmental nucleus (Gudden)
Ent — entorhinal cortex
E/OV — ependymal olfactory ventricle
EPL — external plexiform layer
Fl — flocculus
fmi — forceps minor of the corpus callosum
FrPaM — frontoparietal cortex, motor area
FrPaSS — frontoparietal cortex, somatosensory area
GL — glomerular layer
GP — globus pallidus
Gr — gracile nucleus

H — hippocampus
IC — inferior colliculus
ic — internal capsule
Int — interpositus cerebellar nucleus
IO — inferior olive
IP — interpeduncular nucleus
IPL — internal plexiform layer
LatC — lateral cervical nucleus
LC — locus coeruleus
LH — lateral hypothalamic area
ll — lateral lemniscus
lo — lateral olfactory tract
LS — lateral septal nucleus
Md — reticular nucleus of the medulla
ME — median eminence
Mi — mitral cell layer
MS — medial septal nucleus
MT — medial terminal nucleus of the accessory optic tract
ON — olfactory nerve layer
Pa — paraventricular hypothalamic nucleus
PCg — posterior cingulate cortex
PFl — paraflocculus
Pn — pontine nuclei
PnO — pontine reticular nucleus, oral part
PO — primary olfactory (pyriform) cortex
Pr5 — principal sensory trigeminal nucleus
PY — pyramidal tract
RF — rhinal fissure
RPn — raphe pontis nucleus
RSpl — retrosplenial cortex
S — subiculum
s5 — sensory root of the trigeminal nerve
sol — solitary tract
sp5 — spinal tract of the trigeminal nerve
Sp5C — nucleus of the spinal tract of the trigeminal nerve, caudal part
Str18a — striate cortex, area 18a
TS — triangular septal nucleus (interstitial nucleus of the ventral hippocampal commissure)
Tu — olfactory tubercle
VDB — nucleus of the vertical limb of the diagonal band, dorsal part
Ve — vestibular nucleus
VMH — ventromedial hypothalamic nucleus

ACKNOWLEDGMENTS

The work described in this chapter was conducted in collaboration with Dr. Michael J. Kuhar of the Neuroscience Branch of NIDA Addiction Research Center and Drs. Marilyn Perrin, Jean Rivier, and Wylie Vale of the Clayton Foundation Laboratories for Peptide Biology at the Salk Institute for Biological Studies. We would like to thank Terrie Pierce and Mary Flutka for manuscript preparation.

REFERENCES

1. **De Souza, E. B., Perrin, M. H., Rivier, J. E., Vale, W. W., and Kuhar, M. J.,** Corticotropin-releasing factor receptors in rat pituitary gland: autoradiographic localization, *Brain Res.*, 296, 202, 1984.
2. **De Souza, E. B. and Kuhar, M. J.,** Corticotropin-releasing factor receptors: autoradiographic identification, in *Neuropeptides in Neurologic and Psychiatric Disease*, Martin, J. B. and Barchas, J., Eds., Raven Press, New York, 1986, 179.
3. **De Souza, E. B. and Kuhar, M. J.,** Corticotropin-releasing factor receptors in the pituitary gland and central nervous system: methods in overview, *Methods Enzymol.*, 124, 560, 1986.
4. **Wynn, P. C., Aguilera, G., Morell, J., and Catt, K. J.,** Properties and regulation of high-affinity pituitary receptors for corticotropin-releasing factor, *Biochem. Biophys. Res. Commun.*, 110, 602, 1983.
5. **Holmes, M. C., Antoni, F. A., and Szeintendnei, T.,** Pituitary receptors for corticotropin-releasing factor: no effect of vasopressin on binding or activation of adenylate cyclase, *Neuroendocrinology*, 39, 162, 1984.
6. **Wynn, P. C., Harwood, J. P., Katt, K. J., and Aguilera, G.,** Regulation of corticotropin-releasing factor (CRF) receptors in the rat pituitary gland: effects of adrenalectomy on CRF receptors, *Endocrinology*, 116, 1653, 1985.
7. **De Souza, E. B.,** Corticotropin-releasing factor receptors in the rat central nervous system: characterization and regional distribution, *J. Neurosci.*, 7(1), 88, 1987.
8. **De Souza, E. B., Perrin, M. H., Whitehouse, P. J., Rivier, J. E., Vale, W. W., and Kuhar, M. J.,** Corticotropin-releasing factor receptors in human pituitary gland: autoradiographic localization, *Neuroendocrinology*, 40, 419, 1985.
9. **Britton, D. R., Koob, G. F., Rivier, J., and Vale, W.,** Intraventricular corticotropin-releasing factor enhances behavioral effects of novelty, *Life Sci.*, 31, 363, 1982.
10. **Peterfreund, R. A. and Vale, W. W.,** Ovine corticotropin-releasing factor stimulates somatostatin secretion from cultured brain cells, *Endocrinology*, 112, 1275, 1983.
11. **Valentino, R. J., Foote, S. L., and Aston-Jones, G.,** Corticotropin-releasing factor activates noradrenergic neurons of the locus ceruleus, *Brain Res.*, 270, 363, 1983.
12. **De Souza, E. B., Insel, T. R., Perrin, M. H., Rivier, J., Vale, W. W., and Kuhar, M. J.,** Corticotropin-releasing factor receptors are widely distributed within the rat central nervous system: an autoradiographic study, *J. Neurosci.*, 5, 3189, 1985.
13. **Whitten, W. K. and Bronson, H.,** The role of pheromones in mammalian reproduction, in *Communication by Chemical Signals*, Vol. 1, Johnston, J. W., Jr., Moulton, D. G., and Turk, A., Eds., Appleton-Century-Crofts, New York, 1970, 309.
14. **Morley, J. E. and Levine, A. S.,** Corticotropin-releasing factor, grooming and ingestive behavior, *Life Sci.*, 31, 1459, 1982.
15. **Levine, A. S., Rogers, B., Kneip, J., Grace, M., and Morley, J. E.,** Effect of centrally administered corticotropin-releasing factor (CRF) on multiple feeding paradigms, *Neuropharmacology*, 22, 337, 1983.
16. **Gosnell, B. A., Morley, J. E., and Levine, A. S.,** Adrenal modulation of the inhibitory effect of corticotropin-releasing factor on feeding, *Peptides*, 4, 807, 1983.
17. **Sirinathsinghji, D. J. S., Rees, L. H., Rivier, J., and Vale, W. W.,** Corticotropin-releasing factor is a potent inhibitor of sexual receptivity in the female rat, *Nature*, 305, 232, 1983.
18. **Sirinathsinghji, D. J. S.,** Inhibitory influence of corticotropin-releasing factor on components of sexual behavior in the male rat, *Brain Res.*, 407, 185, 1987.
19. **Olschowka, J. A., O'Donohue, T. L., Mueller, G. P., and Jacobowitz, D. M.,** The distribution of corticotropin-releasing factor-like immunoreactive neurons in rat brain, *Peptides*, 3, 995, 1982.
20. **Swanson, L. W., Sawchenko, P. E., Rivier, J., and Vale, W. W.,** Organization of ovine corticotropin-releasing factor immunoreactive cells and fibers in the rat brain: an immunohistochemical study, *Neuroendocrinology*, 36, 165, 1983.
21. **Chen, F. M., Bilezikjian, L. M., Perrin, M. H., Rivier, J., and Vale, W. W.,** CRF receptor mediated stimulation of adenylate cyclase activity in rat brain, *Brain Res.*, 381, 49, 1986.
22. **Battaglia, G., Webster, E. L., and De Souza, E. B.,** Characterization of corticotropin-releasing factor (CRF) receptor-mediated adenylate cyclase activity in rat central nervous system, *Synapse*, 1, 572, 1987.
23. **Eberly, L. B., Dudley, C. A., and Moss, R. L.,** Iontophoretic mapping of corticotropin-releasing factor (CRF) sensitive neurones in the rat forebrain, *Peptides*, 4, 837, 1983.
24. **De Souza, E. B., Whitehouse, P. J., Kuhar, M. J., Price, D. L., and Vale, W. W.,** Reciprocal changes in corticotropin-releasing factor (CRF)-like immunoreactivity and CRF receptors in cerebral cortex of Alzheimer's disease, *Nature*, 319, 593, 1986.
25. **Nemeroff, C. B., Owens, M. J., Stanley, M., Andorn, A., and Bissette, G.,** Reduced corticotropin-releasing factor (CRF) receptor number in frontal cortex of suicide victims, *Soc. Neurosci. Abstr.*, 13, 216, 1987.
26. **De Souza, E. B. and Battaglia, G.,** Increased corticotropin-releasing factor receptors in rat cerebral cortex following chronic atropine treatment, *Brain Res.*, 397, 401, 1986.

27. **Crawley, J. N., Olschowka, J. A., Diz, D. I., and Jacobowitz, D. M.,** Behavioral investigation of the coexistence of substance P, corticotropin-releasing factor, and acetylcholinesterase in lateral dorsal tegmental neurons projecting to the medial frontal cortex of the rat, *Peptides,* 6, 891, 1985.
28. **Brown, M. R., Fisher, L. A., Rivier, J., Spiess, J., Rivier, C., and Vale, W.,** Corticotropin-releasing factor: effects on the sympathetic nervous system and oxygen consumption, *Life Sci.,* 30, 207, 1982.
29. **Fisher, L. A., Jessen, G., and Brown, M. R.,** Corticotropin-releasing factor (CRF): mechanism to elevate mean arterial pressure and heart rate, *Regul. Pept.,* 5, 153, 1983.
30. **Britton, D. R., Hoffman, D. K., Lederis, K., and Rivier, J.,** A comparison of the behavioral effects of CRF, sauvagine and urotensin I, *Brain Res.,* 304, 201, 1984.
31. **Sutton, R. E., Koob, G. F., LeMoal, M., Rivier, J., and Vale, W. W.,** Corticotropin-releasing factor produces behavioral activation in rats, *Nature,* 297, 331, 1982.
32. **Veldhuis, H. D. and DeWied, D.,** Differential behavioral actions of corticotropin-releasing factor (CRF), *Pharmacol. Biochem. Behav.,* 21, 707, 1984.
33. **Aldenhoff, J. B., Gruol, D. L., Rivier, J., Vale, W., and Siggins, G. R.,** Corticotropin-releasing factor decreases post-burst hyperpolarizations and excites hippocampal pyramidal neurons *in vitro, Science,* 221, 875, 1983.
34. **Ehlers, C. L., Henriksen, S. J., Wang, M., Rivier, J., Vale, W. W., and Bloom, F. E.,** Corticotropin-releasing factor produces increases in brain excitability and convulsive seizures in rats, *Brain Res.,* 278, 332, 1983.
35. **Weiss, S. R. B., Post, R. M., Gold, P. W., Chrousos, G., Sullivan, T. L., Walker, D., and Pert, A.,** CRF-induced seizures and behavior: interaction with amygdala kindling, *Brain Res.,* 372, 345, 1986.
36. **Bernardini, G. L., Richards, D. B., and Lipton, J. M.,** Antipyretic effect of centrally administered CRF, *Peptides,* 5, 57, 1984.
37. **Ono, N., Lumpkin, M. D., Samson, W. K., McDonald, J. K., and McCann, S. M.,** Intrahypothalamic action of corticotropin-releasing factor (CRF) to inhibit growth hormone and LH release in the rat, *Life Sci.,* 35, 1117, 1984.
38. **Rivier, C. R. and Vale, W.,** Corticotropin-releasing factor (CRF) acts centrally to inhibit growth hormone secretion in the rat, *Endocrinology,* 114, 2409, 1984.
39. **Rivier, C. and Vale, W.,** Influence of corticotropin-releasing factor on reproductive functions in the rat, *Endocrinology,* 114, 914, 1984.
40. **Plotsky, P. M., Bruhn, T. O., and Otto, S.,** Central modulation of immunoreactive arginine vasopressin and oxytocin secretion into the hypophysial-portal circulation by corticotropin-releasing factor, *Endocrinology,* 116, 1669, 1985.
41. **Nikolarakis, K. E., Almeida, O. F. X., and Herz, A.,** Stimulation of hypothalamic β-endorphin and dynorphin release by corticotropin-releasing factor *(in vitro), Brain Res.,* 399, 152, 1986.
42. **Nikolarakis, K. E., Almeida, O. F. X., and Herz, A.,** Corticotropin-releasing factor (CRF) inhibits gonadotropin-releasing hormone (GnRH) release from superfused rat hypothalamus *in vitro, Brain Res.,* 377, 388, 1986.
43. **Grossman, S. P.,** Role of the hypothalamus in the regulation of food and water intake, *Psychol. Rev.,* 82, 200, 1975.
44. **Chappell, P. B., Smith, M. A., Kilts, C. D., Bissette, G., Ritchie, J., Anderson, C., and Nemeroff, C. B.,** Alterations in corticotropin-releasing factor-like immunoreactivity in discrete rat brain after acute and chronic stress, *J. Neurosci.,* 6, 2908, 1986.
45. **Cummings, S., Elde, R., Ells, J., and Lindall, A.,** Corticotropin-releasing factor immunoreactivity is widely distributed within the central nervous system of the rat: an immunohistochemical study, *J. Neurosci.,* 3, 1355, 1983.
46. **Cummings, S., Elde, R., and Sharp, B.,** CRF-immunoreactive neurons within the inferior olive project to the flocculus and dorsal and ventral paraflocculi, *Soc. Neurosci. Abstr.,* 11, 683, 1985.
47. **Young, W. S., III, Walker, L. C., Powers, R. E., De Souza, E. B., and Price, D. L.,** Corticotropin-releasing factor mRNA is expressed in the inferior olives of rodents and primates, *Mol. Brain Res.,* 1, 189, 1986.
48. **Powers, R. E., De Souza, E. B., Walker, L. C., Price, D. L., Vale, W. W., and Young, W. S., III,** Corticotropin-releasing factor as a transmitter in the human olivocerebellar pathway, *Brain Res.,* 415, 347, 1987.
49. **Palkovits, M., Leranth, C., Gorcs, T., and Young, W. S., III,** Corticotropin-releasing factor in the olivocerebellar tract of rats: demonstration by light- and electron-microscopic immunohistochemistry and *in situ* hybridization histochemistry, *Proc. Natl. Acad. Sci. U.S.A.,* 84, 3911, 1987.
50. **Schipper, J., Steinbusch, H. W. M., Vermes, I., and Tilders, F. J. H.,** Mapping of CRF-immunoreactive nerve fibers in the medulla oblongata and spinal cord of the rat, *Brain Res.,* 267, 145, 1983.
51. **Merchenthaler, I., Hynes, M. A., Vigh, S., Schally, A. V., and Petrusz, P.,** Immunocytochemical localization of corticotropin-releasing factor (CRF) in the rat spinal cord, *Brain Res.,* 275, 373, 1983.

52. **Skofitsch, G., Hamill, G. S., and Jacobowitz, D. M.,** Capsaicin depletes corticotropin-releasing factor-like immunoreactive neurons in the rat spinal cord and medulla, *Neuroendocrinology*, 38, 514, 1984.
53. **Skofitsch, G., Insel, T. R., and Jacobowitz, D. M.,** Binding sites for corticotropin releasing factor in sensory areas of the rat hindbrain and spinal cord, *Brain Res. Bull.*, 5, 519, 1985.
54. **Bell, J. A. and De Souza, E. B.,** Functional corticotropin-releasing factor (CRF) receptors in the neonatal rat spinal cord: evidence from autoradiographic and electrophysiological studies, *Soc. Neurosci. Abstr.*, 13, 1666, 1987.
55. **Hamel, D. and Tache, Y.,** Intrathecal (I.T.) injection of bombesin and rat CRF inhibits gastric acid secretion in rats, *Soc. Neurosci. Abstr.*, 10, 812, 1984.
56. **Kuhar, M. J.,** The mismatch problem in receptor mapping studies, *Trends Neurosci.*, 8, 190, 1985.
57. **Skofitsch, G. and Jacobowitz, D. M.,** Distribution of corticotropin-releasing factor-like immunoreactivity in the rat brain by immunohistochemistry and radioimmunoassay: comparison and characterization of ovine and rat/human CRF antisera, *Peptides*, 6, 319, 1985.

Chapter 7

BRAIN CORTICOTROPIN-RELEASING FACTOR AND DEVELOPMENT

Thomas R. Insel

TABLE OF CONTENTS

I. INTRODUCTION

Recent studies of the actions of several neuropeptides in development have demonstrated effects that are distinct from their functions in adulthood. For instance, neuropeptides in the developing rat brain have been implicated in cell maturation,[1] protein synthesis,[2] and neuronal differentiation.[3] Both the numbers and distributions of receptors for several neuropeptides also show unique patterns during the first few weeks of postnatal life. With *in vitro* receptor autoradiography, receptors for opiates,[4,5] neurohypophyseal peptides,[6] neurotensin,[7] and substance P[8] all appear transiently in discrete brain areas. In addition to the growing evidence that neuropeptides have unique functions and anatomic distributions in ontogeny, there is an intriguing literature suggesting that certain peptides may have organizational effects during development, analogous to the effects of sex steroids. Administration of exogenous opiates during a sensitive period in development is associated with analgesia, locomotor deficits, and increased receptor number lasting into adulthood.[9,10] This long-term increase in receptors following exogenous administration of ligand may occur only when the regulatory properties of the peptide and its receptor are being established. By studying such processes during ontogeny, the factors essential for normal development and the sources of some long-term abnormalities may become apparent.

Research on corticotropin-releasing factor (CRF) during development has been focused mostly on the pituitary.[11,12] It has been known for more than 2 decades that glucocorticoid release in rat pups is relatively unresponsive to stress (from day 2 until day 12).[13] Studies of the development of the hypothalamic-pituitary-adrenal (HPA) axis have described increased pituitary sensitivity to glucocorticoid feedback[12] during this period. In addition, there is a transient overshoot of CRF receptors in the anterior pituitary during the 1st postnatal week.[11] By contrast, almost nothing is known about the extrahypothalamic development of CRF, although these pathways appear to profoundly influence behavior and autonomic function. In this chapter, we will review the development of CRF receptors and CRF mRNA in the rat brain, then we will describe some unique behavioral effects of CRF in development, and finally we will present some preliminary results regarding possible organizational effects of CRF.

II. CRF RECEPTORS IN THE DEVELOPING RAT BRAIN

CRF receptors have been characterized and mapped in the adults of several species (see Chapter 6). These receptors generally, though not exclusively, use adenylate cyclase as a second messenger. They are not in perfect register with CRF terminals, suggesting to some[14] that CRF may affect its extrahypothalamic receptors in a nonsynaptic "action at a distance" fashion. If this hypothesis were true, then the receptor map of where CRF acts may be functionally more important than the immunohistochemical map of where CRF is released. For this reason, and because of the variability in CRF immunohistochemical maps,[15] we chose to study the ontogeny of CRF receptors rather than CRF immunoreactivity in development.[16]

A. RECEPTOR BINDING IN BRAIN HOMOGENATES

CRF receptor binding was assessed using ^{125}I-Tyr$^{\circ}$-ovine CRF in whole rat brain homogenates. Briefly, whole brains were removed from rat pups of various ages, homogenized in 30 volumes of ice cold buffer (50 mM Tris HCl, 10 mM $MgCl_2$, 2 mM EGTA, pH 7.0) and centrifuged at 48,000 × g for 20 min at 4°C. The resulting pellet was washed once in buffer, recentrifuged at 48,000 × g for 30 min at 4°C, and resuspended in the same buffer to a final concentration of 20 to 40 mg w/w/ml buffer. Incubations included 100 μl of ^{125}I-Tyr$^{\circ}$-ovine CRF (New England Nuclear, Boston, MA; specific activity = 2200 Ci/mmol)

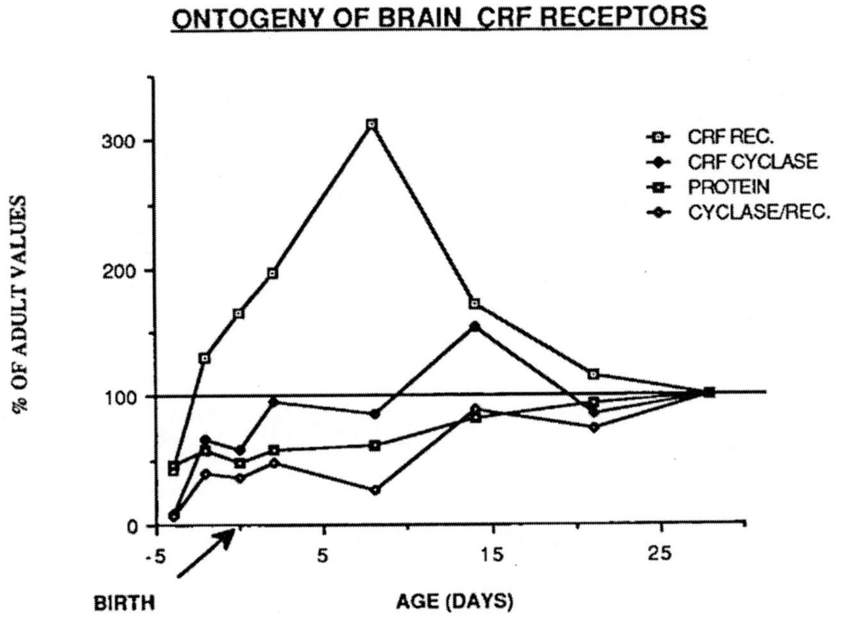

FIGURE 1. Binding of ^{125}I-tyr°-oCRF to whole rat brain homogenates is expressed as percent of day 28 value. Each point represents the mean of three brains at each age. The same brain homogenates were used for measurement of adenylate cyclase by adding 1 m*M* ATP/^{32}P-ATP to 0.8 mg w/w of tissue in 200 μl buffer (100 m*M* Tris HCl, 10 m*M* Mg Cl_2, 0.4 m*M* EGTA, 0.1% BSA, 1 m*M* IBMX, 250 U/ml phosphocreatine kinase, and 5 m*M* creatine phosphate) at 37° with or without 1 *M* rat/human CRF. The reaction was stopped after 10 min by adding 10 μl of 50 m*M* Tris HCl, 45 m*M* ATP, and 2% sodium dodecyl sulfate. The separation of ^{32}P-cAMP from ^{32}P-ATP was accomplished by sequential elution over Dowex and alumina columns. Finally, the ontogeny of CRF receptor linkage to cyclase was calculated by dividing amount of cyclase generated by CRF at each age by number of receptors at this age.

at a concentration of 40,000 cpm/100 μl, 100 μl of tissue, and 100 μl of either incubation buffer or 3 μ*M* unlabeled rat/brain CRF for defining nonspecific binding. Incubation buffer consisted of the Tris buffer used for homogenization with the addition of 0.15% bovine serum albumin, 0.15 m*M* bacitracin, and aprotinin (150 KIU/ml). After a 2-h incubation at room temperature, tissue was separated by centrifugation for 3 min at 12,000 × *g*. The resulting pellet was washed gently with 1 ml of ice cold phosphate buffered saline containing 0.01% Triton and recentrifuged at 12,000 × *g* for 3 min. The supernatant was then aspirated and the pellet radioactivity counted in a gamma counter at 80% efficiency. Protein was measured by the method of Lowry.[17]

In whole rat brain homogenates (Figure 1), CRF receptors can be detected as early as day 17 of gestation (E17), that is, 5 d prior to birth. Receptor number increases steadily, with a peak at postnatal day 8 (P8) when the total number of receptors is 318% of adult values. The number then decreases, reaching final adult values by P28. This overshoot of receptor number is not due to a transient decrease in protein in brain homogenates as the protein concentration increases steadily until P21 (Figure 1).

B. RECEPTOR BINDING IN BRAIN SECTIONS

To localize CRF receptors through ontogeny, we used *in vitro* receptor autoradiography. This technique, which uses the same ligand and buffers as in the homogenate study, employs 16 μm thick cryostat cut sections of unfixed brain. Binding was carried out at room temperature, using approximately 0.1 n*M* of ^{125}I-Tyr°-ovine-CRF. In addition to providing light

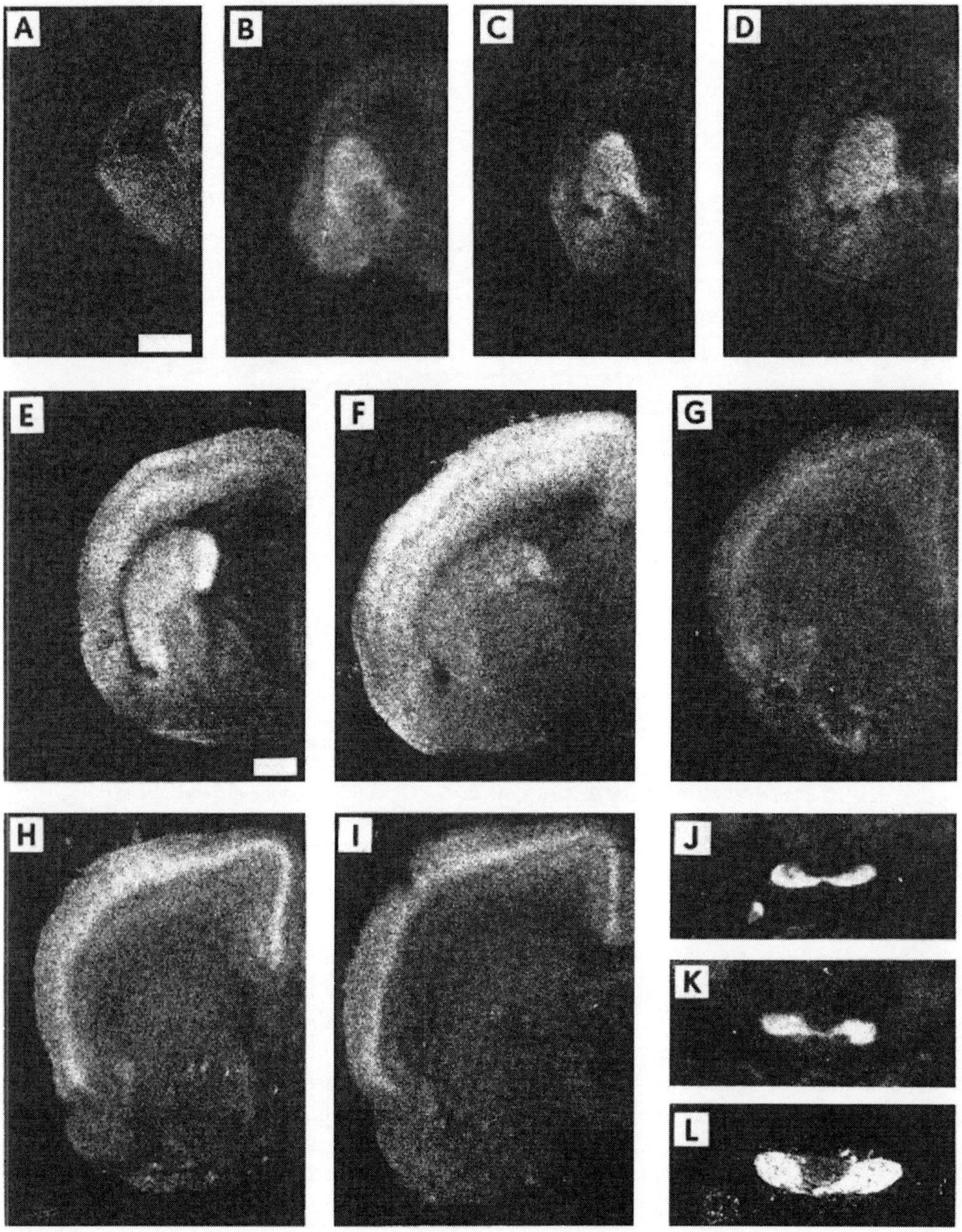

FIGURE 2. The ontogeny of binding of ^{125}I-tyr^{0}-O-CRF to 16 μm rat brain slices is shown in darkfield autoradiograms from E15 (A), E17 (B), E19 (C), E21 (D) all at one magnification and P2 (E), P8 (F), P14 (G), P21 (H), P28 (I) all at one magnification. Note intense binding in striatum until P8 and laminar distribution in cortex developing at P14. For comparison, J to L depicts binding in pituitary at E19, E21, and P14, respectively.

microscopic resolution, this technique uses an iodinated ligand which should not be affected by changes in white matter and brain water content that alter the quenching of radioisotopes such as ^{3}H during development.

By autoradiography, CRF receptors were found to develop along the same time course as described for homogenate binding (Figure 2). Receptors appear initially in the striatum, where they show the highest density from E17 until P8 (Figure 2B to F), after which they decrease markedly. The large overshoot noted with homogenate binding is due to both this striatal component and to an extremely dense distribution across several layers of neocortex (Figure 2F). CRF receptors in amygdala and claustrum and the adult laminar distribution in cortex can be seen by P14 (Figure 2G). By comparison with brain receptors, pituitary receptors appear very dense by E19 (Figure 2J to L) and go through a similar overshoot during the 1st week of postnatal life.[11]

The transient proliferation of receptors in striatum is probably not just a reflection of subsequent decreased density as previously reported with opiate receptors.[4] In the latter case, receptors in striatum appear dense and homogeneous until nigrostriatal afferents arrive postnatally with a resulting patchy distribution.[4] CRF receptors do not form similar patches in adult striatum. In addition, these striatal CRF receptors are probably not on migrating neural elements as cortical neurons are formed earlier and migrate from a more medial, periventricular zone. It is more likely that the transient CRF receptors in striatum are labeling a neural or synaptic population that is eliminated between P8 and P21. It is also possible that these receptors are expressed only transiently in cells that survive.

C. RECEPTOR LINKAGE TO ADENYLATE CYCLASE

When do CRF receptors become functional? CRF can stimulate adenylate cyclase in brain homogenates by E19 and adult levels of cyclase are generated by P2 (Figure 1), suggesting that this receptor system becomes functional very early in brain. However, if one takes into account the large number of CRF receptors present during these first 14 d, the efficiency of this system seems less precocious. Expressed as cyclase generated per receptor, the adult value is not approached until P14 (Figure 1). In other words, although the CRF receptor appears fully linked to adenylate cyclase by P2, relatively few of the receptors show this linkage at this age. Indeed, the striatum with its abundance of receptors until P8 generates relatively little cyclase in response to CRF.[16] Using linkage to cyclase as our criterion, CRF receptors in striatum do not appear to be functional. Of course, these transient receptors may be linked to another second messenger or ion channel, and thus provide a trophic or cell surface marker role in development. Additional studies will be needed to test this hypothesis.

III. DEVELOPMENT OF CRF mRNA

When do neurons begin to express CRF mRNA? Using immunohistochemical techniques, CRF has been identified in the rat median eminence as early as E18, with hypothalamic levels decreasing after birth.[18] In the human hypothalamus, immunoreactive CRF has been detected by 16 weeks of gestation.[19] Using *in situ* hybridization in collaboration with Young (see Chapter 2), we have begun to trace the ontogeny of CRF message in extrahypothalamic sites as well as the paraventricular nucleus (PVN). As shown in Figure 3, this method can clearly discern CRF message in extrahypothalamic sites by P2. Message is particularly dense in cingulate cortex and the bed nucleus of the stria terminalis early in postnatal life. Although it appears that the number of cortical cells expressing CRF mRNA decreases by P28, this difference may in part reflect the concurrent increase in neuropil so that the same number of cells are present in the total structure, although fewer cells are present in any given square millimeter. It is not yet clear whether the average number of grains per cell changes across development.

These data, while preliminary, describe a relatively early development of CRF in both hypothalamic and extrahypothalamic sites. In addition, it does not appear that CRF is selectively present in cells that are eliminated during postnatal sculpting of the brain.

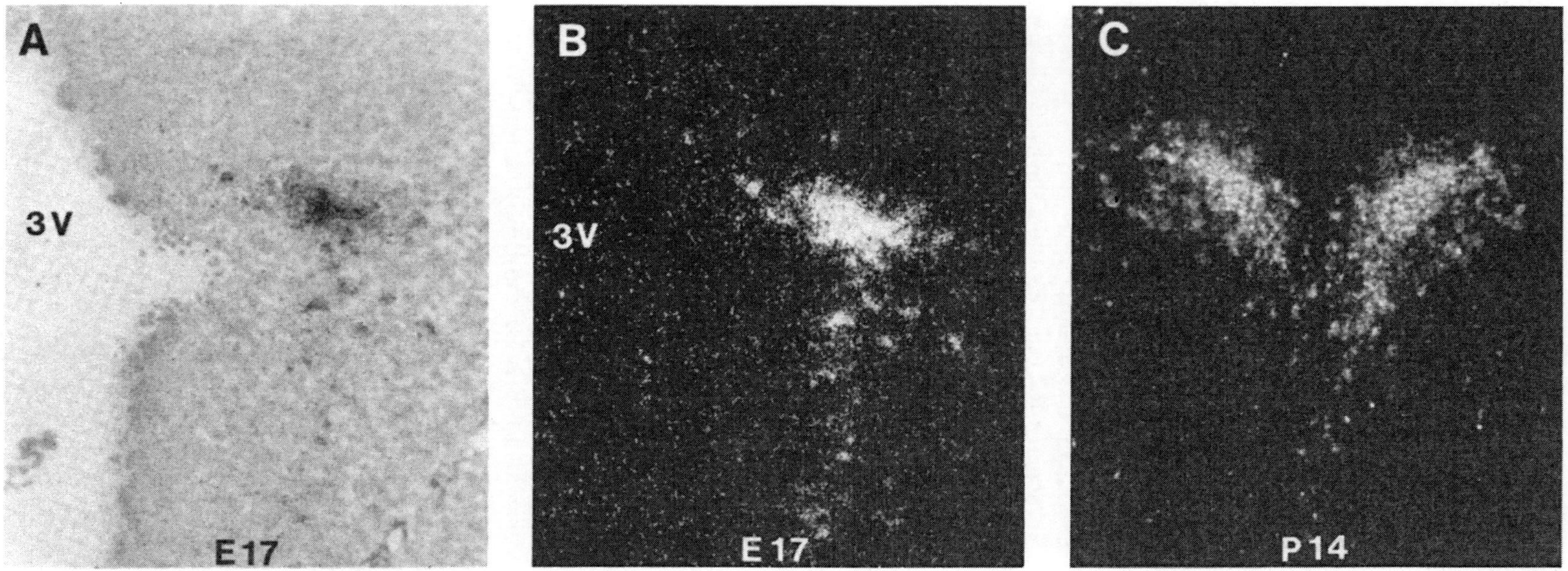

FIGURE 3. *In situ* hybridization with an ^{35}S-labeled synthetic oligonucleotide probe was used to localize cells expressing CRF mRNA during development. Early expression of CRF mRNA was detected in hypothalamic cells on embryonic day 17 as seen in (A) brightfield and (B) darkfield photomicrographs. In (C) this same region on postnatal day 14 appears as the paraventricular nucleus. Scattered extrahypothalamic cells expressing CRF mRNA were also apparent as early as postnatal day 2 in (D) the bed nucleus of the stria terminalis and (E) ngulate cortex. In (F) cells in the same region of the cingulate cortex appear to show less expression of the CRF message on postnatal day 28.

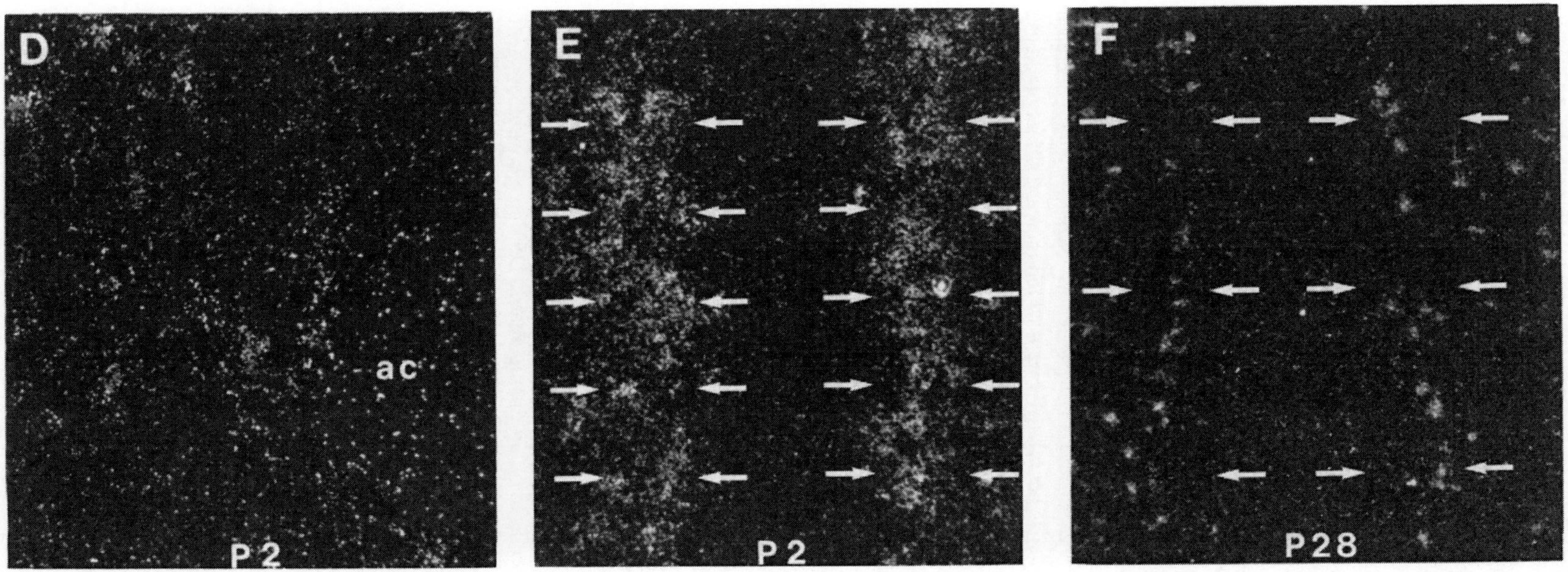

FIGURE 3D—F.

IV. BEHAVIORAL EFFECTS OF CRF IN DEVELOPMENT

The fetus may be exposed to very high levels of CRF, as studies with several species have demonstrated high circulating levels of CRF during pregnancy.[20,21] How much of this peptide can cross the fetal blood-brain barrier is not known. Our own studies in the rat using simultaneous intracarotid injections of ^{125}I-CRF, ^{14}C-inulin, and ^{3}H-water demonstrate significant transport of CRF across the blood-brain barrier as late as P24. This selective permeability to CRF is similar to what has previously been reported with insulin[22] and may involve a similar carrier mechanism.

A. RESPONSE TO SINGLE DOSE ADMINISTRATION OF CRF

Does CRF have behavioral effects in the developing rat? Many of the behaviors affected by CRF in adulthood, such as rearing and exploration (see Chapter 17) are not within the repertoire of pup behavior and thus are not reasonable measures with which to assess CRF's effects in development. When rat pups are stressed, they emit ultrasonic vocalizations (USV). These vocalizations, also known as distress calls, are usually in the 30 to 40 kHz range and are extremely potent stimuli for maternal retrieval. Drugs such as diazepam which decrease anxiety will decrease USV,[23] drugs such as pentylenetetrazol which increase anxiety increase USV,[23] and strains of rats bred for "nervousness" emit more USV when isolated for brief periods.[24] As central administration of CRF to adult rats has been associated with "anxiety-like" behaviors, we hypothesized that CRF would increase rat pup USV.

Our protocol involved a baseline session of 2 min of isolation of 5- to 6-d-old rat pups. Pups were matched for baseline recording rate and then administered 1 μl volumes of either saline of ovine-CRF (0.001, 0.01, or 0.1 μg) by direct intracerebroventricular (ICV) injection. Thirty minutes later, pups were isolated again for 2 min, and USV and locomotor behavior were measured. USV were recorded with a Bruel Kjaer microphone (Model #4385) linked to a sound spectrum analyzer. Sounds were digitized, then analyzed into 100 ms epochs for frequency and power. The number of USV in 2 min was used for data analysis. Locomotor behavior was scored as the number of times the pup crossed into the squares of a 2 × 2-in. grid on the floor of the recording chamber. Each injection volume included 10% India ink. Immediately following the second isolation test, each pup brain was examined to ensure that the injection filled the ventricle. Only pups with injections showing ventricular filling were used for data analysis. Blood was collected from each pup at the time of sacrifice to examine corticosterone.

Contrary to expectations, CRF decreased rather than increased USV in a dose-dependent fashion (Figure 4). Doses as low as 0.01 μg CRF (roughly 1 μg/kg) had significant effects. This decrease in USV was not secondary to decreased arousal as central administration of CRF did not significantly affect locomotor behavior although there was a trend towards fewer crossovers (Figure 4). In addition, CRF given ICV did not significantly increase plasma corticosterone except at the 0.10 μg dose (data not shown). The vocalization effect is thus not mediated by glucocorticoids. Indeed, peripheral (subcutaneous) administration of CRF which increased plasma corticosterone (Figure 5), did not affect USV (Figure 4). Separation for 30 min was also associated with increased plasma corticosterone, as reported by others.[25] As previously noted, 5 d of age is within the stress nonresponsive period. Levels for both baseline and stressed corticosterone (Figure 5) are nearly an order of magnitude below those usually observed in adults.

Thus, it appears that CRF has central effects in 5- to 6-d-old rat pups related to distress behavior. The decrease rather than increase in USV may be due to effects of CRF on thermoregulation (pups call less when they are warm) or on the duration of the calling bout. Generally, pups emit fewer calls with each passing minute of separation. If CRF induced an isolation-like syndrome, as previously described in rhesus monkeys,[26] then the peak

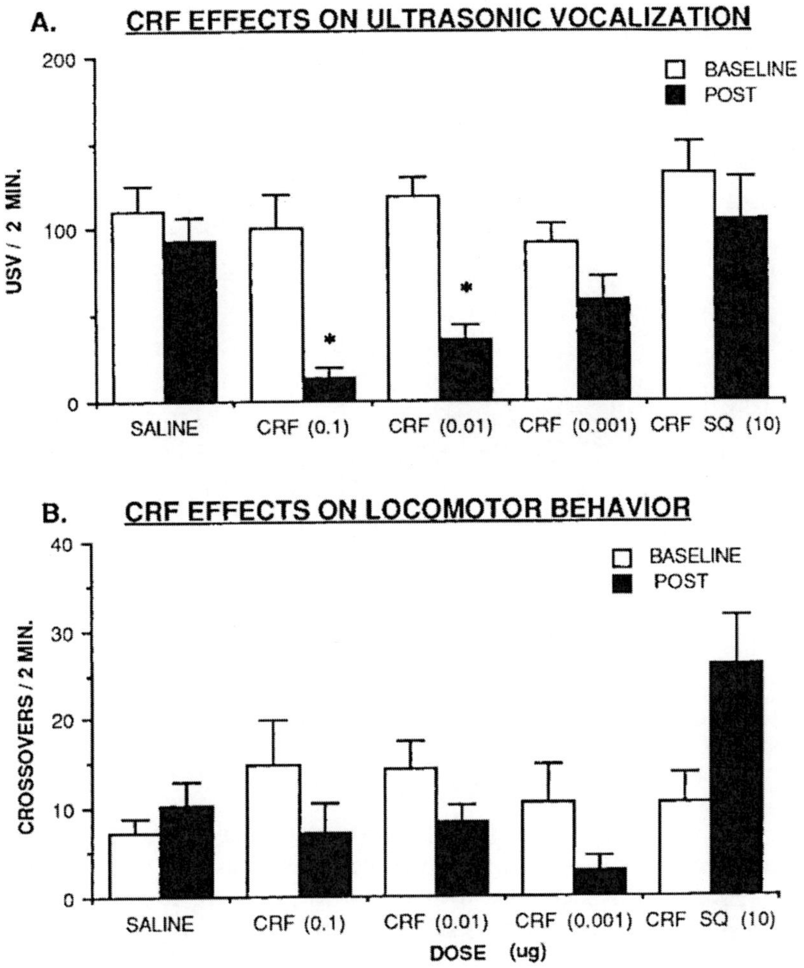

FIGURE 4. Ultrasonic vocalization (A) and locomotor behavior (B) in 5- to 6-d-old rat pups isolated at room temperature for 2 min pre- and 30 min post-ICV or peripheral injection of oCRF. Data represent means from between 6 and 12 pups at each dose and are analyzed by repeated measure ANOVA. In presence of significant main effect for drug, * indicates change from pre- to post-significantly different ($p < 0.05$) from corresponding change with ICV saline administration by post hoc one way ANOVA.

interval of calling may have occurred during the 30-min interval between injection and isolation. Ongoing studies are investigating these two possibilities.

B. EFFECTS OF REPEATED CRF ADMINISTRATION

In studies demonstrating long-term effects of neonatal peptide treatments, Handelmann administered opiates[9] or substance P[27] to rat pups daily from day P1 through P7. The notion that early administration of a neuropeptide could have long-term effects on behavior and even morphology may be particularly relevant to clinical psychiatry. Depression, which has been associated with early loss,[28] is also associated with resistance to dexamethasone suppression and hypersecretion of CRF (see Chapter 17). If early stress permanently alters the sensitivity of CRF receptors, via "organizing" effects at some sensitive period of development, then stressors in adulthood might be expected to elicit altered activity of the HPA axis and potentially abnormal behavior related to extrahypothalamic pathways of CRF.

To test this hypothesis, we administered CRF (10 or 1 μg) daily (days P1 to P7) to rat

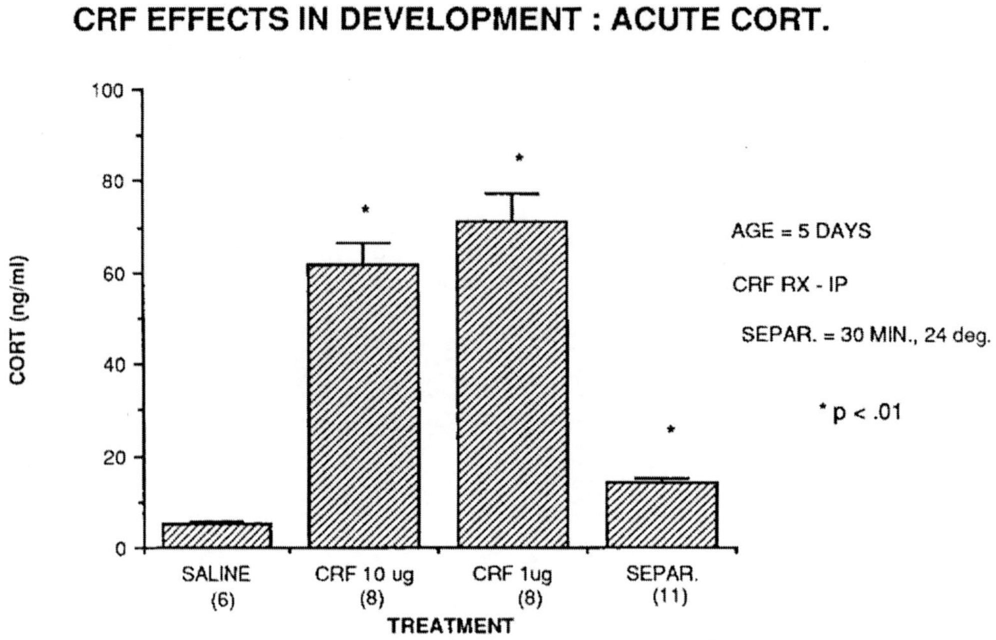

FIGURE 5. Corticosterone increases following peripheral administration of oCRF (1 or 10 μg) or following 30-min period of isolation in 5- to 6-d-old rat pups. Corticosterone was measured with radioimmunoassay (Radioassay Systems Laboratory; Carson, CA) with 5 ng/ml as lower limit of detectability. * signifies $p < 0.01$ difference from saline by Student's t test.

pups by subcutaneous injection (100 μl) at the nape of the neck. Controls were littermates who received saline or, in some cases, were untreated. Pups were then observed for subacute differences on day P8, differences in growth and development (P1 to P21), and differences in behavioral and endocrine responses to stress or CRF (adulthood).

1. Subacute Effects

On day P8, the only significant effect of CRF (10 μg daily) treatment was a sustained increase (approximately 80%) in plasma corticosterone (Table 1). No difference in either the USV or locomotor response to separation was noted. These data suggest that the repeated injections were increasing pituitary activity without altering a behavior which appears sensitive to central administration of CRF. One implication of the increased corticosterone is that apparent CRF effects might be mediated by glucocorticoids and not reflect peptide actions.

2. Growth and Development

The most striking effect of repeated CRF administration was in growth and development. Eye opening, which normally occurs on day P14 to 15 in the rat pup, occurred between P12 and 14 in pups given 10.0 μg of CRF daily (Figure 6). A similar effect has been described by Zadina and Kastin.[29] As CRF appears to have sustained effects on corticosterone, we administered the glucocorticoid dexamethasone as a control. Glucocorticoids alone, even given only on days P1 and 2 also shorten the latency to eye opening (Figure 6) — providing a potential mechanism for the CRF effect.

CRF (10 μg) did not affect weight gain, although the pups receiving 1.0 μg of dexamethasone and the two litters receiving the low dose of CRF (1 μg) showed retarded growth (Figure 7).

TABLE 1
Daily Treatment with CRF — Effects on Day 8

	USV	Locomotor	Plasma cort
Treatment	(calls/2 min)	(crosses/2 min)	(ng/ml)
CRF (10 μg)	99.5 ± 29.2	11.75 ± 3.3	18.7 ± 3.1[a]
Saline	99.0 ± 30.2	11.75 ± 3.3	10.1 ± 1.1

[a] $p < 0.05$ by Student's t test.

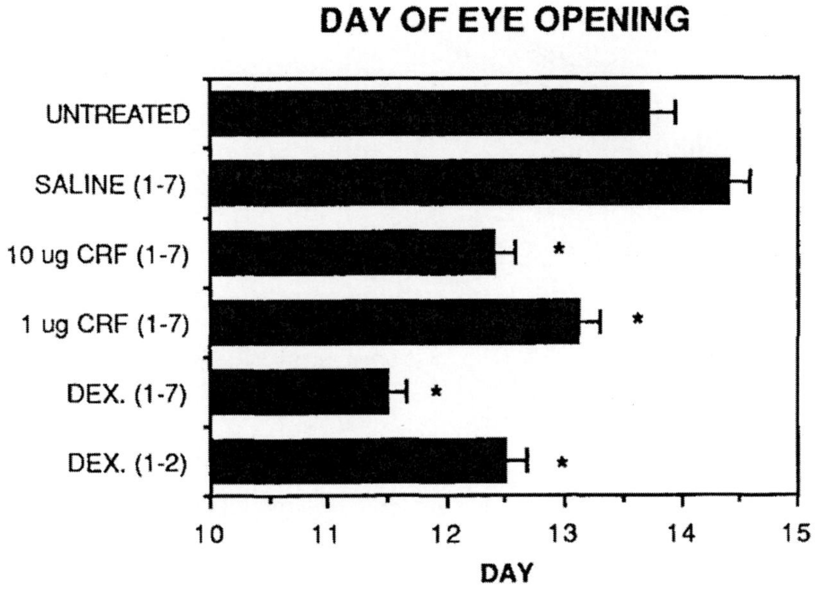

FIGURE 6. Day of eye opening is accelerated by daily peripheral administration of either CRF (1 or 10 μg) or dexamethasone (1 μg) given from P1 to P7 or dexamethasone (1 μg) given on postnatal days 1 and 2 only. Data represent means (± SEM) of at least 17 pups for each treatment. * indicates significant ($p < 0.05$) differences from saline treatment by Student's t test. Saline treatment does not differ from uninjected controls.

3. Long-Term Consequences

Long-term consequences of neonatal CRF administration were studied in several ways. In an open field arena, adults treated with CRF (10 μg) as pups showed increased exploration compared to saline, or dexamethasone treated controls. This effect was evident under both red light (low stress) and white light (high stress) conditions (Table 2 and Figure 8). As dexamethasone appears to show opposite effects to CRF, the peptide effects are probably not mediated by secondary glucocorticoid release.

To assess the endocrine response to a mild stress, group housed animals were socially isolated in a novel cage for 30 min. Blood samples were taken before, at the end of isolation, and 30 min after reunion. There was a tendency for the animals treated with CRF as pups to have higher baseline levels of plasma corticosterone. The plasma level of corticosterone decreased significantly more in the CRF group than the controls during reunion. All groups showed roughly equivalent corticosterone responses to the stressor (Figure 9). To test the HPA axis further, six adults treated with CRF and six treated with saline as pups were given CRF via chronic tail vein cannulae during adulthood. Four doses were given over 2 d and blood was sampled via the cannulae in awake, unrestrained animals. At each dose, the group which had received CRF neonatally, appeared slightly less responsive to CRF administration

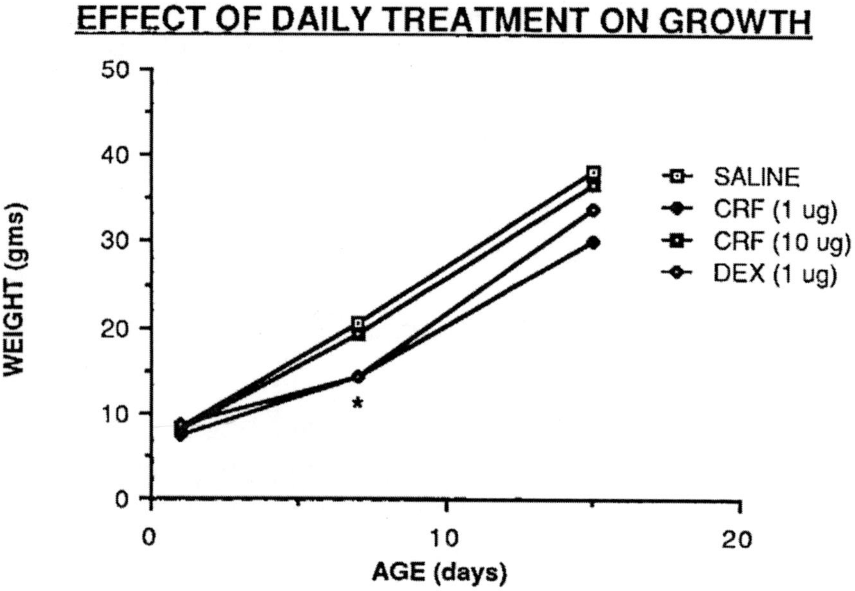

FIGURE 7. Weight gain in pups injected daily with CRF or dexamethasone. Although the high dose CRF treatment did not appear to affect weight gain, the two litters receiving low dose CRF weighed less at birth and gained sluggishly. Dexamethasone (1 μg) daily was also associated with decreased weight gain.

TABLE 2
Adult Open Field Behavior after Neonatal Treatments

Condition: Red Light

Treatment	Inner sq.	Outer sq.	Boli	Rears
Saline	23.3 ± 2.2	41.5 ± 1.4	2.3 ± 0.3	15.7 ± 1.1
CRF (1 μg)	32.0 ± 4.1	38.8 ± 2.0	2.5 ± 0.8	10.6 ± 1.7
CRF (10 μg)	29.1 ± 2.7	27.5 ± 2.6	2.5 ± 0.4	10.5 ± 1.7
Dex.	13.9 ± 1.1	38.5 ± 1.8	3.0 ± 0.4	10.1 ± 0.9

Condition: White Light

Treatment	Inner sq.	Outer sq.	Boli	Rears
Saline	7.2 ± 0.7	19.2 ± 1.6	3.4 ± 0.4	6.0 ± 0.8
CRF (1 μg)	7.3 ± 0.9	20.5 ± 3.0	3.1 ± 0.6	4.9 ± 1.2
CRF (10 μg)	10.7 ± 1.6	15.0 ± 1.8	3.3 ± 0.5	3.0 ± 0.7
Dex.	5.8 ± 0.7	15.9 ± 1.5	3.4 ± 0.5	4.0 ± 0.7

Note: Values represent mean (±SEM) from 2 min test in 1 m diameter open field.

(Figure 10). Consistent with the decreased responsiveness to either stress or exogenous CRF, CRF receptors, using homogenate binding techniques, appeared decreased in pituitaries of the adults who received 10 μg CRF as pups (Figure 11). No change was evident in brain CRF receptors in these animals. Finally, adrenal weights showed a trend towards a decrease in CRF treated animals (454.7 ± 21.4 g). Although this decrease was not significant with respect to saline injected controls (473.0 ± 21.4 g), it was significant ($p < 0.05$) compared to untreated littermate controls (530.1 ± 15.0 g).

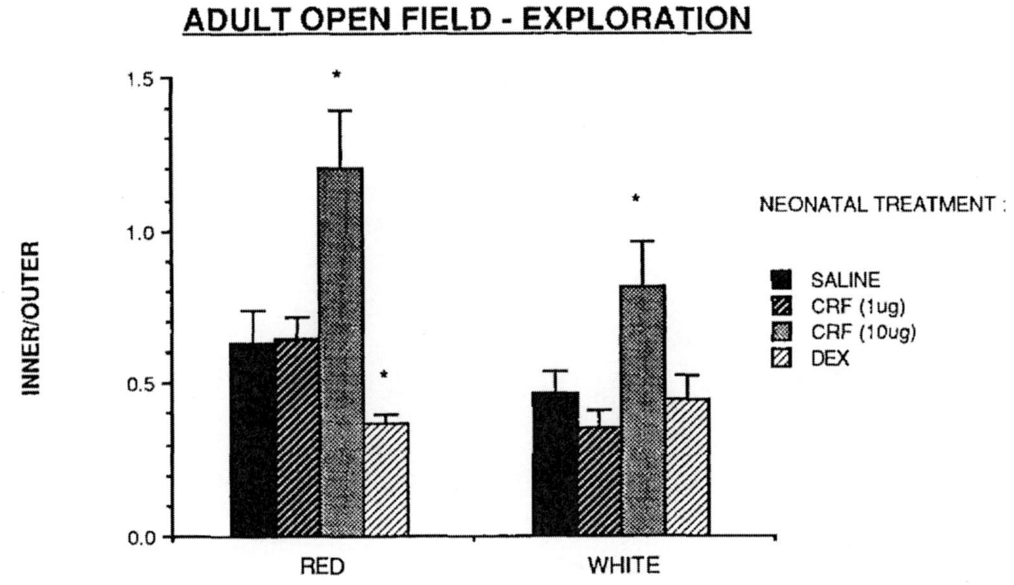

FIGURE 8. Exploratory behavior in adults with neonatal daily treatments of CRF (10 or 1 μg), dexamethasone (1 μg), or saline. Values which represent crossovers into inner squares divided by crossovers into outer squares during a 2-min open field test are calculated from individual results with group means shown in Table 2. Open field testing was done under low stress (red light) and high stress (white light) conditions. * indicates significant ($p < 0.05$) difference from saline by Student's t test.

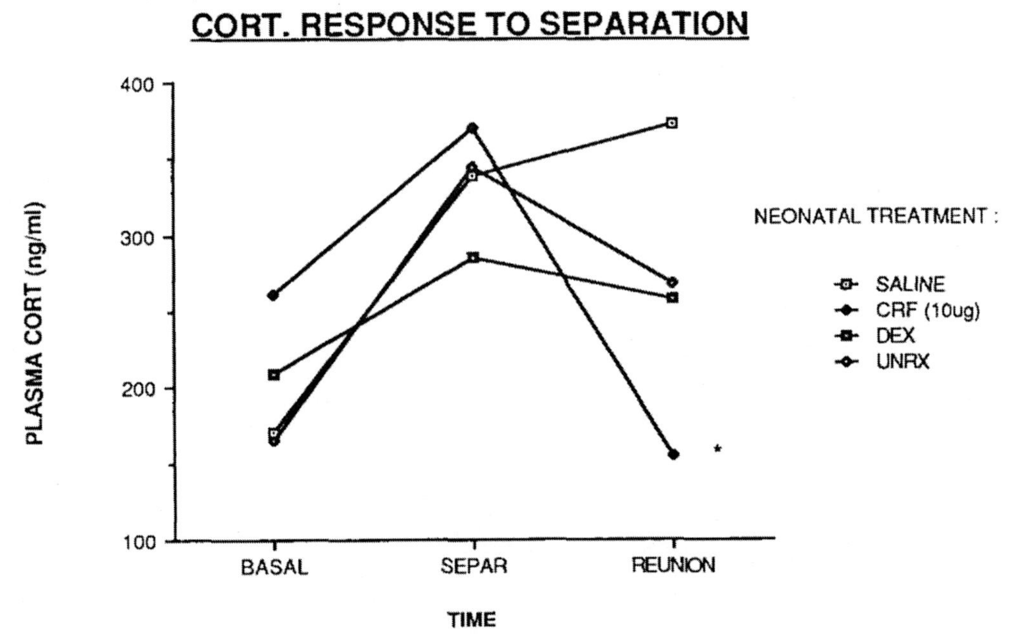

FIGURE 9. Corticosterone response to a 30-min mild environmental stress in adults with neonatal daily treatments of CRF (10 μg), dexamethasone (1 μg), or saline. Socially housed male rats were moved to bare individual plastic cages for 30 min in a novel 20° room with bright light. Each rat was used only once. Blood was collected by cardiac puncture (following brief halothane anesthesia) either prior to separation, at the end of the 30-min isolation, or 30 min following return to the social group. Each point represents the mean of six animals. The only significant difference from saline was the more rapid decrease in corticosterone following reunion in the CRF group.

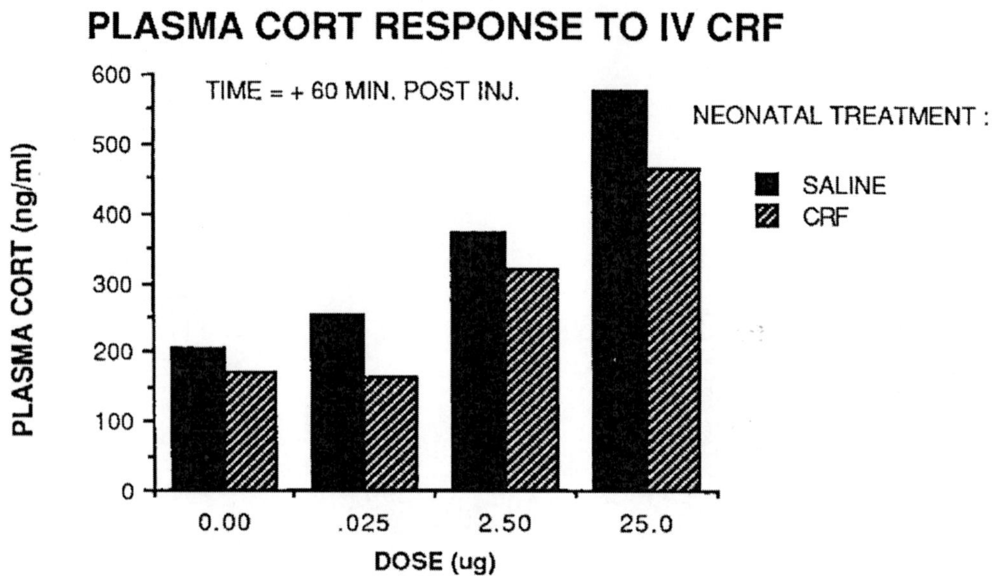

FIGURE 10. Corticosterone increase following exogenous CRF was measured in adult rats with history of either neonatal CRF (10 μg) or saline treatments. oCRF was administered to freely moving rats via an indwelling tail vein cannula inserted at least 24 h before the first dose. To avoid carryover effects, doses of CRF were given in order of potency with at least 4 h between doses. Values represent mean corticosterone at 60 min post injection from six animals under each condition. Although CRF group shows consistently lower corticosterone levels, differences from saline group are not statistically significant.

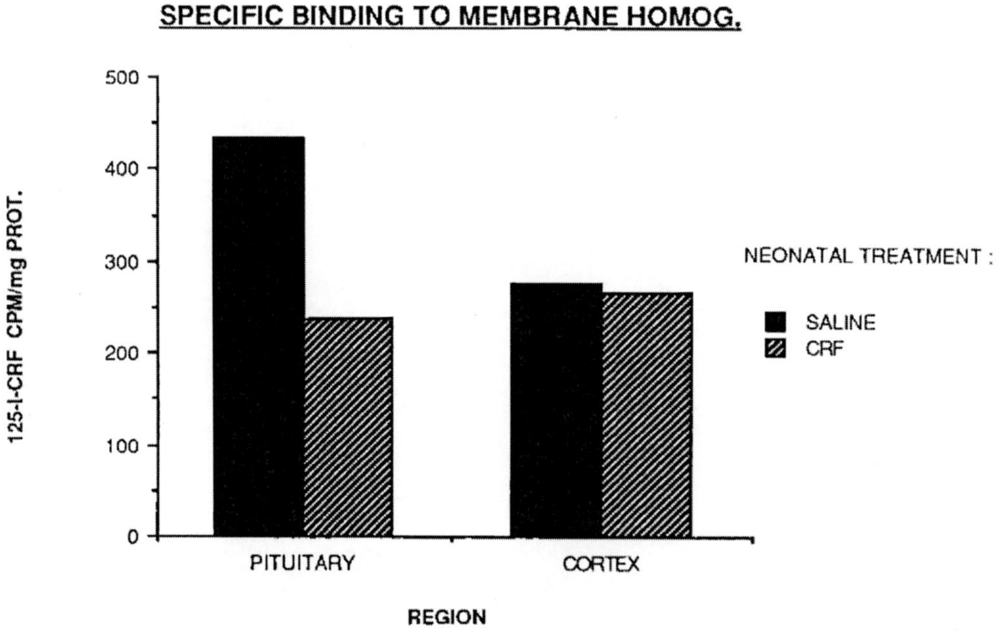

FIGURE 11. CRF receptor binding in pituitary and cortical homogenates from adults with histories of 7-d neonatal treatments with CRF (10 μg) or saline was measured as noted in text. Binding appears significantly decreased in pituitary, but shows no change in brain.

Taken together, it appears that CRF administered to pups has acute effects on growth and development, which may be related to increases in glucocorticoids, and long-term effects on exploratory behavior and HPA axis activity. Contrary to our initial hypothesis, animals receiving CRF showed more exploratory behavior in an open field arena (i.e., appeared less rather than more stress sensitive) and responded with slightly less glucocorticoid output to novelty or exogenous CRF administration. Our experimental results do not clearly model a developmental hypothesis of HPA axis abnormalities in clinical depression. On the other hand, these results are generally consistent with an earlier literature[30,31] documenting long-term effects of handling rat pups. Stress during infancy has long been thought to immunize the rat pup, so that during adulthood, stressors elicit less behavioral and physiologic responses. It is still far from clear however that CRF given either peripherally or centrally to rat pups is the neuroendocrine equivalent of stress.

V. SUMMARY

In this chapter we have examined three aspects of the ontogeny of brain CRF pathways. CRF receptors and CRF mRNA were found to develop by E17 and the receptors appeared to show developmentally restricted expression early in postnatal life. Exogenous CRF administration was associated with decreased USV production. This pharmacologic effect was elicited by central but not peripheral administration of the peptide. Finally, repeated administration of CRF through the first postnatal week was shown to confer effects on growth and development (probably via increased corticosterone) and long-term effects on exploratory behavior (unrelated to glucocorticoid effects). The extent to which peripheral administration of CRF results in increased brain levels of the peptide remains unclear although preliminary data suggest that in the rat a carrier mechanism for transporting this peptide across the blood brain barrier exists until P24. An even more critical question is the extent to which environmental stressors increase CRF production during development. As data accumulate documenting alterations in brain CRF content or receptors in several human disease states, the study of the development of CRF pathways and the process by which their sensitivity becomes regulated during development takes on increasing importance.

ACKNOWLEDGMENTS

The author wishes to acknowledge George Battaglia and Errol De Souza for their generous contribution of adenylate cyclase data and W. Scott Young III for his expert assistance with *in situ* hybridization.

REFERENCES

1. **Bardo, M. T., Schmidt, R. H., and Bhatnagar, R. K.,** Effects of morphine on sprouting of locus coeruleus fibers in the neonatal rat, *Dev. Brain Res.*, 22, 161, 1985.
2. **Bartolome, J. V., Bartolome, M. B., Daltner, L. A., Evans, C. J., Barchas, J. D., Kuhn, C. M., and Schanberg, S. M.,** Effects of β-endorphin on ornithine decarboxylase in tissues of developing rats: a potential role for this endogenous neuropeptide in the modulation of tissue growth, *Life Sci.*, 38, 2355, 1986.
3. **Shyr, S. W., Crowley, W. R., and Grosvenor, C. E.,** Effect of neonatal prolactin deficiency on prepubertal tuberinfundibular and tuberohypophyseal dopaminergic neuronal activity, *Endocrinology*, 119, 1217, 1986.
4. **Kent, J. L., Pert, C. B., and Herkenham, M.,** Ontogeny of opiate receptors in rat forebrain: visualization by *in vitro* autoradiography, *Dev. Brain Res.*, 2, 487, 1982.
5. **Hammer, R. P., Jr.,** Ontogeny of opiate receptors in the rat medial preoptic area: critical periods in regional development, *Int. J. Dev. Neurosci.*, 3, 541, 1985.

6. **Petracca, F. M., Baskin, D. G., Diaz, J., and Dorsa, D. M.,** Ontogenetic changes in vasopressin binding site distribution in rat brain: an autoradiographic study, *Dev. Brain Res.*, 28, 68, 1986.
7. **Kiyama, H., Inagaki, S., Kito, S., and Tohyama, M.,** Ontogeny of [^{3}H]neurotensin binding sites in the rat cerebral cortex: autoradiographic study, *Dev. Brain Res.*, 31, 303, 1987.
8. **Quirion, R. and Dam, T.-V.,** Ontogeny of substance P receptor binding sites in rat brain, *J. Neurosci.*, 6, 2187, 1986.
9. **Handelmann, G. E. and Dow-Edwards, D.,** Modulation of brain development by morphine: effects on central motor systems and behavior, *Peptides*, 6, 29, 1985.
10. **Handelmann, G. E. and Quirion, R.,** Neonatal exposure to morphine increases μ opiate binding in the adult forebrain, *Eur. J. Pharmacol.*, 94, 357, 1983.
11. **Walker, C.-D., Perrin, M., Vale, W., and Rivier, C.,** Ontogeny of the stress response in the rat: role of the pituitary and the hypothalamus, *Endocrinology*, 118, 1445, 1986.
12. **Walker, C.-D., Sapolsky, R. M., Meaney, M. J., Vale, W. W., and Rivier, C. L.,** Increased pituitary sensitivity to glucocorticoid feedback during the stress nonresponsive period in the neonatal rat, *Endocrinology*, 119, 1816, 1986.
13. **Sapolsky, R. M. and Meaney, M. J.,** Maturation of the adrenocortical stress response: neuroendocrine control mechanisms and the stress hyporesponsive period, *Brain Res. Rev.*, 11, 65, 1986.
14. **Herkenham, M.,** Mismatches between neurotransmitter and receptor localizations in brain: observations and implications, *Neuroscience*, 23, 1, 1987.
15. **Skofitsch, G. and Jacobowitz, D. M.,** Distribution of corticotropin releasing factor-like immunoreactivity in the rat brain by immunohistochemistry and radioimmunoassay: comparison and characterization of ovine and rat/human CRF antisera, *Peptides*, 6, 319, 1985.
16. **Insel, T. R., Battaglia, G., Fairbanks, D. W., and De Souza, E. B.,** The ontogeny of brain receptors for corticotropin-releasing factor and the development of their functional association with adenylate cyclase, *J. Neurosci.*, in press.
17. **Lowry, O. H., Rosebrough, N. J., Farr, A. L., and Randall, R. J.,** Protein measurement with the Folin phenol reagent, *J. Biol. Chem.*, 193, 265, 1951.
18. **Bugnon, C., Fellmann, D., Gouget, A., and Cardot, J.,** Ontogeny of the corticoliberin neuroglandular system in rat brain, *Nature*, 298, 159, 1982.
19. **Bresson, J.-L., Clavequin, M.-C., Fellmann, D., and Bugnon, C.,** Human corticoliberin hypothalamic neuroglandular system: comparative immunocytochemical study with anti-rat and anti-ovine corticotropin-releasing factor sera in the early stages of development, *Dev. Brain Res.*, 32, 241, 1987.
20. **Sasaki, A., Shinkawa, O., Margioris, A. N., Liotta, A., Sato, S., Murakami, O., Meigan, G., Shimizu, Y., Hanew, K., and Yoshinaga, K.,** Immunoreactive CRF in human plasma during pregnancy, labor and delivery, *J. Clin. Endocrinol. Metab.*, 64, 224, 1987.
21. **Goland, R. S., Starle, S. L., Brown, L. S., and Frantz, A. G.,** High levels of CRF immunoreactive in maternal and fetal plasma during pregnancy, *J. Clin. Endocrinol. Metab.*, 63, 1199, 1986.
22. **Duffy, K. R. and Pardridge, W. M.,** Blood-brain barrier transcytosis of insulin in developing rabbits, *Brain Res.*, 420, 32, 1987.
23. **Insel, T. R., Hill, J. L., and Mayor, R. B.,** Rat pup ultrasonic isolation calls: possible mediation by the benzodiazepine receptor complex, *Pharmacol. Biochem. Behav.*, 24, 1263, 1986.
24. **Insel, T. R. and Hill, J. L.,** Infant separation distress in genetically fearful rats, *Biol. Psychiatry*, 22, 783, 1987.
25. **Stanton, M. E., Wallstrom, J., and Levine, S.,** Maternal contact inhibits pituitary-adrenal stress responses in preweanling rats, *Dev. Psychobiol.*, 20, 131, 1987.
26. **Kalin, N. H., Shelton, S. E., Kraemer, G. W., and McKinney, W. E.,** Corticotropin-releasing factor administered intraventricularly to rhesus monkeys, *Peptides*, 4, 217, 1983.
27. **Handelmann, G.E., Selsky, J. H., and Helke, C. J.,** Substance P administration to neonatal rats increases adult sensitivity to substance P, *Physiol. Behav.*, 33, 297, 1984.
28. **Roy, A.,** Early parental separation and adult depression, *Arch. Gen. Psychiatry*, 42, 987, 1985.
29. **Zadina, J. E. and Kastin, A. J.,** Neonatal peptides affect developing rats: β-endorphin alters nociception and opiate receptors, corticotropin-releasing factor alters corticosterone, *Dev. Brain Res.*, 29, 21, 1986.
30. **Levine, S. and Mullins, R. F., Jr.,** Hormonal influence on brain organization in infant rats, *Science*, 152, 1585, 1966.
31. **Denenberg, V. H.,** The effects of early experience, in *The Behavior of Domestic Animals*, Hafez, E. S. E., Eds., Williams & Wilkins, Baltimore, 1969, 95.

Chapter 8

NEUROTRANSMITTER REGULATION OF CRF SECRETION *IN VITRO*

Michael J. Owens and Charles B. Nemeroff

A decade prior to elucidation of the sequence of corticotropin-releasing factor (CRF), investigators were already studying the neurotransmitter regulation of CRF release *in vitro* using bioassays for ACTH and adrenal glucocorticoids as a measure of "CRF" activity. Even with the availability of sensitive and specific radioimmunoassays for CRF, considerable controversy exists concerning the role of various neurotransmitters in regulating the secretion of hypothalamic CRF.

Utilizing immunocytochemical methods, Liposits et al.[1] have demonstrated tyrosine hydroxylase immunoreactive nerve terminals, a specific marker for catecholaminergic neurons, innervating CRF-containing perikarya in the paraventricular nucleus (PVN). This same group also found that phenethanolamine-*N*-methyl transferase (PMNT) immunoreactive nerve terminals, a specific marker for epinephrine-containing neurons, arising from the C_1 (ventral lateral medulla) and C_2 (dorsal vagal complex) cell groups, establish direct synaptic contact with PVN CRF-containing perikarya.[2] Recently, evidence for direct serotonergic innervation of the CRF perikarya in the PVN has also been provided.[3] These findings provide direct anatomical evidence for catecholaminergic and serotonergic modulation of CRF secretion.

Data derived from various *in vivo* techniques (plasma ACTH and glucocorticoid concentrations, hypothalamic CRF content, selective brain lesions, etc.), that have been used in the past to infer which transmitters are responsible for CRF secretion, are weakened by the fact that they are always one, two, three, or more steps removed from the actual coupling of the neurotransmitter to its receptor on the CRF perikaryon and the resultant stimulation or inhibition of CRF release into the hypothalamo-hypophysial portal system.

More than 10 years ago, Edwardson and Bennett[4,5] reported that acetylcholine, dopamine, and norepinephrine stimulated the release of CRF-like bioactivity from synaptosomes prepared from rat and sheep hypothalami. More recent *in vitro* studies by our group utilizing minced hypothalami[6] or median eminence alone provides discrepant findings. These conflicting results led several investigators to study CRF release from intact neurons.

In Table 1 we have summarized the literature concerning regulation of hypothalamic CRF release *in vitro*. Unfortunately, the majority of these studies were conducted prior to the advent of specific and sensitive radioimmunoassay methods for measurement of CRF, and therefore utilized a variety of bioassay methods[7,8] to quantify CRF release. While these bioassays are, in general, quite sensitive, they lack specificity because many secretagogues other than CRF itself possess CRF-like activity, e.g., vasopressin, oxytocin, and epinephrine. For this reason, a specific and sensitive radioimmunoassay for quantifying CRF concentrations is clearly the method of choice to elucidate the neurotransmitter regulation of CRF secretion.

The contradictory data in the literature are not solely due to differences between bioassay and radioimmunoassay methods but to differences in incubation techniques (static vs. continuous flow incubations), or other less well-defined variables, as well.

In some studies[9,10] animals were hypophysectomized or adrenalectomized prior to experimentation to increase CRF secretion as much as three to four times the rate in control animals. This documented increase in CRF secretion suggests that certain pathophysiological changes can be investigated *in vitro*. An intact negative feedback loop of glucocorticoids on CRF release has been unequivocally demonstrated in these *in vitro* experiments. Both

TABLE 1
Effects of Neurotransmitters, Hormones, and Drugs on CRF Release *In Vitro*

	Stimulates CRF release	Inhibits CRF release	No effect on CRF release
Acetylcholine (Ach)	10, 13-17, 21, 53, 54		55
Serotonin (5-HT)	8, 15-18, 21, 53, 54, 58		55, 57
Norepinephrine (NE)	55	14, 19, 53	13, 57
Dopamine	55		13, 19, 53
Epinephrine			53
Angiotensin II	9, 53		
Morphine	20-22		
β-Endorphin	21	21, 23	
met-Enkephalin	20-22	23, 56	
leu-Enkephalin	20, 21	23	
Dynorphin		23	
Histamine			13, 53, 55
GABA		17	14, 53, 55
$ACTH_{1-39}$		12	
Corticosterone		11	53
Dexamethasone		9	
Aldosterone		11	
Estrogens	11		
Progestins			11
Androgens			11
cAMP	9		
Glycine		53	
Glutamate			53

Note: Numbers refer to citations in references.

Suda et al.,[9] utilizing dexamethasone, and Buckingham,[11] utilizing corticosterone, have reported a dose-dependent inhibition of CRF secretion by glucocorticoids suggesting a direct long loop negative feedback of adrenal steroids on the hypothalamus itself. Suda's group reported that the effects of dexamethasone on the hypothalamic explant were above the level of the median eminence. A rebound increase in the basal secretion of CRF after removal of dexamethasone was seen, suggesting that short-term incubation with the steroid could decrease release without altering CRF synthesis.

Furthermore, Suda et al.[12] have reported the existence of a short loop negative feedback effect of ACTH on CRF release. By using various ACTH fragments, it appears that the active portion of the ACTH molecule which exerts this effect is in the 1-17 region. The exact anatomical site(s) where ACTH acts within the hypothalamus (e.g., median eminence, paraventricular nucleus, etc.) has not yet been determined.

Other steroid hormones, while not altering basal CRF release, have been reported to modulate CRF secretion in response to other secretagogues.[11] The mineralocorticoid, aldosterone, partially inhibited CRF release in response to both acetylcholine and serotonin. Estrogens, on the other hand, potentiated the effects of these neurotransmitters. Neither progestins nor androgens altered CRF secretion.

The majority of these investigations, and *in vivo* studies as well, demonstrate both stimulatory cholinergic and serotonergic components to hypothalamic CRF release. While there certainly appears to be a stimulatory cholinergic component, it remains to be clarified as to whether it is predominantly muscarinic or nicotinic in nature, or a combination of both. While Hillhouse et al.[13] reported that the effects of acetylcholine were blocked by hexamethonium (a nicotinic antagonist) and only partially antagonized by atropine (a muscarinic antagonist), Suda et al.[14] reported that the effects of acetylcholine on CRF release were completely blocked by atropine and partially antagonized by hexamethonium. There are also reports[13,14] that the stimulatory effects of acetylcholine on CRF release can be attenuated by norepinephrine. Buckingham and Hodges[15] reported that both nicotone itself and bethanechol, a relatively specific muscarinic agonist, stimulate CRF release. These stimulatory effects of acetylcholine and cholinergic agonists on CRF secretion can be antagonized by the serotonergic receptor antagonist cyproheptadine.[15,16] This suggests that there may be a serotonergic interneuron between the cholinergic terminals and the CRF perikarya. However, histochemical verification of this hypothesis is not available.

In general, both *in vivo* and *in vitro* data support a stimulatory role for serotonin on hypothalamic CRF secretion. In one case, this stimulatory effect was blocked by both atropine and hexamethonium[17] suggesting a cholinergic interneuron. Another group,[15] however, could not confirm these findings. Although stimulation of CRF secretion by serotonin could only be demonstrated after an overnight incubation, Calegero et al.[18] reported that the effects of 5-HT were completely blocked by ritanserin, suggesting that the action of 5-HT is mediated by the 5-HT_2 receptor subtype.

The effects of norepinephrine and the opioid peptides are less clear. Recently, Suda et al.[19] reported that norepinephrine has a potent inhibitory effect mediated by α_1 and β-receptors.

Buckingham[20-22] has repeatedly reported that various opioid peptides directly stimulate hypothalamic CRF release *in vitro*. The only exception is the bell-shaped dose-response curve generated by β-endorphin.[21] Concentrations of β-endorphin greater than 10^{-7} *M* were found to inhibit basal CRF release, while concentrations less than 10^{-7} *M* retain their stimulatory activity. In contrast, Suda et al.[23] reported that a variety of opioid peptides including β-endorphin inhibited CRF release at all concentrations tested.

With the availability of a sensitive and specific radioimmunoassay for CRF in our laboratory made possible by Vale's generous contribution of a CRF antiserum, we began by studying CRF release from brain slices. We easily demonstrated potassium-induced, calcium-dependent release of CRF from minced hypothalamus, striatum, and amygdala.[6] However, attempts at whole hypothalamic incubations over the past 2 years, similar to those used by the investigators who generated the data described above, has led us to the inexorable conclusion that the size limitations necessary to keep the hypothalamic CRF pathway intact, result in an explant that quickly becomes hypoxic and nonviable.

Utilizing incubation systems identical to other investigators,[9,21] we were simply unable to generate consistent results with repeated experiments. Although Jones and colleagues[10,24] have described several different methodologies used to establish viability of the explant (O_2 consumption, lactate release, and tissue concentrations of K^+ and ATP), the vast majority of literature from other fields that utilize *in vitro* incubations (cerebral ischemia studies, hippocampal slice physiology, etc.) repeatedly establishes the fact that tissue explants of the size used in these experiments (2 to 3 mm^3, approximately 50 mg w/w) quickly become hypoxic *in vitro*.[25-29] In general, brain tissue slices greater than 600 to 800 μm in thickness develop swelling and neuronal chromatolysis even after short term incubation. Interestingly, while Jones' group[30] purports that the hypothalamus survives in a good functional state for 2 h, and the Japanese investigators'[9] of upwards of 4 h, Calegero et al.[18] require a 24-h preincubation prior to experimental testing. They have, however, purportedly found no histological differences between freshly dissected explants and those incubated overnight.

Because of the tortuous pathway taken by the CRF neuron from the PVN to the median eminence,[31] the explant block cannot feasibly be made much smaller. Medial bisection of the hypothalamus, while decreasing overall explant size and exposing the third ventricle directly to the incubation medium, undoubtedly results in major trauma to the median eminence.[31] Utilizing a dissecting microscope in an attempt to directly expose the PVN without damaging the median eminence, we bisected the hypothalamus in such a way as to remove one of the paraventricular nuclei while leaving the ME untampered with. While theoretically this may remove 50% of the stimulatory or inhibitory input to the ME, it allows for direct exposure of the contralateral PVN to incubation media. This preparation again proved unsatisfactory. We have also added exogenous hydrogen peroxide as a supplementary oxygen source[33,34] without success. Other theoretical possibilities such as addition of oxygen-carrying fluorocarbons[35,36] to the media and incubation in a large sealed diving chamber under higher atmospheric pressures in order to increase oxygen tension in the media are impractical for large numbers of experiments. We have also used neonatal and young rats, in hopes that they may possess greater resistance to hypoxia, without success. It is our belief that current technology is unable to provide a suitable incubation system that allows for the consistent study of CRF release from these large hypothalamic explants.

In vivo techniques in which the viability of the tissue is not a problem include portal vessel cannulation or the use of push-pull cannulae in the median eminence.[37,38] Plotsky et al. have used this approach to examine the effects of noradrenergic and serotonergic agents, as well as opioids, arginine vasopressin, interleukin-1, and CRF itself, on CRF secretion into the portal vessels.[39-45] The results from these studies indicate a stimulatory role for 5-HT,[40] NE,[41] and IL-1,[42] and an inhibitory role for opioids[43] and vasopressin.[44] CRF itself was without effect.[45] Both techniques suffer from the fact that the secretagogue under study (if administered systemically) must be able to cross the blood brain barrier and those that do will certainly be acting at other brain areas in addition to the PVN thus confounding any attempt to study transmitter control at the level of the hypothalamic CRF perikarya themselves.

One technique that may be exploited in the future is the direct perfusion, through the anterior cerebral artery, of an isolated hypothalamic block *in vitro*.[46,47] We have recently applied this technique to simultaneously determine CRF and ACTH release from an isolated hypothalamic-hypophyseal block *in vitro*.[48] In pilot studies we demonstrated an inhibitory effect of the triazalobenzodiazepine, alprazolam, on CRF secretion following stimulation by 5-HT. With the proper perfusion parameters, this technique theoretically allows for the direct investigation of CRF release from isolated hypothalami *in vitro*.

In this chapter thus far, we have focused on CRF release from the median eminence. Releasable pools of CRF are found in other areas of the brain as well. As noted earlier, utilizing a minced tissue preparation, we have previously shown a calcium-dependent, potassium-stimulated release of CRF from the amygdala, midgrain, cerebellum, and striatum,[6] and just recently from slices of prefrontal cortex as well.[49] This comes as no surprise as several other investigators[50-52] have been able to show stimulated release of other neuropeptides such as cholecystokinin, somatostatin, neurotensin, and TRH from various brain regions including the hippocampus, caudate-putamen, and cerebral cortex *in vitro*. These studies have utilized either continuous superfusion of thin slices or static incubations of these slices in small chambers. We have, to date, been unable to obtain a consistent CRF response to various secretagogues (*d*-amphetamine, 5-HT, norepinephrine) other than depolarizing concentrations of potassium in the frontal cortex. This may be due to a masking of the true effects of the transmitters by leakage of CRF from severed neurons or, quite possibly, a small truly functional response obscured by the "noise" of the system. Because of the small size of the CRF neurons in the cortex, viability does not pose a problem, and future investigations of this brain area should be fruitful.

Wider use of sensitive radioimmunoassays, and the use of immunocytochemical techniques and selective brain lesions should allow investigators in the near future to develop a plausible map of the various CRF pathways and their respective neurotransmitter inputs throughout the CNS.

ACKNOWLEDGMENTS

We are grateful to Sheila Walker for preparation of this manuscript. Supported by NIMH MH-42088.

REFERENCES

1. **Liposits, Zs., Sherman, D., Phelix, C., and Paull, W. K.**, A combined light and electron microscopic immunocytochemical method for the simultaneous localization of multiple tissue antigens, *Histochemistry*, 85, 95, 1986.
2. **Liposits, Zs., Phelix, C., and Paull, W. K.**, Adrenergic innervation of corticotropin releasing factor (CRF)-synthesizing neurons in the hypothalamic paraventricular nucleus of the rat, *Histochemistry*, 84, 201, 1986.
3. **Liposits, Zs., Phelix, C., and Paull, W. K.**, Synpatic interaction of serotonergic axons and corticotropin releasing factor (CRF) synthesizing neurons in the hypothalamic paraventricular nucleus of the rat, *Histochemistry*, 86, 541, 1987.
4. **Edwardson, J. A. and Bennett, G. W.**, Modulation of corticotrophin-releasing factor release from hypothalamic synaptosomes, *Nature*, 251, 425, 1974.
5. **Bennett, G. W. and Edwardson, J. A.**, Release of corticotrophin releasing factor and other hypophysiotropic substances from isolated nerve-endings (synaptosomes), *J. Endocrinol.*, 65, 33, 1975.
6. **Smith, M. A., Bissette, G., Slotkin, T. A., Knight, D. L., and Nemeroff, C. B.**, Release of corticotropin-releasing factor from rat brain regions *in vitro*, *Endocrinology*, 118, 1997, 1986.
7. **Vale, W. and Grant, G.**, *In vitro* pituitary hormone secretion assay for hypophysiotropic substances, in *Methods in Enzymology*, Vol. 37, O'Malley, B. W. and Hardman, T. G., Eds., Academic Press, New York, 1974, 82.
8. **Holmes, M. C., Di Renzo, G., Beckford, U., Gillham, B., and Jones, M. T.**, Role of serotonin in the control of secretion of corticotrophin releasing factor, *J. Endocrinol.*, 93, 151, 1982.
9. **Suda, T., Yajima, F., Tomori, N., Demura, H., and Shizume K.**, *In vitro* study of immunoreactive corticotropin-releasing factor release from the rat hypothalamus, *Life Sci.*, 37, 1499, 1985.
10. **Bradbury, M. W. B., Burden, J., Hillhouse, E. W., and Jones, M. T.**, Stimulation electrically and by acetylcholine of the rat hypothalamus *in vitro*, *J. Physiol.*, 239, 269, 1974.
11. **Buckingham, J. C.**, Effects of adrenocortical and gonadal steroids on the secretion *in vitro* of corticotrophin and its hypothalamic releasing factor, *J. Endocrinol.*, 93, 123, 1982.
12. **Suda, T., Yajima, F., Tomori, N., Sumitomo, T., Nakagami, Y., Ushiyama, T., Demura, H., and Shizume, K.**, Inhibitory effect of adrenocorticotropin on corticotropin-releasing factor release from rat hypothalamus *in vitro*, *Endocrinology*, 118, 459, 1986.
13. **Hillhouse, E. W., Burden, J., and Jones, M.T.**, The effect of various putative neurotransmitters on the release of corticotropin releasing hormone from the hypothalamus of the rat *in vitro*. I. The effect of acetylcholine and noradrenaline, *Neuroendocrinology*, 17, 1, 1975.
14. **Suda, T., Yajima, F., Tomori, N., Sumitomo, T., Nakagami, Y., Ushiyama, T., Demura, H., and Shizume, K.**, Stimulatory effect of acetylcholine on immunoreactive corticotropin-releasing factor release from the rat hypothalamus *in vitro*, *Life Sci.*, 40, 673, 1987.
15. **Buckingham, J. C. and Hodges, J. R.**, Hypothalamic receptors influence the secretion of corticotrophin releasing hormone in the rat, *J. Physiol.*, 290, 421, 1979.
16. **Jones, M. T., Birmingham, M., Gillham, B., Holmes, M., and Smith, T.**, The effect of cyproheptadine on the release of corticotrophin releasing factor, *Clin. Endocrinol.*, 10, 203, 1979.
17. **Jones, M. T., Hillhouse, E. W., and Burden, J.**, Effect of various putative neurotransmitters on the secretion of corticotrophin-releasing hormone from the rat hypothalamus *in vitro* — a model of the neurotransmitters involved, *J. Endocrinol.*, 69, 1, 1976.
18. **Calogero, A. E., Bernardini, R., Margioris, A. N., Gallucci, W. T., Munson, P. J., Tamarkin, L., Tomai, T. P., Brady, L., Gold, P. W., and Chrousos, G. P.**, Serotonin stimulates rat hypothalamic corticotropin releasing hormone secretion *in vitro*, *Endocrinology*, in press.
19. **Suda, T., Yajima, F., Tomori, N., Sumitomo, T., Nakagami, Y., Ushiyama, T., Demura, J., and Shizume, K.**, Inhibitory effect of norepinephrine on immunoreactive corticotropin-releasing factor release from the rat hypothalamus *in vitro*, *Life Sci.*, 40, 1645, 1987.
20. **Buckingham, J. C.**, Secretion of corticotrophin and its hypothalamic releasing factor in response to morphine and opioid peptides, *Neuroendocrinology*, 35, 111, 1982.

21. **Buckingham, J. C.,** Stimulation and inhibition of corticotrophin releasing factor secretion by beta endorphin, *Neuroendocrinology*, 42, 148, 1986.
22. **Buckingham, J. C. and Cooper, T. A.,** Interrelationships of opioidergic and adrenergic mechanisms controlling the secretion of corticotrophin releasing factor in the rat, *Neuroendocrinology*, 46, 199, 1987.
23. **Yajima, F., Suda, T., Tomori, N., Sumitomo, T., Nakagami, Y., Ushiyama, T., Demura, H., and Shizume, K.,** Effects of opioid peptides on immunoreactive corticotropin-releasing factor release from the rat hypothalamus *in vitro*, *Life Sci.*, 39, 181, 1986.
24. **Gillham, B., Beckford, U., Insall, R. L., Skelly, A. McL., and Jones, M. T.,** Dynamics of the formation and release of corticotrophin releasing activity by the rat hypothalamus *in vitro*, *J. Endocrinol.*, 90, 201, 1981.
25. **Elliot, K. A. C.,** The use of brain slices, in *Handbook of Neurochemistry*, Vol. 2, 1st ed., Lajtha, A., Ed., Plenum Press, New York, 1969, 103.
26. **Boujon, C. E., Bestetti, G. E., Reymond, M. J., and Rossi, G. L.,** A model for combined morphological and functional investigations on the isolated mediobasal rat hypothalamus, *Neuroendocrinology*, 45, 311, 1987.
27. **Fujii, T., Baumgartl, H. and Lubbers, D. W.,** Limiting section thickness of guinea pig olfactory cortical slices studied from tissue pO_2 values and electrical activities, *Pfluegers Arch.*, 393, 83, 1982.
28. **Schurr, A., Teyler, T. J., and Tseng, M. T., Eds.,** *Brain Slices: Fundamentals, Applications and Implications*, S. Karger, Basel, 1987.
29. **Dingledine, R., Ed.,** *Brain Slices*, Plenum Press, New York, 1984.
30. **Jones, M. T. and Hillhouse, E. W.,** Neurotransmitter regulation of corticotropin-releasing factor *in vitro*, *Ann. N.Y. Acad. Sci.*, 297, 536, 1977.
31. **Merchenthaler, I., Vigh, S., Petrusz, P., and Schally, A. V.,** The paraventriculoinfudibular corticotropin releasing factor (CRF)-pathway as revealed by immunocytochemistry in long-term hypophysectomized or adrenalectomized rats, *Regul. Pept.*, 5, 295, 1983.
32. **Zamora, A. J. and Ramirez, V. D.,** Ultrastructure of the rat median eminence after superfusion, *Cell Tissue Res.*, 226, 27, 1982.
33. **Walton, K. and Fulton, B.,** Hydrogen peroxide as a source of molecular oxygen for *in vitro* mammalian CNS preparations, *Brain Res.*, 278, 387, 1983.
34. **Llinas, R. Yarom, Y., and Sugimori, M.,** Isolated mammalian brain *in vitro:* new technique for analysis of electrical activity of neuronal circuit function, *Fed. Proc.*, 40, 2240, 1981.
35. **Kopp, S. J., Krieglstein, J., Freidank, A., Rachman, A., Seibert, A., and Cohen, M. M.,** P-31 nuclear magnetic resonance analysis of brain. II. Effects of oxygen deprivation on isolated perfused and nonperfused rat brain, *J. Neurochem.*, 43, 1716, 1984.
36. **Dirks, B., Hanke, J., Krieglstein, J., Stock, R., and Wickop, G.,** Studies on the linkage of energy metabolism and neuronal activity in the isolated perfused rat brain, *J. Neurochem.*, 35, 311, 1980.
37. **Levine, J. E. and Ramirez, V. D.,** *In vivo* release of luteinizing hormone-releasing hormone estimated with push-pull cannulae from the mediobasal hypothalami of ovariectomized, steroid-primed rats, *Endocrinology*, 107, 1782, 1980.
38. **Ixart, G., Barbanel, G., Conte-Devolx, B., Grino, M., Oliver, C., and Assenmacher, I.,** Evidence for basal and stress-induced release of corticotropin releasing factor in the push-pull cannulated median eminence of conscious free-moving rats, *Neurosci. Lett.*, 74, 85, 1987.
39. **Gibbs, D. M.,** Measurement of hypothalamic corticotropin-releasing factors in hypophyseal portal blood, *Fed. Proc.*, 44, 203, 1985.
40. **Gibbs, D. M. and Vale, W.,** Effects of the serotonin reuptake inhibitor fluoxetine on corticotropin-releasing factor and vasopressin secretion into hypophysial portal blood, *Brain Res.*, 280, 176, 1983.
41. **Plotsky, P. M.,** Facilitation of immunoreactive corticotropin-releasing factor secretion into the hypophysial-portal circulation after activation of catecholaminergic pathways or central norepinephrine injection, *Endocrinology*, 121, 924, 1987.
42. **Sapolsky, R., Rivier, C., Yamamoto, G., Plotsky, P., and Vale, W.,** Interleukin-1 stimulates the secretion of hypothalamic corticotropin-releasing factor, *Science*, 238, 522, 1987.
43. **Plotsky, P. M.,** Opioid inhibition of immunoreactive corticotropin-releasing factor secretion into the hypophysial-portal circulation of rats, *Regul. Pept.*, 16, 235, 1986.
44. **Plotsky, P. M., Bruhn, T. O., and Vale, W.,** Central modulation of immunoreactive corticotropin-releasing factor secretion by arginine vasopression, *Endocrinology*, 115, 1639, 1984.
45. **Plotsky, P. M., Bruhn, T. O., and Otto, S.,** Central modulation of immunoreactive arginine vasopressin and oxytocin secretion into the hypophysial-portal circulation by corticotropin-releasing factor, *Endocrinology*, 116, 1669, 1985.
46. **Bonrque, C. N. and Renaud, L.P.,** A perfused *in vitro* prep of hypothalamus for electrophysiological studies on neurosecretory neurons, *J. Neurosci. Methods*, 7, 203, 1983.
47. **Campbell, E. A., Chuang, T. T., Gillham, B., and Jones, M. T.,** An 'in vitro' preparation of rat perfused/perifused hypothalamus for studying the secretion of various hypothalamic releasing and inhibiting hormones, *Endocr. Soc. Abstr.*, 506, 1987.

48. **Ritchie, J. C., Liu, P. K., Nemeroff, C. B., and Davis, M. D.,** Simultaneous determination of CRF and ACTH release from hypothalamic-pituitary blocks perfused *in vitro:* responses to physiological and pharmacological manipulation, *Soc. Neurosci. Abstr.*, 13, 1371, 1987.
49. **Owens, M. J., Maynor, B., and Nemeroff, C. B.,** Release of corticotropin-releasing factor (CRF) from rat prefrontal cortex *in vitro*, *Soc. Neurosci. Abstr.*, 13, 1110, 1987.
50. **Thal, L. J., Laing, K., Horowitz, S. G., and Makman, M. H.,** Dopamine stimulates rat cortical somatostatin release, *Brain Res.*, 372, 205, 1986.
51. **Mendez, M., Joseph-Bravo, P., Cisneros, M., Vargas, M. A., and Charli, J.-L.,** Regional distribution of *in vitro* release of thyrotropin releasing hormone in rat brain, *Peptides*, 8, 291, 1987.
52. **Gysling, K. and Beinfeld, M. C.,** The regulation of cholecystokinin release from rat caudatoputamen *in vitro*, *Brain Res.*, 407, 110, 1987.
53. **Buckingham, J. C. and Hodges, J. R.,** Production of corticotrophin releasing hormone by the isolated hypothalamus of the rat, *J. Physiol.*, 272, 469, 1977.
54. **Nicholson, S., Lin, J.-H., Mahmoud, S., Campbell, E., Gillham, B., and Jones, M.,** Diurnal variations in responsiveness of the hypothalamo-pituitary-adrenocortical axis of the rat, *Neuroendocrinology*, 40, 217, 1985.
55. **Fehm, H. L., Voigt, K. H., Lang, R. E., and Pfeiffer, E. F.,** Effects of neurotransmitters on the release of corticotropin releasing hormone (CRH) by rat hypothalamic tissue *in vitro*, *Exp. Brain Res.*, 39, 229, 1980.
56. **Hashimoto, K., Suemaru, S., Ono, N., Hattori, T., Inoue, H., Takao, T., Sugawara, M., Kageyama, J., and Ota, Z.,** Dual effects of (D-Ala^2, Met^5)-enkephalinamide on CRF and ACTH secretion, *Peptides*, 8, 113, 1987.
57. **Tate, P. W., Newell, D. C., Cook, E. E., Martinson, D. R., and Hagan, T. C.,** Lack of effect of serotonin and norepinephrine on CRF release from hypothalami *in vitro*, *Horm. Metab. Res.*, 15, 342, 1983.
58. **Nakagami, Y., Suda, T., Yajima, F., Ushiyama, T., Tomori, N., Sumitomo, T., Demura, H., and Shizume, K.,** Effects of serotonin, cyproheptadine and reserpine on corticotropin-releasing factor release from the rat hypothalamus *in vitro*, *Brain Res.*, 386, 232, 1986.

Chapter 9

CORTICOTROPIN-RELEASING FACTOR (CRF) RECEPTORS IN BRAIN: CHARACTERIZATION AND REGULATION STUDIES

Errol B. De Souza and Dimitri E. Grigoriadis

TABLE OF CONTENTS

I. INTRODUCTION

There is accumulating evidence suggesting that corticotropin-releasing factor (CRF) may act as a neurotransmitter or neuromodulator in the central nervous system (CNS) where it appears to play a role in the integration of an organism's responses to stress. The initial event in the action of a neurotransmitter is binding to a membrane receptor. An understanding of the structural requirements for CRF receptor recognition and subsequent signal transduction mechanisms are crucial to understanding the mechanisms by which CRF produces its varied effects in brain. This chapter provides a summary of the data from radioligand binding studies using iodine-125-labeled analogs of CRF to examine in detail the kinetic and pharmacological characteristics and regional distribution of CRF receptors in brain membrane homogenates.

II. METHODS

A. CHOICE OF RADIOLIGAND

Since native ovine CRF (oCRF) and rat/human CRF (rCRF) do not contain a tyrosine residue, analogs of CRF with additions of tyrosine at the N-terminal region of the peptide or tyrosine substituted analogs of CRF have been used for iodine-125 labeling to produce radioligands suitable for identifying and characterizing CRF receptors in membrane homogenates. The iodine-125-labeled analogs of CRF that have been used in radioligand binding assays include [^{125}I-Tyr^{0}-oCRF][1-3], [^{125}I-Tyr^{0}-rCRF][4-6], [Nle^{21}, ^{125}I-Tyr^{32}-oCRF][7,8] and [^{125}I-Tyr^{3}, Pro^{4}, Nle^{18,21}, α-helical rCRF3-41][9]. While all of the iodine-125-labeled analogs of CRF provide good probes to study CRF receptors, the oCRF radiolabels have lower levels of nonspecific binding and higher levels of specific binding. The differences in the level of specific binding may relate to the structural characteristics and/or stability of the radioligand. At present, we routinely use commercial preparations of ^{125}I-Tyr^{0}-ovine or rat/human CRF (New England Nuclear Corp., Boston, MA; specific activity 2200 Ci/mmol).

B. TISSUE PREPARATION

A schematic of a flowchart for the tissue preparation and CRF binding assay is shown in Figure 1. Brain regions of interest were dissected and homogenized in 20 volumes of cold buffer (50 m*M* Tris HCl, 10 m*M* $MgCl_2$, 2 m*M* EGTA, pH 7.2 at 22°C) using a Brinkman Polytron (setting of 5 for 20 s) and centrifuged at 48,000 × *g* for 10 min at 4°C. The supernatant was discarded and the pellet was rehomogenized in 20 volumes of cold buffer, and centrifuged at 48,000 × *g* for 30 min at 4°C. The resulting pellet was suspended in buffer to a final concentration of 20 to 40 mg original wet weight (o.w.w.) of tissue per milliliter in the same buffer. Protein concentrations were determined according to the method of Lowry et al.[10]

C. ^{125}I-CRF BINDING ASSAY

One hundred microliters of the membrane suspension was added to 1.5 ml polypropylene microtubes (Sarstedt, W. Germany) containing 100 μl of an ^{125}I-CRF solution (approximately 0.15 n*M*) and 100 μl of the incubation buffer or an appropriate concentration of unlabeled rCRF or other competing peptide. Nonspecific binding was determined in the presence of 1 μM rCRF or oCRF. Incubations were carried out for 2 h at room temperature. The tissue was separated from the incubation medium by centrifugation in a Beckman microfuge for 3 min at 12,000 × *g*. The resulting pellet was washed gently with 1 ml of ice-cold phosphate-buffered saline, pH 7.2, containing 0.01% Triton® X-100, and the contents were recentrifuged for 3 min at 12,000 × *g*. The supernatant was aspirated and the radioactivity of the pellet was measured in a gamma counter at 80% efficiency.

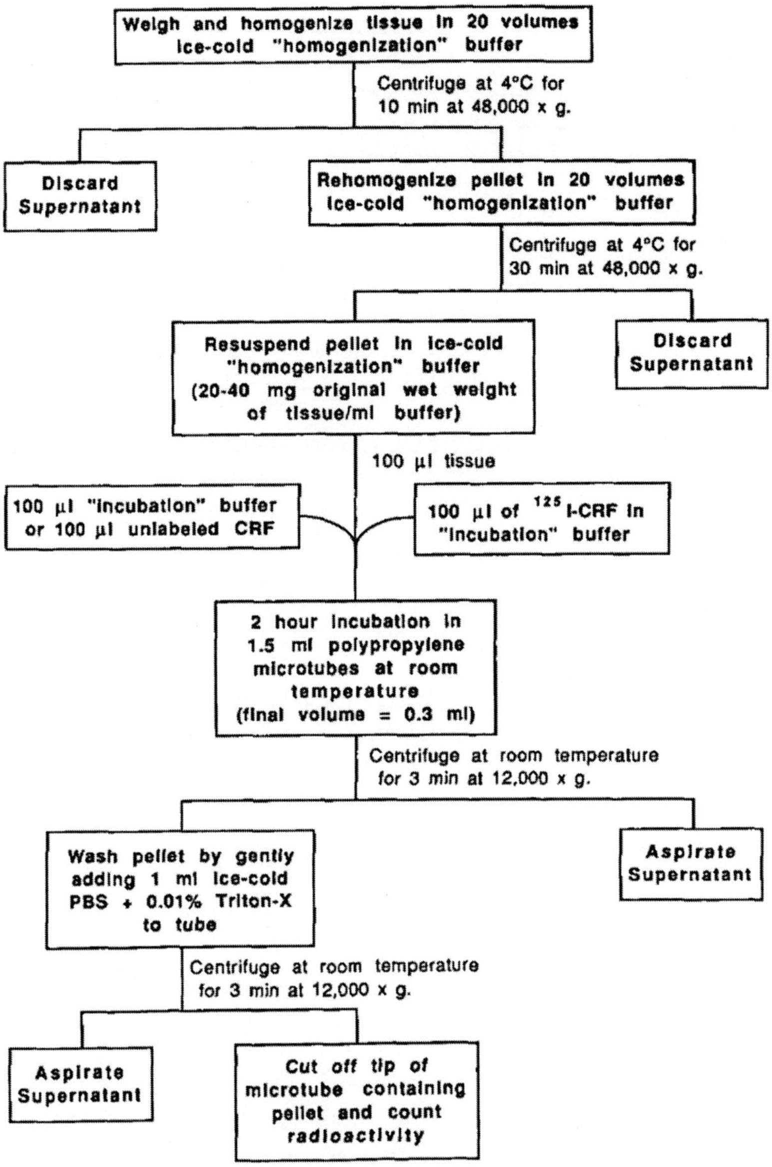

FIGURE 1. Flowchart for tissue preparation and ^{125}I-CRF binding assay. "Homogenization" buffer: 50 m*M* Tris HCl/10 m*M* $MgCl_2$/2 m*M* EGTA, pH 7.0 at 22°C; "Incubation" buffer: 50 m*M* Tris HCl/10 m*M* $MgCl_2$/2 m*M* EGTA/0.15% BSA/150 KIU/ml aprotinin/15 m*M* bacitracin, pH 7.0 at 22°C; PBS: phosphate-buffered saline, pH 7.2. For explanation, see Section II.C.

D. CHEMICAL AFFINITY CROSS-LINKING OF ^{125}I-CRF TO MEMBRANE HOMOGENATES

Tissue homogenates were prepared as described above for the reversible ^{125}I-CRF binding assay with the following modifications. Tissues were placed in Dulbecco's phosphate buffered saline, 10 m*M* $MgCl_2$, 2 m*M* EGTA, 0.15% BSA, 0.15 m*M* bacitracin, 0.1 m*M* phenylmethylsulfonyl fluoride, and 1.5% aprotonin, pH 7.0 at 22°C homogenized as above and resuspended to a final concentration of 30 to 40 mg o.w.w./ml for brain tissues and 10 to 15 mg o.w.w/ml for pituitary tissues. The binding of ^{125}I-oCRF or ^{125}I-r/hCRF to membrane homogenates was performed as described in Section II.C. The procedure for the chemical

chemical cross-linking of ^{125}I-CRF to membrane homogenates followed closely that of Pilch and Czech[11] with minor modifications. The bifunctional cross-linking reagent disuccinimidyl suberate (DSS) was dissolved in dimethyl sulfoxide to yield a final concentration of 1.5 m*M* in the assay. Following the 2 h incubation of ^{125}I-CRF with membrane homogenates, 10 μl of the DSS solution was added to each tube and allowed to react for 20 min at 22°C. The reaction was terminated by the addition of 1.0 ml ice-cold wash buffer (10 m*M* Tris-HCl, 1 m*M* EDTA; pH 7.0 at 0 to 4°C) and subsequent centrifugation at 12,000 rpm for 5 min in a Beckman microfuge. The pellets were washed gently by the addition of 1 ml ice-cold wash buffer and recentrifuged. Final pellets were monitored for bound radioactivity in a gamma counter (LKB) at 80% efficiency and then solubilized in electrophoresis sample buffer (see below) prior to being subjected to sodium dodecyl sulfate polyacrylamide gel electrophoresis (SDS-PAGE).

E. SDS-POLYACRYLAMIDE GEL ELECTROPHORESIS

Samples for electrophoresis were resuspended in SDS-PAGE sample buffer containing 50 m*M* Tris-HCl, 10% glycerol, 5% β-mercaptoethanol and 0.005% bromophenol blue (pH 6.8 at 22°C) and incubated for 45 min at 22°C. Samples were then transferred to a boiling water bath for an additional 15 min before being electrophoresed on a discontinuous slab gel (6% stacking and 12% running) according to the method of Laemmli.[12] An equivalent amount of protein (200 to 300 μg) was loaded onto each lane and the gels were typically run overnight at a constant current of 10 to 15 mA. Prestained protein standards (Sigma) were included on each gel and used to calculate a standard curve from their relative mobilities. The molecular weights of the sample proteins were then determined from the standard curve. Following electrophoresis, the gels were dried using a Bio-Rad slab gel dryer and exposed to Kodak X-AR film using Lightning-Plus intensifying screens (Dupont) at −70°C for 24 to 60 h before being developed. The film optical density values and gel profiles were obtained using the Loats PC-based image analysis system (Amersham).

F. RP-HPLC ANALYSIS OF CRF SOLUTIONS AND TISSUE EXTRACTS

Aliquots of an ^{125}I-rCRF solution (approximately 3000 cpm), the supernatant from the first centrifugation of a sample assayed as described in Section II.C, and an acid extract (20% acetonitrile in 0.1% trifluoroacetic acid) from the final pellet of the same sample were analyzed by reversed-phase high-performance liquid chromatography (RP-HPLC) on a Waters C_{18} μ-Bondapak column (4.6 mm × 30cm) equilibrated with solution A (20% acetonitrile in 0.1% trifluoroacetic acid) at a flow rate of 1.0 ml/min; 10 min after sample injection, a linear gradient to 85% solution B (80% acetonitrile in 0.1% trifluoroacetic acid) over 65 min was initiated. Unlabeled rCRF was detected by absorbance at 220 nm.

G. DATA ANALYSIS

The radioligand binding data were analyzed by the computer program EBDA,[13] which provides initial estimates of equilibrium binding parameters by Scatchard, Hill, and Eadie-Hofstee analysis and then produces a file for the nonlinear curve-fitting program LIGAND,[14] which gives final parameter estimates.

III. CHARACTERIZATION OF CRF RECEPTORS IN BRAIN

A. STUDIES TO OPTIMIZE ^{125}I-CRF BINDING

On the basis of previous autoradiographic localization studies of the distribution of binding sites for iodine-125-labeled analogs of ovine CRF in rat brain (see Chapter 6) and preliminary results of regional homogenate assays, the binding of ^{125}I-rCRF was characterized in membrane preparations of rat olfactory bulb, a region with a high concentration of CRF binding sites.

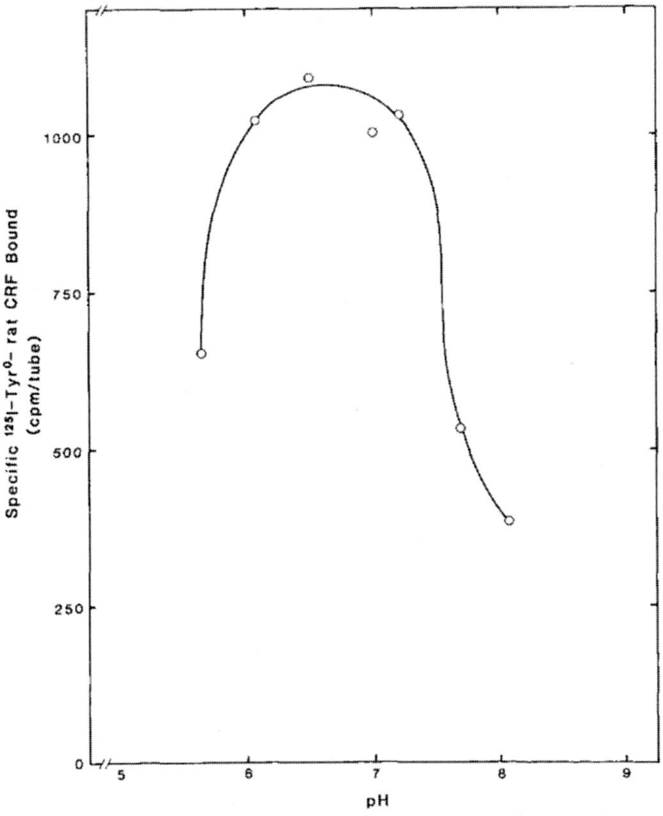

FIGURE 2. Effect of pH on ^{125}I-Tyr° rat/human CRF binding to rat olfactory bulb membranes. ^{125}I-Tyr° rat/human CRF was incubated in 50 m*M* Tris-HCl, 10 m*M* $MgCl_2$, 2 m*M* EGTA, 10^{-4} *M* bacitracin, and 100 kallikrein units/ml aprotinin for 120 min at room temperature with tissue at pH values ranging from 5.5 to 8.2. Each point is the average of three or four determinations, which differed by less than 10%. (From De Souza, E. B., *J. Neurosci.*, 7, 88, 1987. With permission.)

In all experiments, specific binding of ^{125}I-rCRF to rat brain membranes was defined as binding inhibited by 1 μ*M* unlabeled rCRF. ^{125}I-rCRF binding was sensitive to the pH and ionic strength of the incubation buffer. Optimal binding was observed over a relatively broad pH range of 6.1 to 7.3; the binding was decreased by 40 and 65% at pH 5.6 and 8.1, respectively, (Figure 2). The sensitivity of ^{125}I-rCRF binding to the ionic strength of the buffer in the range of 25 to 170 m*M* Tris-HCl buffer was optimal at 50 m*M* Tris-HCl and declined precipitously at higher ionic strengths (data not shown). The substitution of a sodium phosphate buffer was found to have no effect on the binding. The binding of ^{125}I-rCRF was dependent on the presence of magnesium ions; linear increases in specific binding were observed up to a concentration of 10 m*M* $MgCl_2$ (Figure 3). This effect appeared to be specific to magnesium, as the presence of up to 12.5 m*M* NaCl had no effect on binding, and smaller increases were observed in the presence of a comparable concentration of $CaCl_2$ (Figure 3). Incubation of varying concentrations of membranes with ^{125}I-rCRF indicated that the specific binding of the radioligand was linear with protein concentration to approximately 100 μg/tube; higher protein concentrations resulted primarily in an increase in the nonspecifically bound ^{125}I-rCRF (data not shown). Tissue that was boiled for 5 min before incubation showed no detectable specific ^{125}I-rCRF binding. On the basis of these preliminary experiments, all subsequent assays were routinely performed in 50 m*M* Tris-HCl buffer containing

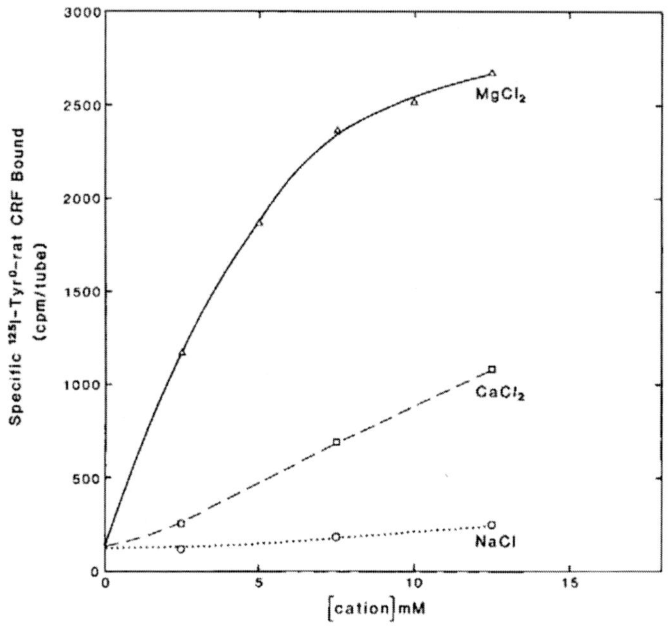

FIGURE 3. Effects of cations on ^{125}I-Tyr° rat/human CRF binding to rat olfactory bulb membranes. ^{125}I-Tyr° rat/human CRF was incubated for 120 min at room temperature in 50 m*M* Tris-HCl containing 2 m*M* EGTA, 10^{-4} *M* bacitracin, 100 kallikrein units/ml aprotinin, pH 7.2, and varying concentrations of $MgCl_2$ (triangles), $CaCl_2$ (squares), and NaCl (circles). Each point is the average of three or four determinations, which varied by less than 10%. (From De Souza, E. B., *J. Neurosci.*, 7, 88, 1987. With permission.)

10 m*M* $MgCl_2$ (pH 7.2 at 23°C), and the incubations were carried out with a membrane protein concentration of 25 to 50 μg/tube. In addition, the protease inhibitors bacitracin (10^{-4} *M*) and aprotinin (100 kallikrein units/ml) were included in the incubation buffer to minimize degradation of the peptide and/or receptor during the course of the assay. Under these assay conditions, specific binding represented 70 to 75% of total ^{125}I-rCRF binding.

To determine whether the assay conditions described above were effective in limiting degradation of ^{125}I-rCRF, and to ascertain that membrane-bound radioactivity represented intact^{125}I-rCRF, analytical RP-HPLC profiles of the added ligand, the postincubation supernatant, and extracted membrane-bound ligand were compared. After 120 min of incubation, the ^{125}I-rCRF in the incubation medium and ^{125}I-rCRF extracted from the membrane pellet eluted as a single peak with "stock" ^{125}I-rCRF (Figure 4), thereby demonstrating the stability of the radioiodinated ligand under routine assay conditions, as well as confirming the identity of the membrane-bound ^{125}I-rCRF.

The subcellular distribution of ^{125}I-rCRF binding sites was examined by comparing the specific binding observed in pellets obtained from successive fractionation steps. The density of specific ^{125}I-rCRF binding sites was greater in the crude mitochondrial/synaptosomal pellet (P_2) than in the P_1 fraction representing nuclei and cellular debris (Table 1). No significant amount of specific ^{125}I-rCRF binding was found in the microsomal pellet (P_3). This crude subcellular distribution of binding sites for ^{125}I-rCRF is consistent with a localization of CRF receptors to membranes derived from nerve terminals.

B. KINETICS OF ^{125}I-CRF BINDING

1. Association Studies

The time courses for the specific binding of ^{125}I-rCRF to rat olfactory bulb membranes

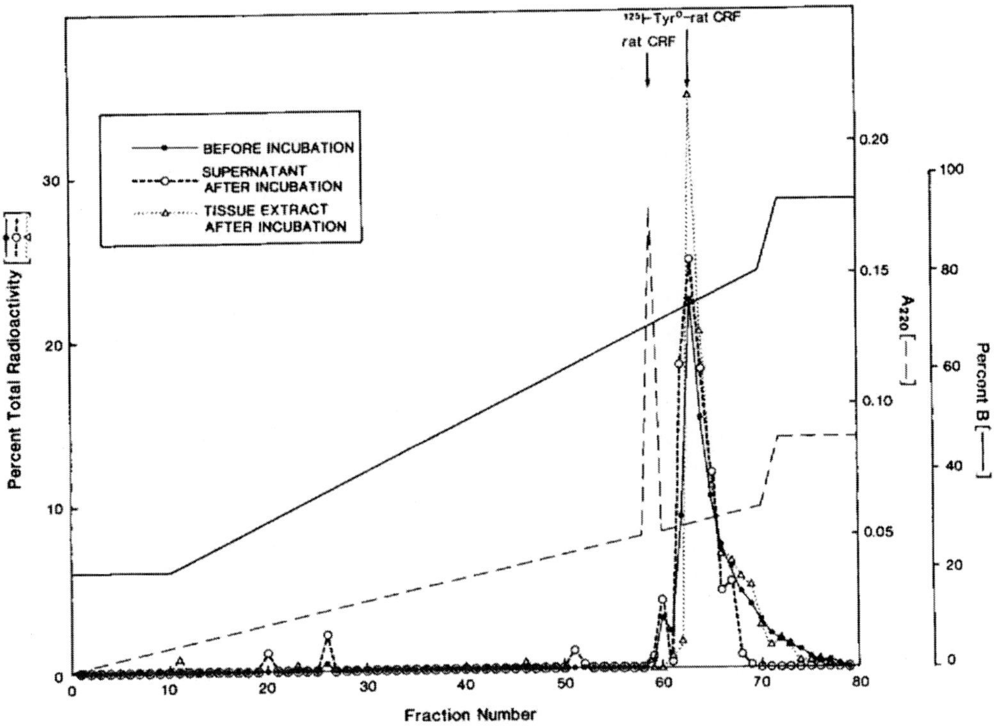

FIGURE 4. RP-HPLC profile of ^{125}I-Tyr° rat/human CRF: Effects of incubation with rat olfactory bulb membranes. Crude membranes were incubated under standard assay conditions with ^{125}I-Tyr° rat /human CRF for 120 min at room temperature. The reaction was terminated by centrifugation at 12,000 × *g* for 3 min. Aliquots of ^{125}I-Tyr° rat/human CRF in the medium before and after, and a 20% acetonitrile in 0.1% trifluoroacetic acid extract of the tissue-bound radioactivity were analyzed by RP-HPLC using conditions described in Methods. The elution profile of unlabeled rat CRF was determined by absorbance at 220 nm. For symbols, see inset. (From De Souza, E. B., *J. Neurosci.*, 7, 88, 1987. With permission.)

TABLE 1
^{125}I-Tyr° Rat/Human CRF Binding to Subcellular Fractions of Rat Olfactory Bulb

Olfactory bulb fraction	Specific ^{125}I-Tyr° rat/human CRF bound (fmol/mg protein)
1,000 × *g* pellet (P_1)	17.1 ± 0.2
20,000 × *g* pellet (P_2)	31.0 ± 1.0
100,000 × *g* pellet (P_3)	0

Note: Aliquots of rat olfactory bulb tissue comprising the various fractions were incubated with 35 p*M* ^{125}I-Tyr° rat/human CRF for 120 min at room temperature in the absence (total) or presence (blank) of 1 μ*M* unlabeled rat CRF. At the concentration of ligand used in the fractionation studies, approximately 15% of the total ^{125}I-Tyr° rat/human CRF binding sites were occupied. The data are based on five determinations from a representative experiment (mean ± SEM), which were confirmed by two independent repetitions of the experiment and were found to vary less than 10% from each other.

From De Souza, E. B., *J. Neurosci.*, 7, 88, 1987. With permission.

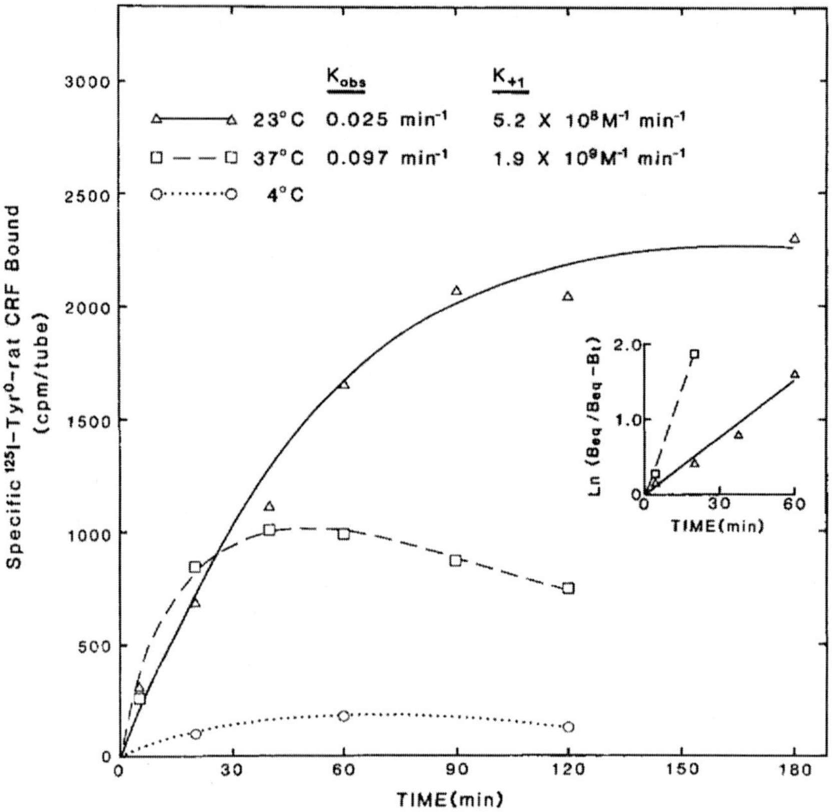

FIGURE 5. Time course association of ^{125}I-Tyr° rat/human CRF to rat olfactory bulb membranes as a function of temperature. Specific binding of ^{125}I-Tyr° rat/human CRF at 4, 23, and 37°C is plotted as a function of time after tissue addition. Nonspecific binding (binding in the presence of 1 μ*M* rat CRF) was subtracted from total binding. Linear conversion of the association data is included as an inset, where B_{eq} represents specific binding at the steady-state level and B_t represents specific binding at time t. Linear regression yielded the slope presented as K_{obs}, which was used in the calculation of the association rate constant, K_{+1}. Each value is the average of three or four determinations, which varied by less than 10%. (From De Souza, E. B., *J. Neurosci.*, 7, 88, 1987. With permission.)

are shown in Figure 5. The association rate of ^{125}I-rCRF was temperature dependent and exhibited pseudo-first-order kinetics. At room temperature (23°C), the binding increased with time to reach a steady state after 90 min and equilibrium binding was maintained for at least 3 h. The rate of association was more rapid at 37°C (K_1 = 1.9 × 10^9 M^{-1} min^{-1}) than at room temperature (K_1 = 5.2 × 10^8 M^{-1} min^{-1}) but this higher rate of association at 37°C was accompanied by a substantially lower level and duration of equilibrium-specific binding of ^{125}I-rCRF. The decrease in ^{125}I-rCRF binding with longer periods of incubation at 37°C probably resulted from a more rapid degradation of the ligand, as evidenced by the proportionally higher level of nonspecific binding at this temperature (data not shown). Negligible amounts of specific ^{125}I-rCRF binding were observed at 4°C, with incubation periods lasting up to 2 h.

2. Dissociation Studies

The time course of ^{125}I-rCRF dissociation from rat olfactory bulb membranes is shown in Figure 6. ^{125}I-rCRF binding decreased as a function of time following the addition of 1 μ*M* unlabeled rat CRF, demonstrating the reversibility of the binding reaction. A plot of

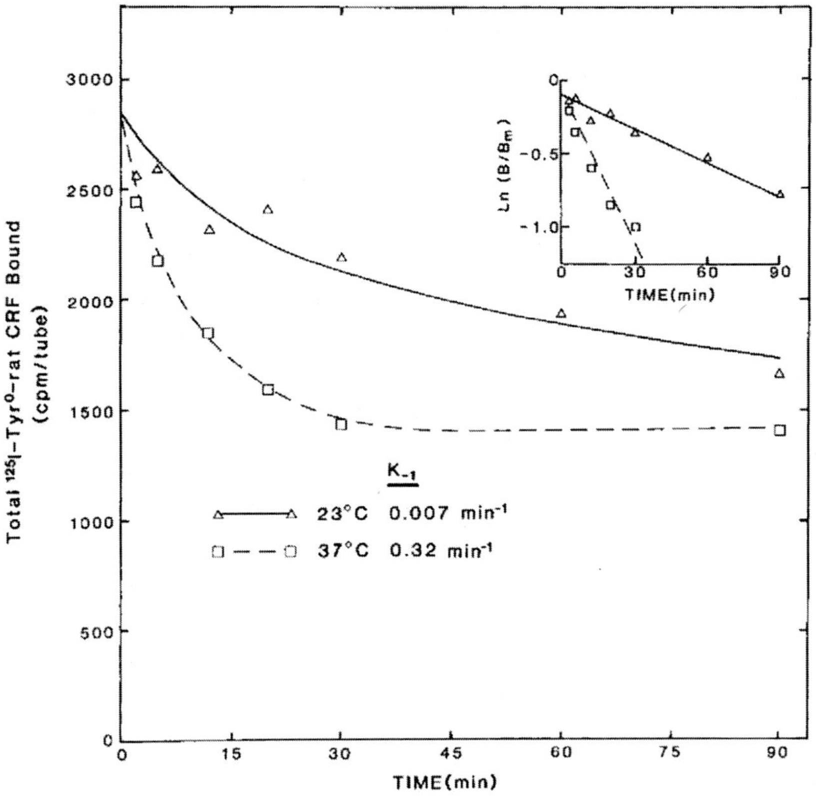

FIGURE 6. Time course of dissociation of ^{125}I-Tyr° rat/human CRF from rat olfactory bulb membranes as a function of temperature. Binding of ^{125}I-rCRF at 23 and 37°C is plotted as a function of time after the addition of excess unlabeled peptide. Homogenates were incubated with ^{125}I-Tyr° rat/human CRF at room temperature for 120 min and dissociation was initiated by the addition of 1 μ*M* rat CRF. Linear conversion of the data is included as an inset, where B represents binding at time t and B_m represents maximal binding; linear regression yielded the slope indicated as K_{-1}. Each value is the average of three or four determinations, which varied by less than 10%. (From De Souza, E. B., *J. Neurosci.*, 7, 88, 1987. With permission.)

$\ln(B/B_m)$ against time (see inset in Figure 6) resulted in a monophasic profile at both 23 and 37°C. The rate of dissociation of ^{125}I-rCRF was temperature dependent, with more rapid dissociation observed at 37°C (k_{-1} = 0.32 min^{-1}) than at 23°C (k_{-1} = 0.007 min^{-1}), corresponding to apparent half-times of 2.2 and 99 min, respectively.

C. ^{125}I-rCRF BINDING CHARACTERISTICS AT EQUILIBRIUM

1. Saturation Studies

The concentration-dependent equilibrium binding of ^{125}I-rCRF to rat olfactory bulb membranes was examined and is shown in Figure 7A. Specific ^{125}I-rCRF binding was saturable and Scatchard analysis of the saturation data (Figure 7B) indicated that the binding exhibited high affinity, with an apparent equilibrium dissociation constant (K_D) of 192 p*M* and a maximum number of binding sites (B_{max}) of 243 fmol/mg protein. In the range of ligand concentrations (0.05 to 1 n*M*) used in our saturation studies, ^{125}I-rCRF bound to a single class of sites, as indicated by both the straight line obtained from the Scatchard plot (correlation coefficient = 0.98) and a Hill coefficient of 1.0. Inhibition studies employing increasing concentrations of unlabeled rCRF (0.1 to 100 n*M*) yielded a curvilinear plot of bound hormone vs. bound/free hormone. Scatchard analysis of these data using the nonlinear

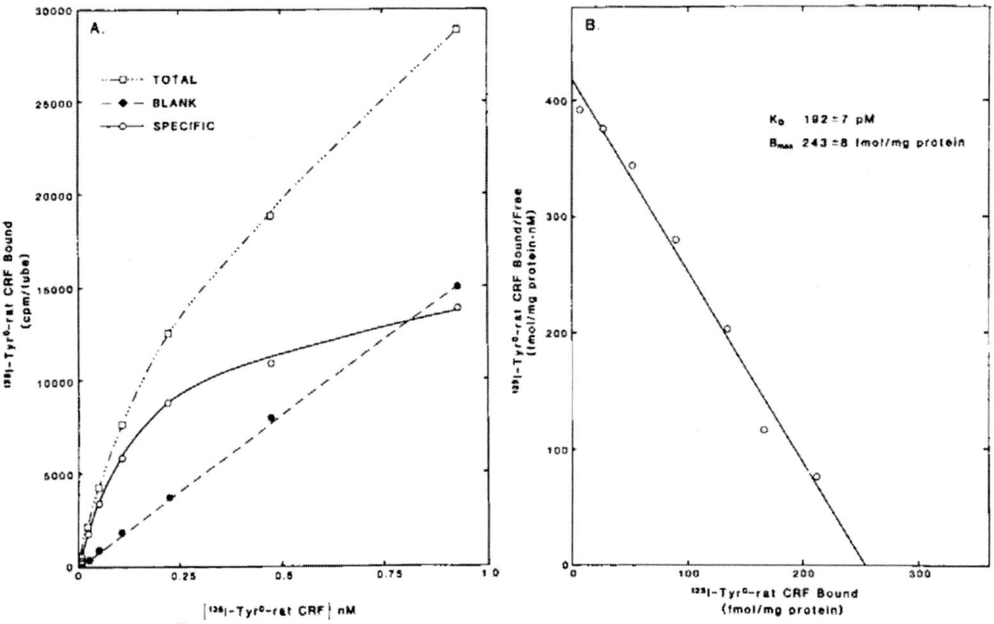

FIGURE 7. The binding of ^{125}I-Tyr^0 rat/human CRF to rat olfactory bulb membranes as a function of increasing ligand concentration. (A) Direct plot of the data showing the total amount of ^{125}I-Tyr^0 rat/human CRF bound, the amount of ^{125}I-Tyr^0 rat/human CRF bound in the presence of 1 μM rat CRF (blank), and specific binding (total minus blank). (B) Scatchard plot of ^{125}I-Tyr^0 rat/human CRF specific binding. Crude membrane preparations of rat olfactory bulb were incubated for 120 min at room temperature with increasing concentrations (0.05 to 1 n*M*) of ^{125}I-Tyr^0 rat/human CRF. The concentration of free ^{125}I-Tyr^0 rat/human CRF was measured directly in aliquots of the supernatant after the binding reaction was terminated by centrifugation. The data shown are from a representative experiment. Each value is the average of three determinations, which varied by less than 10%. (From De Souza, E. B., *J. Neurosci.*, 7, 88, 1987. With permission.)

curve-fitting program LIGAND[14] suggested the presence of a second, lower-affinity CRF binding site with an apparent K_D of approximately 20 n*M*; the Hill coefficient associated with binding under these conditions was 0.75 (data not shown).

2. Pharmacological Characteristics

Pharmacological characteristics of the binding site were examined by determining the relative potencies of a variety of CRF-related and unrelated peptides in displacing specifically bound ^{125}I-rCRF. The results of these studies are summarized in Figure 8 and Table 2. Rat CRF was slightly more potent than the oCRF in displacing the iodinated ligand; the inhibitory binding-affinity constant (K_i) values for rat and ovine CRF were 1 and 2.5 n*M*, respectively. Acetylated oCRF(4-41), a fragment more potent than the endogenous ligand in stimulating ACTH release, exhibited a slightly higher affinity (K_i = 0.5 n*M*) for the ^{125}I-rCRF binding site, whereas oCRF(1-39), a fragment with 1000-fold lower potency in the ACTH bioassay, was approximately 1000 times less potent than the endogenous ligand in displacing ^{125}I-rCRF from its binding site. The putative weak CRF antagonist, alpha-helical oCRF(9-41), displaced the ^{125}I-rCRF with a K_i value of approximately 15 n*M*. A variety of biologically inactive CRF fragments representing the amino terminus [rCRF(1-20)], the carboxy terminus [rCRF(21-41)], and the middle portion [rCRF(6-33)] of the active peptide were found to have no effect on ^{125}I-rCRF binding. Equally ineffective in inhibiting ^{125}I-rCRF binding were the unrelated peptides angiotensin II, arginine vasopressin, vasoactive intestinal peptide, and an analog of growth hormone-releasing factor.

The binding sites for ^{125}I-rCRF in rat olfactory bulb membrane preparations exhibit a

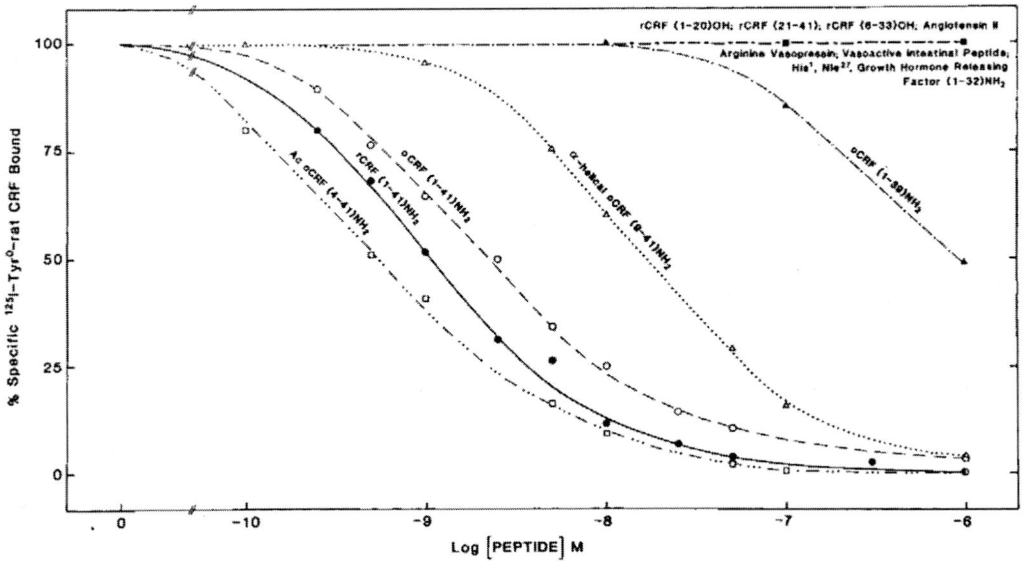

FIGURE 8. Characterization of the pharmacological specificity of ^{125}I-Tyr° rat/human CRF binding in rat olfactory bulb membranes. Crude membranes were incubated for 120 min at room temperature in the presence of 0.1 n*M* ^{125}I-Tyr° rat/human CRF and varying concentrations of CRF-related and unrelated peptides. Nonspecific binding was determined in the presence of 1 μ*M* rat CRF and was subtracted from the total binding. The data shown are from representative experiments. Each point represents the mean of a triplicate determination where the standard error is less than 10% of the mean. Shown are acetyl oCRF(4-41)NH_2; rCRF(1-41)NH_2; oCRF(1-41)NH_2; α-helical oCRF(9-41)NH_2; oCRF(1-39)NH_2 and one of several noncompeting rCRF fragments or unrelated peptides. (From De Souza, E. B., *J. Neurosci.*, 7, 88, 1987. With permission.)

pharmacological specificity for CRF analogs and fragments that correlates extremely well with their relative intrinsic potencies in stimulating or inhibiting anterior pituitary secretion of ACTH *in vitro* (see Table 2). The specificity of the ^{125}I-rCRF binding site is further strengthened by the lack of inhibitory activity of peptides such as angiotensin II, arginine vasopressin, and vasoactive intestinal peptide, all of which have been shown to also stimulate ACTH secretion through their own respective receptors.[15-17] The kinetic and pharmacological characteristics of the ^{125}I-rCRF binding sites in rat olfactory bulb were comparable to the characteristics found in the other brain regions examined and are similar to those previously observed in rat,[1,7,18] bovine[7,8] and human[19] pituitary, and in rat[2,9,20-22] and human[6] brain using a variety of ligands. These data substantiate earlier suggestions[23-25] that some structural requirements for CRF activity are shared by brain and pituitary receptors.

D. REGIONAL DISTRIBUTION OF ^{125}I-CRF BINDING SITES IN RAT CNS

Saturation studies of CRF binding in various CNS regions indicated generally uniform K_D values but marked differences in the density of ^{125}I-rCRF binding sites across 11 rat CNS regions (Figure 9). The highest density of binding sites in the rat CNS was in the olfactory bulb. The density of anterior pituitary binding sites (approximately 2.5 times that of olfactory bulb) is included to indicate the relative abundance of brain binding sites in comparison with this tissue, more commonly employed in CRF binding and bioassays. High levels of ^{125}I-rCRF binding sites were detected in olfactory bulb, cerebellum, cortex, and striatum. Progressively lower but significant receptor densities were observed in cervical spinal cord, hypothalamus, medulla, midbrain, thalamus, pons, and hippocampus.

These results confirm the presence and heterogeneous distribution of specific, high-affinity receptors for CRF throughout the rat CNS. In general, the distribution of CRF binding sites at the gross level of these homogenate binding assays correlates well with

TABLE 2
Pharmacological Specificity of ^{125}I-Tyr° Rat/human CRF Receptor Binding to Olfactory Bulb Membranes

^{125}I-Tyr° Rat/Human CRF Binding Data

Peptide	K_i (n*M*)	Hill coefficient	Relative potency	Bioassay data: intrinsic activity
rCRF	1.01 ± 0.05 (6)	0.84 ± 0.03	1.0	1.0
oCRF	2.39 ± 0.11 (4)	0.66 ± 0.06	0.4	1.0
α-helical oCRF (9-41)	16.73 ± 1.67 (3)	0.83 ± 0.01	0.06	<0.1
Ac oCRF (4-41)	0.47 ± 0.08 (3)	0.86 ± 0.04	2.1	1.22
oCRF (1-39)	828(2)		0.001	<0.01
rCRF (1-20)	>1000 (3)		<0.001	<0.01
rCRF (21-41)	>1000 (3)		<0.001	<0.01
rCRF (6-33)	>1000 (3)		<0.001	<0.01
Angiotensin II	>1000 (3)		<0.001	
AVP	>1000 (3)		<0.001	
VIP	>1000 (3)		<0.001	
His^1, Nle^{27}, GRF	>1000 (3)		<0.001	

Note: Peptides at 3—12 concentrations competed with 30 to 50 p*M* ^{125}I-Tyr° rat/human CRF during 120 min incubation at room temperature. All assays were conducted in triplicate for the number of experiments in parentheses, and the K_i (inhibitory binding affinity constant) and Hill coefficient (mean ± SEM) values were obtained from competition data analyzed by the computer program EBDA, which provided initial estimates of the equilibrium-binding parameters. The bioassay data were obtained from previous studies by Rivier et al. rCRF, Rat/human corticotropin-releasing factor; oCRF, ovine corticotropin-releasing factor; AVP, arginine vasopressin; VIP vasoactive intestinal peptide; GRF, growth hormone-releasing factor.

From De Souza, E. B., *J. Neurosci.*, 7, 88, 1987. With permission.

previous data from higher-resolution autoradiographic localization studies. A detailed discussion of the role of CRF receptors in the brain with regard to the correlation between regional levels of binding sites for ^{125}I-rCRF and the reported distribution of CRF-immunoreactive fibers, and a correspondence between the regional distribution of CRF receptors and the electrophysiological, behavioral, autonomic, and endocrine effects have been discussed in detail in Chapter 6.

E. IDENTIFICATION OF THE LIGAND BINDING SUBUNIT FOR THE CRF RECEPTOR IN BRAIN

To determine the apparent molecular weight of the brain CRF receptor, we cross-linked ^{125}I-oCRF to membrane receptors in rat frontal cortex, olfactory bulb, and cerebellum using the reagent DSS and solubilized the covalently labeled receptors in SDS. The samples were then electrophoresed as described in Section II. In order to characterize the pharmacological specificity of the covalently labeled proteins as the CRF receptor, we inhibited the covalent attachment by the addition in the assay of various related and unrelated CRF analogs. In the frontal cortex, ^{125}I-oCRF was cross-linked to a protein which migrated with an apparent molecular weight of 58,000 Da (Figure 10). The covalent incorporation of ^{125}I-oCRF into this protein could be specifically inhibited by the inclusion of various other peptides. At a 100 n*M* concentration of each of the following peptides, the pharmacological rank-order profile was as follows: Nle^{21},Tyr^{32}-oCRF ≅ rat/human CRF ≅ ovine CRF ≅ α-helical CRF(6-41) > α-helical oCRF(9-41) ≧ oCRF(7-41) > rat/human CRF(1-20) ≅ vasoactive

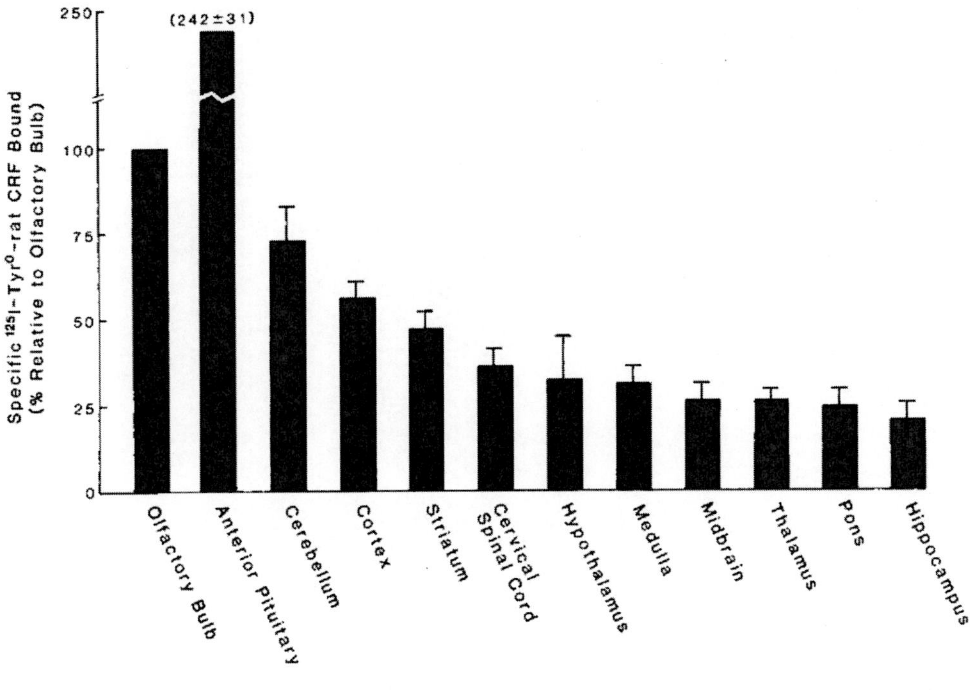

FIGURE 9. Regional distribution of binding sites for ^{125}I-Tyr° rat/human CRF in the rat CNS and anterior pituitary. Crude membrane preparations from each of the indicated regions were incubated for 2 h at room temperature in the presence of 0.1 n*M* ^{125}I-Tyr° rat/human CRF and increasing concentrations of unlabeled rat CRF. The values for each region are presented here as a percentage of the level of binding observed in rat olfactory bulb membranes. Relative values represent the mean ± SEM determined from results of three to five experiments. (From De Souza, E. B., J. *Neurosci.*, 7, 88, 1987. With permission.)

intestinal peptide. This pharmacological profile is similar to that demonstrated for the CRF binding site in membrane homogenates shown in Figure 8. These data provide further evidence that the ligand binding subunit for the brain CRF receptor resides on a peptide of approximately 58,000 Da.

In olfactory bulb and cerebellar homogenates, ^{125}I-oCRF was functionally cross-linked to a membrane protein which also migrated with an apparent molecular weight of 58,000 Da and demonstrated the same specificity and pharmacological rank order for the CRF receptor as the frontal cortex (Figure 10). In rat anterior pituitary however, covalent attachment of ^{125}I-oCRF to membrane homogenates yielded a different pattern of labeling. In the pituitary, affinity cross-linked ^{125}I-oCRF followed by SDS-PAGE and autoradiography revealed four distinctly labeled proteins with apparent molecular weights of 74,800, 44,000, 24,000, and 20,000 Da. Unlabeled rat/human CRF at a concentration of 1 μ*M* inhibited the attachment to the 74,800 and the 20,000 Da protein fragments (Figure 10). While the 74,800 Da protein corresponds to the CRF binding protein previously described by Nishimura et al.[26] in bovine pituitary, it is unclear whether the minor, specifically labeled M_r = 20,000 band observed on SDS-PAGE represents a functional subunit of the CRF receptor only in the pituitary or whether it is a degradation product containing the binding site for CRF. Also evident from Figure 10, there was a greater incorporation of ^{125}I-oCRF into the major protein band in the anterior pituitary when compared to the olfactory bulb, which in turn had a greater incorporation of radiolabel than did the frontal cortex. The increased incorporation of ^{125}I-oCRF in the pituitary reflects the higher density of receptors present in that tissue. Furthermore, the relative density of the receptors observed under these conditions, is in agreement with the homogenate binding data shown in Figure 9 demonstrating highest

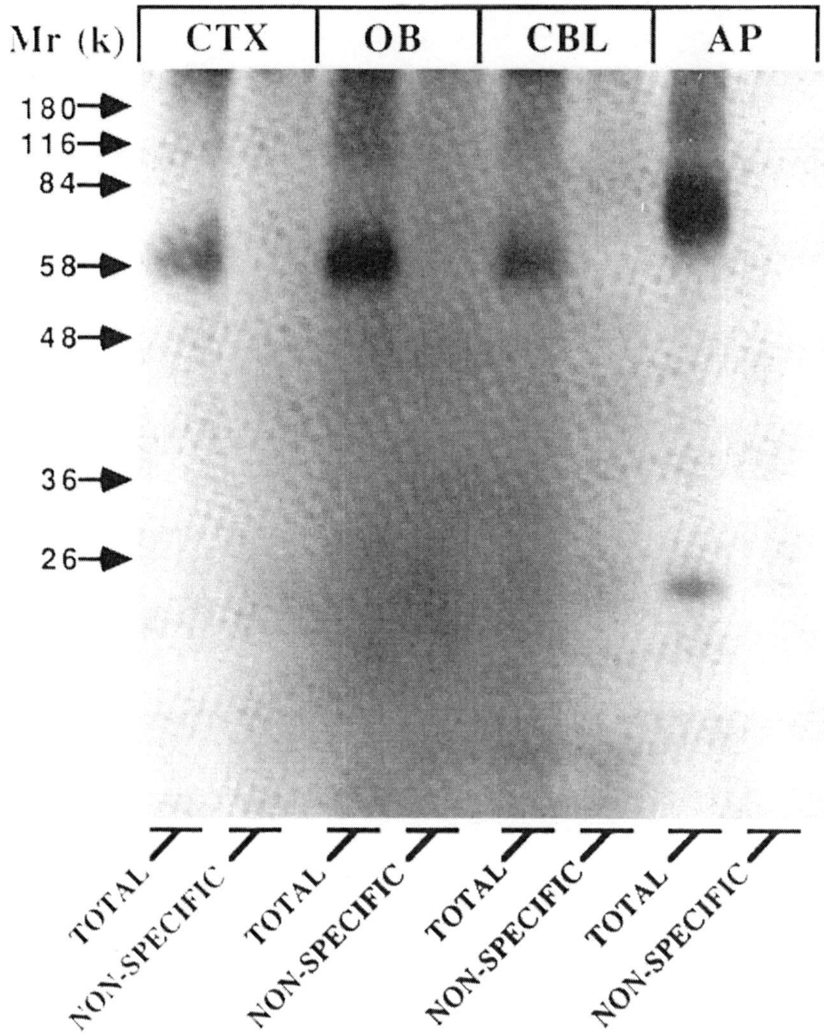

FIGURE 10. Covalent incorporation of ^{125}I-Tyr°-oCRF into rat brain and pituitary membrane homogenates. Labeling patterns of ^{125}I-Tyr°-oCRF cross-linked membranes in rat demonstrate that the specific binding site in brain is different from that of the pituitary. Equivalent amounts of protein (300 to 400 μg) were loaded onto each lane and electrophoresed as described in Section II.E. Molecular weights of known prestained protein standards run in conjunction with the samples are shown on the left. Rat cerebral cortex (CTX; lanes 1 and 2), olfactory bulb (OB; lanes 3 and 4) and cerebellum (CBL; lanes 5 and 6) all exhibit a binding site for CRF with an apparent molecular weight of 58,000 Da while the anterior pituitary (AP; lanes 7 and 8) exhibits a major labeled protein specific for the CRF receptor with an apparent molecular weight of 74,800 Da. In all tissues there also appears to be a faintly labeled protein with a molecular weight of approximately 116,000 Da as well as lower molecular weight peptides which are more apparent in pituitary than in brain. In all tissues examined, nonspecific binding was defined in the presence of 1 μ*M* rat/human CRF.

concentrations of CRF receptors in the anterior pituitary with progressively lower densities present in olfactory bulb and frontal cortex, respectively (refer to Figure 9).

The pharmacological rank order of potencies for the 74,800 Da labeled protein in the anterior pituitary was similar to that obtained in the 58,000 Da protein in the frontal cortex. This rank order in both tissues correlates extremely well with their relative intrinsic potency in stimulating or inhibiting anterior pituitary secretion of adrenocorticotropic hormone secretion *in vitro*[27-30] and adenylate cyclase activity in brain.[3,9] This specificity in the two tissues is further strengthened by the lack of inhibitory potency of fragments (rat/human CRF(1-20)) and vasoactive intestinal peptide. These data therefore strongly suggest that the 58,000 Da protein in brain and the 74,800 and 20,000 Da proteins in anterior pituitary represent the CRF binding subunit in the respective tissues and substantiate earlier suggestions that some of the structural requirements for CRF activity are shared by both brain and pituitary receptors.[23-25]

In order to further establish that the difference observed in the labeling pattern of ^{125}I-Tyr°-oCRF in the cortex and anterior pituitary was not a function of the ligand, we used ^{125}I-Tyr°-rat/human CRF and the cross-linking agent DSS to covalently label receptors in the two tissues. As demonstrated in Figure 11, both ligands exhibited identical labeling patterns in the respective tissues. In rat frontal cortex, (lanes 1, 2 and 5, 6), both ligands were incorporated into proteins with an apparent molecular weight of 58,000 Da. Similarly, in the anterior pituitary (lanes 3, 4 and 7, 8) the pattern of covalent labeling was identical with both ligands being incorporated into the 74,800 Da and the 20,000 Da proteins. Although both ligands identified the identical binding sites in each tissue, ^{125}I-Tyr°-rat/human CRF was observed to incorporate more heavily into the nonspecific fragments present in both tissues (Figure 11; compare lanes 1 to 4 with 5 to 8). The consequence of this binding in homogenates is to increase the level of nonspecific binding of the ligand to membranes. As stated in Section II.A, the choice of radioligands for the labeling of CRF receptors is largely based on the extent of nonspecific binding that a particular probe exhibits. Thus inasmuch as both ligands recognize and can identify the CRF binding subunit in both brain and pituitary, ^{125}I-Tyr°-oCRF is the ligand of choice since it displays less nonspecific binding to membranes than ^{125}I-Tyr°-rat/human CRF.

Previous studies have identified and characterized CRF receptors in rat,[1,7,8] cow,[7,31] monkey,[32] and human[19] anterior pituitary and in rat,[3,4,8,9,20,21] monkey,[8,33] and human[6] brain that have comparable kinetic and pharmacological characteristics. We wished to further determine whether the heterogeneity of CRF receptors observed in the rat anterior pituitary and brain was specific to the species, or if it was consistent across a variety of species. Preliminary data demonstrate that the distinct different patterns of labeling observed in rat anterior pituitary and brain was consistent across the various species examined. Specifically, rhesus monkey, porcine, and bovine anterior pituitary all showed comparable profiles on SDS-PAGE with the incorporation of ^{125}I-oCRF into four major bands that were identical to those observed in rat anterior pituitary. Similarly, the SDS-PAGE profiles of cerebral cortex of human, monkey, and rat were all identical, with incorporation of ^{125}I-oCRF into a single major protein band of 58,000 Da. Thus using the chemical cross-linking technique, we have identified a brain CRF binding protein with a different molecular weight than the CRF binding protein in the pituitary.

IV. REGULATION OF CRF RECEPTORS IN RAT BRAIN

A. *IN VITRO* STUDIES

1. Effects of Magnesium Ions and Guanine Nucleotides

Magnesium ions have been shown to enhance agonist binding to receptors coupled to adenylate cyclase by stabilizing the high-affinity form of the receptor-effector complex. In contrast, guanine nucleotides have been shown to selectively decrease the affinity of agonists for their receptors in these systems. To determine whether CRF receptors in brain might be coupled to adenylate cyclase, we carried out a series of studies to examine the effects of magnesium ions and nucleotides on ^{125}I-rCRF binding in rat olfactory bulb. The addition of increasing concentrations of guanosine 5′-triphosphate (GTP) to the incubation medium resulted in an inhibition of ^{125}I-rCRF binding; the nonhydrolyzable GTP analog 5′-guanylyl-

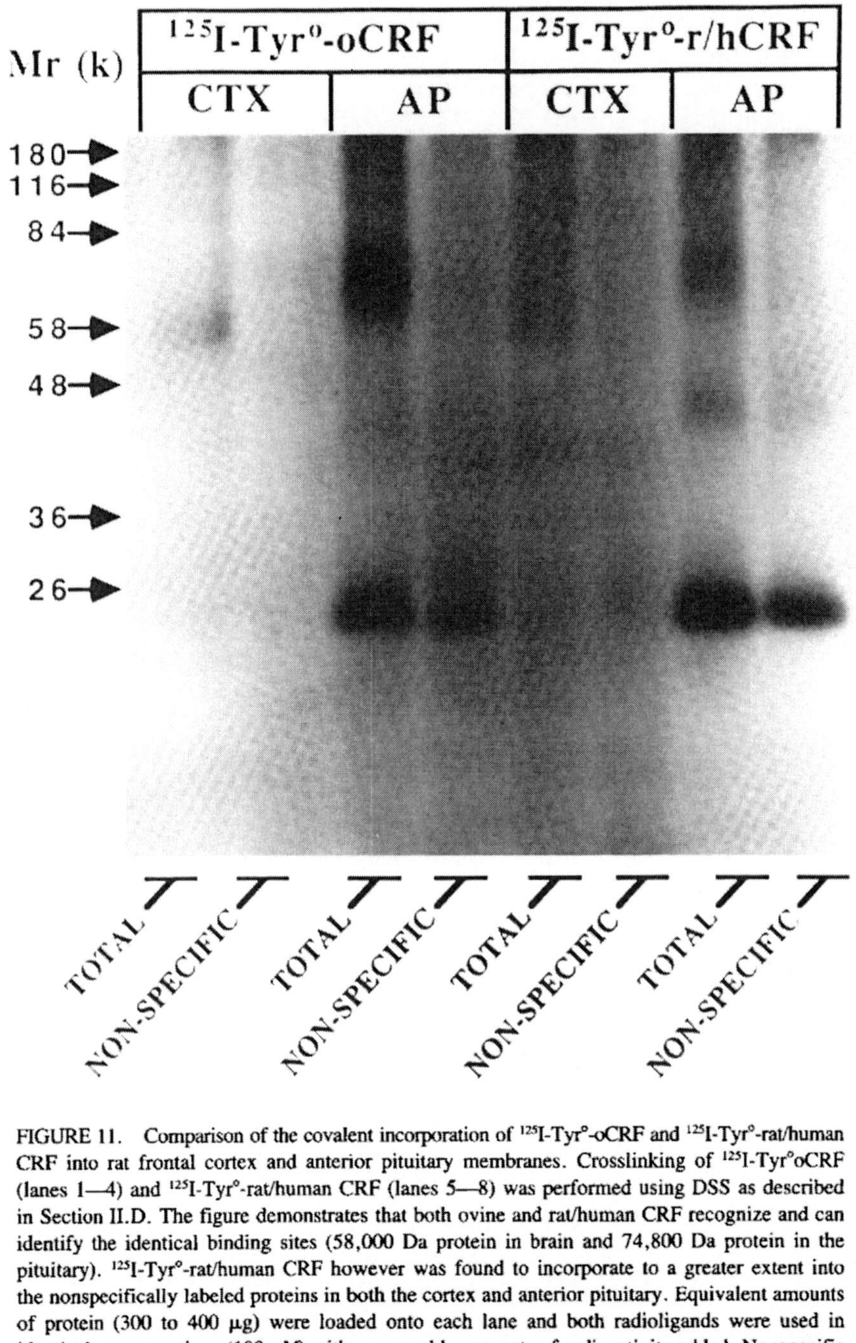

FIGURE 11. Comparison of the covalent incorporation of ^{125}I-Tyr°-oCRF and ^{125}I-Tyr°-rat/human CRF into rat frontal cortex and anterior pituitary membranes. Crosslinking of ^{125}I-Tyr°oCRF (lanes 1—4) and ^{125}I-Tyr°-rat/human CRF (lanes 5—8) was performed using DSS as described in Section II.D. The figure demonstrates that both ovine and rat/human CRF recognize and can identify the identical binding sites (58,000 Da protein in brain and 74,800 Da protein in the pituitary). ^{125}I-Tyr°-rat/human CRF however was found to incorporate to a greater extent into the nonspecifically labeled proteins in both the cortex and anterior pituitary. Equivalent amounts of protein (300 to 400 μg) were loaded onto each lane and both radioligands were used in identical concentrations (100 p*M*) with comparable amounts of radioactivity added. Nonspecific binding was defined in the presence of 1 μ*M* unlabeled rat/human CRF for ^{125}I-Tyr°-oCRF binding and 1 μ*M* unlabeled oCRF for ^{125}I-Tyr°-rat/human CRF binding.

imidodiphosphate (GppNHp) was even more potent (Figure 12). This effect appeared to be specific to the guanine nucleotides, as similar experiments indicated that equimolar concentrations of ATP had no effect on ^{125}I-rCRF binding (data not shown). GTP and GppNHp inhibited a maximum of 50% of the specific ^{125}I-rCRF binding. Conversely, as mentioned previously, magnesium ions appeared to be obligatory for obtaining significant ^{125}I-rCRF binding in olfactory bulb; specific ^{125}I-rCRF binding increased in a linear fashion with

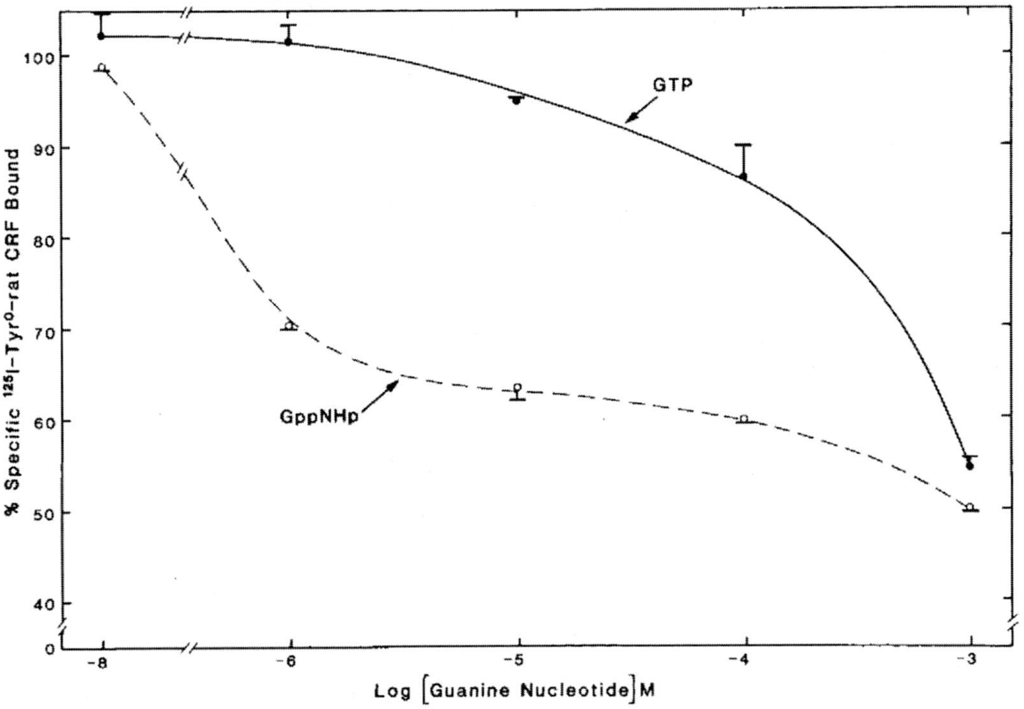

FIGURE 12. Effects of guanine nucleotides on the binding of ^{125}I-Tyr° rat/human CRF to rat olfactory bulb membranes. Crude membrane preparations of olfactory bulb were incubated for 2 h at room temperature in the presence of 0.1 n*M* ^{125}I-Tyr° rat/human CRF and the indicated concentrations of guanosine-5′-triphosphate (GTP) or 5′-guanylylimidodiphosphate (GppNHp). Nonspecific binding was determined in the presence of 1 μ*M* rat CRF. The data are based on quadruplicate determinations; mean ± SEM. (From De Souza, E. B., *J. Neurosci.*, 7, 88, 1987. With permission.)

increasing concentrations of $MgCl_2$ (Figure 3). To determine whether the effects of guanine nucleotides and magnesium ions were mediated by changes in receptor density or affinity, we carried out a series of parallel ^{125}I-rCRF saturation assays in the presence of 1 m*M* $MgCl_2$, 10 m*M* $MgCl_2$, and 10 m*M* $MgCl_2$ + 200 μ*M* NaGTP. The Scatchard analyses of these saturation data are presented in Figure 13. GTP increased the apparent K_D value of ^{125}I-rCRF by 35%; this decrease in affinity was accomplished by a small (10%) reduction in the B_{max} value. Conversely, the effect of magnesium on increased ^{125}I-rCRF binding was accomplished by decreasing both the K_D (50% decrease, from 1 to 10 m*M* $MgCl_2$) and increasing the number of binding sites (Figure 13). These results are, in general, consistent with the characteristic effects of adenylate cyclase modulators on agonist binding to receptors mediating their effects through cAMP production.

Monovalent cations had no effect on ^{125}I-rCRF binding over the range of concentrations examined, whereas divalent cations such as $CaCl_2$ and $MgCl_2$ increased binding affinity. The effects of $MgCl_2$ on the binding of ^{125}I-rCRF are consistent with previous data demonstrating the ability of magnesium ions to enhance agonist binding to receptors coupled to adenylate cyclase.[34] Likewise, the ability of guanine nucleotides to inhibit ^{125}I-rCRF binding is consistent with the hypothesis that receptors for CRF are coupled to adenylate cyclase. There are recent studies reporting the ability of CRF to stimulate adenylate cyclase activity in some brain areas[2,3,9] (see Chapter 10).

B. *IN VIVO* STUDIES

1. Effects of Atropine Treatment in Rats

Recent anatomical and behavioral studies have indicated interactions between CRF and

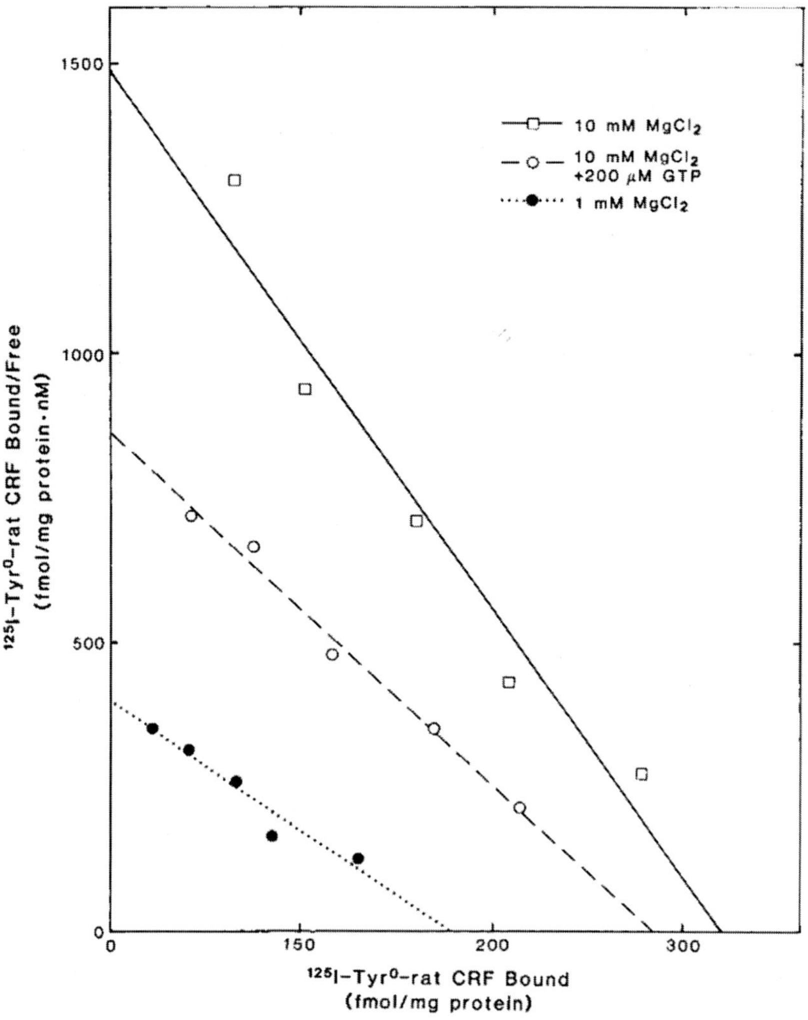

FIGURE 13. Effects of magnesium ions and guanosine-5′-triphosphate (GTP) on the binding of ^{125}I-Tyr° rat/human CRF to rat olfactory bulb membranes. Scatchard plots of the ^{125}I-Tyr° rat/human CRF binding to crude membranes in the presence of 1 m*M* $MgCl_2$, 10 m*M* $MgCl_2$, or 10 m*M* $MgCl_2$ plus 200 μ*M* GTP and increasing concentrations of ligand (0.05 to 1 n*M*). Incubations were for 120 min at room temperature. Nonspecific binding was in the presence of 1 μ*M* rat CRF. The concentration of free ^{125}I-Tyr° rat/human CRF was measured directly in aliquots of the supernatant after the binding reaction was terminated by centrifugation. Points are the means of triplicate determinations in a representative experiment. (From De Souza, E. B., *J. Neurosci.*, 7, 88, 1987. With permission.)

acetylcholine in brain.[35] Specifically, colocalization of CRF and acetylcholinesterase was detected by retrograde tracing and immunocytochemical staining in some brainstem nuclei projecting to the frontal cortex. In behavioral studies, CRF injected into the rat frontal cortex did not initiate, but significantly inhibited the effects of carbachol, a muscarinic cholinergic receptor agonist, on a stereotyped motor "boxing" behavior.[35] We have recently found that CRF content was reduced and there was a reciprocal increase in CRF receptors in affected cerebral cortical areas in Alzheimer's disease.[6] Furthermore, these changes in CRF in Alzheimer's disease correlated significantly with decrements in choline acetyltransferase activity. The demonstration of significant correlations between changes in CRF and choline acetyltransferase activity in human cortex, complemented by similar preliminary studies in rodents described above, suggests that CRF and cholinergic systems may interact.

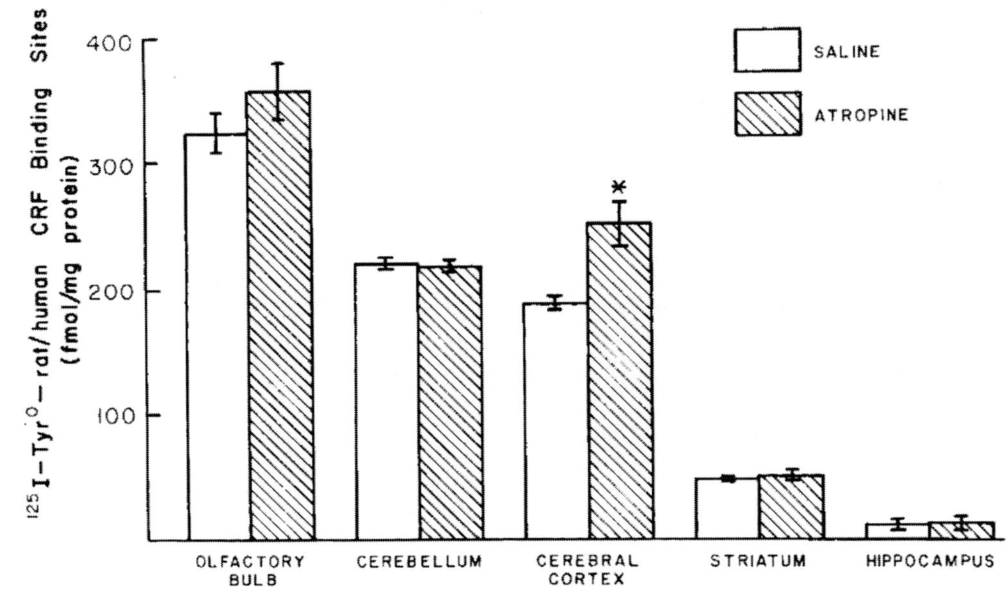

FIGURE 14. Effects of chronic atropine treatment on ^{125}I-Tyr^{0}-rat/human CRF binding sites in discrete regions of rat brain. Values represent the mean ± SEM; n = 5 rats/group. *, Significant difference from saline-treated control group at $p < 0.01$. B_{max} values and details of treatment and assay conditions are described in the text. (From De Souza, E. B., and Battaglia, G., *Brain Res.*, 397, 401, 1986. With permission.)

To investigate the functional consequence of cholinergic deficits on the regulation of brain CRF neuronal activity, we designed a study in which rats were treated chronically with atropine (14 d, 20 mg/kg/d, s.c.), a muscarinic cholinergic receptor antagonist, and CRF receptors were measured in various regions of brain.[5] Chronic atropine treatment produced significant increases in muscarinic cholinergic receptors in the frontoparietal cortex (30% increase) and hippocampus (20% increase). No significant changes in the concentration of CRF binding sites were observed in olfactory bulb, cerebellum, striatum, and hippocampus (Figure 14). In contrast, there was a significant and selective increase in CRF receptors in the frontoparietal cortex of atropine-treated rats (Figure 14). These results demonstrating a selective up-regulation of CRF receptors in the cerebral cortex of atropine-treated rats suggest that the reported reciprocal changes in pre- and postsynaptic markers of CRF in the cerebral cortex of patients with Alzheimer's disease[6] (see Chapter 24) may be, in part, a consequence of deficits in the cholinergic projections to the cerebral cortex in Alzheimer's disease. Additional studies are necessary to determine the precise population(s) of cerebral cortical CRF neurons that may be modulated by acetylcholine, and to determine the functional significance of the interaction between CRF and cholinergic systems.

V. SUMMARY AND CONCLUSIONS

In summary, we have used the iodine-125-labeled analogs of CRF to identify, characterize, and localize receptors for CRF in discrete areas of the rat CNS. In addition, we have identified brain and pituitary CRF binding proteins with identical pharmacologic characteristics but different apparent molecular weights (brain: 58,000 Da; anterior pituitary: 74,800 Da). The effects of guanine nucleotides in decreasing and of magnesium ions in increasing ^{125}I-rCRF binding are consistent with CRF receptors being coupled to guanine nucleotide regulatory protein(s) and support data demonstrating that some populations of CRF receptors are coupled to adenylate cyclase; however, whether there are additional CRF-mediated signal transduction mechanisms remains to be determined. The effects of chronic atropine treatment to selectively up-regulate CRF receptors in the cerebral cortex suggest an interaction between CRF and acetylcholine. These data provide further support for the proposed role of CRF as a neurotransmitter in the CNS. This study demonstrating the characteristics of CRF receptors in brain provides a means for better understanding the various functions of this neuropeptide under physiological and pathological conditions.

VI. ACKNOWLEDGMENTS

We would like to thank George Battaglia and Elizabeth L. Webster for helpful discussions and comments regarding the manuscript. In addition, we would like to thank Terrie Pierce and Mary Flutka for manuscript preparation. D.E.G. is a Medical Research Council of Canada Postdoctoral Fellow.

REFERENCES

1. **Wynn, P. C., Aguilera, G., Morell, J., and Catt, K. J.,** Properties and regulation of high-affinity pituitary receptors for corticotropin-releasing factor, *Biochem. Biophys. Res. Commun.*, 110, 602, 1983.
2. **Wynn, P. C., Hauger, R. L., Holmes, M. C., Millan, M. A., Catt, K. J., and Aguilera, G.,** Brain and pituitary receptors for corticotropin releasing factor: localization and differential regulation after adrenalectomy, *Peptides*, 5, 1077, 1984.
3. **Battaglia, G., Webster, E. L., and De Souza, E. B.,** Characterization of corticotropin-releasing factor (CRF) receptor-mediated adenylate cyclase activity in rat central nervous system, *Synapse*, 1, 572, 1987.
4. **De Souza, E. B.,** Corticotropin-releasing factor receptors in the rat central nervous system: characterization and regional distribution, *J. Neurosci.*, 7, 88, 1987.
5. **De Souza, E. B. and Battaglia, G.,** Increased corticotropin-releasing factor receptors in rat cerebral cortex following chronic atropine treatment, *Brain Res.*, 397, 401, 1986.
6. **De Souza, E. B., Whitehouse, P. J., Kuhar, M. J., Price, D. L., and Vale, W. W.,** Reciprocal changes in corticotropin-releasing factor (CRF)-like immunoreactivity and CRF receptors in cerebral cortex of Alzheimer's disease, *Nature*, 319, 593, 1986.
7. **De Souza, E. B., Perrin, M. H., Rivier, J. E., Vale, W. W., and Kuhar, M. J.,** Corticotropin-releasing factor receptors in rat pituitary gland: autoradiographic localization, *Brain Res.*, 296, 202, 1984.
8. **De Souza, E. B. and Kuhar, M. J.,** Corticotropin-releasing factor receptors in the pituitary gland and central nervous system: methods and overview, *Methods Enzymol.*, 124, 560, 1986.
9. **Chen, F. M., Bilezikjian, L. M., Perrin, M. H., Rivier, J., and Vale, W. W.,** Corticotropin-releasing factor receptor-mediated stimulation of adenylate cyclase activity in the rat brain, *Brain Res.*, 381, 49, 1986.
10. **Lowry, O. H., Rosebrough, N. J., Farr, A. L., and Randall, R. J.,** Protein measurement with the Folin phenol reagent, *J. Biol. Chem.*, 265, 1983, 1951.
11. **Pilch, P. F. and Czech, M. P.,** Interaction of cross-linking agents with the insulin effector system of isolated fat cells. Covalent linkage of ^{125}I-insulin to a plasma membrane receptor protein of 140,000 daltons, *J. Biol. Chem.*, 254, 3375, 1979.
12. **Laemmli, U. K.,** Cleavage of structural proteins during the assembly of the head of bacteriophage T4, *Nature*, 227, 680, 1970.
13. **McPherson, G. A.,** A practical computer-based approach to the analysis of radioligand binding experiments, *Comput. Programs Biomed.*, 17, 107, 1983.
14. **Munson, P. J. and Rodbard, D.,** LIGAND: a versatile computerized approach for characterization of ligand-binding systems, *Anal. Biochem.*, 297, 220, 1980.
15. **Vale, W. W. and Rivier, C.,** Substances modulating the secretion of ACTH by cultured anterior pituitary cells, *Fed. Proc.*, 36, 2094, 1977.
16. **Steele, M. K, Negro-Vilar, A., and McCann, S. W.,** Effect of angiotensin II on in vivo and in vitro release of anterior pituitary hormones in the female rat, *Endocrinology*, 109, 893, 1981.
17. **Westendorf, J. M., Philips, M. A., and Schonrunn, A.,** Vasoactive intestinal peptide stimulates hormone release from corticotropic cells in culture, *Endocrinology*, 112, 550, 1983.

18. **Holmes, M. C., Antoni, F. A., and Szeintendnei, T.,** Pituitary receptors for corticotropin-releasing factor: no effect of vasopressin on binding or activation of adenylate cyclase, *Neuroendocrinology*, 39, 1001, 1984.
19. **De Souza, E. B., Perrin, M. H., Whitehouse, P. J., Rivier, J. E., Vale, W. W., and Kuhar, M. J.,** Corticotropin-releasing factor receptors in human pituitary gland: autoradiographic localization, *Neuroendocrinology*, 40, 419, 1985.
20. **De Souza, E. B., Perrin, M. H., Insel, T. R., Rivier, J., Vale, W. W., and Kuhar, M. J.,** Corticotropin-releasing factor receptors in rat forebrain: autoradiographic identification, *Science*, 224, 1449, 1984.
21. **De Souza, E. B., Insel, T. R., Perrin, M. H., Rivier, J., Vale, W. W., and Kuhar, M. J.,** Corticotropin-releasing factor receptors are widely distributed within the rat central nervous system: an autoradiographic study, *J. Neurosci.*, 5, 3189, 1985.
22. **De Souza, E. B., Insel, T. R., Perrin, M. H., Rivier, J., Vale, W. W., and Kuhar, M. J.,** Differential regulation of corticotropin-releasing factor receptors in anterior and intermediate lobes of pituitary and in brain following adrenalectomy in rats, *Neurosci. Lett.*, 56, 121, 1985.
23. **Peterfreund, R. A. and Vale, W. W.,** Ovine corticotropin releasing factor stimulates somatostatin secretion from cultured brain cells, *Endocrinology*, 112, 1275, 1983.
24. **Valentino, R. J., Foote, S. L., and Aston-Jones, G.,** Corticotropin-releasing factor activates noradrenergic neurons of the locus ceruleus, *Brain Res.*, 270, 363, 1983.
25. **Britton, D. R., Hoffman, D. K., Lederis, K., and Rivier, J.,** A comparison of the behavioral effects of CRF, sauvagine and urotensin I, *Brain Res.*, 304, 201, 1984.
26. **Nishimura, E., Billestrup, N., Perrin, M., and Vale, W.,** Identification and characterization of a pituitary corticotropin-releasing factor binding protein by chemical cross-linking, *J. Biol. Chem.*, 262, 12893, 1987.
27. **Vale, W., Spiess, J., Rivier, C., and Rivier, J.,** Characterization of a 41-residue ovine hypothalamic peptide that stimulates secretion of corticotropin and β-endorphin, *Science*, 213, 1394, 1981.
28. **Rivier, J., Spiess, J., Rivier, C., Galyean, R., and Vale, W.,** Solid phase synthesis of amunine (CRF), sauvagine, and neurotensin I, in *Peptides*, Blaha, K. and Malon, P., Eds., 1982, 597.
29. **Rivier, J., Spiess, J., and Vale, W.,** Characterization of rat hypothalamic corticotropin-releasing factor, *Proc. Natl. Acad. Sci. U.S.A.*, 80, 4851, 1983.
30. **Rivier, J., Rivier, C., and Vale, W.,** Synthetic competitive antagonists of corticotropin-releasing factor: effect on ACTH secretion in the rat, *Science*, 224, 889, 1984.
31. **Perrin, M. H., Haas, Y., Rivier, J., and Vale, W.,** Corticotropin-releasing factor binding to the anterior pituitary receptor is modulated by divalent cations and guanyl-nucleotides, *Endocrinology*, 118, 1171, 1986.
32. **Millan, M. A., Samra, A. B., Wynn, P. C., Catt, K. J., and Aguilera, G.,** Receptor and action of corticotropin-releasing hormone in the primate pituitary gland, *J. Clin. Endocrinol. Metab.*, 64, 1036, 1987.
33. **Millan, M. A., Jacobowitz, D. M., Hauger, R. L., Catt, K. J., and Aguilera, G.,** Distribution of corticotropin releasing factor receptor in primate brain, *Proc. Natl. Acad. Sci. U.S.A.*, 83, 1921, 1986.
34. **DeLean, A., Stadel, J. M., and Lefkowitz, R. J.,** A ternary complex model explains the agonist-specific binding properties of the adenylate cyclase-coupled β-adrenergic receptor, *J. Biol. Chem.*, 255, 7108, 1980.
35. **Crawley, J. N., Olschowka, J. A., Diz, D. I., and Jacobowitz, D. M.,** Behavioral investigation of the coexistence of substance P, corticotropin-releasing factor, and acetylcholinesterase in lateral dorsal tegmental neurons projecting to the medial frontal cortex of the rat, *Peptides*, 6, 891, 1985.

Chapter 10

CHARACTERIZATION OF SECOND MESSENGERS COUPLED TO CORTICOTROPIN-RELEASING FACTOR (CRF) RECEPTORS IN BRAIN

George Battaglia, Elizabeth L. Webster, and Errol B. De Souza

TABLE OF CONTENTS

I. INTRODUCTION

CRF receptors have recently been identified, characterized, and localized in the pituitary gland.[1-4] CRF-stimulated proopiomelanocortin (POMC)-derived peptide secretion is mediated by the activation of adenylate cyclase in both anterior[5,6] and intermediate[7] lobes of the pituitary. The effects of CRF, however, are not restricted to the pituitary. Recently, we and others have characterized and localized specific binding sites for CRF in rat,[8-11] monkey,[4,12] and human[4,13] central nervous system (CNS). As reported for the pituitary, CRF receptors in brain can stimulate cAMP production.[14,15]

Since increasing evidence suggests a neurotransmitter or neuromodulatory role for CRF in CNS, an understanding of its mechanism of signal transduction is essential toward understanding the varied effects of CRF in brain. This chapter provides a summary of the characteristics of CRF-stimulated adenylate cyclase activity in brain with respect to its pharmacology, regional distribution, and sensitivity to *in vitro* regulation.

II. METHODS

A. TISSUE PREPARATION

Male Sprague-Dawley rats (150 to 180 g) were killed by decapitation and brain regions of interest were dissected as described previously[10] and homogenized in 30 volumes of cold buffer (50 m*M* Tris HCl, 10 m*M* $MgCl_2$, 2 m*M* EGTA, pH 7.2 at 22°C) using a Brinkman polytron (setting of 5 for 20 s) and centrifuged (38,000 × *g*, 30 min). The pellet was suspended in buffer, washed one more time and resuspended to a final concentration of 20 mg original wet weight (o.w.w.) of tissue per milliliter in the same buffer. Protein concentrations were determined according to the method of Lowry et al.[16]

B. ADENYLATE CYCLASE ASSAY

In routine studies, adenylate cyclase assays were carried out at 37°C for 10 min in 200 μl of buffer containing 100 m*M* Tris HCl, 10 m*M* $MgCl_2$, 0.4 m*M* EGTA, 0.1% bovine serum albumin (BSA), 1 m*M* isobutylmethylxanthine (IBMX), 250 units/ml phosphocreatine kinase, 5 m*M* creatine phosphate, 100 μ*M* guanosine 5′-triphosphate (GTP) and 0.8 mg o.w.w. of tissue (approximately 40 to 60 μg protein) unless otherwise specified. Reactions were initiated by the addition of 1 m*M* ATP/[^{32}P]ATP (approximately 2 to 4 μCi/tube) and terminated by the addition of 100 μl of 50 m*M* Tris HCl, 45 m*M* ATP and 2% sodium dodecyl sulfate (SDS). One milliliter of [^{3}H]cAMP (approximately 40,000 dpm) in distilled water was added to each tube to monitor the recovery of cAMP. The separation of [^{32}P]cAMP from [^{32}P] ATP was accomplished by sequential elution over Dowex and alumina columns.[17] Recovery was consistently greater than 80%. CRF receptor binding assays were carried out as described in Chapter 9.

III. CHARACTERIZATION OF CRF-STIMULATED ADENYLATE CYCLASE ACTIVITY IN BRAIN

A. STUDIES TO OPTIMIZE CRF-STIMULATED ADENYLATE CYCLASE ACTIVITY IN RAT BRAIN

In order to elucidate the biochemical parameters that would provide a reliable stimulatory response for both pharmacologic characterization and regional studies of CRF receptor-mediated cAMP production, the stimulation by 1 μ*M* rat/human CRF (rCRF) was investigated in homogenates of cerebral cortex with respect to tissue protein concentration, assay time and temperature. Preliminary studies indicated that there was a substantial CRF stimulation of adenylate cyclase in this brain region.

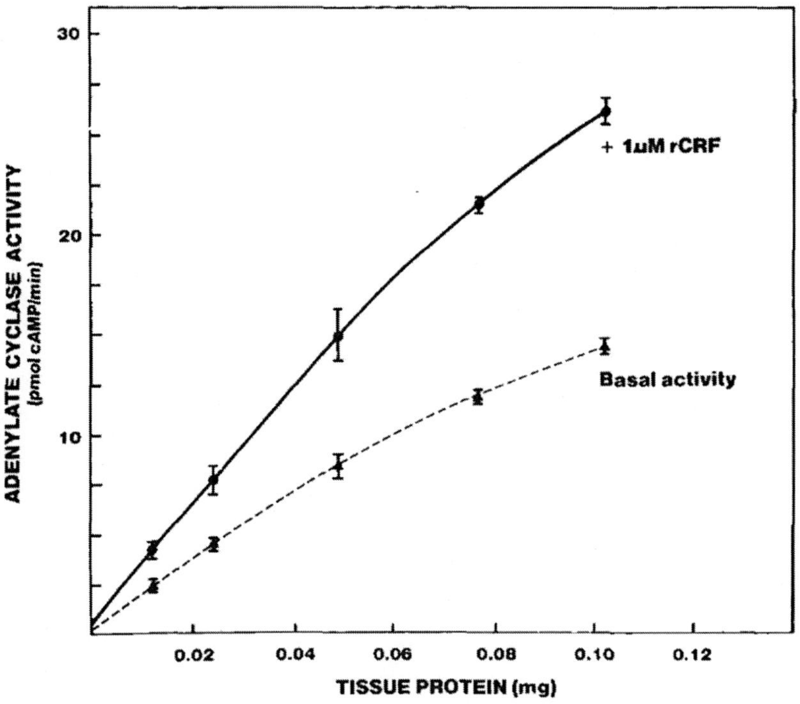

FIGURE 1. Effect of increasing tissue protein concentration on basal and rCRF-stimulated adenylate cyclase activity in brain. Homogenates of rat frontoparietal cortex (0.015 to 0.12 mg protein) were added to tubes containing either 100 μ*M* GTP (-▲-) or 100 μ*M* GTP plus 1 μ*M* rCRF (-●-) and incubations were allowed to proceed for 10 min at 37°C. Data represent the mean ± SEM of values from three independent experiments in which each point was determined in triplicate. (From Battaglia, G., Webster, E. L., and De Souza, E. B., *Synapse*, 1, 572, 1987. With permission.)

1. Tissue Linearity

The correspondence between protein content and CRF-stimulated cAMP production was assessed at 37°C, using a 10-min incubation. As shown in Figure 1, increases in basal and rCRF-stimulated adenylate cyclase activity were linear with tissue protein concentration up to approximately 0.08 mg protein; at 0.1 mg protein, some deviation from linearity was observed for both basal and rCRF-stimulated enzyme activity. Therefore, a concentration of 0.05 mg protein/tube (range 0.04 to 0.06 mg protein) was used in subsequent studies since this amount of tissue provided a reliable stimulation which was midway on the linear portion of the curve.

2. Incubation Time

In determining an incubation time, it is appropriate to choose a maximal period of time during which CRF stimulation of cAMP increases linearly with time. As deviations from linearity might reflect a desensitization of the system, choosing a maximal time prior to such an occurrence serves to maximize the reliability and degree of the stimulatory response. With respect to incubation time (Figure 2), both basal and 1 μ*M* rCRF-stimulated adenylate cyclase activity increased linearly up to 20 min at 37°C. Since some small deviation from linearity occurred at the 20-min time point, an incubation time of 10 min was chosen, as this time period provided both a reliable absolute and percent stimulation by CRF above basal activity.

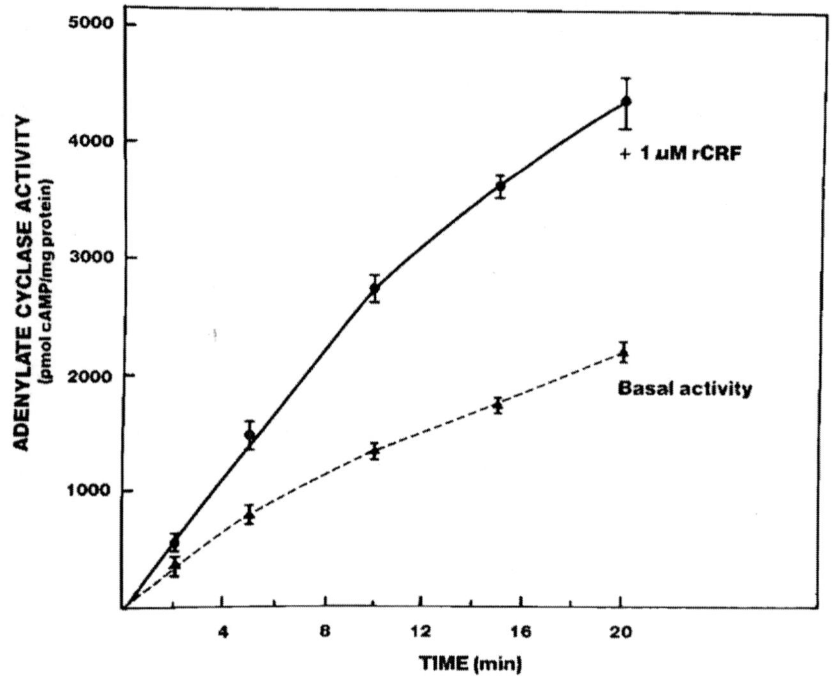

FIGURE 2. Time course of basal and rCRF-stimulated adenylate cyclase activity in homogenates of rat frontoparietal cortex. Tubes containing tissue (50 μg protein) and either 100 μ*M* GTP (-▲-) or 100 μ*M* GTP plus 1 μ*M* rCRF (-●-) were incubated at 37°C for various periods of time (2 to 20 min). Data represent the mean ± SEM of values from three independent experiments in which each point was determined in triplicate. (From Battaglia, G., Webster, E. L., and De Souza, E. B., *Synapse*, 1, 572, 1987. With permission.)

3. Temperature Dependence

Since the optimal temperature for rCRF binding is 22°C, with decreases observed at 37°C,[10] it was necessary to confirm that 37°C was indeed an optimal temperature for rCRF-stimulated adenylate cyclase activity. As shown in Figure 3, progressively lower adenylate cyclase stimulation by rCRF was obtained at 30 and 22°C, respectively. At 30°C, CRF stimulation of enzyme activity was only 60% of that at 37°C, while at 22°C, more marked reductions in CRF-stimulated as well as basal adenylate cyclase activity were observed. Thus, all subsequent studies were carried out at 37°C for 10 min using approximately 40 to 60 μg protein/tube.

4. Effects of Guanine Nucleotides

The role of the guanine nucleotide regulatory protein in CRF-stimulated adenylate cyclase activity was investigated with respect to the concentrations of guanosine 5′-triphosphate (GTP) and its nonhydrolyzable analog, 5′-guanylylimidodiphosphate (Gpp(NH)p), that were required to elicit maximal responses to CRF in terms of an absolute (i.e., pmol cAMP/min/mg protein) and percent stimulation above basal activity.

In cerebral cortex homogenates, both GTP and Gpp(NH)p stimulated baseline enzyme activity (i.e., enzyme activity in the absence of nucleotides) and facilitated the stimulation of adenylate cyclase activity by a maximally effective concentration (1 μ*M*) of rCRF (Figure 4). For comparison purposes, data are presented as a percent of maximal stimulation. With respect to baseline enzyme activity in the absence of nucleotides (67 ± 12 pmol cAMP/min/mg protein), the maximal stimulation by Gpp(NH)p above baseline (598 ± 10 pmol cAMP/min/mg protein) was significantly greater than that by GTP (43 ± 8 pmol cAMP/

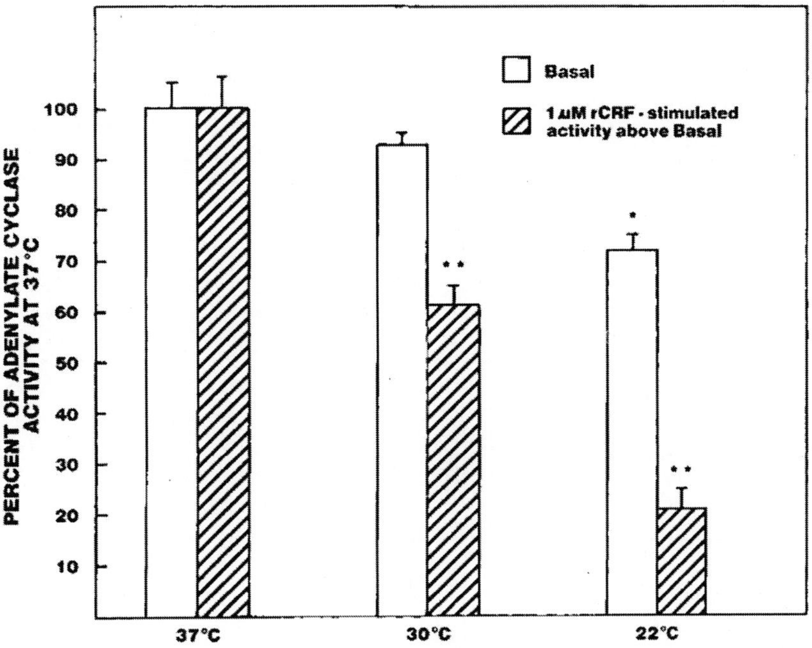

FIGURE 3. Effect of temperature on basal adenylate cyclase activity (□) and CRF-stimulated adenylate cyclase activity above basal (▨) in homogenates of rat frontoparietal cortex. Data are calculated as a percent of the enzyme activity at 37°C; basal activity defined as that in the presence of 100 μM GTP was 133 ± 24 pmol cAMP/min/mg protein. Data represent the mean ± SEM of values from three independent experiments in which each point was determined in triplicate. * and **, significant differences at $p < 0.05$ and $p < 0.01$ from corresponding adenylate cyclase activity at 37°C, respectively; Student's t-test. (From Battaglia, G., Webster, E. L., and De Souza, E. B., *Synapse*, 1, 572, 1987. With permission.)

min/mg protein). Gpp(NH)p was also more potent than GTP with respect to the concentration required to elicit half maximal stimulation of baseline activity (i.e., EC_{50} values of 500 and 1400 nM for Gpp(NH)p and GTP, respectively). Similarly, Gpp(NH)p was more potent than GTP in facilitating the stimulation of adenylate cyclase activity by 1 μM rCRF with respect to the concentration required to elicit half maximal stimulation (EC_{50} values of 40 and 700 nM for Gpp(NH)p and GTP, respectively) as well as the maximal absolute stimulation elicited by 1 μM rCRF (278 ± 25 and 115 ± 7 pmol cAMP/min/mg protein for Gpp(NH)p and GTP, respectively) above basal (i.e., activity in the presence of the respective concentration of nucleotide alone). In the absence of any added guanine nucleotide, 1 μM rCRF-stimulated adenylate cyclase activity was negligible (15 pmol cAMP/min/mg protein). This small stimulation, presumably facilitated by endogenous nucleotides, was subtracted from the amount of CRF-stimulated enzyme activity at each of the various concentrations of nucleotide. These data confirm the obligatory role of a guanine nucleotide regulatory protein in CRF receptor-mediated stimulation of cAMP production and indicate that both GTP and Gpp(NH)p can facilitate CRF stimulation.

B. PHARMACOLOGIC PROFILE OF CRF RECEPTOR MEDIATED cAMP PRODUCTION

1. Stimulation of cAMP by Analogs of CRF

Upon establishing appropriate assay conditions, the pharmacology of the receptor mediating the stimulation of adenylate cyclase by CRF was investigated with respect to the activity of various biologically active and inactive CRF-related peptides (Figures 5 and 6;

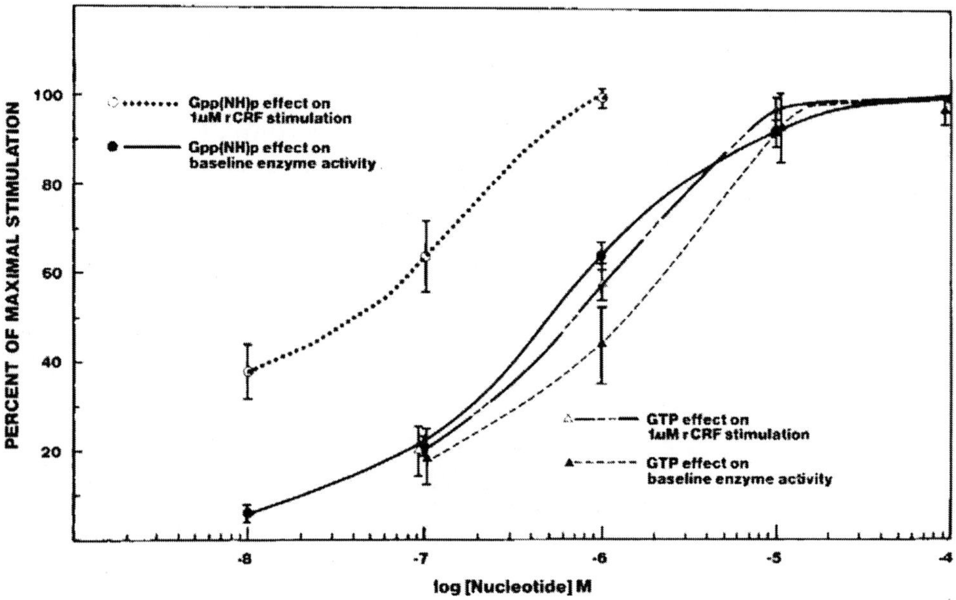

FIGURE 4. Concentration-dependent effects of GTP and its nonhydrolyzable analog, Gpp(NH)p, on the stimulation above baseline adenylate cyclase activity (no nucleotide) and the stimulation by 1 μM rCRF above that elicited by various concentrations of the respective nucleotides. The stimulation by 1 μM rCRF in the absence of nucleotides (~15 pmol/cAMP/min/mg protein) was subtracted from the CRF stimulation at all concentrations of nucleotides. Baseline enzyme activity defined as the activity in the absence of any nucleotides was 67 ± 12 pmol cAMP/min/mg protein while the activity in the presence of 100 μM GTP was 126 ± 6 pmol cAMP/min/mg protein. Enzyme activity in the presence of 100 μM Gpp(NH)p was 723 ± 20 pmol cAMP/min/mg protein. Stimulation by 1 μM rCRF above that in the presence of 1 μM Gpp(NH)p or 100 μM GTP was 278 ± 25 and 115 ± 7 pmol cAMP/min/mg protein, respectively. Data represent the mean ± SEM of values from three independent experiments in which each point was determined in triplicate. (From Battaglia, G., Webster, E. L., and De Souza, E. B., *Synapse*, 1, 572, 1987. With permission.)

Table 1). These studies were carried out in cerebral cortical homogenates in the presence of 100 μM GTP. The rank order of potency of various biologically active compounds in stimulating adenylate cyclase activity was [$Nle^{21,38}$] rCRF > rCRF ≃ acetyl oCRF(4-41) ≃ alpha helical oCRF > oCRF (Figure 5; Table 1). Nearly a sixfold difference in potency was observed between the most potent ([$Nle^{21,38}$]rCRF) and least potent (oCRF) peptide. Two biologically weak compounds, oCRF(1-39) and oCRF(7-41) did not significantly stimulate enzyme activity at concentrations up to 1 μM (Figure 5). The rank order of potency in stimulating adenylate cyclase activity by the various CRF-related peptides was consistent with their affinities for CRF receptors in brain[10] and their relative potencies in stimulating pituitary ACTH secretion *in vitro*.[18,19]

It was of interest to note that although rCRF, acetyl oCRF(4-41), and alpha helical oCRF had comparable half-maximal effective concentration (EC_{50}) values (Table 1), differences in the shape of their concentration-dependent stimulation curves were observed (Figure 5). The response for alpha-helical oCRF appeared to be more shallow than that for either rCRF or acetyl oCRF (4-41). While most of the biologically active CRF compounds elicited their maximal stimulation of adenylate cyclase at 1 μM concentrations and had comparable maximum initial velocity (V_{max}) values, the maximal stimulation of adenylate cyclase elicited by alpha-helical oCRF was markedly lower (Table 1). These data indicate that not all CRF-related peptides exhibit comparable intrinsic activities with respect to their stimulation of adenylate cyclase activity.

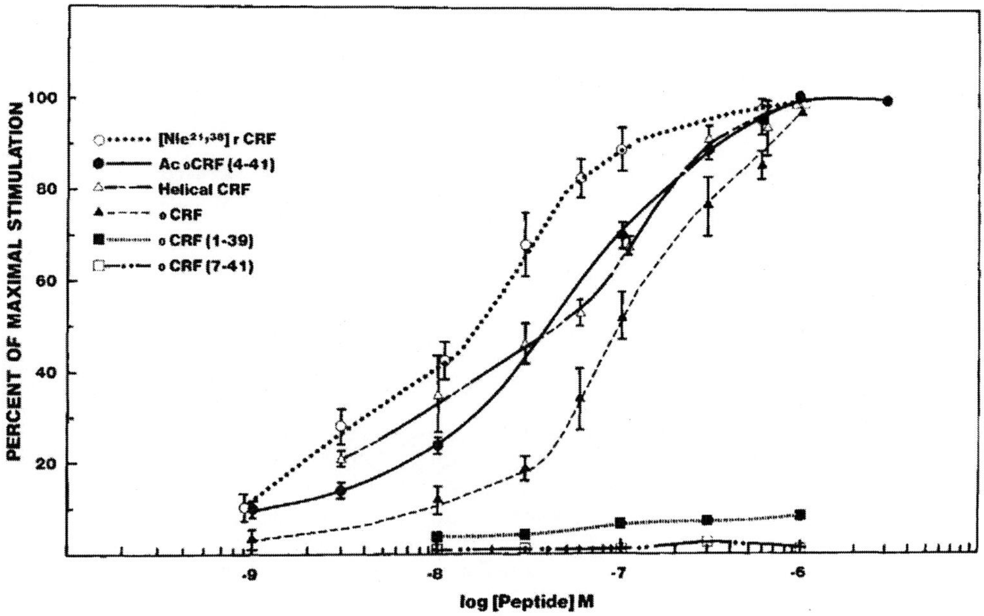

FIGURE 5. Pharmacologic profile of CRF receptor-mediated stimulation of adenylate cyclase activity in homogenates of rat frontoparietal cortex. Data are expressed as a percent of the maximal stimulation above basal enzyme activity elicited by 1 to 3 μ*M* of CRF or the respective CRF-related peptides. Basal activity (activity in the presence of 100 μ*M* GTP) was approximately 110 pmol cAMP/min/mg protein. Each point represents the mean ± SEM of data from three independent experiments in which each concentration point was determined in triplicate. (From Battaglia, G., Webster, E. L., and De Souza, E. B., *Synapse*, 1, 572, 1987. With permission.)

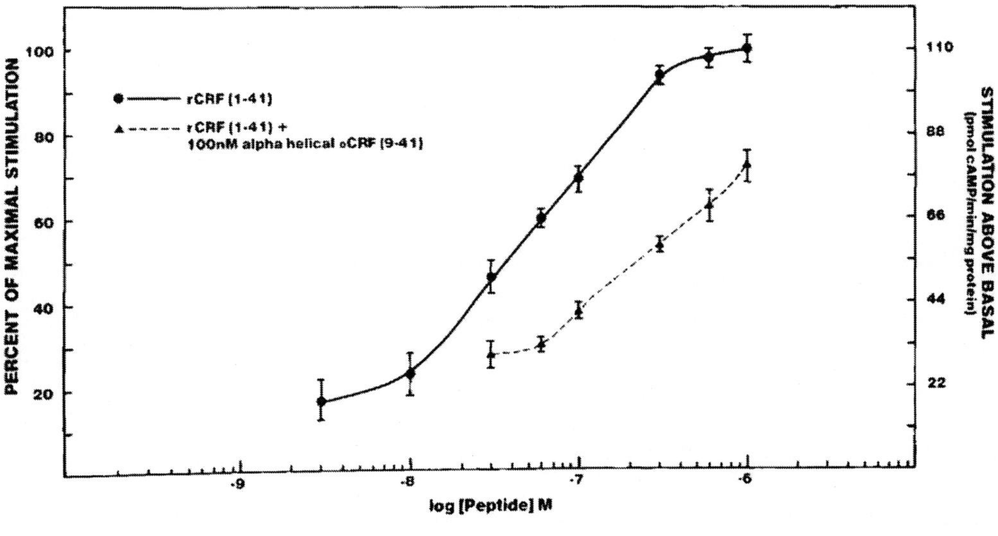

FIGURE 6. Rat CRF stimulation of adenylate cyclase activity in frontoparietal cortex homogenates in the absence and presence of 100 n*M* of the CRF receptor antagonist alpha helical oCRF(9-41). Data are expressed as a percent of the maximal stimulation by 1 μ*M* rCRF above basal and represent the mean ± SEM of values from three independent experiments in which incubations in the absence and presence of antagonist were carried out in parallel. (From Battaglia, G., Webster, E. L., and De Souza, E. B., *Synapse*, 1, 572, 1987. With permission.)

TABLE 1
Stimulation of Adenylate Cyclase Activity in Rat Frontoparietal Cortex by Various CRF-Related Peptides

Peptide	EC_{50} (n*M*)	V_{max} (pmol cAMP/min/mg protein)
[$Nle^{21,38}$] rat CRF	16 ± 5	110 ± 6
rat CRF	36 ± 4	115 ± 7
Acetyl ovine CRF(4-41)	38 ± 2	98 ± 9
Alpha-helical ovine CRF	43 ± 12	36 ± 7
ovine CRF	100 ± 21	121 ± 22
ovine CRF(1-39)	>1000	ND
ovine CRF(7-41)	>1000	ND

Note: Data represent the mean and SEM of values calculated from at least three independent concentration-response curves in which each concentration was determined in triplicate. ND indicates that there was no significant detectable stimulation at peptide concentrations up to 1 μ*M*. Basal activity (in the presence of 100 μ*M* GTP) was approximately 110 pmol cAMP/min/mg protein. V_{max} values represent the stimulation above basal activity.

From Battaglia, G., Webster, E. L., and De Souza, E. B., *Synapse*, 1, 572, 1987. With permission.

2. Effects of a CRF Receptor Antagonist

In order to examine the receptor specificity of these peptides, we investigated the effects of the weak receptor antagonist, alpha-helical oCRF(9-41), on the stimulation of enzyme activity by rCRF. A concentration of 100 n*M* alpha-helical oCRF(9-41) was without effect on adenylate cyclase activity in the absence of rCRF. As shown in Figure 6, the percent stimulation by various concentrations of rCRF in the presence of 100 n*M* alpha helical oCRF(9-41) was attenuated with respect to the percent stimulation by rCRF alone. The attenuation of rCRF-stimulated adenylate cyclase activity by alpha-helical oCRF(9-41) was evidenced by an apparent fivefold shift in the EC_{50} for rCRF stimulation, from 36 to 190 n*M* in the absence vs. the presence of 100 n*M* alpha-helical oCRF(9-41), respectively (Figure 6). These data indicate that both compounds interact in a competitive manner with the same receptor population.

C. REGIONAL CORRESPONDENCE BETWEEN CRF-STIMULATED ADENYLATE CYCLASE ACTIVITY AND CRF RECEPTORS

In order to assess the correspondence between receptor number and cAMP production, CRF-stimulated adenylate cyclase activity and ^{125}I-Tyr^{0}-ovine CRF binding were measured in parallel in various regions of rat CNS.

1. CRF-Stimulated cAMP Production

As shown in Table 2, region-specific differences in the maximal stimulation of adenylate cyclase activity were observed with both rat/human and ovine CRF. While basal activity varied by up to nearly threefold, differences in basal enzyme activities in the brain areas examined did not correspond to the regional differences observed for maximal CRF-stimulated activities. For example, basal activities were comparable in cerebral cortex and spinal cord, while CRF-stimulated cAMP production was greatest in cerebral cortex but negligible in spinal cord (Table 2; Figure 7A). With respect to the percent stimulation above basal enzyme activity, CRF stimulation was likewise greatest in cerebral cortex and represented nearly a twofold increase (100%). The smallest percent increase over basal activity, observed

TABLE 2
CRF Receptor-Mediated Adenylate Cyclase Activity in Various Regions of Rat CNS

	Adenylate cyclase activity (pmol cAMP/min/mg protein)		
Region	**Basal**	**+ 1μ*M* rCRF**	**+ 1μ*M* oCRF**
Frontoparietal cortex	114 ± 9	209 ± 18 (95 ± 10)	206 ± 16 (92 ± 12)
Olfactory bulb	106 ± 13	161 ± 18 (54 ± 6)	156 ± 19 (48 ± 7)
Cerebellum	135 ± 10	182 ± 15 (47 ± 5)	182 ± 15 (47 ± 7)
Midbrain	192 ± 16	217 ± 17 (25 ± 1)	218 ± 17 (26 ± 2)
Hippocampus	116 ± 4	136 ± 4 (20 ± 1)	134 ± 5 (18 ± 2)
Striatum	162 ± 5	178 ± 7 (17 ± 3)	178 ± 8 (17 ± 5)
Hypothalamus	276 ± 3	288 ± 2 (13 ± 3)	290 ± 4 (14 ± 4)
Spinal cord	110 ± 19	118 ± 18 (8 ± 4)	119 ± 20 (9 ± 3)

Note: Adenylate cyclase activity in various regions of the CNS in the presence of 100 μ*M* GTP (basal), 100 μ*M* GTP plus 1 μ*M* rat CRF or 100 μ*M* GTP plus 1 μ*M* ovine CRF. Data represent the mean ± SEM of values from three or four experiments in which triplicate determinations of each point in all regions were assayed in parallel. Data in parentheses represent the mean and SEM of four or five individual values of net CRF stimulation above basal activity calculated from all experiments.

From Battaglia, G., Webster, E. L., and De Souza, E. B., *Synapse*, 1, 572, 1987. With permission.

in hypothalamus (~6%), reflected both the small degree of absolute stimulation by CRF (15 pmol cAMP/min/mg protein) and the high basal enzyme activity (276 ± 3 pmol cAMP/min/mg protein) in this region. For comparative purposes, the stimulation by 1 μ*M* rCRF above basal activity in various regions of rat CNS was plotted as a percent of that observed in frontoparietal cortex, a region with the greatest CRF-stimulated enzyme activity. As shown in Figure 7A, olfactory bulb and cerebellum exhibited the next greatest degree of rCRF-stimulated adenylate cyclase activity followed by progressively lower stimulation in midbrain, hippocampus, striatum, hypothalamus, and spinal cord, respectively.

2. CRF RECEPTOR DENSITIES

In contrast with the CRF stimulation of adenylate cyclase activity noted in various regions of rat CNS, a somewhat different profile in ^{125}I-Tyr°-ovine CRF specific binding was observed in the brain regions described above. The highest concentration of CRF receptors was localized to olfactory bulb (~twofold greater than cerebral cortex) followed by cerebral cortex and cerebellum (with comparable numbers of receptors), and progressively lower binding in hypothalamus, striatum, midbrain, hippocampus, and spinal cord, respectively (Figure 7B). Although these studies were carried out using a single concentration of ^{125}I-Tyr°-ovine CRF, saturation studies of CRF binding in these areas indicated little regional difference in receptor affinity for CRF (see Chapter 9).

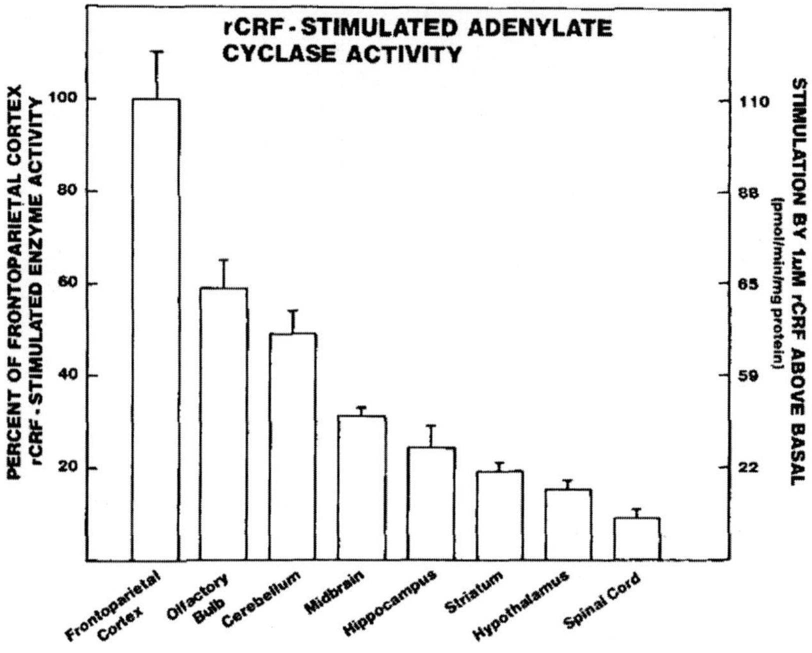

A

FIGURE 7. (A) Stimulation of adenylate cyclase activity by 1 µ*M* rCRF above basal enzyme activity in discrete regions of rat CNS. Incubations were carried out for 10 min at 37°C as described in Methods and the data are calculated as a percent of the maximal stimulation obtained in frontoparietal cortex. (B) Distribution of CRF receptors labeled by ^{125}I-Tyr^{o}-oCRF in discrete regions of rat CNS. Specific binding of ^{125}I-Tyr^{o}-oCRF was determined as outlined in Chapter 9, and data are calculated as a percent of the amount of binding in frontoparietal cortex homogenates. (C) Regional differences in CRF Activity Indices in rat CNS. CRF Activity Indices were calculated from CRF-stimulated adenylate cyclase activities (pmol cAMP/min/mg protein) and CRF receptor densities (fmol/mg protein) in respective regions of CNS. Data are plotted as a percentage of the CRF Activity Index determined in frontoparietal cortex. (From Battaglia, G., Webster, E. L., and De Souza, E. B., *Synapse*, 1, 572, 1987. With permission.)

3. CRF Activity Index

An additional means of assessing the degree of association between CRF receptors and the stimulatory response to CRF in various regions of CNS can be elucidated by calculation of the stimulation of cAMP production per unit CRF receptor rather than per milligram of protein since both adenylate cyclase activities and CRF receptor densities were originally calculated per unit protein in each region. As shown in Figure 7C such an index which we refer to as the CRF Activity Index (i.e., fmol cAMP/min/fmol CRF receptor) reveals a large disparity between cerebral cortex and olfactory bulb, two regions with less marked differences in CRF stimulation of cAMP per unit protein. Other regions with far smaller degrees of cAMP production than olfactory bulb (e.g., midbrain, hippocampus, and striatum) reveal CRF Activity Indices greater than olfactory bulb with some being closer to that observed in cerebral cortex. These data suggest that in some brain regions such as cerebral cortex CRF receptors may be exclusively or predominantly associated with a stimulatory response for cAMP while in other brain regions (e.g., olfactory bulb) a subpopulation of CRF receptors may be more closely associated with an alternative second messenger function (e.g., inhibition of adenylate cyclase or phosphoinositide metabolism).

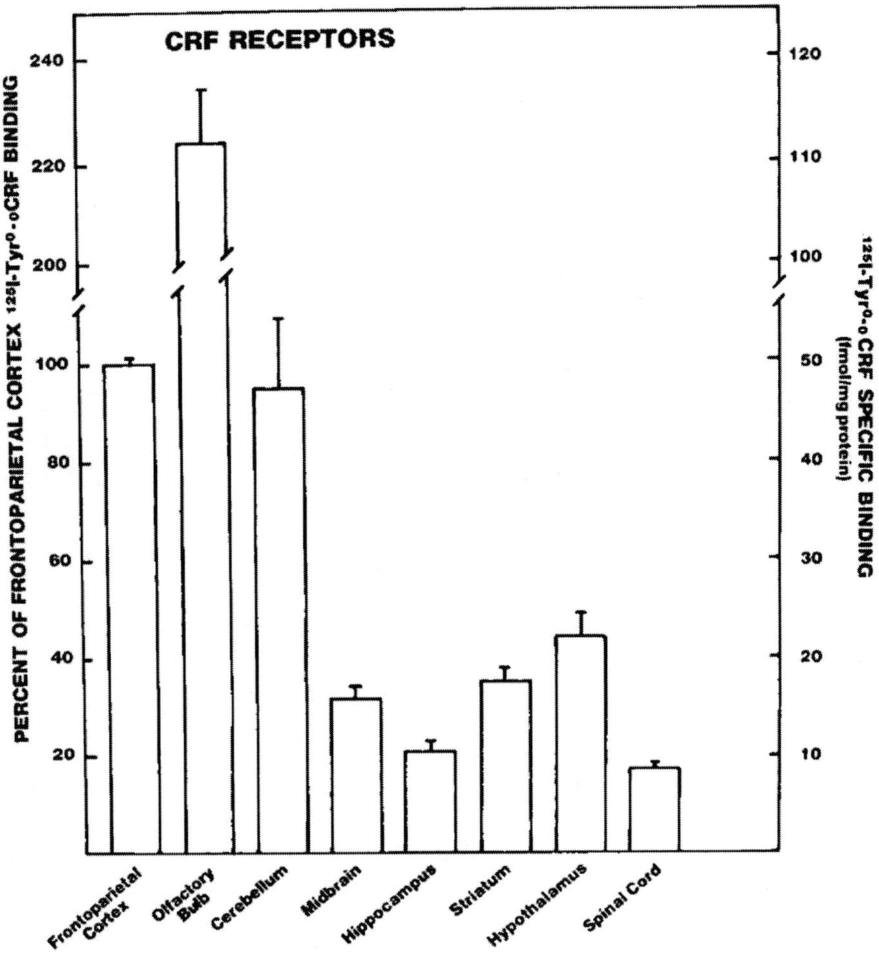

FIGURE 7B.

IV. REGULATION OF CRF-STIMULATED ADENYLATE CYCLASE ACTIVITY IN RAT BRAIN

A. *IN VITRO* POTENTIATION BY FORSKOLIN AND ATTENUATION BY Gpp(NH)p OF CRF-STIMULATED cAMP PRODUCTION

Since regional studies indicated some degree of noncorrespondence between the density of CRF receptors and the degree of CRF-stimulated enzyme activity, it was of interest to investigate whether CRF receptors and/or CRF-stimulation of cAMP in various brain regions were subject to similar regulatory responses *in vitro*. One means for assessing similarities or differences in the functional coupling of CRF receptors with the catalytic unit of adenylate cyclase can be elucidated by examining the effects of the potent adenylate cyclase modulator, forskolin. Forskolin is a unique diterpene activator of adenylate cyclase activity which acts directly at the catalytic subunit and has also been shown in some systems to facilitate and potentiate hormonal stimulation.[20] As shown in Table 3 and Figure 8, we observed region-specific differences in the stimulation of catalytic subunit activity and in the ability of forskolin to potentiate CRF-stimulated adenylate cyclase activity. In the absence of CRF, forskolin stimulation of adenylate cyclase activity was comparable in most of the CNS regions examined with markedly greater stimulation observed only in the striatum (Table 3). In contrast, with respect to receptor-mediated enzyme activity, forskolin potentiated the

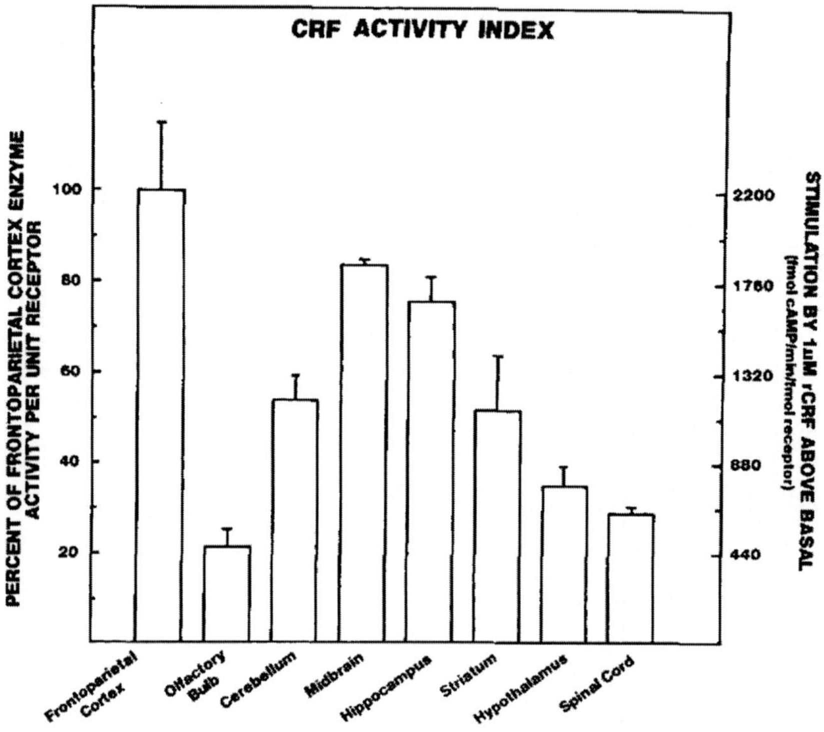

FIGURE 7C.

TABLE 3
Effects of Forskolin and CRF on Adenylate Cyclase Activity in Discrete Areas of Rat CNS

	Activity (pmol cAMP/min/mg protein)			
Region	**Basal**	**+10 μ*M* Forskolin**	**+10 μ*M* Forskolin + 1μ*M* rCRF**	**Stimulation by forskolin alone (% basal activity)**
Frontoparietal cortex	114 ± 9	1000 ± 83	1247 ± 99	854 ± 48
Olfactory bulb	101 ± 9	941 ± 52	1020 ± 70	953 ± 53
Cerebellum	140 ± 11	1018 ± 82	1268 ± 110	729 ± 14
Midbrain	198 ± 19	985 ± 130	1087 ± 113	490 ± 38
Hippocampus	116 ± 6	867 ± 70	897 ± 67	897 ± 67
Striatum	167 ± 5	3752 ± 247	3900 ± 232	2245 ± 149
Hypothalamus	281 ± 6	969 ± 19	1017 ± 34	344 ± 5
Spinal cord	134 ± 15	623 ± 144	670 ± 134	494 ± 33

Note: Comparison of adenylate cyclase activity in various regions of rat CNS in the absence (basal), presence of 10 μ*M* forskolin and 10 μ*M* forskolin plus 1 μ*M* rCRF. 100 μ*M* GTP was included in all assay tubes. Data represent the mean ± SEM of values from three to five experiments in which each point was determined in triplicate.

From Battaglia, G., Webster, E. L., and De Souza, E. B., *Synapse*, 1, 572, 1987. With permission.

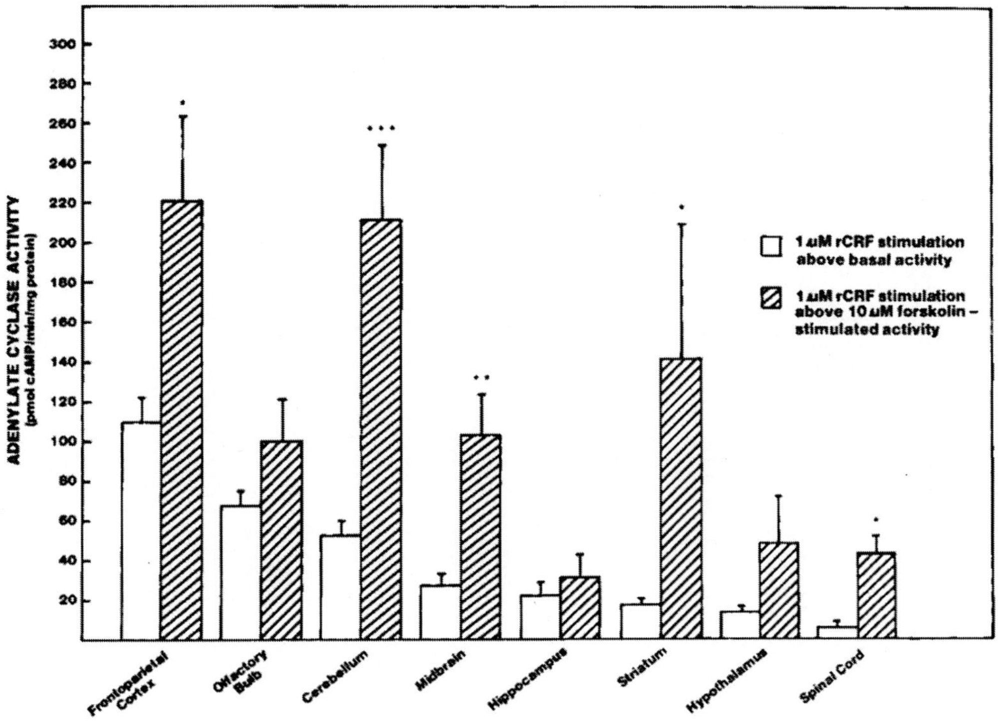

FIGURE 8. Potentiation by forskolin of 1 μ*M* rCRF-stimulated adenylate cyclase activity in discrete regions of rat CNS. Data represent the stimulation of adenylate cyclase activity by 1 μ*M* rCRF above basal (+100 μ*M* GTP) and that in the presence of 10 μ*M* forskolin (+100 μ*M* GTP). Values represent mean ± SEM of data from three or four experiments. Cyclic AMP production by 1 μ*M* rCRF was significantly potentiated by forskolin in frontoparietal cortex, cerebellum, midbrain, striatum and spinal cord as determined by a paired Student's t-test. The stimulation of adenylate cyclase activity by 10 μ*M* forskolin alone (in the absence of CRF) varied from 3.5 times basal activity in hypothalamus to 26 times basal activity in striatum. *, **, and ***, significant differences from corresponding control group (i.e., no forskolin) at $p < 0.05$, $p < 0.025$, and $p < 0.005$, respectively. (From Battaglia, G., Webster, E. L., and De Souza, E. B., *Synapse,* 1, 572, 1987. With permission.)

stimulation by CRF in only some regions of rat CNS; significant potentiation was observed in cerebral cortex, cerebellum, midbrain, striatum, and spinal cord (Figure 8). In these regions, the effect of forskolin was to increase the intrinsic activity (i.e., V_{max}) of rCRF-stimulated adenylate cyclase activity while it was ineffective in the other regions examined. These data, demonstrating regional differences in the effects of forskolin on CRF stimulation of adenylate cyclase activity, suggest that there may be subtle differences in the interaction between CRF receptors and the stimulatory catalytic subunit in different regions of rat CNS.

Since CRF receptor-mediated stimulation of adenylate cyclase activity is greater in the presence of maximally effective concentration of Gpp(NH)p, it appears that like forskolin, Gpp(NH)p can increase the intrinsic activity of CRF. However, the effects of Gpp(NH)p on CRF stimulation of cAMP are somewhat different from those of forskolin in that at a higher concentration (100 μ*M*) of Gpp(NH)p, the absolute stimulation by CRF is attenuated by 35% compared with the absolute CRF stimulation observed at 1 μ*M* Gpp(NH)p. This attenuated response was specific for CRF as the stimulation by 1 μ*M* vasoactive intestinal peptide (VIP) was not decreased by increasing concentrations of Gpp(NH)p (Figure 9). These data indicate that while GTP and Gpp(NH)p can each facilitate CRF stimulation of adenylate cyclase activity, some significant differences may exist in the mechanism by which each of the nucleotides promote the interaction between CRF receptors and guanine nucleotide regulatory proteins which affect catalytic unit activity.

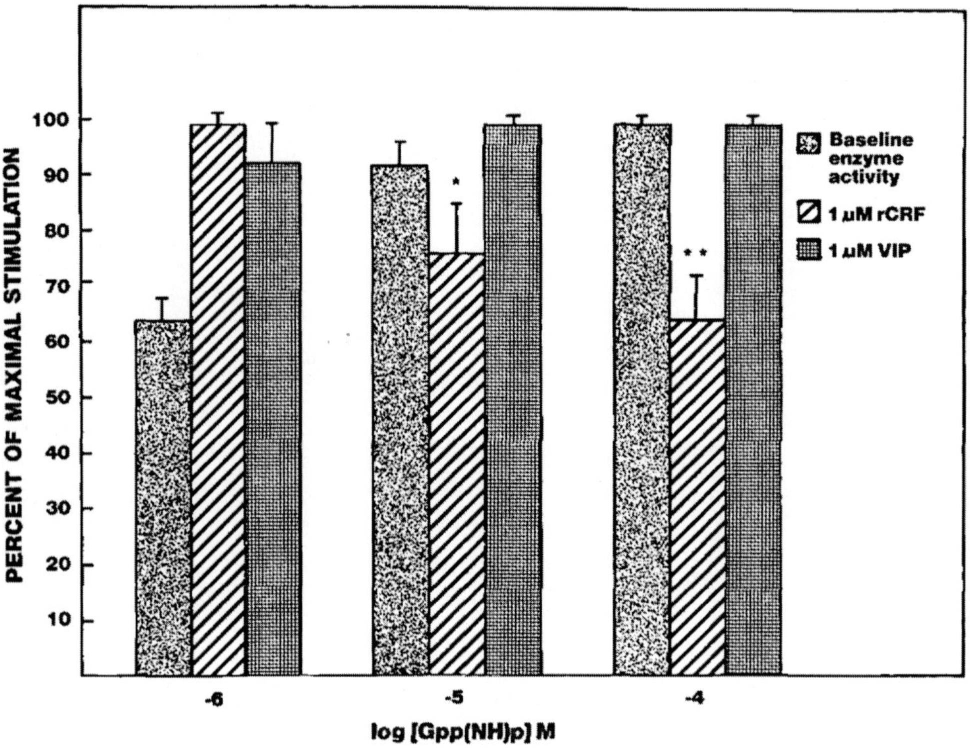

FIGURE 9. Selective attenuation of 1 µ*M* rCRF-stimulated cAMP production in rat frontoparietal cortex homogenates by high concentrations of Gpp(NH)p. Data are presented as a percent of the maximal stimulation elicited by Gpp(NH)p above baseline activity (no nucleotide) and as a percent of the maximal stimulation by 1 µ*M* rCRF or 1 µ*M* VIP above that elicited by Gpp(NH)p alone. Maximal stimulation by 100 µ*M* Gpp(NH)p above baseline activity was 656 ± 28 pmol cAMP/min/mg protein while the stimulation by 1 µ*M* VIP above that in the presence of 100 µ*M* Gpp(NH)p was 283 ± 32 pmol cAMP/min/mg protein. Maximal stimulation by 1 µ*M* rCRF (278 ± 25 pmol cAMP/min/mg) above that of the nucleotide alone was observed at 1 µ*M* Gpp(NH)p. At higher Gpp(NH)p concentrations (10 and 100 µ*M*) 1 µ*M* rCRF-stimulated cAMP production was significantly [($p < 0.05$ (*) and $p < 0.01$ (**), respectively; Student's t-test)] attenuated. The data represent the mean ± SEM of values from four experiments in which each point was determined in triplicate. (From Battaglia, G., Webster, E. L., and De Souza, E. B., *Synapse*, 1, 572, 1987. With permission.)

Gpp(NH)p can also facilitate receptor-mediated inhibition of adenylate cyclase activity for receptors which are coupled in an inhibitory fashion to the catalytic subunit.[21] Since higher concentrations of Gpp(NH)p (1 µ*M*) attenuate CRF stimulation of adenylate cyclase activity, one can speculate that the attenuation in enzyme activity may be due to Gpp(NH)p facilitating the interaction of CRF with receptors coupled to inhibitory components of guanine nucleotide regulatory proteins. This hypothesis presumes the existence of a subpopulation of "inhibitory" CRF receptors in the cerebral cortex. Alternatively, higher concentrations of Gpp(NH)p may exert effects on a guanine nucleotide regulatory protein involved in phosphoinositide metabolism which in turn may alter the ability of CRF to increase cAMP production. However, such a mechanism for the effect of Gpp(NH)p would likely result in nonspecific membrane effects which would also attenuate VIP stimulation. The specificity of the Gpp(NH)p attenuation of CRF-stimulated adenylate cyclase activity strongly indicates that this is not due to a generalized membrane effect and suggests that there may be a subpopulation of CRF receptors not coupled to a stimulatory guanine nucleotide regulatory protein.

V. SUMMARY AND CONCLUSIONS

In summary, CRF stimulates adenylate cyclase activity in rat CNS. Stimulation by CRF is dependent on time, temperature, tissue protein concentration, and guanine nucleotides. The rank order of potency for CRF analogs and fragments in stimulating adenylate cyclase activity [($Nle^{21,38}$) rCRF > rCRF ≃ acetyl oCRF(4-41) ≃ alpha-helical oCRF > oCRF > oCRF(1-39) ≃ oCRF(7-41)] is consistent with their affinities for CRF receptors in brain and their relative potencies in stimulating pituitary ACTH secretion *in vitro*. The regional distribution of ^{125}I-Tyr^{0}-ovine CRF binding (olfactory bulb > cerebral cortex ≃ cerebellum > hypothalamus > striatum ≥ midbrain > hippocampus ≥ spinal cord) does not correspond with the regional degree of CRF receptor-mediated stimulation of adenylate cyclase (cerebral cortex > olfactory bulb ≥ cerebellum > midbrain ≥ hippocampus > striatum ≥ hypothalamus > spinal cord). Regional differences in the ability of forskolin to potentiate CRF-stimulated enzyme activity suggest region-specific differences in the sensitivity to *in vitro* regulation. Marked regional differences in CRF Activity Indices and the attenuation of CRF stimulation by high concentrations of Gpp(NH)p suggest that in addition to the stimulatory effect of CRF on cAMP production, CRF receptors may mediate some effects via other mechanisms. These data provide further support for the putative neurotransmitter role of CRF in the CNS and should provide the basis for understanding how CRF may mediate its effects under physiologic or pathologic conditions.

ACKNOWLEDGMENTS

We would like to thank Dr. Paul Manis for providing us with a computer program for the calculation of adenylate cyclase activities. We are especially grateful to Dr. Jean Rivier of the Salk Institute, Peptide Biology Laboratory for providing CRF analogs and fragments. We would also like to thank Brian Kuyatt for excellent technical assistance and Terrie Pierce and Mary Flutka for manuscript preparation.

REFERENCES

1. **Wynn, P. C., Aguilera, G., Morell, J., and Catt, K. J.,** Properties and regulation of high-affinity pituitary receptors for corticotropin-releasing factor, *Biochem. Biophys. Res. Commun.*, 110, 602, 1983.
2. **De Souza, E. B., Perrin, M. H., Rivier, J. E., Vale, W. W., and Kuhar, M. J.,** Corticotropin-releasing factor receptors in rat pituitary gland: autoradiographic localization, *Brain Res.*, 296, 202, 1984.
3. **De Souza, E. B., Perrin, M. H., Whitehouse, P. J., Rivier, J. E., Vale, W. W., and Kuhar, M. J.,** Corticotropin-releasing factor receptors in human pituitary gland: autoradiographic localization, *Neuroendocrinology*, 40, 419, 1985.
4. **De Souza, E. B. and Kuhar, M. J.,** Corticotropin-releasing factor receptors in the pituitary gland and central nervous system: methods and overview, *Methods Enzymol.*, 124, 560, 1986.
5. **Giguere, V., Labrie, F., Cote, J., Coy, D. H., Sueiras-Diaz, Jr., and Schally, A. V.,** Stimulation of cyclic AMP accumulation and corticotropin release by synthetic ovine corticotropin-releasing factor in rat anterior pituitary cells: site of glucocorticoid action, *Proc. Natl. Acad. Sci. U.S.A.*, 79, 3466, 1982.
6. **Bilezikjian, L. M. and Vale, W.,** Glucocorticoids inhibit corticotropin-releasing factor-induced production of adenosine 3′,5′-monophosphate in cultured anterior pituitary cells, *Endocrinology*, 113, 657, 1983.
7. **Meunier, H., Lefevre, G., Dumont, D., and Labrie, F.,** CRF stimulates alpha-MSH secretion and cyclic AMP accumulation of rat pars intermedia cells, *Life Sci.*, 31, 2129, 1982.
8. **De Souza, E. B., Perrin, M. H., Insel, T. R., Rivier, J., Vale, W. W., and Kuhar, M. J.,** Corticotropin-releasing factor receptors in rat forebrain: autoradiographic identification, *Science*, 224, 1449, 1984.
9. **De Souza, E. B., Insel, T. R., Perrin, M. H., Rivier, J., Vale, W. W., and Kuhar, M. J.,** Corticotropin-releasing factor receptors are widely distributed within the rat central nervous system: an autoradiographic study, *J. Neurosci.*, 5, 3189, 1985.

10. **De Souza, E. B.,** Corticotropin-releasing factor receptors in the rat central nervous system: characterization and regional distribution, *J. Neurosci.*, 7, 88, 1987.
11. **Wynn, P. C., Hauger, R. L., Holmes, M. C., Millan, M. A., Catt, K. J., and Aguilera, G.,** Brain and pituitary receptors for corticotropin-releasing factor: localization and differential regulation after adrenalectomy, *Peptides*, 5, 1077, 1984.
12. **Millan, M. A., Jacobowitz, D. M., Hauger, R. L., Catt, K. J., and Aguilera, G.,** Distribution of corticotropin-releasing factor receptors in primate brain, *Proc. Natl. Acad. Sci. U.S.A.*, 83, 1921, 1986.
13. **De Souza, E. B., Whitehouse, P. J., Kuhar, M. J., Price, D. L., and Vale, W. W.,** Reciprocal changes in corticotropin-releasing factor (CRF)-like immunoreactivity and CRF receptors in cerebral cortex of Alzheimer's disease, *Nature*, 319, 593, 1986.
14. **Chen, F. M., Bilezikjian, L. M., Perrin, M. H., Rivier, J., and Vale, W.,** CRF receptor-mediated stimulation of adenylate cyclase activity in rat brain, *Brain Res.*, 381, 49, 1986.
15. **Battaglia, G., Webster, E. L., and De Souza, E. B.,** Characterization of corticotropin-releasing factor (CRF) receptor-mediated stimulation of adenylate cyclase activity in rat brain, *Synapse*, 1, 572, 1987.
16. **Lowry, O. H., Rosebrough, N. J., Farr, A. L., and Randall, R. J.,** Protein measurement with the Folin phenol reagent, *J. Biol. Chem.*, 193, 265, 1951.
17. **Salomon, Y.,** Adenylate cyclase assay, *Adv. Cyclic Nucleotide Res.*, 10, 35, 1979.
18. **Vale, W., Spiess, J., Rivier, C., and Rivier, J.,** Characterization of a 41-residue ovine hypothalamic peptide that stimulates secretion of corticotropin and beta-endorphin, *Science*, 213, 1394, 1981.
19. **Rivier, J., Rivier, C., and Vale, W.,** Synthetic competitive antagonists of corticotropin-releasing factor: effect on ACTH secretion in the rat, *Science*, 224, 889, 1984.
20. **Seamon, K. B. and Daly, J. W.,** Forskolin: a unique diterpene activator of cyclic AMP-generating systems, *J. Cyclic Nucleotide Res.*, 7, 201, 1981.
21. **Steer, M. L. and Wood, A.,** Regulation of human platelet adenylate cyclase by epinephrine, prostaglandin E_1, and guanine nucleotides, *J. Biol. Chem.*, 254, 10791, 1979.

Chapter 11

RECEPTORS FOR CORTICOTROPIN-RELEASING FACTOR

Greti Aguilera, Mirza Flores, Pilar Carvallo, James P. Harwood, Monica Millan, and Kevin J. Catt

TABLE OF CONTENTS

I. INTRODUCTION

Since the characterization of corticotropin releasing factor (CRF) in 1981, evidence has accumulated to indicate that the hypothalamic peptide plays an important role in the regulation of ACTH secretion as well as in mediating visceral and behavioral responses to stress.[1,2] The peptide is a potent stimulus of ACTH and β-endorphin secretion from the anterior pituitary and also of the POMC derived peptides α-MSH, β-endorphin, and corticotropin-like intermediate lobe peptide (CLIP) from the intermediate lobe of the pituitary. In addition, synthetic CRF has been shown to have direct actions in the brain that mimic physiological responses to stress. The initial event in the action of peptide hormones is the binding to a plasma membrane receptor, and knowledge of the CRF receptor properties is crucial in understanding the mode of interaction of the peptide with its target tissues. Receptors for CRF were first identified in rat pituitary membranes using radioiodinated Tyr-oCRF, and subsequent studies by radioassays in membranes, autoradiographic procedures, and cytochemical techniques using biotinylated or fluorescein-conjugated CRF analogs have facilitated the identification and characterization of CRF receptors in the central and peripheral nervous systems, which has contributed to our understanding of the physiological actions of CRF.[3-10]

In all target tissues studied CRF receptors are coupled to adenylate cyclase, and at least in the pituitary, increases in cyclic AMP are recognized as the major mediator of the cell response.[11,12] Other intracellular messenger systems, such as calcium and arachidonic acid metabolites have also been implicated in the mechanism of action of CRF, and their importance relative to the cyclic AMP system may vary in different target tissues.[13-16]

Although the binding characteristics of the CRF receptor are similar in all tissues, it is not clear whether the receptor is structurally the same. Little is known about the physicochemical characteristics of the CRF receptor and further studies in this area are needed to define specific properties of the receptor in the individual tissues. This chapter will review the current knowledge of the properties, distribution, mechanism of action, and regulation of CRF receptors in the different target tissues.

II. PITUITARY CRF RECEPTORS

A. CELLULAR LOCALIZATION AND BINDING PROPERTIES

Receptors for CRF have been identified and characterized in the pituitary gland from rat and a number of primates.[3-5,17,18] In all species, autoradiographic analysis of the binding of radiolabeled CRF have shown the receptors to be localized in the anterior and intermediate lobes, with no staining in the neural lobe (Figure 1). In rat pituitary, CRF binding is associated with dense clusters homogeneously distributed throughout the anterior lobe. In primates there are marked differences in the concentration and distribution of receptors within species. In the cynomolgus monkey, confluent clusters of high binding density are homogeneously distributed throughout the anterior lobe while in the human pituitary, such clusters are found preferentially in the anteromedial portion of the anterior pituitary. In contrast to human and cynomolgus monkey, the pituitary gland of the marmoset, a new world primate, exhibits very high binding density in the intermediate lobe and sparsely scattered clusters in the anterior lobe.

In spite of differences in distribution, the binding properties of CRF receptors are similar in pituitary membranes from rat, marmoset, and cynomolgus monkey. The specific binding of ^{125}I-Tyr-oCRF to anterior pituitary membrane-rich fractions is time- and temperature-dependent, and reaches a steady state within 30 to 40 min at 22°C. Scatchard analysis of equilibrium binding data shows a single class of CRF receptors with K_d of 10^{-9} *M*. The number of sites, after correction for the binding activity of the tracer, is about 500 fmol/mg

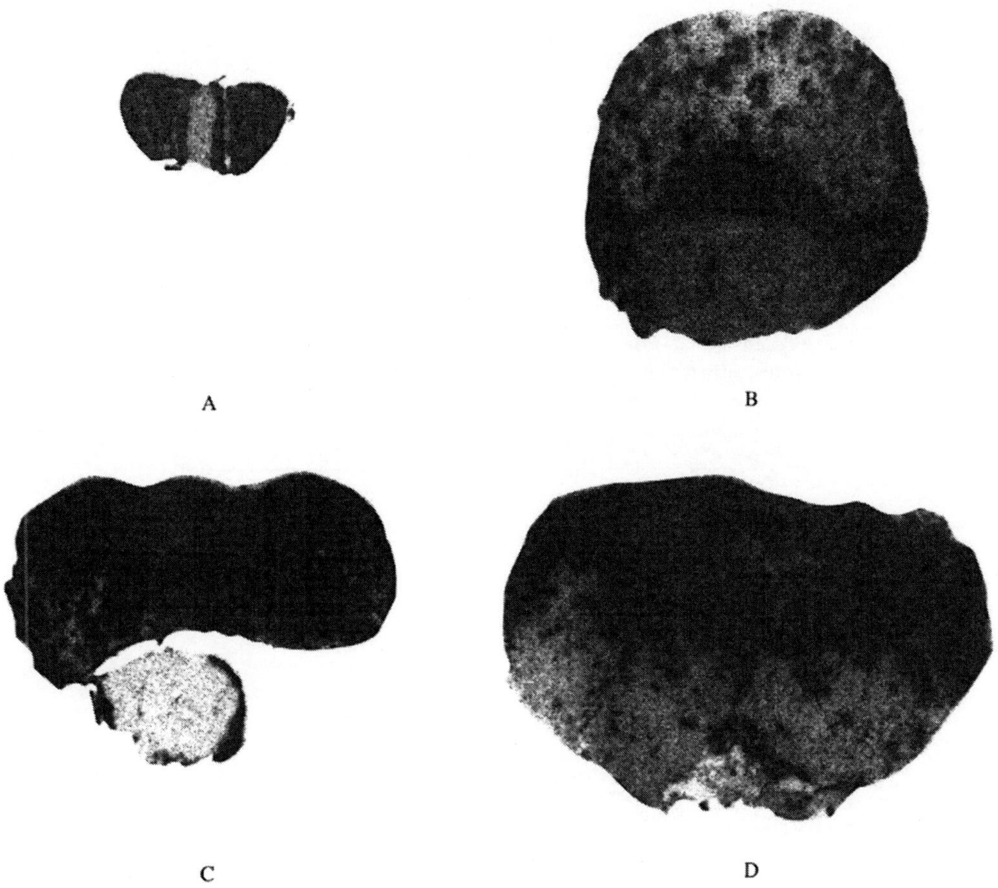

FIGURE 1. Autoradiographic analysis of ^{125}I-Tyr-oCRF binding to 20 μm frozen sections of rat (A), marmoset (B), cynomolgus monkey (C), and human (D) pituitaries.

in the rat and cynomolgus monkey, while it is considerably lower in the marmoset (less than 50%), which is consistent with the autoradiographic data (Table 1). ^{125}I-Tyr-oCRF binding is enhanced by the divalent cations magnesium, calcium, and manganese in the 2 to 5 m*M* range, while it is unaffected by sodium concentrations up to 140 m*M*. Addition of dithiothreitol (up to 1 m*M*), aprotinin (100 KIU/ml), and EGTA (2 m*M*) also increases CRF binding, probably by inhibiting tracer and/or receptor degradation. Binding is markedly decreased by freezing and thawing of pituitaries, or by keeping tissue or membranes at 4°C for more than 2 h.

The binding of CRF to pituitary membranes is markedly influenced by guanyl nucleotides.[4,17,19] Addition of the nondegradable GTP analog, Gpp(NH)p, inhibits the binding of ^{125}I-Tyr-oCRF by 60% with ID_{50} of 3 μ*M* (Figure 2). This effect of Gpp(NH)p on CRF receptor affinity is also seen during kinetic studies on CRF binding, when addition of the guanyl nucleotide causes rapid dissociation of the bound peptide. Such guanyl nucleotide action is characteristic of receptors that are coupled to adenylate cyclase by a guanyl nucleotide regulatory protein, as is likely to be the case for the CRF receptor.[11,12]

The specificity of the binding sites for ^{125}I-Tyr-oCRF is indicated by the ability of CRF peptides to inhibit tracer binding, with potencies similar to their biological activities. Thus, oCRF, rat CRF and Tyr-oCRF are equipotent, with ID_{50}s of about 1 n*M*, while the CRF 15-41 fragment is 1000 times less potent, and CRF 21-41, vasopressin, angiotensin II, ACTH, and GnRH are completely inactive.[3]

TABLE 1
Binding Constants of Pituitary CRF Receptors in the Rat and Primate

Species	Pituitary CRF receptors	
	Concentration (fmol/mg)	K_d (n*M*)
Rat (n = 5)	324 ± 16	1.2 ± 0.8
Cynomolgus monkey (n = 3)	605 ± 121	1.9 ± 0.23
Marmoset (n = 2)	200 ± 15	5.6 ± 2.7

Note: Values are the mean ± SE of the number of experiments indicated in parenthesis for rat and cynomolgus monkey, and mean ± SD of two experiments for the marmoset.

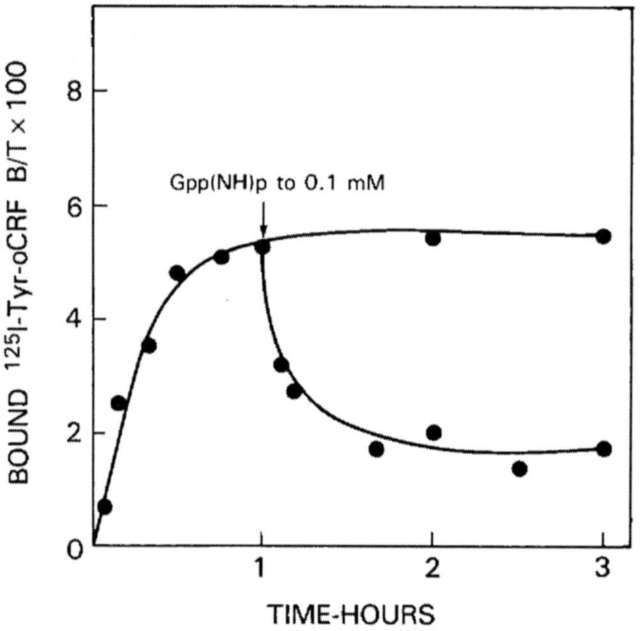

FIGURE 2. Steady-state binding of ^{125}I-Tyr-oCRF to anterior pituitary membrane-rich fractions and its dissociation by the guanosine triphosphate analog Gpp(NH)p.

Validation of a peptide hormone binding site as a receptor which mediates hormone action requires that receptor occupancy results in the characteristic biological response of the cell. In the corticotroph, such a relationship was sought by comparison of the binding activities of a series of CRF peptides with their ACTH releasing potencies. In cultured pituitary cells, both CRF and Tyr-oCRF stimulate ACTH release eightfold, with half-maximum effective concentrations (ED_{50}) of 3.2 and 2.8 × 10^{-10} *M*, respectively. The 15-41 CRF fragment is a weak agonist with full intrinsic activity (ED_{50} 5 × 10^{-7} *M*), whereas the 21-41 CRF fragment is completely inactive (Figure 3). Also, in cultured pituitary cells from cynomolgus monkey CRF stimulated ACTH release with and ED_{50} in the nanomolar range.[18] The maximal ACTH response to CRF occurs with about 50% receptor occupancy, indicating the presence of spare receptors. Such excess of CRF receptors in the corticotroph could be relevant to the effects of adrenalectomy, in which high rates of ACTH secretion

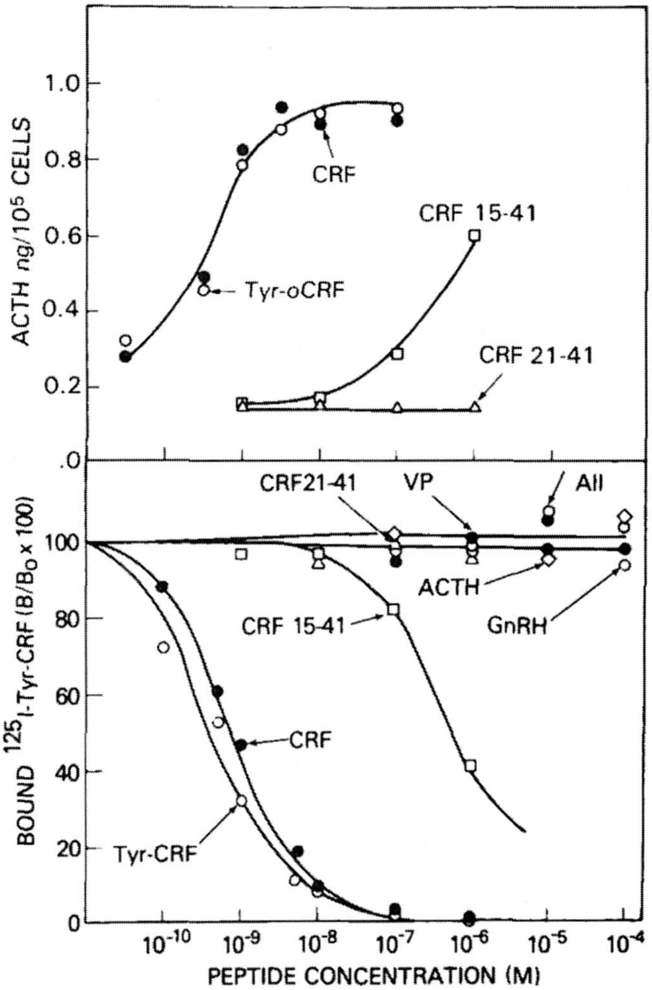

FIGURE 3. Dose response curves of the stimulation of ACTH release from dispersed rat anterior pituitary cells by CRF peptides (upper panel), and binding inhibitory activities of CRF related and unrelated peptides on the binding of ^{125}I-Tyr-oCRF to rat pituitary particles (lower panel). VP = vasopressin; AII = angiotensin II; GnRH = gonadotropin-releasing hormone.

are maintained in the presence of marked down-regulation of CRF receptors. The concentrations of CRF reported in portal blood are in the range of the pituitary CRF receptor affinity, supporting the view that the binding sites detected by radioligand assay represent the functional receptors through which CRF regulates ACTH secretion.[2,20]

Studies of the physicochemical properties of the CRF receptor have been difficult because of the lability of the receptor. However, the use of bifunctional crosslinking agents had made possible the identification of a CRF binding protein in solubilized membranes from rat and bovine pituitaries and ATt-20 cells. SDS polyacrylamide gel electrophoresis of rat anterior pituitary membrane-rich fractions crosslinked to ^{125}I-Tyr-oCRF with disuccinimidyl suberate reveal the presence of specifically radiolabeled proteins by autoradiography (Figure 4). A main band of approximate molecular weight 70 kDa which is abolished when the crosslinking is performed in the presence of an excess of unlabeled CRF is observed in all experiments. Minor bands with molecular weights of about 40 and 28 kDa are also blocked

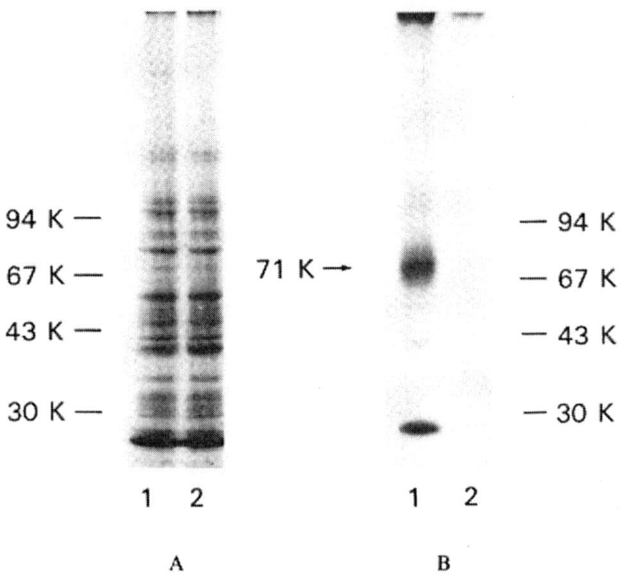

FIGURE 4. SDS/polyacrylamide gel electrophoresis of solubilized rat anterior pituitary membranes crosslinked with radioiodinated Tyr-oCRF using disuccinimidyl suberate. (A) Coomassie blue staining; (B) Autoradiographs of the gel indicating a major radiolabeled band with molecular weight of 71 kDa. Lane 1 shows specific binding and lane 2 the nonspecific binding in the presence of 1 μ*M* CRF.

by CRF, but their presence is less consistent from experiment to experiment. The characteristics of the 70 kDa protein identified in rat pituitary is similar to those described from bovine pituitary and ATt-20 cells using similar crosslinking techniques.[21,22]

B. MECHANISM OF ACTION OF CRF

Several studies in rat and monkey pituitary cells have shown that stimulation of ACTH release by CRF is accompanied by parallel increases in cyclic AMP production.[11,12,18] As in many cAMP-dependent endocrine systems, the ED_{50} for CRF stimulation of cyclic AMP production is higher than that for ACTH release. However, in all studies a minor increase in cyclic AMP is observed with concentrations of CRF at the threshold for stimulation of ACTH release. The role of cAMP as the second messenger for CRF action is also indicated by the ability of isobutylmethylxanthine to potentiate cAMP production and ACTH release, and the lack of additivity between the stimulatory effects of CRF and those of 8-bromo-cAMP or cholera toxin on ACTH release.[12] The stimulation of cAMP production by CRF is due to increased synthesis of the nucleotide, since CRF directly stimulates adenylate cyclase in pituitary homogenates and membrane fractions.[12] CRF stimulation of adenylate cyclase is enhanced by guanyl nucleotides, suggesting that the CRF receptor is coupled to the enzyme by the guanyl nucleotide binding protein, N_s. The role of cAMP in the stimulation of ACTH release by CRF is also indicated by the ability of the hypothalamic peptide to stimulate cAMP-dependent protein kinase at doses similar to those required to activate adenylate cyclase in pituitary homogenates and ACTH release in cultured pituitary cells.[12]

The interaction of CRF with other ACTH secretagogues plays an important role during the physiological control of ACTH secretion.[2] VP, NE, and AII alone are minor stimuli of ACTH secretion *in vitro*, but they markedly potentiate the stimulatory action of CRF, an effect that is likely mediated by protein kinase C. VP has been shown to increase inositol phosphate formation in pituitary cells (Figure 5) and recent studies have shown that the

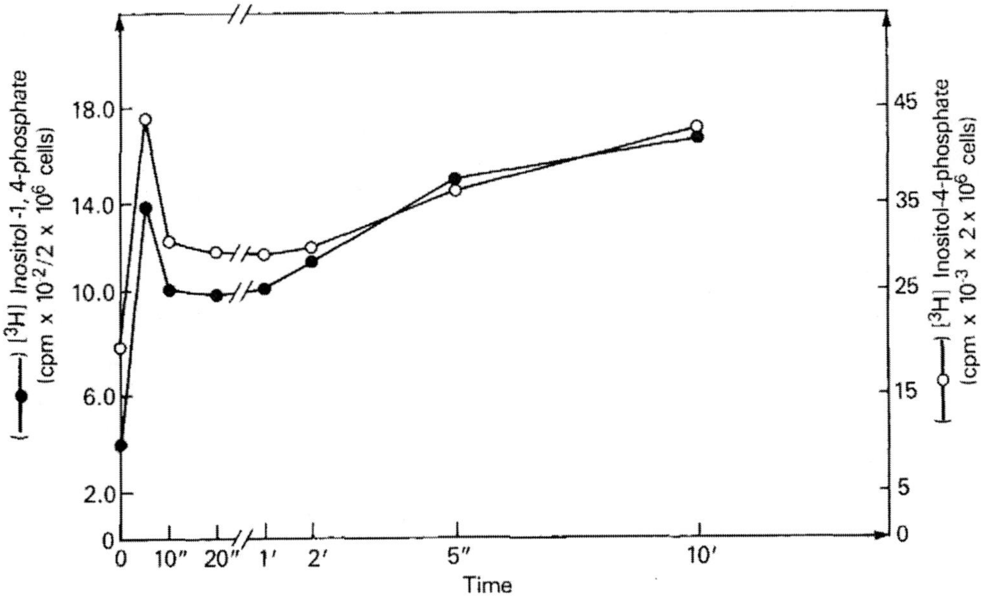

FIGURE 5. Effect of vasopressin on inositol phosphate formation in rat pituitary cells. Cultured cells were prelabeled with ^{3}H inositol for 24 h, washed and incubated with VP for the periods of time indicated in the presence of 10 m*M* LiCl. Inositol phosphates were measured by HPLC after extraction with EDTA tri-*N*-octylamine/freon.

peptide causes rapid translocation of protein kinase C from cytosol to membranes in the anterior pituitary.[24]

In contrast to CRF, these stimuli are ineffective in increasing cAMP production when added alone to pituitary cells and have no direct stimulatory effect in adenylate cyclase assays. However, similar to the effect on ACTH release, VP and NE, but not AII, have been shown to potentiate the stimulatory effect of CRF on cAMP production.[25,26] Such enhancement of the ability of CRF to increase cAMP production probably contributes to the potentiation of CRF-stimulated ACTH release by VP and NE, but does not account for the synergistic effect of VP on ACTH release by maximally effective concentrations of CRF.

There is evidence that non-cAMP mediators such as calcium and phospholipid metabolites are also involved in postreceptor mediation of the action of CRF. Protein carboxymethylation and phospholipid methylation occur during CRF stimulation of ACTH release in ATt-20 cells and phospholipid methylation is frequently followed by calcium-dependent activation of a membrane-bound phospholipase A.[27,28] There is evidence for the involvement of arachidonic acid in the mechanism of action of ACTH regulators, including CRF. The ability of CRF to release arachidonic acid and the stimulatory effect of arachidonic acid on ACTH release in isolated pituitary cells in a nonadditive manner with CRF, suggests that arachidonic acid metabolism is part of the mechanism of action of CRF. Studies using inhibitors of the different pathways of arachidonic acid metabolism indicate that arachidonic acid products may have a dual modulatory role on ACTH release, with cyclooxygenase metabolites being inhibitory and lipoxygenase products being stimulatory. Thus, lipoxygenase inhibitors such as nordihydroguairetic acid (NDGA) and 5,8,11,14-eicosatetraynoic acid (ETYA) partially inhibit CRF-stimulated ACTH secretion, whereas cyclooxygenase inhibition by indomethacin has the opposite effect.[13]

Calcium is not only important as a messenger system for cAMP-independent hormones, but also has a role in CRF-stimulated ACTH release. In the absence of extracellular calcium, the effect of CRF on ACTH release from normal rat anterior pituitary cells is partially inhibited.[29,30] Kinetic studies have shown that the initial rate of CRF stimulated ACTH

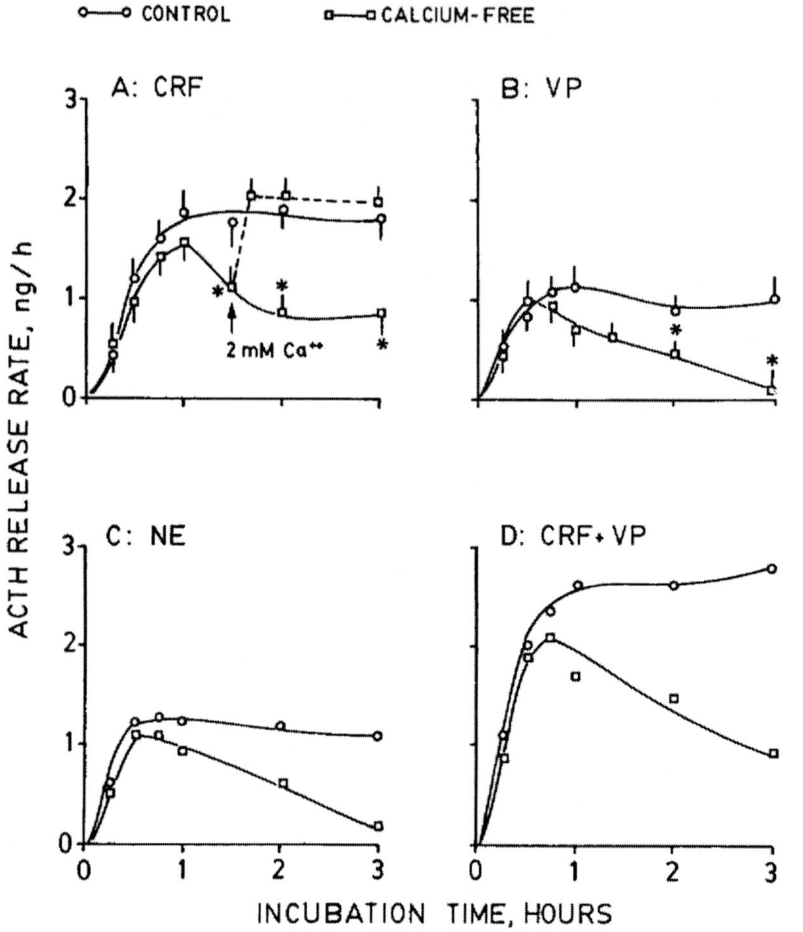

FIGURE 6. Effect of calcium-free medium on ACTH release rate. Cultures were washed with PBS and incubated with the secretagogue (10 n*M*) in the absence and presence of calcium. At each time point, 10 μl of medium was removed, diluted 50-fold with PBS and rapidly frozen until measurement of ACTH by RIA. ACTH release rate was derived by calculating the amount of ACTH released between two successive time points divided by the intervening time period. The data are the means ± SEM of three experiments (A and B) or means of two experiments (C and D). In panel A, 2 m*M* $CaCl_2$ was added 90 min after the initial incubation in calcium-free medium (dotted lines). $^*p < 0.05$ compared to control.

secretion is unaffected by the absence of calcium, but after 45 min the secretion rate decreases by 40% (Figure 6). There is evidence indicating that calcium influx through voltage-dependent calcium channels is involved during activation of the corticotroph by CRF. Thus, the voltage-dependent calcium channel agonist, BK 8644, increases basal and maximal ACTH release stimulated by CRF, while the calcium channel antagonist, nitrendipine, is inhibitory. Also, CRF and cAMP analogs have been shown to increase cytosolic calcium in ATt-20 cells, an effect which is blocked by calcium channel antagonists.[14]

The absence of extracellular calcium has no effect on CRF-stimulated cAMP accumulation, indicating that adenylate cyclase activity can be sustained by intracellular calcium levels. On the other hand, depletion of intracellular calcium by incubation of the cells with EGTA inhibits CRF-stimulated cAMP production, and this effect is reversed by addition of calcium to permeabilized cells, consistent with the recognized calcium/calmodulin dependence of adenylate cyclase.[31,32]

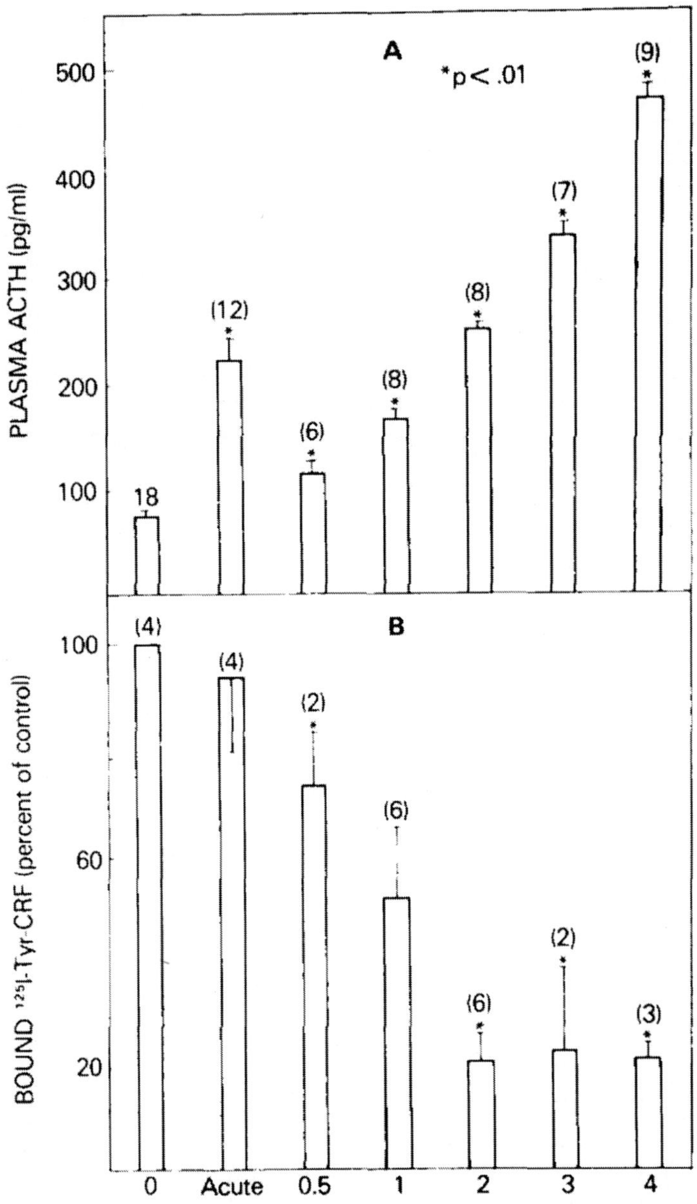

FIGURE 7. Effect of adrenalectomy on plasma ACTH levels (A) and pituitary CRF receptors (B) in the rat. CRF receptor concentration was calculated by Scatchard analysis. The number of experiments is indicated in parenthesis.

C. REGULATION OF PITUITARY CRF RECEPTORS

The interaction of peptide hormones with the target cell is frequently accompanied by regulatory changes of their receptors. In many cases, changes in receptors parallel responsiveness of the cell to the hormone, and receptor regulation has an important role in modulating hormone responses. A number of studies have shown that pituitary CRF receptors are modulated following alterations of the hypothalamic-pituitary-adrenal axis. Following adrenalectomy, despite progressively increasing plasma ACTH levels, pituitary CRF binding is significantly reduced by 24 h, followed by a sustained 80% decrease after 48 h (Figure 7). The reduction in binding is due to a decrease in receptor concentration with no change in binding activity.[33]

The reduction of CRF receptors after adrenalectomy is accompanied by a 60% decrease in CRF-stimulated adenylate cyclase activity at 24 h, with no further change at later times. The decreases in CRF receptors and adenylate cyclase activity following adrenalectomy are prevented by dexamethasone treatment. Consistent with the decreased capacity of CRF to activate adenylate cyclase in membranes, CRF-stimulated cyclic AMP production is significantly reduced in cultured pituitary cells from adrenalectomized rats. However, in contrast to the decrease in CRF receptors and cyclic AMP production, CRF-stimulated ACTH release in cultured pituitary cells is increased, with no change in sensitivity to CRF. Such increased ACTH release from corticotrophs with reduced CRF receptors and impaired activation of adenylate cyclase indicates that elevated ACTH secretion can be maintained by occupancy of few receptors and the generation of only small quantities of cyclic AMP. It is likely that synergistic interactions between CRF and other ACTH secretagogues contribute to the sustained increase in ACTH secretion that follows adrenalectomy.[33]

Increased exposure of target tissues to peptide hormones is commonly associated with down-regulation and densensitization of the homologous receptors.[34] CRF levels in the portal blood are elevated after adrenalectomy.[35,36] Thus, studies with prolonged CRF infusion is normal rats have shown decreases in pituitary CRF receptors of only 30 to 40% compared with the 70 to 80% loss observed after adrenalectomy and therefore other factors are likely to be involved in this process. However, increases in CRF are only partially responsible for the effect of adrenalectomy on pituitary CRF receptor levels.[37]

The secretion of VP into the portal circulation is increased after adrenalectomy, and recent studies have shown that in addition to potentiating CRF-stimulated ACTH secretion, VP also influences the regulation of CRF receptors in the anterior pituitary.[38-41] In di/di Brattleboro rats which lack endogenous brain VP, adrenalectomy only causes a slight decrease in CRF receptors (compared with the 80% loss in Sprague-Dawley or Long-Evans rats) but the effect could be reproduced by infusion of VP during the postadrenalectomy period (Figure 8). Also in Sprague-Dawley rats, in contrast to the minor receptor loss obtained with CRF infusion alone, the simultaneous infusion of CRF and VP mimicked the effect of adrenalectomy on pituitary CRF receptors.[39]

Glucocorticoids also regulate CRF receptor concentration in the pituitary *in vitro* and *in vivo*. In isolated pituitary cells, incubation with corticosterone results in CRF receptor loss measured using biotinylated or fluorescein labeled CRF.[9,10] *In vivo*, corticosterone administration (0.5 to 150 mg/day) for 1 to 4 d in adult male rats causes a dose-dependent decrease in the number of CRF receptors in the anterior pituitary, in parallel with the reduction in ACTH secretion.[42] Low doses of corticosterone, which produce only transient increases in plasma steroid levels, caused significant reductions in both plasma ACTH and pituitary CRF receptors. The latter effect of low doses of the naturally occurring glucocorticoid suggests that receptor down-regulation in the anterior pituitary may be of physiological importance, and could participate in the inhibitory action of glucocorticoids on ACTH release. However, glucocorticoids inhibit not only CRF-stimulated ACTH release, but the response to other hormones such as VP, norepinephrine, and angiotensin II, as well as to postreceptor stimulants such as 8-Br-cyclic AMP and phorbol esters.[43] Therefore, the decrease in cell-surface CRF receptors can only partially account for the inhibitory effects of glucocorticoids on ACTH release.

The decrease in CRF receptors may reflect a general inhibitory effect of glucocorticoids upon the corticotroph. Since glucocorticoids have been shown to cause a rapid decrease in POMC gene transcription, it is possible that the synthesis of cellular proteins including the CRF receptor could also be reduced. The direct down-regulatory effect of glucocorticoids on pituitary CRF receptors differs from the effects of glucocorticoid replacement in adrenalectomized rats. In the latter situation, glucocorticoids are likely to block ligand-induced CRF receptor down-regulation by inhibiting the increased release of CRF and VP into the portal circulation secondary to adrenalectomy.[33]

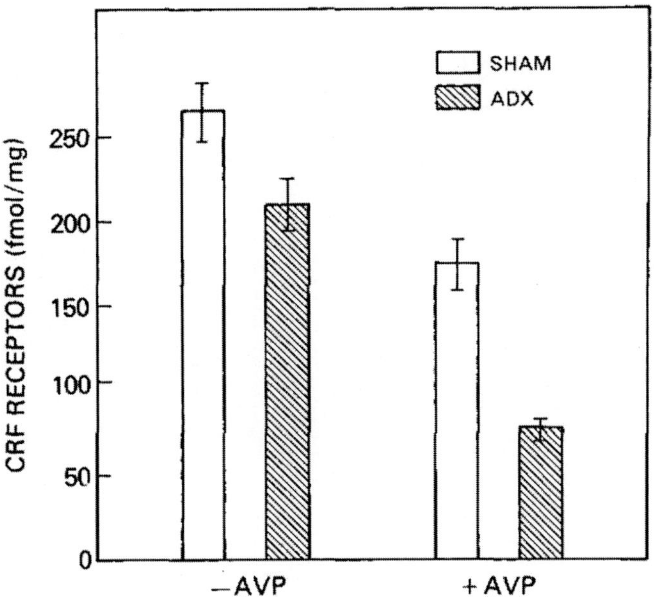

FIGURE 8. Effect of vasopressin infusion on anterior pituitary CRF receptor concentration in sham-operated and adrenalectomized di/di Brattleboro rats. Bars represent the mean and SE of the values obtained by Scatchard analysis of the binding data in four experiments.

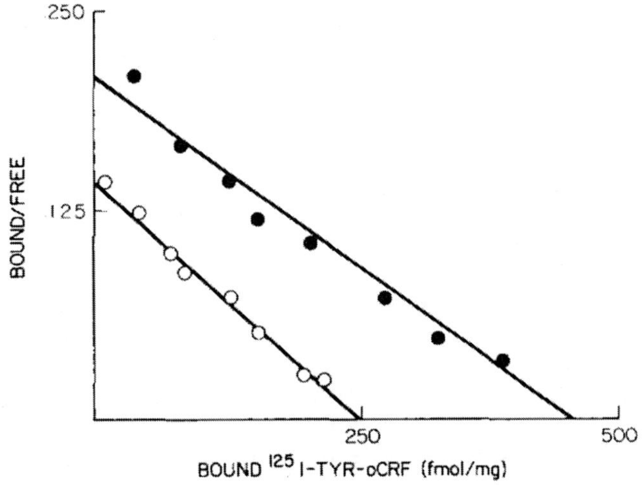

FIGURE 9. Scatchard analysis of the binding of ^{125}I-Tyr-oCRF to anterior pituitary particles from control rats (●) and rats subjected to continuous immobilization stress for 48 h (O).

Prolonged immobilization stress for more than 12 h also results in reduction of CRF receptor number in the anterior pituitary (Figure 9.)[44] Several factors may contribute to the pituitary CRF receptor down-regulation during stress such as elevated plasma glucocorticoids levels and increased secretion of hypothalamic CRF and VP. The decrease in pituitary CRF receptors is accompanied by decreased CRF stimulated cyclic AMP and ACTH release in cultured pituitary cells from 48 h restrained rats. However, concomitant incubation of the cells with CRF and VP restores cyclic AMP and ACTH responses to levels observed in

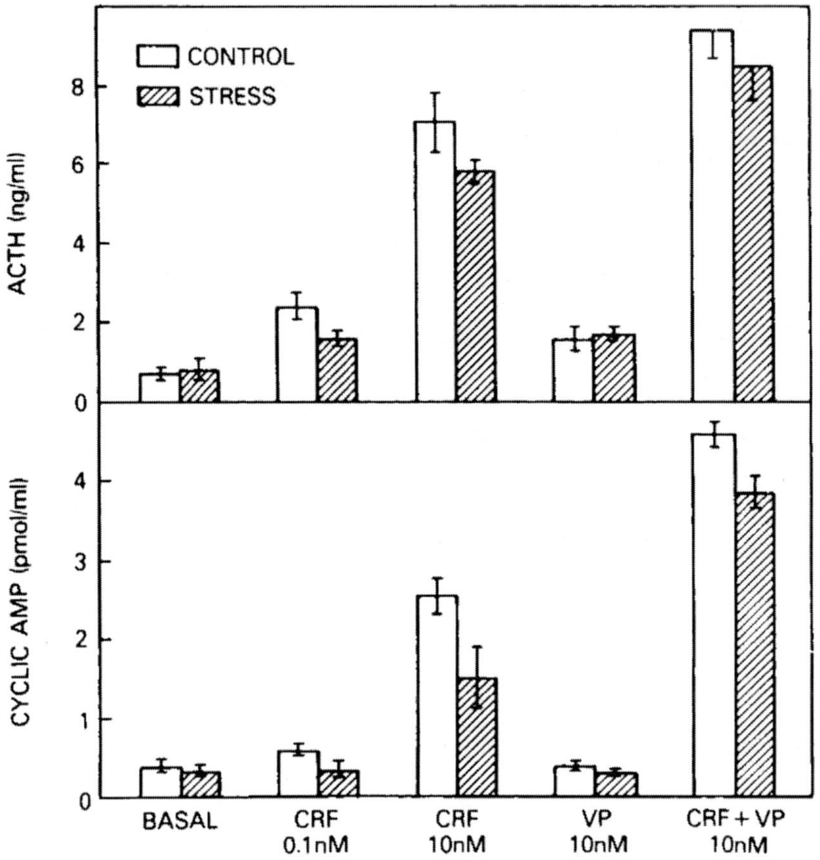

FIGURE 10. ACTH and cyclic AMP responses to CRF and VP or their combination in 24 h cultured pituitary cells from control rats and rats subjected to immobilization stress for 48 h. For ACTH determination, cells were incubated for 3 h with stimulants, and for cAMP production for 30 min in the presence of 1 m*M* of the phosphodiesterase inhibitor, methylisobutylxanthine. Bars are the mean and SE from the data in three experiments.

control cells, suggesting that the simultaneous release of both regulators from the hypothalamus determines the ACTH response (Figure 10). In contrast to the desensitization of the cells to CRF *in vitro*, ACTH responses to CRF injection *in vivo* are increased following 48 h immobilization, probably due to interaction of CRF with other endogenous regulators. On the other hand, after 6 d of daily 3 h foot shock stress, plasma ACTH responses to CRF are decreased suggesting the involvement of different neural pathways and releasing factors in different types of stress.[45] ACTH responses to novel stimuli are usually enhanced during chronic stress which is consistent with the concept that the responsiveness of the corticotroph is preserved despite pituitary CRF receptor desensitization due to the coordinated action of CRF and other regulators.[44,46-49]

III. CORTICOTROPIN-RELEASING FACTOR RECEPTORS IN THE NERVOUS SYSTEM

The biological role of CRF is not limited to the pituitary, and several studies have shown visceral and behavioral effects following central administration of the peptide. The availability of synthetic CRF and antisera against the peptide have permitted detailed studies on the localization and actions of the peptide in the brain and other tissues. In addition to the paraventricular-median eminence neuronal system responsible for CRF release to the portal circulation, CRF-containing cell bodies have been described in many extrahypothalamic sites including several regions of the limbic system and central structures associated with the control of autonomic function.[50-54] Intracerebroventricular administration of CRF to rat, dog, and monkey results in activation of the hypothalamic-pituitary-adrenal axis and the sympathetic nervous system, resulting in visceral, metabolic, and behavioral responses.[1,2,55-57] These observations suggest that CRF coordinates the responses to stress by actions within the nervous system, as well as in the pituitary gland. This hypothesis has been supported by the demonstration of receptors for CRF widely distributed in cortical and limbic areas of the brain, as well as in the adrenal medulla and sympathetic ganglia of the peripheral nervous system.[5-7]

A. BRAIN CRF RECEPTORS

1. Binding Properties of Brain CRF Receptor Sites

The ligand binding characteristics of brain CRF receptors have been studied in membrane-rich fractions from rat and monkey brain, using the radioiodinated Tyr-oCRF derivative previously employed to characterize pituitary CRF receptor sites.[6] Receptor binding properties are similar in all brain areas in the species studied.[6,7] Specific binding of ^{125}I-Tyr-oCRF to membrane-rich particles of rat olfactory bulb is temperature dependent, linear with protein, and reaches equilibrium within 30 min at 22°C. Addition of excess unlabeled CRF at the time of equilibrium binding causes rapid dissociation of the bound tracer with a half-time of 15 min followed by a slower component with a half-time of 75 min. Computer analysis of the equilibrium binding data shows a single high-affinity site with 1 n*M* and a concentration of 50 to 100 fmol/mg protein. As in the pituitary gland, ovine CRF and rat CRF compete with ^{125}I-Tyr-oCRF with equal potencies, while the 15 to 41 CRF fragment is 1000-fold less potent, and unrelated peptides including angiotensin II, arginine-vasopressin, and $ACTH^{1-24}$ are inactive. Similar findings have been obtained in autoradiographic studies where the binding to slide-mounted sections has an affinity in the nanomolar range, and as observed in particulate brain fractions, bound radioactivity is displaced by CRF analogs and not by unrelated peptides.[5,6]

2. Mechanism of Action of CRF in the Brain

As in the pituitary gland, CRF receptors in the brain are linked to adenylate cyclase and cAMP production.[6,7,58] It has been shown that CRF increases adenylate cyclase activity in membrane-rich particles prepared from rat frontal cortex and amygdala, areas rich in CRF receptors, but not in areas where no CRF binding is detectable. Similar stimulatory effects of CRF on adenylate cyclase have been observed in cynomolgus monkey orbital cortex and amygdala, areas with abundant CRF receptors.[7] Other reports have shown a dissociation between the degree of activation of adenylate cyclase by CRF and CRF receptor content in brain particles from different brain areas of the rat.[58] Although such studies are difficult to interpret because of varying levels of basal activity and other adenylate cyclase linked systems, the results could reflect regional differences in CRF receptor-messenger coupling in the brain.

3. CRF Receptor Distribution in Rat and Primate Brain

Autoradiographic mapping of CRF receptors in rat and monkey brain reveal predominant localization in two functionally distinct systems, the neocortex and the limbic system.[5-7] In rat brain (Table 2), the receptors are highly concentrated throughout the cerebral cortex, with relatively higher densities in the anterior cingulate cortex, frontoparietal (somatosensory area) and temporal cortex (auditory area). The highest concentration is found in layer IV of the cerebral cortex, followed by layers I to III. Two cortical structures related to the limbic

TABLE 2
Regional Distribution of CRF Receptors in the Rat Brain

	^{125}I-Tyr-oCRF binding (optical density × 10^3)		
Region	**Layer 1—3**		**Layer 4**
Cerebral			
Prefrontal	240		470
Anterior cingulate	220		440
Posterior cingulate	260		290
Frontoparietal (motor)	260		320
Frontoparietal (sensory)	310		400
Striate			370
Temporal auditory area			610
Entorhinal			220
Subiculum		260	
Hippocampus		170	
Basal telencephalon			
Olfactory bulb (EPI)		510	
Lateral intermediate and medial septal nuclei		230	
Nucleus accumbens		220	
Caudate putamen		154	
Bed nucleus stria terminalis		300	
Amygdala			
Anterior amygdaloid area		290	
Medial amygdaloid area		230	
Basolateral amygdaloid area		370	
Claustrum		160	
Diencephalon			
Paraventricular thalamic nuclei anterior		90	
Dorsolateral thalamic nuclei		130	
Supramammillary nuclei		150	
Dorsomedial hypothalamic nucleus		150	
Arcuate nucleus		100	
Paraventricular nucleus		0	
Periventricular nucleus		0	
Median eminence		0	
Brain stem			
Superior colliculus		300	
Interpeduncular nucleus		410	
Periaqueductal gray		80	
Locus coeruleus		120	
Inferior olive		270	
Pontine nucleus		320	
Nucleus of the solitary tract		90	
Dorsal tegmental nucleus		210	
Spinal trigeminal tract		150	
Cerebellum granular layer		460	

Note: Data are the means of optical density values in two to three determinations after subtracting the nonspecific background.

system, the subiculum and the hippocampus, also contain moderate receptor densities. In the extracortical areas, higher receptor concentrations are found in the external plexiform layer of the olfactory bulb, bed nucleus of the stria terminalis, basolateral amygdaloid area, superior colliculus, interpeduncular nucleus, and pontine nucleus.

Moderate concentrations of receptors are found in the lateral intermediate and medial septal nuclei, nucleus accumbens, anterior and medial amygdaloid areas, claustrum, supra-

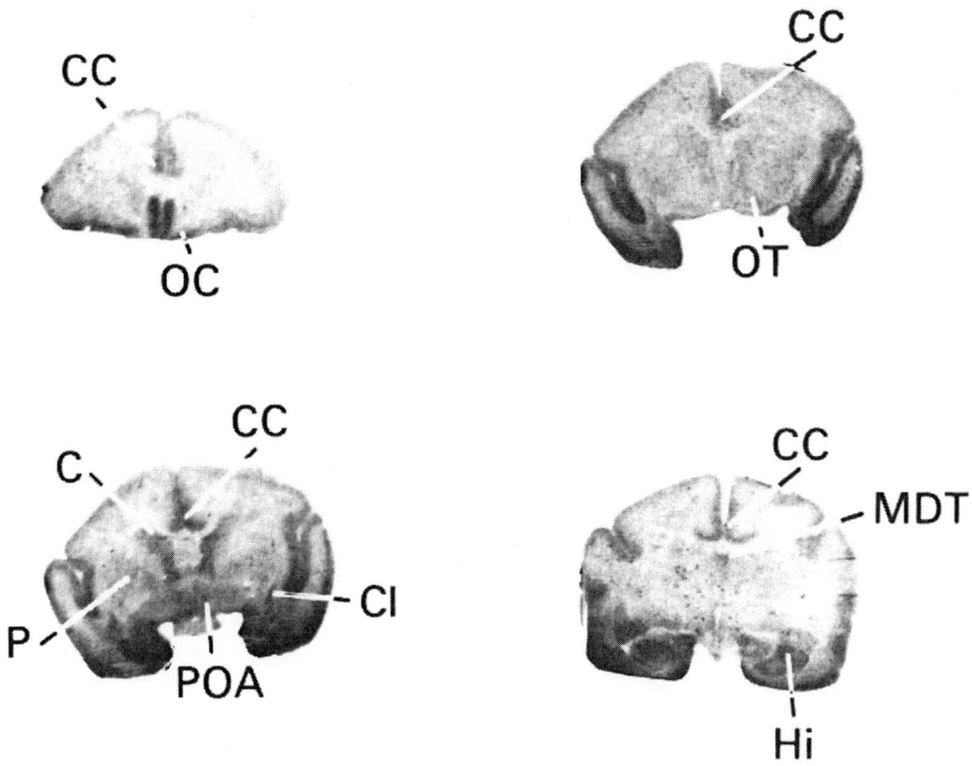

FIGURE 11. Autoradiographic analysis of ^{125}I-Tyr-oCRF binding to coronal sections, from rostral to caudal, of marmoset brain. CC, cingulate cortex; OC, orbital cortex; OT, olfactory tubercle; C, caudate; P, putamen; Cl, claustrum; POA, preoptic area; MDT, medial dorsal thalamus; Hi, hippocampus.

mammillary nuclei, dorsomedial hypothalamic nucleus, inferior olive, dorsal tegmental nucleus, and spinal trigeminal tract. Low receptor concentrations are found in the caudate putamen, paraventricular thalamic nuclei anterior, dorsolateral thalamic nuclei, arcuate nucleus, periaquaductal gray, locus coeruleus, and nucleus of the solitary tract. The cerebellar cortex also contains a high density of receptors that are mainly concentrated in the granular layer. In autoradiographic studies using ^{125}I[NLe^{21},Tyr^{32}]CRF as the labeled ligand, similar CRF receptor localization has been observed in some areas of the forebrain including neocortex, thalamus, hippocampus, striatum, and lateral nucleus of the amygdala and external layer of the median eminence.[5]

In the monkey, CRF receptor distribution in the brain cortex and limbic system-related structures resembles that observed in the rat brain (Figures 11 and 12). Whereas in the rat and marmoset the receptors are evenly distributed throughout the cortex, in the cynomolgus monkey there are marked differences between the cortical areas. The highest binding is in the prefrontal, orbital, and insular cortices, regions of relatively late phylogenetic acquisition that are well developed only in primates and particularly man.[59] These areas receive abundant innervation from the dorsomedial nucleus of the thalamus, which relays impulses from several autonomic centers.[59] This area of the thalamus, which plays an important role in emotional responses, also contains abundant CRF receptors (Figure 12d).

In rat and monkey, CRF receptors are also found in two hypothalamic areas involved in the control of gonadotropin secretion, the preoptic area and the arcuate nucleus (Figure 12b and d). A possible role for CRF in the control of sexual function is supported by studies showing inhibition of sexual behavior in the female rat, and decreases in plasma luteinizing hormone and gonadotropin-releasing hormone in portal blood after CRF injection into the

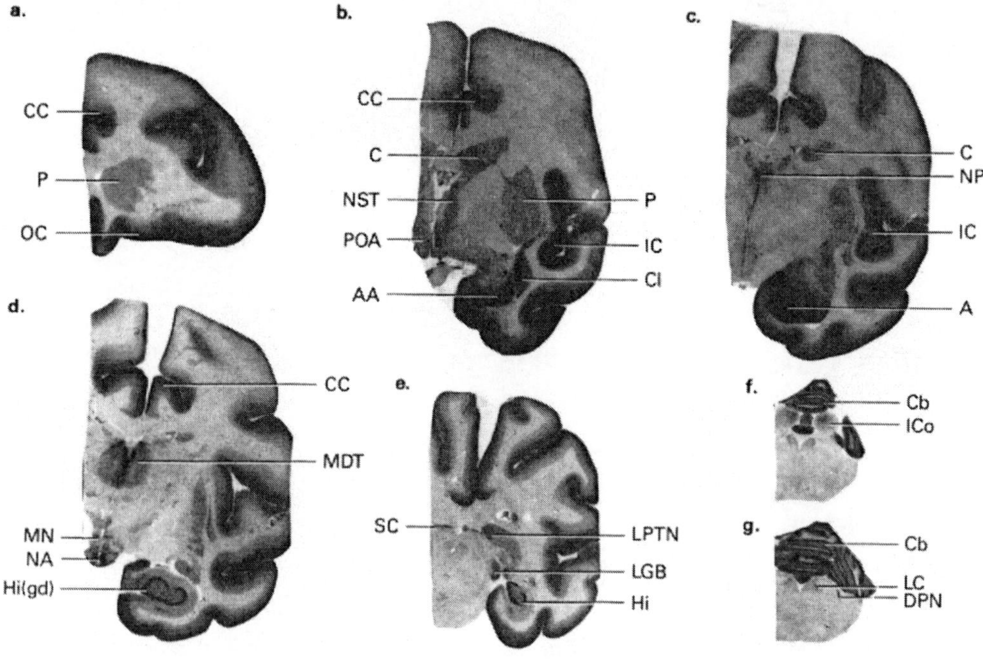

FIGURE 12. Autoradiographic distribution of CRF receptors in cynomolgus monkey brain. CC, cingulate cortex; P, putamen; OC, orbital cortex; C, caudate; OT, olfactory tubercle; Cl, claustrum; TC, temporal cortex; NST, nucleus of stria terminalis; IC, insular cortex; POA, preoptic area; AA, anterior amygdala; NPT, paraventricular nucleus of thalamus; A, amygdala; MDT, medial dorsal thalamus; MN, mamillary nucleus; Hi(gd), hippocampus (gyrus dentatus); SC, superior colliculus; LPTN, lateral posterior thalamic nucleus; LGB, lateral geniculate body; Hi, hippocampus; Cb, cerebellum; ICo, inferior colliculus; LC, locus coeruleus; DPN, dorsal parabrachial nucleus.

arcuate-ventromedial hypothalamic regions.[37,60-62] The presence of CRF receptors in these areas of the primate brain suggests that CRF influences gonadotropin secretion in man and could be involved in the mechanism of altered gonadal function during prolonged stress. The arcuate nucleus has projections to a number of hypothalamic and limbic structures that are rich in CRF, ACTH, and β-endorphin.[63] Coexistence of CRF and POMC-derived peptides has been described in many other structures that contain CRF receptors, such as the nucleus accumbens, stria terminalis, preoptic area, amygdala, geniculate bodies, locus coeruleus, and parabrachial nucleus.[63] The similar anatomical distribution of CRF and its receptors and of opiocortin peptides suggests that, as in the pituitary gland, both systems are functionally related in the brain.

Also in the hypothalamus, it is interesting to note the high concentration of CRF receptors in the nucleus tuberis lateralis, a structure of unknown function at the present time. This nucleus is well developed in primates, and the presence of CRF receptors may provide a basis for the study of its function.[64]

Local neuronal activation has been observed after microinjection of CRF in the locus coeruleus of the rat and this effect may be mediated by the CRF receptors observed in this area.[65] Studies using retrograde staining techniques have shown that the locus coeruleus has connections to the hypothalamus, hippocampus, brain cortex, olfactory bulb, cerebellum, and spinal cord. These multiple connections suggest that the locus coeruleus has an integrative role in the central nervous system relating to endocrine, autonomic, and behavioral control. It should be noted that all these structures, including the locus coeruleus and connected structures, contain immunoreactive CRF and CRF receptors (Figure 12g), suggesting that the peptide has a role as mediator of neural transmission in this pathway. A high density of CRF receptors is also observed in the amygdala (Figure 12b and c). This important component of the limbic system has both efferent and afferent connections with the cortical areas that contain the highest CRF receptor concentrations, namely, the frontal, orbital, cingulate, temporal, and insular cortices. The amygdaloid nuclei also receive projections from the locus coeruleus, hypothalamus, and dorsomedial thalamic nucleus and have efferent projections to the dorsomedial thalamic nucleus, nucleus stria terminalis, preoptic area, septal regions, and arcuate nucleus. All of these areas connected to the amygdala are found to contain CRF receptors. It should be noted that the connections of the amygdala to the septal and preoptic area are unique for the higher mammals,[66] a feature that correlates with the presence of CRF receptors in these regions in the monkey but not in the rat.[6] Electrical stimulation of the amygdala in experimental animals has been shown to cause arousal, attention, fear, and rage reactions associated with sympathetic activation.[67] These reactions are similar to those observed during stress and after intracerebroventricular administration of CRF.[1] Since CRF and its receptors are present in the amygdala, it is likely that the peptide is important in the generation of some of these responses.

In the primate, there is a high CRF receptor concentration throughout the limbic lobe, a structure composed of the cingulate and parahippocampal cortex, and the hippocampus. The limbic lobe is connected with the hypothalamus, other limbic structures and the neocortex.[68] The limbic system has a primary role in the control of behavior, emotion, and autonomic and endocrine functions, and a number of these limbic system-mediated responses can be mimicked by central administration of CRF in the rat, dog, and monkey.[1,56]

There is evidence suggesting that excessive CRF activity in the central nervous system is related to certain types of depression in humans. In unrestrained monkeys, large doses of CRF injected into the brain causes a depressed state with huddling and lying down behavior, similar to that observed during isolation of monkeys.[56] Many patients with depression show hyperactivity of the pituitary-adrenal axis with elevated basal cortisol levels, and with reduced ACTH responses to exogenous CRF, suggestive of increased endogenous CRF production.[69,70] This possibility is supported by the demonstration of elevated concentrations of CRF in the spinal fluid of depressed patients.[71] The presence of CRF receptors in the cerebral cortex and limbic system provides a mechanism through which increased activity of CRF neurons could be expressed as the behavioral and visceral components of depression.

Although the mechanisms by which CRF modulates neuronal activity and the exact physiological role of the peptide in the nervous system will require further study, the presence of CRF and functional CRF receptors in discrete structures of the brain is a strong indication that CRF modulates central nervous system function.

4. Regulation of Brain CRF Receptors

In contrast to the marked reduction in CRF receptors in the anterior pituitary, CRF receptor concentrations in the brain and intermediate pituitary lobe are not affected by adrenalectomy, stress, or glucocorticoid administration.[5,6,42,44] Comparison of brain and pituitary CRF receptors 48 h after adrenalectomy shows the expected reduction in the anterior pituitary, but no significant difference in CRF receptor concentration or affinity in the olfactory bulbs.[6] Autoradiographic studies confirm the decrease in receptors in the anterior pituitary and no difference in optical densities in different brain areas including the prefrontal cortex, frontal parietal cortex, basolateral amygdaloid nuclei, lateral amygdaloid nucleus, and bed nucleus stria terminalis. In contrast to the decrease in optical density of the autoradiography in the anterior pituitary of adrenalectomized rats, the intermediate lobe, which is less sensitive to negative glucocorticoid feedback, showed no change in receptor density after adrenalectomy.[6]

Similar to the effects of adrenalectomy, chronic stress, and glucocorticoid administration, which also cause marked reduction in CRF receptors in the anterior pituitary, have no detectable effect on CRF receptors in areas of the brain including amygdala, cortex, hippocampus, septal area, and olfactory bulb.[42,44] The mechanism by which pituitary and brain CRF receptors are differentially regulated have not been elucidated, but it is likely that the interaction of CRF with its brain receptor involves mechanisms different from the processing of the hormone receptor complex by endocytosis and receptor desensitization that occurs in the pituitary gland.

B. CRF RECEPTORS IN THE PERIPHERAL SYMPATHETIC NERVOUS SYSTEM

Several studies have demonstrated that immunoreactive CRF is present in peripheral tissues including the chromaffin tissue of the adrenal medulla.[72-76] Since the peptide is known to act centrally to modulate the activity of the autonomic nervous system, it is likely that CRF could also exert regulatory effects upon the sympathetic system in the periphery. In support of this possibility are observations on the presence of CRF receptors in the adrenal medulla and sympathetic ganglia, and the actions of the peptide upon chromaffin cell function.[53]

Binding studies of ^{125}I-Tyr-oCRF to membrane-rich fractions from monkey adrenal reveal that specific and high affinity CRF receptors with K_d of 3 n*M* are present in the adrenal medulla. As previously demonstrated for CRF receptors in other tissues, those present in the adrenal medulla are also coupled to adenylate cyclase. CRF stimulates the conversion of [^{32}P]ATP to cyclic AMP in membrane-rich medullary particles at concentrations in the range of the CRF binding affinity for the adrenal medullary receptors.[77]

Binding sites for ^{125}I-Tyr-oCRF in the adrenal medulla are also demonstrable by autoradiographic analysis of radioligand binding in slide-mounted frozen sections of rhesus monkey adrenal glands. Consistent with the results in adrenal membranes, CRF receptors are confined to the adrenal medulla, with no specific autoradiographic staining in the adrenal cortex. Autoradiographic analyses also have shown specific CRF binding in several sympathetic ganglia including those of the lumbar, thoracic, cervical, and celiac groups. In all sympathetic ganglia, complete inhibition of the specific binding is observed with 100 n*M* CRF, indicating that the binding sites are of high affinity.[77]

CRF receptor activation in the adrenal medulla also results in changes in secretory activity, as shown by studies in which the effects of CRF on catecholamine and enkephalin release were analyzed in cultured bovine chromaffin cells. In short-term studies, incubation of the cells with CRF for a few hours has no effect on catecholamine secretion. However, after 24 h of incubation with CRF there is a dose-dependent stimulation of epinephrine, norepinephrine, and met-enkephalin release into the culture medium.[77] Although the physiological role of CRF in the autonomic nervous system will require further study, the presence of the peptide and its functional receptors in the brain, adrenal medulla, and sympathetic ganglia suggest that CRF can regulate the activity of the autonomic nervous system at both central and peripheral levels.

IV. CONCLUSIONS

Binding sites for CRF with characteristics of receptors which mediate the biological actions of the peptide have been identified in the anterior and intermediate lobes of the pituitary, and in the nervous system at sites related to the stress response. In keeping with the characteristics of most peptide hormone receptors, CRF binding sites are saturable, high affinity and specific for CRF-related peptides. In the anterior pituitary the affinity of CRF analogs for the CRF receptor parallel the ACTH secretory activity of the peptides. In pituitary and brain membrane fractions the binding is enhanced by divalent cations and is inhibited by guanylnucleotides, which is consistent with this receptor being coupled to adenylate cyclase. Crosslinking and gel electrophoresis studies suggest that the pituitary CRF receptor is a single protein of molecular weight about 70 kDa.

The binding of CRF to its membrane receptor triggers the formation of intracellular signals that lead to synthesis and release of ACTH. Several lines of evidence indicate that cAMP is the main mediator of CRF action. CRF increases cAMP accumulation in pituitary cells, stimulates adenylate cyclase activity, and activates cAMP-dependent protein kinase at concentrations in the range of its ACTH stimulating activity. In addition to cAMP, the mechanisms of action of CRF may involve other messenger systems including calcium and arachidonic acid and its metabolites. Of particular importance is the interaction of CRF with cyclic AMP-independent regulators mainly vasopressin, which are weak stimulators alone, but markedly potentiate the effect of CRF in the corticotroph. The effect of vasopressin involves calcium and phospholipid-dependent mechanisms with stimulation of protein kinase C. Autoradiographic studies of CRF binding have provided valuable new information about the distribution and potential functions of CRF receptors especially in the brain. The distribution of CRF receptors in cortical and limbic areas of the brain correlates with the immunohistochemical localization of immunoreactive CRF pathways and the effects of centrally administered CRF on behavioral and autonomic responses. The presence of CRF receptors in sympathetic ganglia and adrenal medulla, the actions of CRF on adenylate cyclase, and the release of catecholamines and enkephalin from chromaffin cells, all support a regulatory role for CRF on autonomic activity at both central and peripheral levels.

CRF receptors in the pituitary but not in the brain undergo regulatory changes following alterations in the hypothalamo-pituitary-adrenal axis. Following adrenalectomy and chronic stress, the increased ACTH secretion is accompanied by a loss of pituitary CRF receptors, a result of the coordinated actions of CRF and vasopressin in the corticotroph. During glucocorticoid administration, CRF receptors decrease in parallel with inhibition of ACTH secretion suggesting that CRF receptor down-regulation contributes to the physiological effect of the steroid. The widespread distribution of CRF binding sites in the pituitary, cortical and limbic areas of the brain and peripheral nervous system emphasize the importance of CRF as a mediator of the endocrine as well as the behavioral and autonomic responses to stress.

REFERENCES

1. **Vale, W. W., Rivier, C., Brown, M. R., Spiess, J., Koob, G., Swanson, L., Bilezikjian, L., Bloom, F., and Rivier, J.,** Chemical and biological characterization of corticotropin releasing factor, *Rec. Prog. Horm. Res.*, 39, 245, 1983.
2. **Rivier, C. L. and Plotsky, P. M.,** Mediation by corticotropin releasing factor (CRF) of adenohypophysial hormone secretion, *Annu. Rev. Physiol.*, 48, 475, 1986.
3. **Wynn, P. C., Aguilera, G., Morell, J., and Catt, K. J.,** Properties and regulation of high-affinity pituitary receptors for corticotropin-releasing factor, *Biochem. Biophys. Res. Commun.*, 110, 602, 1983.
4. **Holmes, M. C., Antoni, F. A., and Szentendrei, T.,** Pituitary receptors for corticotropin-releasing factor: no effect of vasopressin on binding or activation of adenylate cyclase, *Neuroendocrinology*, 39, 162, 1984.
5. **De Souza, E. B. and Kuhar, M. J.,** Corticotropin-releasing factor receptors in the pituitary gland and central nervous system: methods and overview, *Methods Enzymol.*, 124, 560, 1986.
6. **Wynn, P. C., Hauger, R. L., Holmes, M. C., Millan, M. A., Catt, K. J., and Aguilera, G.,** Brain and pituitary receptors for corticotropin releasing factor: localization and differential regulation after adrenalectomy, *Peptides*, 5, 1077, 1984.
7. **Millan, M., Jacobowitz, D., Catt, K. J., and Aguilera, G.,** Distribution of corticotropin releasing factor receptors in primate brain, *Proc. Natl. Acad. Sci. U.S.A.*, 83, 1921, 1986.
8. **Leroux, P. and Pelletier, G.,** Radioautographic study of binding and internalization of corticotropin-releasing factor by rat anterior pituitary corticotrophs, *Endocrinology*, 114, 14, 1984.
9. **Childs, G. V., Morell, J. L., and Aguilera, G.,** Cytochemical studies of CRH receptors in anterior lobe corticotrophs: binding, glucocorticoid regulation and endocytosis of [Biotinyl-Ser^1]CRH, *Endocrinology*, 119, 2129, 1986.

10. **Schwart, J., Billestrup, N., Perrin, M., Rivier, J., and Vale, W. W.,** Identification of corticotropin-releasing factor (CRF) target cells and effects of dexamethasone on binding in anterior pituitary using a fluorescent analog of CRF, *Endocrinology,* 119, 2376, 1986.
11. **Giguere, V., Labrie, F., Cote, J., Coy, D. H., Sueiras-Diaz, J., and Schally, A. V.,** Stimulation of cyclic AMP accumulation and corticotropin release by synthetic ovine corticotropin-releasing factor in rat anterior pituitary cells: site of glucocorticoid action, *Proc. Natl. Acad. Sci. U.S.A.,* 79, 3466, 1982.
12. **Aguilera, G., Harwood, J. P., Wilson, J. X., Morell, J., Brown, J. H., and Catt, K. J.,** Mechanisms of action of corticotropin-releasing factor and other regulators of corticotropin release in rat pituitary cells, *J. Biol. Chem.,* 258, 8039, 1983.
13. **Abou-Samra, A.-B., Catt, K. J., and Aguilera, G.,** Role of arachidonic acid in the regulation of adrenocorticotropin release from rat anterior pituitary cell cultures, *Endocrinology,* 119, 1427, 1986.
14. **Abou-Samra, A.-B., Catt, K. J., and Aguilera, G.,** Calcium-dependent control of corticotropin release in rat anterior pituitary cell cultures, *Endocrinology,* 121, 965, 1987.
15. **Vlaskovska, M. and Knepel, W.,** Beta-endorphin and adrenocorticotropin release from rat adrenohypophysis *in vitro:* evidence for local modulation by arachidonic acid metabolites of the cyclooxygenase and lipoxygenase pathway, *Neuroendocrinology,* 39, 334, 1984.
16. **Luini, G., Lewis, D., Guild, S., Corda, D., and Axelrod, J.,** Hormone secretagogues increase cytosolic calcium by increasing cAMP in corticotropin-secreting cells, *Proc. Natl. Acad. Sci. U.S.A.,* 82, 8034, 1985.
17. **Perrin, M. H., Haas, Y., Rivier, J. E., and Vale, W. W.,** Corticotropin-releasing factor binding to the anterior pituitary receptor is modulated by divalent cations and guanyl nucleotides, *Endocrinology,* 118, 1171, 1986.
18. **Millan, M. A., Abou-Samra, A.-B., Wynn, P. C., Catt, K. J., and Aguilera, G.,** Receptors and actions of corticotropin releasing hormone in the primate pituitary gland, *J. Clin. Endocrinol. Metab.,* 64, 1036, 1987.
19. **Aguilera, G., Wynn, P. C., Harwood, J. P., Hauger, R. L., Millan, M. A., Grewe, C., and Catt, K. J.,** Receptor-mediated actions of corticotropin-releasing factor in pituitary gland and nervous system, *Neuroendocrinology,* 43, 79, 1986.
20. **Gibbs, D. M. and Vale, W. W.,** Presence of corticotropin releasing factor-like immunoreactivity in hypophysial portal blood, *Endocrinology,* 111, 1418, 1982.
21. **Nishimura, E., Billestrup, N., Perrin, M., and Vale, W. W.,** Identification and characterization of a pituitary corticotropin-releasing factor binding protein by chemical cross-linking, *J. Biol. Chem.,* 262, 2893, 1987.
22. **Rosendale, B. E., Jarrett, D. B., and Robinson, A. G.,** Identification of a corticotropin-releasing factor-binding protein in the plasma membrane of ATt-20 mouse pituitary tumor cells and its regulation by dexamethasone, *Endocrinology,* 120, 2357, 1987.
23. **Abou-Samra, A.-B., Catt, K. J., and Aguilera, A.,** Involvement of protein kinase C in the regulation of adrenocorticotropin release from rat anterior pituitary cells, *Endocrinology,* 118, 212, 1986.
24. **Carvallo, P. and Aguilera, G.,** Protein kinase C mediates the effect of vasopressin in pituitary corticotrophs, *70th Endocr. Soc. Annu. Meet.,* New Orleans, 1988, abstr. 440.
25. **Giguere, V. and Labrie, F.,** Vasopressin potentiates cyclic AMP accumulation of ACTH release induced by corticotropin-releasing factor (CRF) in rat anterior pituitary cells in culture, *Endocrinology,* 111, 1752, 1982.
26. **Abou-Samra, A.-B., Harwood, J. P., Manganiello, V. C., Catt, K. J., and Aguilera, G.,** Phorbol 12-myristate 13-acetate and vasopressin potentiate the effect of corticotropin-releasing factor on cyclic AMP production in rat anterior pituitary cells, *J. Biol. Chem.,* 262, 1129, 1987.
27. **Hook, V. Y. H., Heisler, S., and Axelrod, J.,** Corticotropin-releasing factor stimulates phospholipid methylation and corticotropin secretion in mouse pituitary tumor cells, *Proc. Natl. Acad. Sci. U.S.A.,* 79, 6220, 1982.
28. **Heisler, S., Hook, V. Y. H., and Axelrod, J.,** Corticotropin releasing factor stimulation of protein carboxymethylation in mouse pituitary tumor cells, *Biochem. Pharmacol.,* 32, 1295, 1983.
29. **Giguere, V., Lefevre, G., and Labrie, F.,** Site of calcium requirement for stimulation of ACTH release in rat anterior pituitary cells in culture by synthetic ovine corticotropin-releasing factor, *Life Sci.,* 31, 3057, 1982.
30. **Murakami, K., Hashimoto, K., and Ota, Z.,** The effects of nifedipine on CRF-41 and AVP-induced ACTH released *in vitro, Acta Endocrinol. (Copenhagen),* 109, 32, 1985.
31. **Piascik, M. T., Bibich, M., and Rush, M. E.,** Calmodulin stimulation and calcium regulation of smooth muscle adenylate cyclase activity, *J. Biol. Chem.,* 258, 10913, 1983.
32. **Resink, T. J., Stucki, S., Grigorian, G. Y., Zschauer, A., and Buhler, F. R.,** Biphasic Ca^{2+} response of adenylate cyclase. The role of calmodulin in its activation by Ca^{2+} ions, *Eur. J. Biochem.,* 154, 451, 1986.

35. **Wynn, P. C., Harwood, J. P., Catt, K. J., and Aguilera, G.,** Regulation of corticotropin-releasing factor (CRF) receptors in the rat pituitary gland: effect of adrenalectomy on CRF receptors and corticotroph responses, *Endocrinology,* 116, 1653, 1985.
34. **Catt, K. J., Harwood, J. P., Aguilera, G., and Dufau, M. L.,** Hormonal regulation of peptide receptors and target cells responses, *Nature,* 280, 109, 1979.
35. **Oliver, C., Conte-Devolx, B., Rey, M., Boudouresque, F., Giraud, P., Castanas, E., and Porter, J. C.,** Immunoreactive 41-CRF in hypophysial portal blood of intact and adrenalectomized rats, *Acta Endocrinol.,* 103, 98, 1983.
36. **Plotsky, P. M. and Sawchenko, P. E.,** Hypophysial-portal plasma levels, median eminence content, and immunohistochemical staining of corticotropin-releasing factor, arginine vasopressin, and oxytocin after pharmacological adrenalectomy, *Endocrinology,* 120, 1361, 1987.
37. **Wynn, P. C., Harwood, J. P., Grewe, C., and Aguilera, G.,** Corticotropin-releasing factor (CRF) induces desensitization of the rat pituitary CRF receptor-adenylate cyclase complex, *Endocrinology,* 122, 351, 1988.
38. **Holmes, M. C., Catt, K. J., and Aguilera, G.,** Involvement of vasopressin in the downregulation of pituitary corticotropin releasing factor receptors following adrenalectomy, *Endocrinology,* 121, 2093, 1987.
39. **Holmes, M. C., Antoni, F. A., Catt, K. J., and Aguilera, G.,** Predominant release of vasopressin vs. corticotropin-releasing factor from the isolated median eminence after adrenalectomy, *Neuroendocrinology,* 43, 245, 1985.
40. **Kiss, J. Z., Mezey, E., and Skirboll, L.,** Corticotropin-releasing factor-immunoreactive neurons of the paraventricular nucleus become vasopressin positive after adrenalectomy, *Proc. Natl. Acad. Sci. U.S.A.,* 81, 1854, 1984.
41. **Koenig, J. I., Meltzer, H. Y., Devane, G. D., and Gudelsky, G. A.,** The concentration of arginine vasopressin in pituitary stalk plasma of the rat after adrenalectomy or morphine, *Endocrinology,* 118, 2534, 1986.
42. **Hauger, R. L., Millan, M. A., Catt, K. J., and Aguilera, G.,** Differential regulation of brain and pituitary corticotropin-releasing factor receptors by corticosterone, *Endocrinology,* 120, 1527, 1987.
43. **Abou-Samra, A.-B., Catt, K. J., and Aguilera, G.,** Biphasic inhibition of adrenocorticotropin release by corticosterone in cultured anterior pituitary cells, *Endocrinology,* 119, 972, 1986.
44. **Hauger, R. L., Millan, M. A., Lorang, M., Harwood, J. P., and Aguilera, G.,** Corticotropin releasing factor receptors and pituitary adrenal responses during immobilization stress, *Endocrinology,* 123, 396, 1988.
45. **Rivier, C. and Vale, W. W.,** Diminished responsiveness of the hypothalamic-pituitary-adrenal axis of the rat during exposure to prolonged stress: a pituitary-mediated mechanism, *Endocrinology,* 121, 1320, 1987.
46. **Sakellaris, P. C. and Vernikos-Danellis, J.,** Increased rate of response of the pituitary-adrenal system in rats adapted to chronic stress, *Endocrinology,* 97, 597, 1975.
47. **Gann, D. S., Bereiter, D. A., Carlson, D. E., and Thrivikraman, K. V.,** Neural interaction in control of adrenocorticotropin, *Fed. Proc,* 44, 161, 1985.
48. **Garcia-Marquez, C. and Armario, A.,** Chronic stress depresses exploratory activity and behavioral performance in the forced swimming test without altering ACTH response to a novel acute stressor, *Physiol. Behav.,* 40, 33, 1987.
49. **Raff, H., Sandri, R., and Segerson, T. P.,** Renin, ACTH and adrenocortical function during hypoxia and hemorrhage in conscious rats, *Am. J. Physiol.,* 250, R240, 1986.
50. **Swanson, L. W., Sawchenko, P. E., Rivier, J., and Vale, W. W.,** Organization of ovine corticotropin-releasing factor immunoreactive cells and fibers in the rat brain: an immunohistochemical study, *Neuroendocrinology,* 36, 165, 1983.
51. **Antoni, F. A., Palkovits, M., Makara, G. B., Linton, E. A., Lowry, P. J., and Kiss, J. Z.,** Immunoreactive corticotropin-releasing hormone in the hypothalamo-infundibular tract, *Neuroendocrinology,* 36, 415, 1983.
52. **Fischman, A. J. and Moldow, R. L.,** Extrahypothalamic distribution of CRF-like immunoreactivity in the rat brain, *Peptides,* 3, 149, 1982.
53. **Merchenthaler, I., Vigh, S., Petrusz, P., and Schally, A. V.,** Immunocytochemical localization of corticotropin-releasing factor (CRF) in the rat brain, *Am. J. Anat.,* 165, 385, 1982.
54. **Olschokwa, J. A., O'Donohue, T. L., Mueller, G. P., and Jacobowitz, D. M.,** The distribution of corticotropin releasing factor-like immunoreactive neurons in rat brain, *Peptides,* 3, 995, 1982.
55. **Brown, M.,** Corticotropin releasing factor: central nervous system sites of action, *Brain Res.,* 399, 10, 1986.
56. **Kalin, N. H.,** Behavioral effects of ovine corticotropin-releasing factor administered to rhesus monkeys, *Fed. Proc.,* 44, 249, 1985.
57. **Koob, G. F. and Bloom, F. E.,** Corticotropin-releasing factor and behavior, *Fed. Proc.,* 44, 259, 1985.

58. **Chen, F. M., Bilezikjian, L. M., Perrin, M. H., Rivier, J., and Vale, W.,** Corticotropin releasing factor receptor-mediated stimulation of adenylate cyclase activity in the rat brain, *Brain Res.*, 381, 49, 1986.
59. **Carpenter, M. B. and Sutin, J.,** in *Human Neuroanatomy*, Carpenter, M. B. and Sutin, J., Eds., Williams & Wilkins, Baltimore, 1983, 643.
60. **Rivier, C. and Vale, W.,** Influence of corticotropin-releasing factor on reproductive functions in the rat, *Endocrinology*, 114, 914, 1984.
61. **Sirinathsinghji, D. J. S., Rees, L. H., Rivier, J., and Vale, W.,** Corticotropin-releasing factor is a potent inhibitor of sexual receptivity in the female rat, *Nature*, 305, 1983.
62. **Petraglia, F., Sutton, S., Vale, W., and Plotsky, P.,** Corticotropin releasing factor decreases plasma LH levels in female rats by inhibiting GnRH release into hypophyseal portal circulation, *Endocrinology*, 120, 1083, 1987.
63. **Knigge, K. M. and Joseph, S. A.,** Anatomy of the opioid systems of the brain, *Can. J. Neurol. Sci.*, 11, 14, 1984.
64. **Nauta, W. J. H. and Haymaker, W.,** Hypothalamic nuclei and fiber connections, in *The Hypothalamus*, Haymaker, W. X., Anderson, X., and Nauta, W. J. H., Eds., Charles C. Thomas, Springfield, IL, 1969, 136.
65. **Valentino, R. J., Foote, S. L., and Aston-Jones, G.,** Corticotropin-releasing factor activates noradrenergic neurons of the locus coeruleus, *Brain Res.*, 270, 363, 1983.
66. **Nauta, W. J. H.,** Fibre degeneration following lesions of the amygdaloid complex in the monkey, *J. Anat.*, 95, 515, 1986.
67. **Kaada, B. R.,** Stimulation and ablation of the amygdaloid complex with reference to functional representations, in *The Neurobiology of the Amygdala*, Eleftheriou, B. E., Ed., Plenum Press, New York, 1972, 205.
68. **MacLean, P. D.,** Some psychiatric implications of physiological studies on frontotemporal portions of the limbic system, *Electroencephalogr. Clin. Neurophysiol.*, 4, 407, 1952.
69. **Carroll, B. J.,** Neuroendocrine regulation in depression. I. Limbic system-adrenocortical dysfunction, *Arch. Gen. Psychiatry*, 37, 737, 1980.
70. **Gold, P. W., Loriaux, D. L., Roy, A., Kling, M. A., and Calabrese, J. R.,** Responses to corticotropin-releasing hormone in the hypercortisolism of depression and Cushing's disease, *N. Engl. J. Med.*, 314, 1986.
71. **Tomori, N., Suda, T., Tozawa, F., Demura, H., Shizume, K., and Mouri, T.,** Immunoreactive corticotropin-releasing factor concentrations in cerebrospinal fluid from patients with hypothalamic-pituitary-adrenal disorders, *J. Clin. Endocrinol. Metab.*, 57, 1305, 1983.
72. **Nemeroff, C. B., Widerlov, E., Bissette, G., Walleus, H., Karlsson, I., Eklund, K., Kilts, C. D., and Loosen, P. T.,** Elevated concentrations of CSF corticotropin-releasing factor-like immunoreactivity in depressed patients, *Science*, 226, 1342, 1984.
73. **Hashimoto, K., Murakami, K., Hattori, T., Nimi, M., Fujino, K., and Ota, Z.,** Corticotropin-releasing factor (CRF)-like immunoreactivity in the adrenal medulla, *Peptides*, 5, 707, 1983.
74. **Suda, T., Tomori, N., Tozawa, F., Demura, H., Shizume, K., Mouri, T., Miura, Y., and Sasano, N.,** Immunoreactive corticotropin and corticotropin-releasing factor in human hypothalamus, adrenal, lung cancer, and pheochromocytoma, *J. Clin. Endocrinol. Metab.*, 58, 919, 1984.
75. **Petrusz, P., Merchenthaler, I., Maderdrut, J. L., and Heitz, P. U.,** Central and peripheral distribution of corticotropin-releasing factor, *Fed. Proc.*, 44, 229, 1985.
76. **Bruhn, T. O., Engeland, W. C., Anthony, E. L. P., Gann, D. S., and Jackson, I. M. D.,** Corticotropin-releasing factor in the adrenal medulla, in *The Hypothalamic-Pituitary-Adrenal Axis Revisited*, Vol. 512, Ganong, W. F., Dallman, M. F., and Roberts, J. L., Eds., Ann. of The New York Academy of Sciences, New York, 1988, 115.
77. **Udelsman, R., Harwood, J. P., Millan, M. A., Chrousos, G. P., Goldstein, D. S., Zimlichman, R., Catt, K. J., and Aguilera, G.,** Functional corticotropin releasing factor receptor in the primate peripheral sympathetic nervous system, *Nature*, 319, 147, 1986.

Chapter 12

REGULATION OF ADRENOCORTICOTROPIC HORMONE (ACTH) SECRETION BY CORTICOTROPIN-RELEASING FACTOR (CRF)

Catherine Rivier, Mark Smith, and Wylie Vale

TABLE OF CONTENTS

I. INTRODUCTION

It is now well established that the mechanisms which modulate adrenocorticotropic hormone (ACTH) secretion by the pituitary involve the interaction of several secretagogues. In this multifactorial regulation, corticotropin-releasing factor (CRF), first isolated from sheep and later characterized in a variety of species,[1-8] represents the predominant physiological stimulus. In addition, arginine-vasopressin, catecholamines, and angiotensin-II also act on the pituitary to increase the release of ACTH, but their effect is highly dependent upon the presence of endogenous CRF with which they exert an additive or synergetic action. This review will focus on the effect of CRF, alone or combined with other secretagogues, to modulate corticotropin secretion. For additional discussion of the role and function of CRF, as well as secretagogues, in controlling ACTH release, the reader is referred to a number of previously published reviews.[9-16]

II. RESULTS AND DISCUSSION

A. PITUITARY EFFECT OF CRF

1. Basal Secretion

The high potency and intrinsic activity exhibited by CRF to stimulate the release of ACTH has been established in a number of *in vitro* and *in vivo* models. Using fetal rat pituitary glands, Dupouy and Chatelain have observed that the response to CRF increased with gestational age.[17] When pituitary cells from adult donor rats or cows were maintained in culture, the effect of CRF was dose dependent, with an ED_{50} of approximately 0.5×10^{-9} *M* CRF and a maximum ACTH stimulation observed at about 5×10^{-9} *M* CRF.[18-22] Larger concentrations of CRF are reported to be necessary for maximum ACTH release by AtT-20 mouse pituitary tumor cells[23] and in short-term incubations of pituitary gland fragments.[24-26] The long-term exposure of anterior pituitary cells in culture[21] or intact rats,[27] to CRF also results in an increased ACTH synthesis, or POMC gene transcription, respectively. Childs and Burke, using a reverse hemolytic plaque assay, (a technique which allows the study of the secretion of individual cells,[28]) have shown that CRF (0.1 to 10 n*M*) increased the percentage of ACTH plaques to 9.8% of the total population of anterior pituitary cells.[29] Because ACTH cells are believed to represent approximately 10% of the anterior lobe cell population,[30] these results suggest that the secretion of most, if not all, corticotrophs, can be stimulated by CRF.

The *in vivo* ability of CRF administered by intravenous, subcutaneous or cerebroventricular routes to increase ACTH secretion has also been established in several species, including rat,[26,31,32] sheep,[33,34] primates,[35-37] and man.[38-48] While in man and rhesus monkeys the potency and duration of action of ovine and rat/human CRF are different,[41,42,45] they are indistinguishable in the rat (Figures 1 and 2). In this species, plasma ACTH levels reached a peak at 5 to 10 min following the i.v. injection of CRF, and stayed elevated for a period of time which was in direct relationship with the dose of the peptides. In contrast, CRF induces a biphasic pattern of ACTH secretion in man.[37,42,43] Because CRF is structurally related to other peptides such as urotensin and sauvagine,[49-51] a number of studies have compared the stimulation of these homologous compounds on ACTH secretion. Using either an *in vitro* or in an *in vivo* model, we have shown that CRF, sauvagine, and urotensin I were equipotent to release ACTH.[52] In contrast, urotensin I is several times more potent than the two other peptides to increase ACTH secretion by the fish pituitary.[50,53]

2. Effect of Glucocorticoids

The negative feedback exerted by glucocorticoids on ACTH secretion is well established

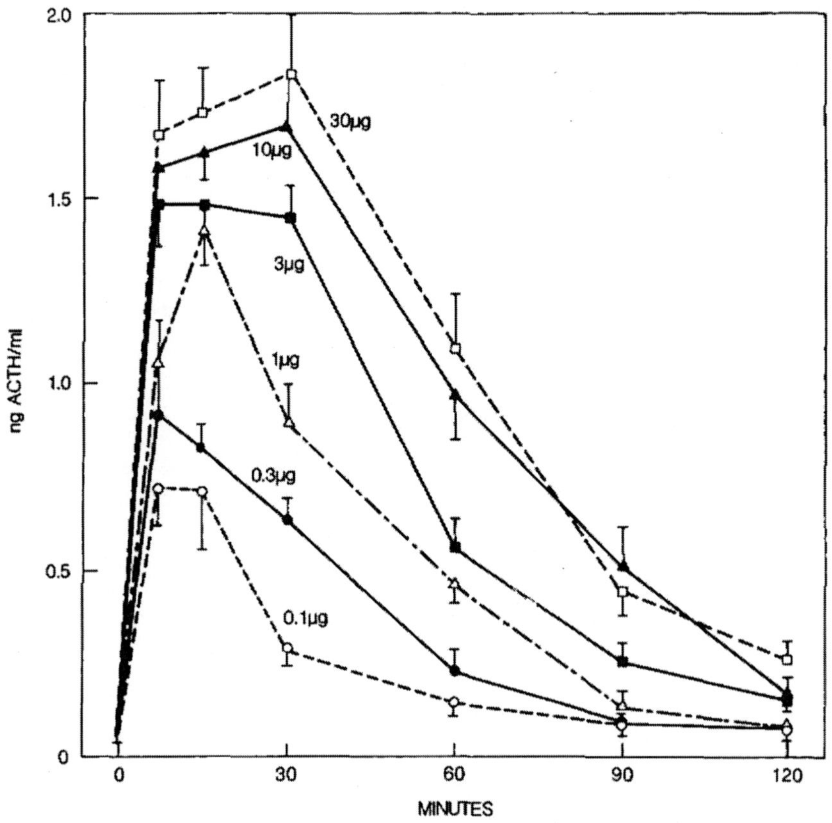

FIGURE 1. Effect of the i.v. injection of ovine CRF to intact, freely moving rats. Each point represents the mean ± SEM of five animals.

and has been extensively discussed.[54] In cultured pituitary cells,[21] as well as in the intact rat,[32] corticosteroids interfere with the stimulatory action of CRF in a dose-dependent manner. Corticosteroid feedback is believed to operate in at least three time domains, fast, intermediate, and slow.[54] *In vivo*, fast feedback occurs during the initial period of increasing plasma corticoid concentration, and is rate sensitive as well as short-lived.[32] This rapid inhibition also occurs *in vitro*.[55,56] Delayed feedback appears to depend on the levels of circulating steroids achieved, as well as the total dose of steroid administered to rats. This inhibition occurs several hours after exposure of dispersed pituitary cells or lesioned rats[54,55] to corticoids. All three are demonstrable at the level of the anterior pituitary using *in vitro* or *in vivo* systems.[22,32,55,57] In contrast with what has been observed in *in vitro* experiments,where high concentrations of CRF can still elicit the release of ACTH even at maximally effective doses of dexamethasone[21] (thus suggesting that this inhibition was noncompetitive), in intact rats large doses of dexamethasone were capable of totally abolishing the stimulatory action of up to 30 µg CRF, which represents a pharmacological amount of peptide (Figure 3A). One hundred µg CRF induced a very small, but detectable rise in plasma ACTH levels measured 10 min later. Measurement of plasma corticosterone levels indicated that these small increases in ACTH levels were, however, sufficient to stimulate adrenal function (Figure 3B). When ACTH and corticosterone levels were measured 30 min after the injection of CRF (Figure 4A and B), the ability of CRF to elevate circulating levels of corticoids even in the presence of large amounts of dexamethasone was even more apparent—a finding which probably reflects the cumulative action of even small elevations of ACTH levels over a prolonged period of time. This finding is consistent with the obser-

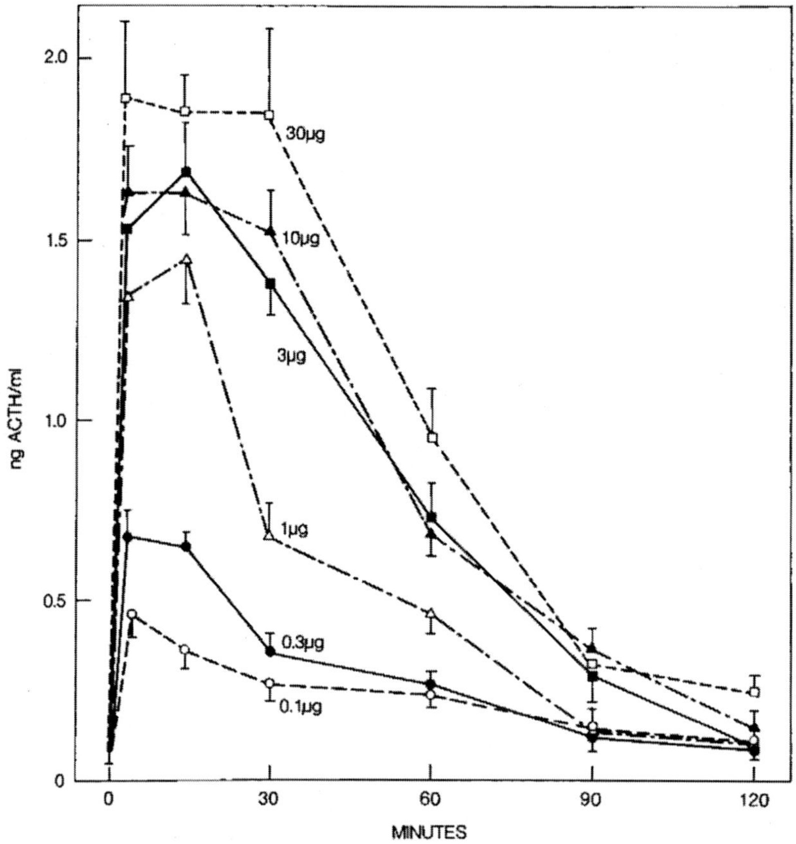

FIGURE 2. Effect of the i.v. injection of rat/human CRF to intact freely moving rats. Each point represents the mean ± SEM of five animals.

vation that very small increases in ACTH release, which may not be easily detectable by RIA, are capable of stimulating corticoid secretion.[54]

3. Interaction with other Secretagogues

A number of *in vitro* and *in vivo* experiments have shown that arginine-vasopressin, catecholamines, or angiotensin-II (A-II) represent weak secretagogues of ACTH secretion. Using cultured pituitary cells, Vale et al. have shown that arginine vasopressin (AVP), oxytocin, catecholamines, or A-II stimulated ACTH secretion, but had lower potency and intrinsic activity than CRF.[21] CRF and AVP, oxytocin, A-II, or epinephrine, exhibited effect additivity, because the coaddition of AVP or oxytocin to maximally effective doses of CRF caused additional increases in ACTH release.[21] The same results have been obtained by other investigators using similar or comparable *in vitro* systems,[58-71] pharmacologically blocked rats,[31,75,76] or human volunteers.[43,72-74] In contrast, the peripheral injection of AVP, catecholamines or A-II to intact rats elicit large increases in plasma ACTH levels.[31,52,77] We therefore asked the following question: Is the stimulatory action of these compounds due to their interaction with endogenous CRF? Two models were used to test this hypothesis. First, rats were injected with chlorpromazine, morphine, and Nembutal to block the release of endogenous CRF.[78] In these animals, AVP, catecholamines and A-II produced only marginal elevations of plasma ACTH levels.[31,77,79] In the second model, normal rabbit serum or an antiserum against CRF[14] were administered immediately before the compound to be tested, and we observed that the immunoneutralization of endogenous CRF markedly decreased the

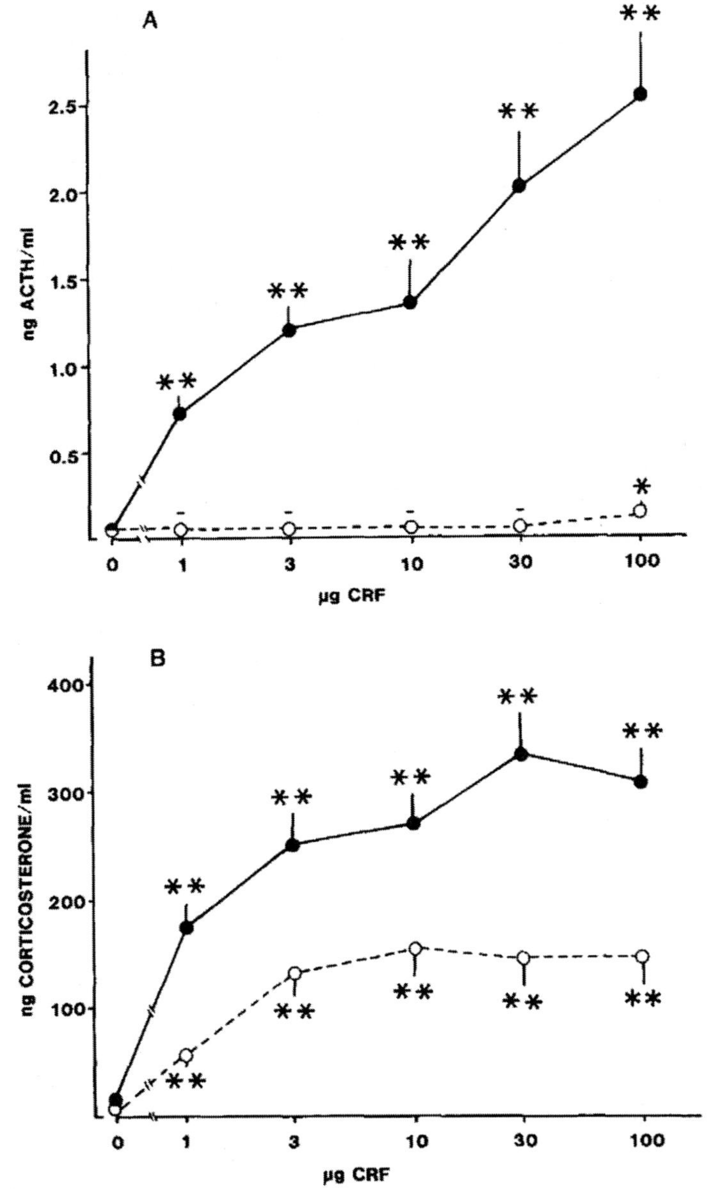

FIGURE 3. Effect of the i.v. administration of CRF in the absence (●) or presence (○) of 100 µg dexamethasone (DEX) on ACTH and corticosterone secretion by intact freely moving rats. Blood samples were obtained 10 min after the injection of CRF. DEX was administered s.c. 6 h before the assay. Each point represents the mean ± SEM of five animals. ⁻, $p > 0.05$; *, $p < 0.05$; **, $p < 0.01$. ACTH was measured in 50 µl plasma, and the limit of sensitivity of the RIA was 2.1 pg tube, or 42 pg/ml.

stimulatory effect of all three compounds.[31,77] The results from these two sets of experiments support the hypothesis that the *in vivo* ability of AVP, catecholamines or A-II to increase ACTH secretion is due to their additive or potentiating effects with CRF.It should be noted that the diminished ability of Brattleboro rats (which lack endogenous vasopressin) to release ACTH when compared to normal rats also suggests that vasopressin exerts a synergic role with CRF.[80]

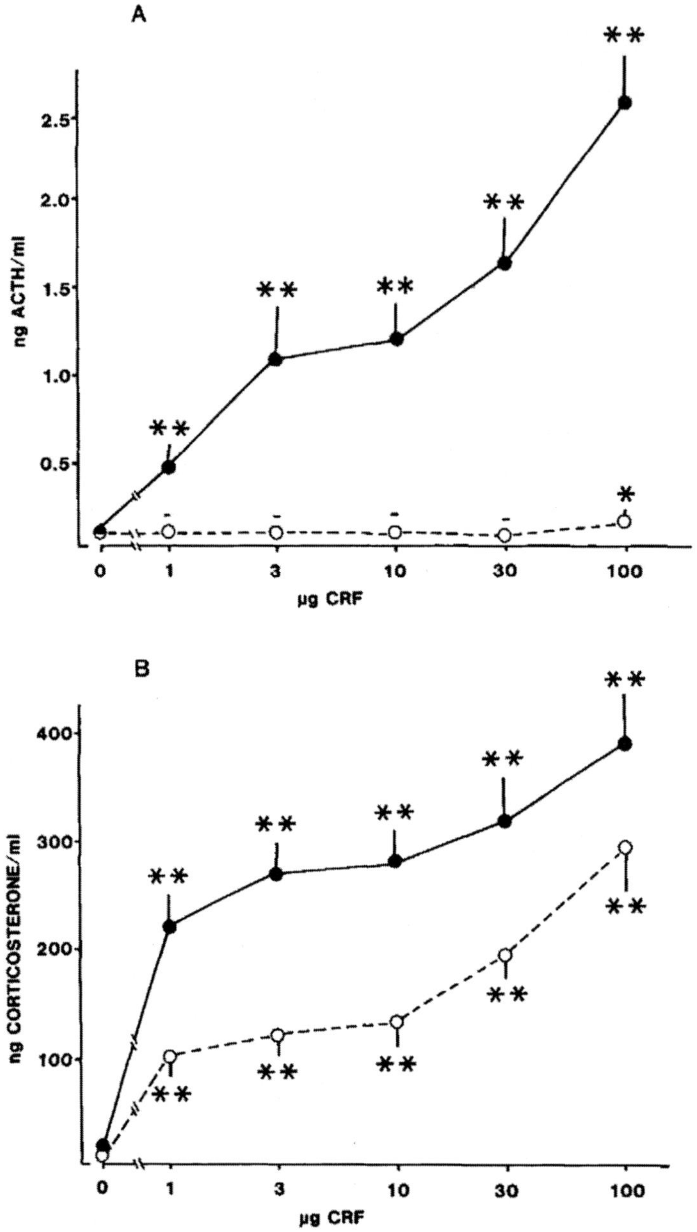

FIGURE 4. Same as in Figure 3. Blood samples were obtained 30 min after the injection of CRF.

The nature of the interaction between AVP and CRF on the corticotroph has been the subject of controversy. Because AVP effects diuresis and vasoconstriction through two different types of receptors (respectively, V_2, which are coupled to adenylate cyclase, and V_1, which are not),[81] a number of investigators have attempted to determine which class of receptors mediated the ACTH-releasing action of AVP. Some studies using vasopressin analogs with selective pressor or antidiuretic activity have suggested that the ability of AVP to induce ACTH release was primarily correlated with its pressor effect.[82-84] We have shown that *in vitro* as well as *in vivo* studies that V_1, but not V_2, AVP antagonists inhibited the ACTH-releasing and the CRF-potentiating effects of AVP.[76] These results have suggested

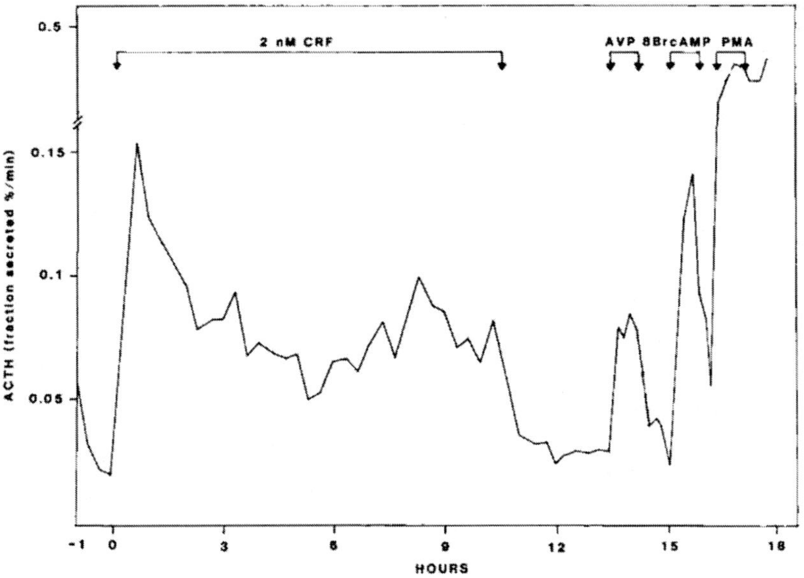

FIGURE 5. Effect of the continuous superfusion of pituitary cells with 2 n*M* CRF. About 8×10^6 cells were superfused at a rate of 10 ml/h. After washing out the CRF, the cells were challenged with 100 n*M* [Arg^8]-vasopressin, 3 m*M* 8-Br cAMP or 100 n*M* PMA. The ACTH content at the start was 1050 ng, 370 ng remained at the end while about 750 ng was secreted, indicating that 60 ng was synthesized during the 18-h experiment.

that the action of AVP on the corticotroph was primarily mediated through V_1 (pressor-like) receptors. Other results, however, do not agree with this conclusion. For example, some vasopressin analogs exhibit a CRF-like activity with a rank order which is different from that for vasopressor activity,[86-89] which suggests that their ACTH-releasing activity may not be wholly related to their pressor activity. Additionally, [1-deamino-8-D-arginine]vasopressin (dDAVP), a vasopressin analog with high antidiuretic activity and low pressor activity,[82] also stimulates ACTH release and potentiates the effect of CRF in both *in vivo* and *in vitro* experiments.[31,76,87] However, whether these effects can be blocked with V_1 or V_2 receptor antagonists, is not resolved.[76,89-91] These observations have led to the hypothesis that the anterior pituitary vasopressin receptors resemble, but are not identical to, V_1 receptors, and might be classified as V_3 (pituitary) receptors.[92-94]

4. Effect of Prolonged Exposure to CRF

Exposure of cultured anterior pituitary cells to CRF for 24 h caused an increase in ACTH synthesis.[21] However, pretreatment of these cells with CRF is known to reduce the ability of CRF to restimulate ACTH release.[95-98] Using rat pituitary cells attached to cytodex beads, which were superfused with 2 n*M* CRF for 10 h, we have observed a biphasic pattern of ACTH secretion in which the initial maximal secretion was followed by a rapid decline (Figure 5). ACTH release, however, was maintained at about 30% of the initial rate, an observation that was also made in intact animals infused with CRF for 24 h.[99] This indicates that only partial desensitization to CRF occurs during prolonged exposure to the peptide. Although the cultured cells were partially refractory to CRF, they remained fully sensitive to AVP, 8-Br cAMP and phorbol myristate acetate (PMA).

The repeated or continuous administration of CRF to human volunteers[100-102] or intact rats[99] also causes some degree of pituitary desensitization. In the rat, we have observed that the acute i.v. injection of CRF caused a period of transient pituitary refractoriness,[103] and the infusion of CRF for 1 to 7 d also caused some degree of desensitization of the pituitary-

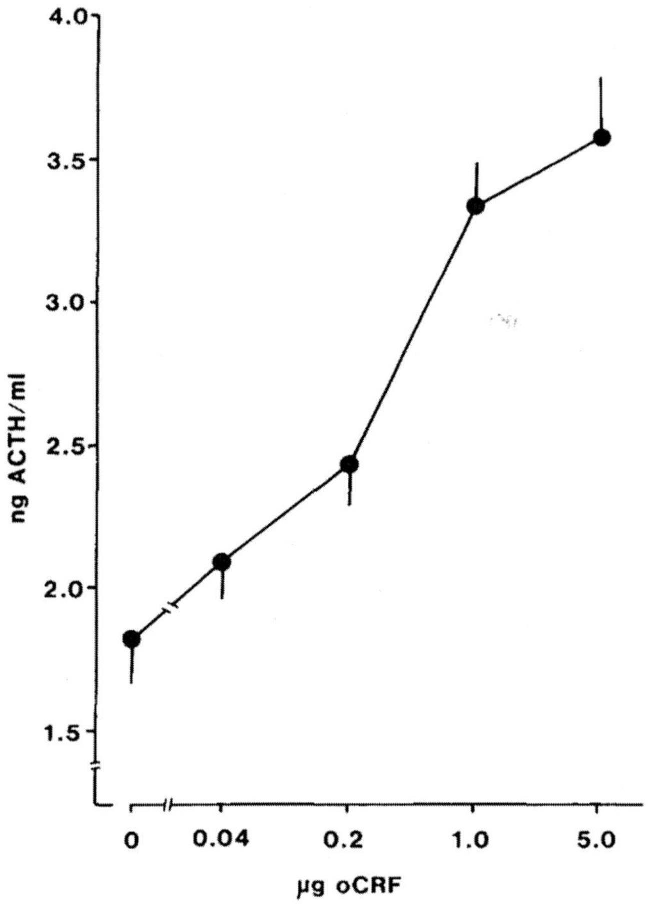

FIGURE 6. Effect of the i.v. administration of ovine CRF (oCRF) on plasma ACTH levels in adrenalectomized, freely moving rats. Blood samples were obtained 5 min after injection of the peptide. Each point represents the mean ± SEM of five animals.

adrenal axis.[99] However, the observation that adrenalectomized rats can maintain constantly elevated plasma ACTH levels, and respond to CRF in a manner similar to that of intact animals (Figure 6), has suggested that the decreased effectiveness of CRF to stimulate pituitary secretion over prolonged periods of time was at least partially mediated through steroid feedback.

Because either intact or adrenalectomized rats exposed to a 3-h stress cannot maintain sustained elevations in their plasma ACTH levels, we examined a possible role of pituitary desensitization as well as steroid feedback in mediating this phenomenon.[104] The observation that CRF, norepinephrine, vasopressin, or phorbol ester all caused significantly smaller increases in the plasma ACTH levels of intact rats subjected to electroshocks, when compared to control animals, suggested that there was no specific desensitization to CRF. Interestingly, the recovery of the responsiveness to CRF administration paralleled pituitary ACTH content: this content was markedly depleted at the end of the 3-h shock session, at a time when the effect of CRF was significantly blunted. Measurement of CRF-induced ACTH release, and pituitary ACTH content, from 1 to 24 h following cessation of the stress, showed a comparable pattern of recovery. These results therefore suggested that the temporary decrease of a readily releasable pituitary ACTH pool represents one of the mechanisms through which continuous exposure to CRF diminishes ACTH secretion.

B. ROLE OF ENDOGENOUS CRF

Having shown that CRF acted both *in vitro*[21] and *in vivo*[32] to stimulate ACTH secretion, we asked the following question: Is endogenous CRF involved in modulating basal and/or stimulated ACTH release? Hypophysial portal blood contains concentrations of CRF in the range of those shown to stimulate ACTH secretion *in vitro*,[105,106] and exposure to the stress of hemorrhage significantly increases the release of CRF.[107,108] Using a CRF-antiserum[15] or a specific CRF antagonist,[98] we[109-111] as well as other investigators,[112-118] have shown that removal of endogenous CRF markedly decreased basal plasma ACTH levels in adrenalectomized rats, as well as stress-induced ACTH secretion in intact animals. In addition to the above studies, support for the important physiological role of CRF has come from experiments showing that lesions of the mediobasal hypothalamus and destruction of CRF fibers in the rat brain blocked the corticosterone response to stress.[23,118] Also, Moldow and Fishman have reported that the production of CRF antibodies by active immunization in rabbits produced adrenal atrophy and a decrease in pituitary ACTH content.[119]

We consistently observed that ACTH values of stressed rats in which endogenous CRF had been immunoneutralized were significantly lowered, but did not however completely return to baseline values.[110] This suggested that endogenous CRF was a very important physiological regulator of ACTH secretion, but that there were also other factors which played a role. Because exposure to stress is accompanied by the release of not only CRF, but also of those secretagogues which act synergically with it,[107,108] we investigated the relative role of CRF, vasopressin, and catecholamines in mediating stress-induced ACTH secretion. We observed that pretreatment with a vasopressin antagonist decreased the plasma ACTH levels of ether-stressed rats.[107,111] Similar results were obtained by other investigators[114,116] with an antiserum against vasopressin. These results strongly suggest that vasopressin also represents a physiologically relevant peptide involved in stress-induced ACTH release. It should be noted, however, that this may not be true of all stresses because Mormede found no effect of a vasopressin antagonist in rats exposed to physiological stress.[120]

The physiological role played by catecholamines has also given rise to controversy. Because catecholamines act both at the level of the pituitary and the brain to exert stimulatory as well as inhibitory effects on ACTH secretion,[122] their precise role in modulating the corticotropins response to stress has been difficult to ascertain. We have observed that the ganglionic blocker chlorisondamine markedly inhibited stress-induced ACTH release.[111] This suggested that peripheral catecholamines mediated stress-induced ACTH release. By contrast, Tilders et al. have reported that catecholamines did not play a significant role.[122,123] These investigators based their conclusion on the observation that extirpation of the adrenal medulla or sympathectomy did not interfere with the pituitary adrenal response to stress. Because the respective role of factors other than CRF in modulating stress-induced ACTH release can vary (for example, endogenous CRF appears to be the sole mediator of ACTH secretion during tail-hanging,[109] and as discussed above, the role of AVP during stress appears to depend on the nature of stress), it is possible that differences in the type of stress used may explain the differences observed with the use of catecholamine blockers.

A-II stimulates ACTH secretion,[125-129] but its role during stress is not clear. Some investigators have proposed that at least in the sheep, A-II might act within the central nervous system to release ACTH under basal as well as stressful circumstances[129] while in the rat, this secretagogue has not been found to mediate the effect of stress.[130] Consequently, the respective role of the pituitary vs. brain sites of action of A-II in modulating ACTH release has not yet been resolved.

III. CONCLUSION

In conclusion, there is a plurality of factors which modulate ACTH secretion during stress.[31,107,111,131,132] Among these, endogenous CRF plays a predominant role, but it also interacts in important ways with other factors such as vasopressin, catecholamines and possibly A-II.

IV. ACKNOWLEDGMENTS

Research was supported by NIH Grants AM-26741, HD-13527, and AA-03504. Research conducted in part by The Clayton Foundation, California Division, C. R. and W. V. are Clayton Foundation Investigators. The authors are grateful to Greta Berg, Robert Galyean, Georgia Morgan, David Hutchinson, Ron Kaiser, Richard McClintock, and Mary Tam for their expert technical assistance and to Bethany Connor for manuscript preparation.

REFERENCES

1. **Esch, F., Ling, N., Bohlen, P., Baird, A., Benoit, R., and Guillemin, R.,** Isolation and characterization of the bovine hypothalamic corticotropin-releasing factor, *Biochem. Biophys. Res. Commun.*, 122, 899, 1984.
2. **Furutani, Y., Morimoto, Y., Shibahara, S., Noda, M., and Takahashi, H.,** Cloning and sequence analysis of cDNA for ovine corticotropin-releasing factor precursor, *Nature*, 301, 537, 1983.
3. **Jingami, H., Mizuno, N., Takahashi, H., Shibahara, S., Furutani, Y., Imura, H., and Numa, S.,** Cloning and sequence analysis of cDNA for rat corticotropin-releasing-factor precursor, *Fed. Eur. Biochem. Soc.*, 191, 63, 1985.
4. **Ling, N., Esch, F., Bohlen, P., Baird, A., and Guillemin, R.,** Isolation and characterization of caprine corticotropin-releasing factor, *Biochem. Biophys. Res. Commun.*, 122, 1218, 1984.
5. **Patthy, M., Schlesinger, D. H., Horvath, J., Mason-Garcia, M., Szoke, B., and Schally, A. V.,** Isolation and amino acid sequence of corticotropin releasing factor from pig hypothalami, *Proc. Natl. Acad. Sci. U.S.A.*, 82, 8762, 1985.
6. **Rivier, J., Spiess, J., and Vale, W.,** Characterization of rat hypothalamic corticotropin-releasing factor, *Proc. Natl. Acad. Sci. U.S.A.*, 80, 4851, 1983.
7. **Spiess, J., Rivier, J., and Vale, W.,** Sequence analysis of rat hypothalamic corticotropin-releasing factor with the orthophthalaldehyde strategy, *Biochemistry*, 22, 4341, 1983.
8. **Vale, W., Spiess, J., Rivier, C., and Rivier, J.,** Characterization of a 41-residue ovine hypothalamic peptide that stimulates secretion of corticotropin and β-endorphin, *Science*, 213, 1394, 1981.
9. **Antoni, F. A.,** Hypothalamic control of adrenocorticotropin secretion: advances since the discovery of 41-residue corticotropin-releasing factor, *Endocr. Rev.*, 7, 351, 1986.
10. **Brodish, A.,** Control of ACTH secretion by cortocitropin-releasing factors(s), *Vitam. Horm.*, 37, 111, 1979.
11. **Buckingham, J. C.,** Corticotrophin releasing factor, *Pharmacol. Rev.*, 31, 253, 1980.
12. **Howlett, T. A. and Rees, L. H.,** Endogenous opioid peptides and hypothalamo-pituitary function, *Annu. Rev. Physiol.*, 48, 527, 1986.
13. **Rivier, C. L. and Plotsky, P. M.,** Mediation by corticotropin-releasing factor (CRF) of adenohypophysial hormone secretion, *Annu. Rev. Physiol.*, 48, 475, 1986.
14. **Vale, W., Rivier, C. Brown, M. R., Spiess, J. Koob, G., Swanson, L., Bilezikjian, L., Bloom, F., and Rivier, J.,** Chemical and biological characterization of corticotropin releasing factor, *Rec. Prog. Horm. Res.*, 39, 245, 1983.
15. **Vale, W. W., Rivier, C., Spiess, J., and Rivier, J.,** Corticotropin releasing factor, in *Brain Peptides*, Krieger, D., Brownstein, M., and Martin, J., Eds., John Wiley & Sons, New York, 1983, chap. 38.
16. **Yasuda, N., Greer, M. A., and Aizawa, T.,** Corticotropin-releasing factor, *Endocr. Rev.*, 3, 123, 1982.
17. **Dupouy, J. P. and Chatelain, A.,** In-vitro effects of corticosterone, synthetic ovine corticotrophin releasing factor and arginine vasopressin on the release of adrenocorticotrophin by fetal rat pituitary glands, *J. Endocrinol.*, 101, 339, 1984.
18. **Aguilera, G., Harwood, J. P., Wilson, J. X., Morell, J., Brown, J. H., and Catt, K. J.,** Mechanisms of action of corticotropin-releasing factor and other regulators of corticotropin release in rat pituitary cells, *J. Biol. Chem.*, 258, 8039, 1983.
19. **Aguilera, G., Wynn, P. C., Harwood, J. P., Hauger, R. L., Millan, M. A., Grewe, C., and Catt, K. J.,** Receptor-mediated actions of corticotropin-releasing factor in pituitary gland and nervous system, *Neuroendocrinology*, 43, 79, 1986.

20. **Baird, A., Wehrenberg, W. B., Shibasaki, T., Benoit, R., Chong-Li, Z., Esch, F., and Ling, N.,** Ovine corticotropin-releasing factor stimulates the concomitant secretion of corticotropin, β-lipotropin, β-endorphin and γ-melanotropin by the bovine adenohypophysis *in vitro*, *Biochem. Biophys. Res. Commun.*, 108, 959, 1982.
21. **Vale, W., Vaughan, J., Smith, M., Yamamoto, G., Rivier, J., and Rivier, C.,** Effects of synthetic ovine corticotropin-releasing factor, glucocorticoids, catecholamines, neurohypophysial peptides, and other substances on cultured corticotropi cells, *Endocrinology*, 113, 1121, 1983.
22. **Hook, V. Y. H., Heisler, S., Sabol, S. L., and Axelrod, J.,** Corticotropin releasing factor stimulates adrenocorticotropin and β-endorphin release from AtT-20 mouse pituitary tumor cells, *Biochem. Biophys. Res. Commun.*, 106, 1364, 1982.
23. **Tilders, F. J. H., Schipper, J., Lowry, P. J., and Vermes, I.,** Effect of hypothalamus lesions on the presence of CRF-immunoreactive nerve terminals in the median eminence and on the pituitary-adrenal response to stress, *Regul. Pept.*, 5, 77, 1982.
24. **Durand, P., Cathiard, A-M., Dacheux, F., Naaman, E., and Saez, J. M.,** *In vitro* stimulation and inhibition of adrenocorticotropin release by pituitary cells from ovine fetuses and lambs, *Endocrinology*, 118, 1387, 1986.
25. **Knepel, W., Homolka, L., Vlaskovska, M., and Nutto, D.,** Stimulation of adrenocorticotropin/β-endorphin release by synthetic ovine corticotropin-releasing factor in vitro, *Neuroendocrinology*, 38, 344, 1984.
26. **Turkelson, C. M., Arimura, A., Culler, M. D., Fishback, J. B., Groot, K., Kanda, M., Luciano, M., Thomas, C. R., Chang, T. D., Chang, J. K., and Shimizu, M.,** In vivo and in vitro release of ACTH by synthetic CRF, *Peptides*, 2, 425, 1981.
27. **Bruhn, T. O., Sutton, R. E., Rivier, C. L., and Vale, W. W.,** Corticotropin-releasing factor regulates proopiomelanocortin messenger ribonucleic acid levels in vivo, *Neuroendocrinology*, 39, 170, 1984.
28. **Frawley, L. S. and Neill, J. D.,** A reverse hemolytic plaque assay for microscopic visualization of growth hormone release from individual cells; evidence for somatotrope heterogeneity, *Neuroendocrinology*, 39, 484, 1984.
29. **Childs, G. V. and Burke, J. A.,** Use of the reverse hemolytic plaque assay to study the regulation of anterior lobe adrenocorticotropin (ACTH) secretion by ACTH-releasing factor, arginine vasopressin, angiotensin II, and glucocorticoids, *Endocrinology*, 120, 439, 1987.
30. **Westlund, K. M., Aguilera, G., and Childs, G. V.,** Quantification of morphological changes in pituitary corticotropes produced by *in vivo* corticotropin releasing-factor stimulation and adrenalectomy, *Endocrinology*, 116, 439, 1985.
31. **Rivier, C. and Vale, W.,** Effects of corticotropin-releasing factor, neurohypophyseal peptides, and catecholamines on pituitary function, *Fed. Proc.*, 44, 189, 1985.
32. **Rivier, C., Brownstein, M., Spiess, J., Rivier, J., and Vale, W.,** *In vivo* corticotropin-releasing factor-induced secretion of adrenocorticotropin, β-endorphin, and corticosterone, *Endocrinology*, 110, 272, 1982.
33. **Donald, R. A., Redekipp, C., Cameron, V., Nicholls, M. G., Bolton, J., Livesey, J., Espiner, E. A., Rivier, J., and Vale, W.,** The hormonal actions of corticotropin-releasing factor in sheep: effect of intravenous and intracerebroventricular injection, *Endocrinology*, 113, 866, 1983.
34. **Kalin, N. H., Conder, J. C., and Shelton, S. E.,** Effects of synthetic ovine CRF on ACTH, cortisol and blood pressure in sheep, *Peptides*, 4, 221, 1983.
35. **Insel, T. R., Aloi, J. A., Goldstein, D., Wood, J. H., and Jimerson, D. C.,** Plasma cortisol and catecholamine responses to intracerebroventricular administration of CRF to rhesus monkeys, *Life Sci.*, 34, 1873, 1984.
36. **Schulte, H. M., Chrousos, G. P., Oldfield, E. H., Gold, P. W., Cutler, B. G., Jr., and Loriaux, D. L.,** The effects of corticotropin releasing factor on the anterior pituitary function of stalk-sectioned cynomolgus macaques: dose response of cortisol secretion, *J. Clin. Endocrinol. Metab.*, 55, 810, 1982.
37. **Schurmeyer, T. H., Gold, P. W., Gallucci, W. T., Tomai, T. P., Cutler, G. B., Jr., Loriaux, D. L., and Chrousos, G. P.,** Effects and pharmacokinetic properties of the rat/human corticotropin-releasing factor in rhesus monkeys, *Endocrinology*, 117, 300, 1985.
38. **Chrousos, G. P., Calabrese, J. R., Avgerinos, P., Kling, M. A., Rubinow, D., Oldfield, E. H., Schuermeyer, T., Kellner, C. H., Cutler, G. B., Jr., Loriaux, D. L., and Gold, P. W.,** Corticotropin releasing factor: basic studies and clinical applications, *Prog. Neuro-Psychopharmacol. Biol. Psychiatry*, 9, 349, 1985.
39. **Grossman, A., Perry, L., Schally, A. V., Rees, L. H., Nieuwenhuyzen Kruseman, A. C., Tomlin, S., Coy, D. H., Comaru-Schally, A.-M., and Besser, G. M.,** New hypothalamic hormone, corticotropin-releasing factor, specifically stimulates the release of adrenocorticotropic hormone and cortisol in man, *Lancet*, 1, 921, 1982.
40. **Hermus, A. R. M. M., Pieters, G. F. F. M., Smals, A. G. H., Benraad, T. J., and Kloppenborg, P. W. C.,** Plasma adrenocorticotropin, cortisol, and aldosterone responses to corticotropin-releasing factor: modulatory effect of basal cortisol levels, *J. Clin. Endocrinol. Metab.*, 58, 187, 1984.

41. **Hollander, C. S., Audhya, T., Nakane, T., Kanie, N., Kageyama, N., Kuwayama, A., Golbe, L., and Schlesinger, D.,** Distribution, biosynthesis, and physiological role of corticotropin-releasing factor in the human: an overview, *Am. Phys.*, 96, 122, 1983.
42. **Nicholson, W. E., DeCherney, G. S., Jackson, R. V., DeBold, C. R., Uderman, H., Alexander, A. N., Rivier, J., Vale, W., and Orth, D. N.,** Plasma distribution, disappearance half-time, metabolic clearance rate, and degradation of synthetic ovine cortocotropin-releasing factor in man, *J. Clin. Endocrinol. Metab.*, 57, 1263, 1983.
43. **Orth, D. N., DeBold, C. R., DeCherney, G. S., Jackson, R. V., Sheldon, W. R., Jr., Nicholson, W. E., Uderman, H., Alexander, N. A., and Island, D. P.,** Clinical studies with synthetic ovine corticotropin-releasing factor, *Fed. Proc.*, 44, 197, 1985.
44. **Schulte, H. M., Chrousos, G. P., Booth, J. D., Oldfield, E. H., Gold, P. W., Cutler, G. B., Jr., and Loriaux, D. L.,** Corticotropin-releasing factor: pharmacokinetics in man, *J. Clin. Endocrinol. Metab.*, 58, 192, 1984.
45. **Schurmeyer, T. H., Avgerinos, P. C., Gold, P. W., Gallucci, W. T., Tomai, T. P., Cutler, G. B., Jr., Loriaux, D. L., and Chrousos, G. P.,** Human corticotropin-releasing factor in man: pharmacokinetic properties and dose-response of plasma adrenocorticotropin and cortisol secretion, *J. Clin. Endocrinol. Metab.*, 59, 1103, 1984.
46. **Sheldon, W. R., Jr., deBold, C. R., Evans, W. S., DeCherney, G. S., Jackson, R. V., Island, D. P., Thorner, M. O., and Orth, D. N.,** Rapid sequential intravenous administration of four hypothalamic releasing hormones as a combined anterior pituitary function test in normal subjects, *J. Clin. Endocrinol. Metab.*, 60, 623, 1985.
47. **Stalla, G. K., Hartwimmer, J., Schopohl, J., von Werder, K., and Muller, O. A.,** Intravenous application of ovine and human corticotropin releasing factor (CRF): ACTH, cortisol and CRF levels, *Neuroendocrinology*, 42, 1, 1986.
48. **Tsukada, T., Nakai, Y., Koh, T., Tsujii, S., and Imura, H.,** Plasma adrenocorticotropin and cortisol responses to intravenous injection of corticotropin-releasing factor in the morning and evening, *J. Clin. Endocrinol. Metab.*, 57, 869, 1983.
49. **Lau, S. H., Rivier, J., Vale, W., Kaiser, E. T., and Kezdy, F. J.,** Surface properties of an amphiphilic peptide hormone and of its analog: corticotropin-releasing factor and sauvagine, *Proc. Natl. Acad. Sci. U.S.A.*, 80, 7070, 1983.
50. **Lederis, K., Letter, A., McMaster, D., Ichikawa, T., MacCannell, K. L., Kobayashi, Y., Rivier, J., Rivier, C., Vale, W., and Fryer, J.,** Isolation, analysis of structure, synthesis, and biological actions of urotensin I neuropeptides, *Can. J. Biochem. Cell Biol.*, 61, 602, 1983.
51. **Montecucchi, P. C., Anastasi, A., de Castiglione, R., and Erspamer, V.,** Isolation and amino acid composition of sauvagine, *Int. J. Pept. Protein Res.*, 16, 191, 1980.
52. **Rivier, C., Rivier, J., Lederis, K., and Vale, W.,** In vitro and in vivo ACTH-releasing activity of ovine CRF, sauvagine and urotensin I, *Regul. Pept.*, 5, 139, 1983.
53. **Fryer, J., Lederis, K., and Rivier, J.,** Urotensin I, A CRF-like neuropeptide, stimulates ACTH release from the teleost pituitary, *Endocrinology*, 113, 2308, 1983.
54. **Keller-Wood, M. E. and Dallman, M. F.,** Corticosteroid inhibition of ACTH secretion, *Endocr. Rev.*, 5, 1, 1984.
55. **Mahmoud, S. N., Scaccianoce, S., Scraggs, P. R., Nicholson, S. A., Gillham, B., and Jones, M. T.,** Characteristics of corticosteroid inhibition of adrenocorticotrophin release from the anterior pituitary gland of the rat, *J. Endocrinol.*, 102, 33, 1984.
56. **Widmaier, E. P. and Dallman, M. F.,** The effects of corticotropin-releasing factor on adrenocorticotropin secretion from perifused pituitaries *in vitro:* rapid inhibition by glucocorticoids, *Endocrinology*, 115, 2368, 1984.
57. **Fryer, J., Lederis, K., and Rivier, J.,** Cortisol inhibits the ACTH-releasing activity of urotensin I, CRF and sauvagine observed with superfused goldfish pituitary cells, *Peptides*, 5, 925, 1984.
58. **Antoni, F. A., Holmes, M. C., Makara, G. B., Karteszi, M., and Laszlo, F. A.,** Evidence that the effects of arginine-8-vasopressin (AVP) on pituitary corticotropin (ACTH) release are mediated by a novel type of receptor, *Peptides*, 5, 519, 1984.
59. **Beny, J.-L. and Baertschi, A. J.,** Synthetic corticoliberin needs arginine vasopressin for full corticotropin releasing activity, *Experientia*, 38, 1078, 1982.
60. **Gaillard, R. C., Schoenenberg, P., Favrod-Coune, C. A., Muller, A. F., Marie, J., Bockaert, J., and Jard, S.,** Properties of rat anterior pituitary vasopressin receptors: relation to adenylate cyclase and the effect of corticotropin-releasing factor, *Proc. Natl. Acad. Sci. U.S.A.*, 81, 2907, 1984.
61. **Gibbs, D. M., Vale, W., Rivier, J., and Yen, S. S. C.,** Oxytocin potentiates the ACTH-releasing activity of CRF(41) but not vasopressin, *Life Sci.*, 34, 2245, 1984.
62. **Giguere, V. and Labrie, F.,** Additive effects of epinephrine and corticotropin-releasing factor (CRF) on adrenocorticotropin release in rat anterior pituitary cells, *Biochem. Biophys. Res. Commun.*, 110, 456, 1983.

63. **Giguere, V. and Labrie, F.**, Vasopressin potentiates cyclic AMP accumulation and ACTH release induced by corticotropin-releasing factor (CRF) in rat anterior pituitary cells in culture, *Endocrinology*, 111, 1752, 1982.
64. **Gillies, G. and Lowry, P.**, Corticotrophin releasing factor may be modulated vasopressin, *Nature*, 278, 463, 1979.
65. **Gillies, G., Puri, A., Hodgkinson, S., and Lowry, P. J.**, Involvement of rat corticotrophin-releasing factor-41-related peptide and vasopressin in adrenocorticotrophin-releasing activity from superfused rat hypothalami *in vitro*, *J. Endocrinol.*, 103, 25, 1984.
66. **Gillies, G. E., Linton, E. A., and Lowry, P. J.**, Corticotropin releasing activity of the new CRF is potentiated several times by vasopressin, *Nature*, 299, 355, 1982.
67. **Murakami, K., Hashimoto, K., and Ota, Z.**, Interaction of synthetic ovine corticotropin releasing factor and arginine vasopressin on in vitro ACTH release by the anterior pituitary of rats, *Neuroendocrinology*, 39, 49, 1984.
68. **Nicholson, S., Adrian, T. E., Gillham, B., Jones, M. T., and Bloom, S. R.**, Effect of hypothalamic neuropeptides on corticotrophin release from quarters of rat anterior pituitary gland *in vitro*, *J. Endocrinol.*, 100, 219, 1984.
69. **Norman, L. J. and Challis, J. R. G.**, Synergism between systemic corticotropin-releasing factor and arginine vasopressin on adrenocorticotropin release *in vivo* varies as a function of gestational age in the ovine fetus, *Endocrinology*, 120, 1052, 1987.
70. **Spinedi, E. and Negro-Vilar, A.**, Angiotensin II and ACTH release: site of action and potency relative to corticotropin releasing factor and vasopressin, *Neuroendocrinology*, 37, 446, 1983.
71. **Turkelson, C. M., Thomas, C. R., Arimura, A., Chang, D., Chang, J. K., and Shimizu, M.**, In vitro potentiation of the activity of synthetic ovine corticotropin-releasing factor by arginine vasopressin, *Peptides*, 1, 111, 1982.
72. **deBold, C. R., Sheldon, W. R., DeCherney, G. S., Jackson, R. V., Alexander, A. N., Vale, W., Rivier, J., and Orth, D. N.**, Arginine vasopressin potentiates adrenocorticotropin release induced by ovine corticotropin-releasing factor, *J. Clin. Invest.*, 73, 533, 1984.
73. **Liu, J. H., Muse, K., Contreras, P., Gibbs, D., Vale, W., Rivier, J., and Yen, S. S. C.**, Augmentation of ACTH-releasing activity of synthetic corticotropin releasing factor (CRF) by vasopressin in women, *J. Clin. Endocrinol. Metab.*, 57, 1087, 1983.
74. **von Bardeleben, U., Holsboer, F., Stalla, G. K., and Müller, O. A.**, Combined administration of human corticotropin-releasing factor and lysine vasopressin induces cortisol escape from dexamethasone suppression in healthy subjects, *Life Sci.*, 37, 1613, 1985.
75. **Fischman, A. J. and Moldow, R. L.**, In vivo potentiation of corticotropin releasing factor activity by vasopressin analogues, *Life Sci.*, 35, 1311, 1984.
76. **Rivier, C., Rivier, J., Mormede, P., and Vale, W.**, Studies of the nature of the interaction between vasopressin and corticotropin-releasing factor on adrenocorticotropin release in the rat, *Endocrinology*, 115, 882, 1984.
77. **Rivier, C. and Vale, W.**, Interaction of corticotropin-releasing factor and arginine vasopressin on adrenocorticotropin secretion *in vivo*, *Endocrinology*, 113, 939, 1983.
78. **Arimura, A., Saito, T., and Schally, A. V.**, Assays for corticotropin-releasing factor (CRF) using rats treated with morphine, chlorpromazine, dexamethasone and nembutal, *Endocrinology*, 81, 235, 1967.
79. **Rivier, C. and Vale, W.**, Effect of angiotensin II on ACTH release in vivo: role of corticotropin releasing factor, *Regul. Pept.*, 7, 253, 1983.
80. **Conte-Devolx, B., Oliver, C., Giraud, P., Castanas, E., Boudouresque, F., Gillioz, F., and Millet, Y.**, Adrenocorticotropin, β-endorphin, and corticosterone secretion in Brattleboro rats, *Endocrinology*, 110, 2097, 1982.
81. **Jard, S.**, Les isorecepteurs de la vasopressine dans le foie et dans le rein: relation entre fixation d'hormone et response biologique, *J. Physiol. (Paris)*, 77, 621, 1981.
82. **Aizawa, T., Yasuda, N., Greer, M. A., and Sawyer, W. H.**, *In vivo* adrenocorticotropin-releasing activity of neurohypophysial hormones and their analogs, *Endocrinology*, 110, 98, 1982.
83. **Anderson, K. E., Arner, B., Hedner, P., and Mulder, J. L.**, Effects of 8-lysine-vasopressin and synthetic analogues on release of ACTH, *Acta Endocrinol. (Copenhagen)*, 69, 640, 1972.
84. **Arimura, A., Schally, A. V., and Bowers, C. Y.**, Corticotropin releasing activity of lysine vasopressin analogues, *Endocrinology*, 84, 579, 1969.
85. **Portanova, R. and Sayers, G.**, Isolated pituitary cells: CRF-like activity of neurohypophysial and related polypeptides, *Exp. Biol. Med.*, 143, 661, 1973.
86. **Knepel, W., Homolka, L., and Vlaskovska, M.**, *In vitro* CRF activity of vasopressin analogs is not related to pressor activity, *Eur. J. Pharmacol.*, 91, 115, 1983.
87. **Mormede, P., LeMoal, M., and Dantzer, R.**, Analysis of the dual mechanism of ACTH release by arginine vasopressin and its analogs in conscious rats, *Regul. Pept.*, 12, 175, 1985.

88. **Knepel, W., Homolka, L., Vlaskovska, M., and Nutto, D.,** *In vitro* adrenocorticotropin/β-endorphin-releasing activity of vasopressin analogs is related neither to pressor nor to antidiuretic activity, *Endocrinology*, 114, 1797, 1984.
89. **Knepel, W., Homolka, L., and Vlaskovska, M.,** In vitro CRF activity of vasopressin analogs is not related to pressor activity, *Eur. J. Pharmacol.*, 91, 115, 1983.
90. **Buckingham, J. C.,** Vasopressin receptors influence the secretion of ACTH by the rat adenohypothesis, *J. Endocrinol.*, 113, 289, 1987.
91. **Buckingham, J. C.,** Two distinct corticotrophin releasing activities of vasopressin, *Br. J. Pharmacol.*, 84, 213, 1985.
92. **Antoni, F. A., Holmes, M. C., and Jones, M. T.,** Oxytocin as well as vasopressin potentiate ovine CRF *in vitro*, *Peptides*, 4, 411, 1983.
93. **Baertschi, A. J., Gahwiler, B., Antoni, F. A., Holmes, M. C., and Makara, G. B.,** No role of vasopressin in stress-induced ACTH secreton?, *Nature*, 308, 85, 1984.
94. **Baertschi, A. J. and Friedli, M.,** A novel type of vasopressin receptor on anterior pituitary corticotrophs?, *Endocrinology*, 116, 499, 1985.
95. **Ceda, G. P. and Hoffman, A. R.,** Glucocorticoid modulation of corticotropin-releasing factor desensitization in cultured rat anterior pituitary cells, *Endocrinology*, 118, 58, 1986.
96. **Hoffman, A. R., Ceda, G., and Reisine, T. D.,** Corticotropin-releasing factor desensitization of adrenocorticotropic hormone release is augmented by arginine vasopressin, *J. Neurosci.*, 5, 234, 1985.
97. **Reisine, T. and Hoffman, A.,** Desensitization of corticotropin-releasing factor receptor, *Biochem. Biophys. Res. Commun.*, 111, 919, 1983.
98. **Rivier, J., Rivier, C., and Vale, W.,** Synthetic competitive antagonists of corticotropin releasing factor: effect of ACTH secretion in the rat, *Science*, 224, 889, 1984.
99. **Rivier, C. and Vale, W.,** Effect of the long-term administration of corticotropin releasing factor on the pituitary-adrenal and pituitary-gonadal axis in the male rat, *J. Clin. Invest.*, 75, 689, 1985.
100. **Schopohl, J., Hauer, A., Kaliebe, T., Stalla, G. K., von Werder, K., and Müller, O. A.,** Repetitive and continuous administration of human corticotropin releasing factor to human subjects, *Acta Endocrinol.*, 112, 157, 1986.
101. **Schulte, H. M., Chrousos, G. P., Gold, P. W., Booth, J. D., Oldfield, E. H., Cutler, G. B., Jr., and Loriaux, D. L.,** Short term continuous corticotropin releasing factor infusion in man: evidence of a restrained corticotroph response, *J. Clin. Endocrinol.Med.*, submitted.
102. **Schulte, H. M., Chrousos, G. P., Gold, P. W., Booth, J. D., Oldfield, E. H., Cutler, G. B., Jr., and Loriaux, D. L.,** Continuous administration of synthetic ovine corticotropin-releasing factor in man, *J. Clin. Invest.*, 75, 1781, 1985.
103. **Rivier, C. and Vale, W.,** Influence of the frequency of ovine corticotropin-releasing factor administration on adrenocorticotropin and corticosterone secretion in the rat, *Endocrinology*, 113, 1422, 1983.
104. **Rivier, C. and Vale, W.,** Diminished responsiveness of the hypothalamic-pituitary-adrenal axis of the rat during exposure to prolonged stress: a pituitary-mediated mechanism, *Endocrinology*, in press.
105. **Gibbs, D. M.,** Measurement of hypothalamic corticotropin-releasing factors in hypophyseal portal blood, *Fed. Proc.*, 44, 203, 1985.
106. **Gibbs, D. M. and Vale, W.,** Presence of corticotropin releasing factor-like immunoreactivity in hypophysial portal blood, *Endocrinology*, 111, 1418, 1982.
107. **Plotsky, P. M.,** Hypophysiotropic regulation of adenohypophyseal adrenocorticotropin secretion, *Fed. Proc.*, 44, 207, 1985.
108. **Plotsky, P. M., Bruhn, T. O., and Vale, W.,** Evidence for multifactor regulation of the adrenocorticotropin secretory response to hemodynamic stimuli, *Endocrinology*, 116, 633, 1985.
109. **Brown, M. R., Rivier, C., and Vale, W.,** Central nervous system regulation of adrenocorticotropin secretion: role of somatostatins, *Endocrinology*, 114, 1546, 1984.
110. **Rivier, C., Rivier, J., and Vale, W.,** Inhibition of adrenocorticotropic hormone secretion in the rat by immunoneutralization of corticotropin-releasing factor (CRF), *Science*, 218, 377, 1982.
111. **Rivier, C. and Vale, W.,** Modulation of stress-induced ACTH release by corticotropin-releasing factor, catecholamines and vasopressin, *Nature*, 305, 325, 1983.
112. **Conte-Devolx, B., Rey, M., Boudouresque, F., Giraud, P., Castanas, E., Millet, Y., Codaccioni, J. L., and Oliver, C.,** Effect of 41-CRF antiserum on the secretion of ACTH, β-endorphin and α-MSH in the rat, *Peptides*, 4, 301, 1983.
113. **Linton, E. A., Tilders, F. J. H., Hodgkinson, S., Berkenbosch, F., Vermes, I., and Lowry, P. J.,** Stress-induced secretion of adreno-corticotropin in rats is inhibited by administration of antisera to ovine corticotropin-releasing factor and vasopressin, *Endocrinology*, 116, 966, 1985.
114. **Nakane, T., Audhya, T., Kanie, N., and Hollander, C. S.,** Evidence for a role of endogenous corticotropin-releasing factor in cold, ether, immobilization, and traumatic stress, *Proc. Natl. Acad. Sci. U.S.A.*, 82, 1247, 1985.

115. **Ono, N., Bedran de Castro, J. C., Khorram, O., and McCann, S. M.,** Role of arginine vasopressin in control of ACTH and LH release during stress, *Life Sci.*, 36, 1779, 1985.
116. **Ono, N., Samson, W. K., McDonald, J. K., Lumpkin, M. D., Bedran de Castro, J. C., and McCann, S. M.,** Effects of intravenous and intraventricular injection of antisera directed against corticotropin-releasing factor on the secretion of anterior pituitary hormones, *Proc. Natl. Acad. Sci. U.S.A.*, 82, 7787, 1985.
117. **Tilders, F. J. H., Berkenbosch, F., and Smelik, P. A.,** Control of secretion of peptides related to adrenocorticotropin, melanocyte-stimulating hormone and endorphin, *Front. Horm. Res.*, 14, 161, 1985.
118. **Bruhn, T. O., Plotsky, P. M., and Vale, W. W.,** Effect of paraventricular lesions on corticotropin-releasing factor (CRF)-like immunoreactivity in the stalk-median eminence: studies on the adrenocorticotropin response to ether stress and exogenous CRF, *Endocrinology*, 114, 57, 1984.
119. **Moldow, R. L. and Fischman, A. J.,** Production of antiserum to CRF associated with adrenal atrophy in a rabbit, *Peptides*, 3, 989, 1982.
120. **Mormede, P.,** The vasopressin receptor antagonist dPTyr(Me)AVP does not prevent stress-induced ACTH and corticosterone release, *Nature*, 302, 345, 1983.
121. **Weiner, R. I. and Ganong, W. F.,** Role of brain monoamines and histamine in regulation of anterior pituitary secretion, *Physiol. Rev.*, 58, 905, 1978.
122. **Tilders, F. J. H. and Berkenbosch, F.,** CRF and catecholamines; their place in the central and peripheral regulation of the stress response, *Acta Endocrinol. (Suppl.)*, 276, 63, 1986.
123. **Tilders, F. J. H., Berkenbosch, F., Vermes, I., Linton, E. A., and Smelik, P. G.,** Role of epinephrine and vasopressin in the control of the pituitary-adrenal response to stress, *Fed. Proc.*, 44, 155, 1985.
124. **Beuers, U., Hertting, G., and Knepel, W.,** Release of β-lipotropin-and β-endorphin-like material induced by angiotensin in the conscious rat, *Br. J. Pharmacol.*, 76, 579, 1982.
125. **Gaillard, R. C., Grossman, A., Gillies, G., Rees, L. H., and Besser, G. M.,** Angiotensin II stimulates the release of ACTH from dispersed rat anterior pituitary cells, *Clin. Endocrinol.*, 15, 573, 1981.
126. **Rayyis, S. S. and Horton, R.,** Effect of angiotensin II on adrenal and pituitary function in man, *J. Clin. Endocrinol.*, 32, 539, 1971.
127. **Sobel, D. O.,** Characterization of angiotensin-mediated ACTH release, *Neuroendocrinology*, 36, 249, 1983.
128. **Sobel, D. O. and Vagnucci, A.,** Angiotensin II mediated ACTH release in rat pituitary cell culture, *Life Sci.*, 30, 1281, 1982.
129. **Cameron, V. A., Espiner, E. A., Nicholls, G., and MacFarlane, M. R.,** Intracerebroventricular captopril reduces plasma ACTH and vasopressin responses to hemorrhagic stress, *Life Sci.*, 38, 553, 1986.
130. **Buckner, F. S., Chen, F.-N., Wade, C. E., and Ganong, W. F.,** Centrally administered inhibitors of the generation and action of angiotensin II do not attenuate the increase in ACTH secretion produced by ether stress in rats, *Neuroendocrinology*, 42, 97, 1986.
131. **Dallman, M. F., Makara, G. B., Roberts, J. L., Levin, N., and Blum, M.,** Corticotrope response to removal of releasing factors and corticosteroids *in vivo*, *Endocrinology*, 117, 2190, 1985.
132. **Makara, G. B.,** Mechanisms by which stressful stimuli activate the pituitary-adrenal system, *Fed. Proc.*, 44, 149, 1985.

Chapter 13

CORTICOTROPIN-RELEASING FACTOR (CRF) ACTIONS ON ADENOHYPOPHYSEAL HORMONE SECRETION BY DIRECT EFFECTS ON THE PITUITARY AND BY INDIRECT EFFECTS ON THE HYPOTHALAMUS

S. M. McCann

TABLE OF CONTENTS

I. OVERVIEW

There is not one but several CRFs which act directly on the corticotrophs to stimulate the release of adrenocorticotropic hormone (ACTH). The peptidic 41 amino acid CRF is the most important of the CRFs; however, vasopressin is very potent to release ACTH and augments the response to CRF. Since both peptides are present in portal blood in high concentration, the bulk of the evidence indicates that they cooperate in evoking the stress-induced release of ACTH. Under certain conditions epinephrine may also play a role as well as other peptides, such as angiotensin II. The actions of all these peptides and of epinephrine are mediated via specific receptors in the gland, i.e., CRF, vasopressin and beta 2 adrenergic receptors. The mechanisms of release of ACTH are complex and differ to some extent with respect to the different secretogogues.

In addition to these powerful effects at the pituitary level, vasopressin and CRF also have important intrahypothalamic actions which help to bring about the stress-induced pattern of adenohypophyseal hormone secretion. Vasopressin directly stimulates ACTH release, but it also potentiates ACTH release by hypothalamic action and appears to be a mediator of the rapid increase in luteinizing hormone (LH) release which occurs immediately after onset of stress. The latter actions appear to be physiologically significant based on immunoneutralization studies with intraventricular injection of highly specific antivasopressin serum.

On the other hand CRF also acts intrahypothalamically to mediate the stress-induced pattern of pituitary hormone release. It augments its own release by ultrashort-loop positive feedback operative in stress. It inhibits growth hormone release to produce the stress pattern of growth hormone release, at least in part by stimulation of somatostatin release and possibly by inhibition of growth hormone-releasing factor (GRF) release. It brings about suppression of LH release by inhibiting the discharge of luteinizing hormone-releasing hormone (LHRH). These actions may involve an intermediary step mediated via beta endorphin neurons. On the basis of immunoneutralization studies and studies with CRF antagonists, it appears that these actions on growth hormone and LH release may be of physiologic significance.

The release of CRF from the hypothalamus is also under the control of several neurotransmitters which include acetylcholine, serotonin, norepinephrine, and GABA.

II. INTRODUCTION

Although CRF was the first peptide to be postulated to control the release of an anterior pituitary hormone, it was one of the last to be characterized and its structure determined. The fundamental idea in this field was that of neurohumoral control of the anterior pituitary via delivery of neurohormones to the gland via the recently discovered hypophyseal portal system of veins.[1,2] According to this concept, a neurohumor would be released from axon terminals in juxtaposition to the primary capillary plexus of the portal system in the median eminence of the tuber cinereum and would then be delivered to the anterior lobe via the portal veins to modify the release of particular pituitary hormones.[2]

III. EPINEPHRINE AS CRF

The first corticotropin-releasing factor to be postulated was the adrenomedullary hormone, epinephrine, which was shown to release ACTH in animals[3-6] and in man.[7] By microinjecting epinephrine into pituitary grafts placed in the anterior chamber of the eye, Long and Fry found that this would evoke ACTH release and therefore claimed that there was an action of epinephrine directly on the pituitary gland.[4] McCann and co-workers[8-10] showed that hypothalamic lesions which did not block the ACTH release induced by vasopressin and corticotropin-releasing factor (CRF) blocked the response to epinephrine and concluded that the action of epinephrine was mediated via the hypothalamus. Recently, the concept that epinephrine could act directly to release ACTH has been resurrected since it has been reported that it would release ACTH from pituitaries incubated *in vitro*[11,12] and in rats with median eminence lesions or pituitary stalk sections.[13] More recently however, Xu et al.[14] have shown that median eminence lesions indeed block the release of ACTH from epinephrine. It was postulated that the lesions of Mezey et al.[13] may not have been complete since the evidence for completeness was only a decline in dopamine content in the gland. Thus, it would appear that epinephrine cannot act on the gland which has been deprived of the other two main CRFs, i.e., vasopressin and CRF itself.

Nevertheless, it has been repeatedly shown that certain doses of epinephrine and other beta agonists can act on pituitary cells *in vitro* to release ACTH.[11,12] Furthermore, beta adrenoreceptors have been found in the pituitary gland.[15] Therefore, it is possible that epinephrine may act *in vivo* in concert with CRF and/or vasopressin to stimulate directly release of ACTH in stress. It would be of interest to perform additional experiments in which the effect of epinephrine on responsiveness of the gland to CRF and vasopressin is determined *in vivo* to evaluate this possibility. Epinephrine released from the adrenal medulla could act in the presence of these factors to potentiate ACTH release by direct pituitary action; however, it is still not certain if circulating epinephrine levels are ever high enough to activate the gland directly.

Injections of epinephrine probably evoke their major ACTH-releasing action *in vivo* by other changes induced by the hormone such as alterations in blood pressure which would be sensed in the brain and result in alterations in release of CRF and/or vasopressin.[14]

Recently, epinephrine has been detected in high concentrations in portal blood.[16] Consequently, another possible avenue for epinephrine to activate ACTH directly would be via its release in high concentrations into portal blood of stressed animals. The answer to this question would require comparison of the epinephrine levels in portal blood in undisturbed and in stressed rats. This cannot be done by the technique of portal blood cannulation because of the stress involved in the procedure; however, it could be done now using a push-pull cannula located in the anterior pituitary gland with measurement of epinephrine in the effluent from the cannula under various conditions. These studies are eagerly awaited.

IV. VASOPRESSIN AS CRF

The ability of stress to release concurrently vasopressin and ACTH was known as early as the 1940s.[17,18] Furthermore, studies with median eminence lesions showed a correlation between the deficiency of vasopressin (antidiuretic hormone) release as indicated by the severity of diabetes insipidus and the release of ACTH as measured by bioassay.[19] Consequently the ability of Pitressin, a partially purified vasopressin preparation, to release ACTH in animals with hypothalamic lesions in which stress-induced ACTH release was blocked was evaluated. Relatively high intravenous doses of vasopressin (MED, 0.5 U) were needed.[19] Later experiments indicated that this was due to loss of sensitivity to vasopressin with time following lesions since in animals with median eminence lesions used only 48 h after operation the MED was fivefold lower.[9] This was still high in terms of the dose which could reach the anterior lobe via the systemic circulation.[17,18] However, it was thought that high concentrations of the peptide might reach the pituitary after its release into the hypophyseal portal vessels in the median eminence.[19] The fact that median eminence lesions in hypophysectomizied animals resulted in the development of severe diabetes insipidus which was not present following hypophysectomy alone suggested that in the hypophysectomized animal, water balance was maintained by vasopressin secreted from the median eminence and pituitary stalk and supported the concept of high concentration of vasopressin in portal blood.[20] Much later, it was indeed shown that the vasopressin titers are high in portal blood and similar to those which activate the release of ACTH from the pituitary directly.[21]

It was next shown that synthetic lysine vasopressin, recently synthesized by du Vigneaud, released ACTH in these animals with lesions.[22] Subsequent studies have shown that vasopressin is also active to release ACTH *in vitro*.[23] Vasopressin could account for all of the activity in posterior pituitary extracts of varying degrees of purity such as protopituitrin and Pitressin.[10]

At this early time it was also shown that the stress-induced increase in plasma vasopressin could be inhibited by cortisol which also inhibited the release of ACTH.[24] Consequently, it appeared that vasopressin might play a physiological role in the control of stress-induced ACTH release and that it was even involved in the negative feedback of corticoids to suppress this release. At that point, the rat with hereditary diabetes insipidus was discovered by Valtin and Schroeder. In collaboration with Valtin we showed that these animals exhibited only a small defect in ACTH release in response to stress as indicated by plasma corticosterone changes; however, adrenal weight was very subnormal.[25] These animals appeared to contain CRF in the hypothalamus. The relatively minor impairment in ACTH release in these rats suggested that CRF was more important than vasopressin as an ACTH-releasing agent; however, the possibility remained that the absence of vasopressin since birth had allowed for compensation.

By this time evidence was accruing for another CRF which was peptidic in nature but whose structure was elusive (see below). In collaborative experiments with Yates, we showed that a dose of vasopressin too small to evoke ACTH release itself potentiated the response to a partially purified CRF.[26] This potentiation was even found in dehydrated rats which were presumably secreting increased amounts of endogenous vasopressin.[26] We postulated that vasopressin might potentiate the action of CRF to release ACTH. The ability of dehydration to also potentiate suggested that vasopressin might play a physiologically significant role in the control of ACTH release. Much later such a potentiating action was found by Gillies and Lowry[27] *in vitro*, and they also reported that vasopressin antibodies would block the ACTH-releasing activity of median eminence extracts.[28]

In more recent studies with hypothalamic lesions ıt has becoıne apparent that most lesions which interfered with ACTH release also produce diabetes insipidus as a result of decreased vasopressin release, a result in agreement with our earlier studies; however, a few rats were found which had impairment of ACTH release without diabetes insipidus.[29] The defect in these rats was almost certainly caused by an interruption of CRF pathways. It is now known that most CRF-containing neurons are located in the paraventricular nucleus with some few also located in the supraoptic nucleus.[39] Most of these lesions presumably damaged not only the paraventriculohypophyseal tract but also the supraopticohypophyseal tract, the damage to which was reflected in the induction of diabetes insipidus.

It has been reported that paraventricular nucleus lesions can completely block the release of ACTH from stress.[31] This result does not agree with our recent[29] and much earlier results[8] in which we found that complete destruction of the paraventricular nucleus was associated with stress-induced release of ACTH. The reason for the discrepancy between the results is probably the duration of time following lesions. In the studies in which paraventricular lesions were shown to interrupt ACTH release, the experiments were carried out 5 to 7 d after induction of lesions, whereas in those in which paraventricular lesions were not effective, the experiments were performed 2 or more weeks after lesion placement. We hypothesized that the failure of paraventricular lesions to block completely is due to either recovery or take over of function by the remaining CRF neurons outside the paraventricular nucleus or the ability of vasopressin to compensate for the deficiency in CRF.

Further evidence supports the role of vasopressin as a CRF. Vasopressin-containing neurons in the paraventricular nucleus project to the external layer of the median eminence,[21] presumably to release vasopressin into the hypophyseal portal vessels as we had postulated many years before. As indicated above, high concentrations of vasopressin have been measured in hypophyseal portal blood of both rat[32] and monkey.[21] On the basis of immunocytochemical evidence as well as radioimmunoassay of plasma vasopressin, it appears that vasopressin release is indeed enhanced after adrenalectomy and decreased by corticoids[21] as reported many years earlier by bioassay.[24] Furthermore many CRF neurons in the paraventricular nucleus also contain vasopressin and it is likely that both peptides are coreleased into portal blood,[33] thereby favoring the potentiating action of vasopressin on CRF action.

From all of this evidence it would appear that vasopressin was the first releasing factor to be characterized and synthesized, although not on the basis of its ACTH-releasing activity. There is now no doubt that another factor is very important, namely the peptidic CRF, which has been synthesized and proven to be highly active *in vivo* and *in vitro*.[34] These two agents may act together to promote ACTH release in stress. The evidence for a role for vasopressin is now conclusive. It can be summarized as follows: (1) vasopressin is released in stressful situations when ACTH release is increased; (2) its release is inversely related to plasma corticoid levels and in parallel with changes in ACTH secretion; (3) it is found in high concentrations in portal blood which should be sufficient to release ACTH; and (4) it is active to release ACTH in all assay systems which include *in vivo* assays in which the ubiquitous stress response is blocked by median eminence lesions, or pharmacological blockade and *in vitro* systems utilizing hemipituitaries or dispersed pituitary cells. Therefore, it would appear that vasopressin is a very important stress hormone. It is involved in the brain, perhaps in consolidation of memory,[35] at the pituitary to stimulate the release of ACTH, at the kidney to diminish renal water excretion and in supporting blood pressure as revealed by recent experiments.[36]

In addition to its ACTH releasing action, various reports over the years have postulated that vasopressin may be important in the release of other pituitary hormones, for example, growth hormone.[37] There appears to be no action of this peptide to release most of the other anterior pituitary hormones; however, we[38] have been able to confirm the early claims of LaBella[39] that vasopressin will release thyrotropin-releasing hormone (TSH). There is a dose-related increase in TSH release by pituitary cells *in vitro* but at higher doses the response declines producing a bell-shaped dose response curve. Vasopressin may be physiologically significant in the release of TSH, although this remains to be determined.

Vasopressin receptors have been described on pituitary cell membrane preparations which appear to be of the pressor and not the antidiuretic type.[40] The action of vasopressin on the gland differs from that of CRF in that it is apparently principally mediated by increased availability of intracellular calcium and the phosphatidylinositol cycle rather than by cyclic AMP which may be one of the principal mediators of CRF action.[41]

V. CORTICOTROPIN-RELEASING FACTOR

In 1955, Saffran and Schally[42] incubated anterior pituitaries *in vitro* and found that pressor neurohypophyseal extracts would release ACTH, but only in the presence of norepinephrine and that neither substance was active by itself. This was evidence in favor of a role for not only vasopressin but also catecholamines and confirmed the earlier *in vivo* studies indicating a role for vasopressin.[19] Later that year Saffran et al.[43] purified posterior pituitary extracts and reported the separation of a substance distinct from vasopressin, which they named corticotropin-releasing factor (CRF) that released ACTH in their *in vitro* system. In the same year Guillemin and Rosenberg[44] cultured hypothalami and pituitaries together and reported that hypothalami in organ culture stimulated ACTH release. To my knowledge these experiments have never been repeated. In 1956, Guillemin and Hearn[45] reported that posterior pituitary extract could release ACTH *in vitro* but that synthetic vasopressin had no significant releasing action. Interestingly there was no significant difference between the results with vasopressin and posterior pituitary extract, and it has subsequently been shown that vasopressin does indeed release ACTH *in vitro* (see above).

Early studies *in vivo* in animals with hypothalamic lesions had failed to reveal any ACTH-releasing activity of crude median eminence extracts, a result probably due to the relative insensitivity to ACTH releasing factors of these animals with chronic lesions. Later studies with animals with acute lesions revealed an ACTH-releasing activity of hypothalamic extracts.[10] This activity could be purified and shown not to be due to vasopressin.[46,10] Therefore a CRF distinct from vasopressin had to exist.

At the same time studies with posterior pituitary extracts revealed that all of the ACTH releasing activity in those extracts could be accounted for by vasopressin.[10] Thus, vasopressin was presumably the major CRF in neural lobe extracts. If it were secreted from vasopressinergic neural terminals in the neural lobe, it could readily reach the anterior lobe via the short portal vessels. On the other hand hypothalamic CRF would be primarily this undefined peptidic CRF.[10] It was easy to purify hypothalamic CRF by gel filtration on Sephadex® G-25 followed by carboxymethylcellulose chromatography. Gel filtration readily separated vasopressin from CRF and other releasing factors.[47] At this time several structures were proposed by Guillemin and Schally as neurohypophyseal CRFs.[48] There is now little doubt that these structures were erroneous.

Little progress was made in the elucidation of the structure of CRF between 1965 and 1981, at which time Vale, Rivier, and associates[34] elucidated the structure of the 41 amino acid peptide, CRF. The activity was confirmed by synthesis of the molecule. The delay in determining the structure of CRF was probably related to the fact that it was a much larger peptide than the previously determined ones.

There is now little doubt that CRF is the principle ACTH-releasing peptide. It is extremely active and gives a steep dose-response curve in assays in contrast to vasopressin which gives a flatter dose-response curve.[49,50] Furthermore, the potentiating action of vasopressin on CRF-induced ACTH release has now been repeatedly confirmed with the synthetic peptide.[49,50] Inactivation of endogenous CRF by the use of antiserum directed against it or an antagonist to the peptide produces an almost complete block of stress-induced ACTH release.[49,51] CRF neurons with cell bodies predominantly in the paraventricular nucleus project to the external layer of the median eminence[30] where they release the peptide into portal blood and high concentrations have indeed been found there.[52]

The peptide interacts with specific receptors on the cell membrane (see Chapters 9 and 11 on CRF receptors) and activates ACTH release at least in part via activating adenylcyclase with resultant generation of cyclic AMP. There are probably roles also of calcium and the phosphatidylinositol cycle, calmodulin, and arachidonic acid metabolites; however, these latter pathways seem to play a lesser role in the activation of ACTH release achieved by CRF on comparison to the activation achieved by vasopressin.[41]

VI. OTHER CRFs

Oxytocin was earlier ruled out as a CRF since it had no action in either *in vivo* or *in vitro* systems;[48] however, recently some evidence has accrued which suggests that oxytocin also may potentiate the response to CRF in rats.[52,53] In the human, if anything it appears to inhibit ACTH release; however, these studies are confounded by possible peripheral actions of the peptide which could alter ACTH release.[53] Angiotensin II (AII) is also present in the circulating blood and may also be released into portal vessels in the median eminence from AII-containing neurons which terminate there.[54] It will release ACTH *in vivo* and *in vitro*.[54] Receptors for the peptide exist on anterior pituitary cells[55] and it therefore may be involved in certain types of stress-induced ACTH release. Studies with saralysin, the AII antagonist, have shown that it failed to alter ACTH release from ether stress even though the saralysin blocked the effects of exogenous AII.[56] This does not negate the possibility that AII may play some role in stress-induced ACTH release but indicates at least in the release induced by ether its role is not essential since there is no modification in ACTH release when its action is eliminated. It would appear that angiotensin II plays a relatively modest role in stress-induced ACTH release.

In summary, it is apparent that ACTH release is not controlled by a single factor but that its control is multifactorial. This is becoming a rule for the control of pituitary hormones. CRF plays the principle role. Vasopressin by itself has ACTH-releasing activity and potentiates the action of CRF. There is a possible small role for epinephrine, AII and possibly even other peptides. The release of ACTH may be determined by different factors in different types of stress. Further studies will be needed to answer this question. Some work using portal blood collectons has already suggested differential patterns of control of ACTH release depending on the stress;[57] however, these studies are clouded by the use of anesthesized animals in highly unphysiological conditions for portal blood collection which renders the results suspect. There will need to be intensive studies using push-pull cannulae in the pituitary to determine whether there are indeed stress specific patterns of release of CRF and other ACTH-releasing substances, such as vasopressin into portal blood in conscious animals.

VII. INTRAHYPOTHALAMIC ACTIONS OF VASOPRESSIN AND CRF TO BRING ABOUT THE STRESS PATTERN OF ANTERIOR PITUITARY HORMONE RELEASE

In this section we will review what is known about the intrahypothalamic actions of these two principal CRFs to alter pituitary hormone secretion.

A. VASOPRESSIN

It now appears that nearly all peptides which act to stimulate or inhibit pituitary hormone secretion by direct action on the gland have an opposite action within the hypothalamus which was early termed ultrashort-loop negative feedback by Martini and co-workers.[58] Vasopressin is no exception and as indicated it stimulates the release of TSH from the thyrotrophs directly but following its intraventricular injection it has an opposite action to inhibit TSH release, an example of ultrashort-loop negative feedback.[38]

On the other hand, in the case of the stress-induced release of ACTH one might visualize that it would be self defeating to have a negative ultrashort-loop feedback and one might expect to have a positive ultrashort-loop feedback. This is the case apparently with vasopressin which was shown many years ago by Hedge and co-workers[59] to augment the release of ACTH by an intrahypothalamic action which was postulated to be by stimulating the release of CRF which in turn released ACTH. No further work on this has been done recently and so it is not certain whether this action is a stimulatory one on the CRF neurons themselves and/or an ultrashort-loop positive feedback of vasopressin to stimulate its own release into portal vessels and thereby promote enhanced ACTH release.

That the actions of vasopressin intrahypothalamically to promote ACTH release may be of physiological significance is indicated by studies employing the microinjection of highly specific vasopressin antiserum into the third cerebral ventricle prior to the administration of ether stress. The antiserum resulted in a highly significant reduction in the ACTH release from ether anesthesia; however, some release still took place. This antiserum had no effect on the stress pattern of prolactin and growth hormone release; however, plasma LH was significantly lowered.[50] It had earlier been shown that following imposition of ether stress there was initial stimulation of LH release as indicated by a rise in plasma LH followed by inhibition associated with a fall in the levels of the hormone.[60,61] Following the intraventricular injection of vasopressin antiserum, there was no increase following ether, instead only a fall in plasma LH.[50] Therefore, these results are consistent with the hypothesis that vasopressin release during stress is responsible for the early rise in LH which follows some stresses. Presumably the vasopressin released within the hypothalamus would stimulate the release of LHRH to in turn release LH.

B. CRF

To determine the intrahypothalamic action of CRF, this peptide was injected into the third ventricle of conscious, freely moving rats and the effect on pituitary hormone release was evaluated. Under resting conditions there was no effect on plasma ACTH except in very high doses which presumably reached the pituitary via the portal circulation. On the other hand plasma growth hormone declined with relatively low doses of CRF given intraventricularly. Higher doses produced an equivalent decline. LH release was not affected by the lower dose that decreased plasma growth hormone but was lowered only by the higher dose.[62] These actions are almost certainly within the hypothalamus since there is no action of CRF to alter growth hormone or LH release by direct action on the pituitary gland.

These findings raised the possibility that intrahypothalamic release of CRF might bring about the stress pattern of growth hormone and LH release in the rat which is characterized by declines in both growth hormone and LH.[61] The behavior of growth hormone in the rat contrasts with that in most species in which plasma growth hormone is elevated by stress. Similar results have been obtained by Rivier and Vale[63,64] following the lateral ventricular injection of the peptide. The disadvantage with this technique is that it spreads the peptide throughout the nervous system with a preferential localization to the side of injection. Therefore, this procedure is not nearly as effective for localization as the injection of the peptide into the third ventricle.

We were surprised that we did not observe an ultrashort-loop negative feedback of CRF to suppress ACTH release following its intraventricular injection and consequently decided to examine the effect of CRF given intraventricularly on the stress-induced release of ACTH evoked by ether. Surprisingly, minute doses of CRF injected intraventricularly, 1000-fold less than sufficient to act directly on the gland by intraventricular injection, resulted in an augmented release of ACTH following ether anesthesia. We take these results to mean that in stress CRF exerts an ultrashort-loop positive feedback to stimulate the further release of ACTH, presumably by actions on the CRF neurons to promote further release of CRF into portal vessels.[65]

The ability of CRF to act within the hypothalamus to alter the release of other hypothalamic peptides has been demonstrated since the peptide will cause a dose-related release of somatostatin from median eminence fragments[66] and will also inhibit LHRH release from such fragments.[67] These actions may at least in part be responsible for the suppression of growth hormone and LH release observed in the *in vivo* experiments.

GRF also stimulates the release of somatostatin *in vitro*.[68] This action of GRF appears to be brought about via a beta endorphinergic step on the basis of studies with naloxone, the opiate antagonist, and highly specific antisera against beta endorphin, both of which block the GRF-induced release of somatostatin.[69,70] Recent experiments by Petraglia[67] suggest that beta endorphin is also involved in mediating the CRF-induced inhibition of LHRH release.

To evaluate the physiologic significance of CRF in invoking these intrahypothalamic responses, we employed highly specific antisera raised against CRF which were again injected into the third ventricle. These antisera almost completely blocked the release of ACTH following stress when injected intravenously as described above.[51] When injected into the third ventricle the antiserum also largely blocked the release of ACTH in ether stressed rats.[51] We believe that these results indicate that CRF antisera in the hypothalamus can block the ultrashort-loop positive feedback of CRF to augment its own release and that this is the major reason for the marked reduction in the ether-induced ACTH release following the injection of the antisera into the ventricle; however, it is impossible to exclude an action of these antisera directly at the pituitary following their uptake by portal vessels and arrival at the corticotrophs.

The intraventricular CRF antisera also blocked the ether-induced decline in plasma growth hormone which occurred in normal rabbit serum-injected animals indicating a physiologically significant intrahypothalamic action of CRF to block growth hormone release,[51] probably mediated at least in part via release of somatostatin.[66] However, a reduction in GRF release induced by CRF cannot be ruled out. In fact this is probably part of the story.

In contrast to the ability of the antiserum to block the stress-induced lowering of plasma growth hormone, it had no effect on plasma LH in the animals.[51] This suggested to us that this effect may not be of physiological significance; however, in the meantime, using an antagonist to CRF at high doses, Rivier and Vale[71] have reported that this antagonist can block the stress-induced inhibition of LH release. Therefore, this action may also have physiologic significance.

VIII. OTHER PUTATIVE SYNAPTIC TRANSMITTERS AND NEUROMODULATORS INVOLVED IN CONTROL OF ACTH SECRETION AT THE HYPOTHALAMIC LEVEL

Activity in this area was relatively quiescent until the description of the structure and synthesis of corticotropin releasing factor, a 41 amino acid peptide.[34] With the availability of the synthetic peptide it has been possible to reassess the role of CRF in the control of ACTH secretion and to develop antisera directed against the peptide which have made it possible to study its localization. CRF neurons are concentrated in the paraventricular nucleus and have the major projection of their axons to the median eminence with some which progress into the neural lobe.[30] The distribution is consistent with the results of lesion and stimulation studies. It has been shown that lesions destroying the paraventricular nucleus have a profound inhibitory effect on ACTH secretion; however, there is frequently recovery with time, probably because all of the CRF system has not been eliminated.[29,31]

A complex interplay of putative synaptic transmitters regulates the release of CRFs from the hpothalamus, which in turn activate ACTH secretion. Much work in the older literature indicates that serotonin has a stimulatory action.[72] This has led to the use of serotonin receptor blockers in the treatment of Cushing's syndrome. There is little doubt also that acetylcholine can stimulate the release of CRF from the hypothalamus *in vitro*.[73] Although early work from Ganong's laboratory appeared to indicate that ACTH secretion is under inhibitory control by central noradrenergic pathways,[74] more recent experiments in both rat and man appear to indicate that the influence is stimulatory rather than inhibitory. For example, Reese et al.[75] showed that amphetamine increased ACTH secretion in man and that this effect was blocked by the highly selective α-adrenergic receptor blocking drug, thymoxamine, which suggests that noradrenergic influences are stimulatory on CRF and ACTH secretion. This work has recently been confirmed using methoxamine, an α-receptor stimulator, which causes ACTH and cortisol release. This effect was blocked by thymoxamine.[76] Additionally administration of sufficient norepinephrine to increase the pulse and blood pressure to the same extent as that produced by methoxamine had absolutely no effect whatsoever on ACTH and cortisol release in man.[76] These data suggest that in man, noradrenergic, α-receptor-mediated influences are stimulatory to ACTH release through CRF mechanisms. Similarly, in the rat recent experiments point to a stimulatory noradrenergic component to hypothalamic factors releasing ACTH.[14] Prostaglandins may mediate the various actions of putative transmitters on ACTH release as revealed by the excellent *in vivo* studies of Hedge.[77]

The endogenous opioid peptides undoubtedly affect ACTH secretion. Although the evidence from rats has suggested dual effects, both inhibitory and stimulatory of the opioid peptides on CRF release, administration of opioid peptides suppresses ACTH and related peptide secretion as well as cortisol release in man and these effects are blockable by naloxone.[78] Furthermore, experiments with naloxone indicate that there is probably a chronic inhibitory opiate tone on ACTH secretion operative in man. Thus, administration of a high dose of naloxone results in an increase in ACTH and cortisol levels in plasma under basal conditions.[79] The high dose required suggests that this response is mediated by kappa or delta receptors. This increase in ACTH secretion following naloxone appears to be mediated by α-adrenoreceptors since the naloxone-induced ACTH and cortisol increase can itself be blocked by the specific α-adrenoreceptor blocking drug, thymoxamine.[79]

REFERENCES

1. **McCann, S. M.,** Discovery of hypothalamic releasing and inhibiting hormones, in *Endocrinology: People and Ideas,* McCann, S. M., Ed., American Physiological Society, Washington, 1988, p. 41.
2. **Harris, G. W.,** The pituitary stalk and ovulation, in *Control of Ovulation,* Villee, C. A., Ed., Pergamon Press, New York, 1961, 56.
3. **Vogt, M.,** Observations on some conditions affecting the rate of hormone output by the suprarenal cortex, *J. Physiol. (London),* 103, 317, 1944.
4. **Long, C. N. H. and Fry, E. G.,** Effect of epinephrine on adrenal cholesterol and ascorbic acid, *Proc. Soc. Exp. Biol. Med.,* 59, 67, 1945.
5. **Sayers, G. and Sayers, M. A.,** Regulation of pituitary adrenocorticotrophic activity during the response to stress of the rat, *Endocrinology,* 40, 265, 1947.
6. **Farrell, G. L. and McCann, S. M.,** Detectable amounts of adrenocorticotrophic hormone in blood following epinephrine, *Endocrinology,* 50, 274, 1952.
7. **Vernikos-Danellis, J. and Marks, B. H.,** Epinephrine-induced release of ACTH in normal human subjects, *Endocrinology,* 70, 525, 1962.
8. **McCann, S. M.,** Effect of hypothalamic lesions on the adrenal cortical responses to stress in the rat, *Am. J. Physiol.,* 175, 13, 1953.
9. **McCann, S. M.,** The ACTH-releasing activity of extracts of the posterior lobe of the pituitary *in vivo, Endocrinology,* 60, 664, 1957.
10. **McCann, S. M. and Haberland, P.,** Further studies on the regulation of pituitary ACTH in rats with hypothalamic lesions, *Endocrinology,* 66, 217, 1960.
11. **Giguere, V., Cote, J., and Labrie, F.,** Characteristics of the α-adrenergic of stimulation of ACTH secretion in rat anterior pituitary cells, *Endocrinology,* 109, 757, 1981.
12. **Tilders, F. J. H., Berkenbosch, F., Vermes, I., Linton, E. A., and Smelik, P. G.,** Role of epinephrine and vasopressin in the control of the pituitary-adrenal response to stress, *Fed. Proc.,* 44, 155, 1985.
13. **Mezey, E., Reisine, T.D., Palkovits, M., Brownstein, J. J., and Axelrod, J.,** Direct stimulation of β_2 adrenergic receptors in rat anterior pituitary induces release of ACTH *in vivo, Proc. Natl. Acad. Sci. U.S.A.,* 80, 6728, 1983.
14. **Xu, R. K., Antunes-Rodrigues, J., and McCann, S. M.,** Evidence for hypothalamic action of epinephrine to stimulate ACTH secretion in the rat, *Neuroendocrinol. Lett.,* 9, 339, 1987.
15. **Petrovic, S. L., McDonald, J. K., Snyder, G. D., and McCann, S. M.,** Characterization of beta-adrenergic receptors in rat brain and pituitary using a new high-affinity ligand, (^{125}I)iodocyanopindolol, *Brain Res.,* 261, 249, 1983.
16. **Johnston, C. A., Gibbs, D. M., and Negro-Vilar, A.,** High concentrations of epinephrine derived from a central source in hypophysial portal plasma, *Endocrinol.,* 113, 819, 1983.
17. **Verney, E. B.,** The antidiuretic hormone and factors which determine its release, *Proc. R. Soc. Lond. Ser. B,* 135, 25, 1947.
18. **Mirsky, I. A., Stein, N., and Paulisch, G.,** Secretion of antidiuretic substance into circulation of adrenalectomized and hypophysectomized rats exposed to noxious stimuli, *Endocrinology,* 55, 28, 1954.

19. **McCann, S. M. and Brobeck, J. R.**, Evidence for a role of the supraopticohypophyseal system in regulation of adrenocorticotrophin secretion, *Proc. Soc. Exp. Biol. Med.*, 87, 318, 1954.
20. **Gale, C. C., Taleisnik, S., and McCann, S. M.**, Production of diabetes insipidus in hypophysectomized rats by hypothalamic lesions, *Am. J. Physiol.*, 201, 811, 1961.
21. **Zimmerman, E.A., Stillman, M. A., Recht, L. D., Antunes, J. L., and Carmel, P. W.**, Vasopressin and corticotropin releasing factor: an axonal pathway to portal capillaries in the zona externa of the median eminence containing vasopressin and its interaction with adrenal corticoids, *Ann. N.Y. Acad. Sci.*, 249, 405, 1977.
22. **McCann, S. M. and Fruit, A.**, Effect of synthetic vasopressin on release of adrenocorticotrophin in rats with hypothalamic lesions, *Proc. Soc. Exp. Biol. Med.*, 96, 556, 1957.
23. **Saffran, M.**, Discussion of paper by R. Guillemin, *Recent Prog. Horm. Res.*, 20, 126, 1964.
24. **McCann, S. M., Fruit, A., and Fulford, B. D.**, Studies on the loci of action of cortical hormones in inhibiting the release of adrenocorticotrophin, *Endocrinology*, 63, 29, 1958.
25. **McCann, S. M., Antunes-Rodrigues, J., Nallar, R., and Valtin, H.**, Pituitary adrenal function in the absence of vasopressin, *Endocrinology*, 79, 1058, 1966.
26. **Yates, F. E., Russell, S. M., Dallman, M. F., Hedge, G. A., McCann, S. M., and Dhariwal, A. P. S.**, Potentiation by vasopressin of corticotropin release induced by corticotropin-releasing factor, *Endocrinology*, 88, 3, 1971.
27. **Gillies, G. E., Linton, E. A., and Lowry, P. J.**, Vasopressin and the corticoliberin complex, in *Neuroendocrinology of Vasopressin, Corticoliberin and Opiomelanocortins*, Baertschi, A. J. and Dreifuss, J. J., Eds., Academic Press, New York, 1982, 457.
28. **Lowry, P.**, In discussion of vasopressin and corticotropin releasing factor: an axonal pathway to portal capilaries in the zona externa of the median eminence containing vasopressin and its interaction with adrenal corticoids, *Ann. N.Y. Acad. Sci.*, 249, 417, 1977.
29. **Murabayashi, U., McCann, S. M., and Antunes-Rodrigues, J.**, Factors controlling adrenal weight and corticosterone secretion in male rats as revealed by median eminence lesions and pharmacological alteration of prolactin secretion, *Brain Res. Bull.*, 19(5), 511, 1987.
30. **Merchenthaler, L., Vigh, S., Petrusz, S., and Schally, A. V.**, Immunocytochemical localization of corticotropin-releasing factor (CRF) in the rat brain, *Am. J. Anat.*, 165, 385, 1982.
31. **Makara, G. B.**, Mechanisms by which stressful stimuli activate the pituitary adrenal system, *Fed. Proc.*, 44, 149, 1985.
32. **Porter, J. C., Barnea, A., Cramer, O. M., and Parker, C. R.**, Hypothalamic peptide and catecholamine secretion: roles for portal and retrograde blood flow in the pituitary stalk in the release of hypothalamic dopamine and pituitary prolactin and LH, *Clin. Obstet. Gynecol.*, 5, 271, 1978.
33. **Whitnall, M. H., Smyth, D., and Gainer, H.**, Vasopressin coexists in half of the CRF axons present in the external zone of the median eminence in normal rats, *Neuroendocrinology*, 45, 420, 1987.
34. **Vale, W., Spiess, J., Rivier, C., and Rivier, J.**, Characterization of a 41-residue ovine hypothalamic peptide that stimulates secretion of corticotropin and beta-endorphin, *Science*, 213, 1394, 1981.
35. **DeWied, D.**, Pituitary and brain peptides and behavior, in *Brain Peptides: A New Endocrinology*, Gotto, A. M., Jr., Peck, E. J., Jr., and Boyd, A. E., III, Eds., Elsevier/North Holland, Amsterdam, 1979, 307.
36. **Mohring, J., Arbogast, R., Dusing, R., Glanzer, K., Kintz, J., Liard, J.-F., Maciel, J. A., Jr., Montani, J. P., and Schoun, J.**, Vasopressor role of vasopressin in hypertension, in *Brain and Pituitary Peptides*, Wuttke, W., Weindl, A., Voigt, K. H., and Dries, R.-R., Eds., S. Karger, Basel, 1980, 157.
37. **Martini, L.**, Neurohypophysis and anterior pituitary activity, in *The Pituitary Gland*, Vol. 3, Harris, G. W. and Donovan, B. T., Eds., Butterworths, London, 1966, 535.
38. **Lumpkin, M. D., Samson, W. K., and McCann, S. M.**, Arginine vasopressin as a thyrotropin-releasing hormone, *Science*, 235, 1070, 1987.
39. **LaBella, F. S.**, Release of TSH *in vivo* and *in vitro* by synthetic neurohypophyseal hormones, *Can. J. Biochem. Physiol.*, 42, 75, 1964.
40. **Gaillard, R. C., Schoenenberg, P., Favrod-Coune, Muller, A. F., Marie, J., Bockaert, J., and Jard, S.**, Properties of rat anterior pituitary vasopressin receptors: relation to adenylate cyclase and the effect of corticotropin-releasing factor, *Proc. Natl. Acad. Sci. U.S.A.*, 81, 2907, 1984.
41. **McCann, S. M.**, The hypothalamus and anterior pituitary, in *Textbook of Endocrine Physiology*, Ojeda, S. R. and Griffin, J. E., Eds., Oxford University Press, in press.
42. **Saffran, M. and Schally, A. V.**, The release of corticotropin by anterior pituitary tissue *in vitro*, *Can. J. Biochem. Physiol.*, 33, 408, 1955.
43. **Saffran, M., Schally, A. V., and Benfey, B. G.**, Stimulation of the release of corticotropin from the adenohypophysis by a neurohypophsial factor, *Endocrinology*, 57, 439, 1955.
44. **Guillemin, R. and Rosenberg, B.**, Humoral hypothalamic control of anterior pituitary: a study with combined tissue cultures, *Endocrinology*, 57, 599, 1955.

45. **Guillemin, R. and Hearn, W. R.,** ACTH release by *in vitro* pituitary. Effect of pitressin and purified arginine-vasopressin, *Proc. Soc. Exp. Biol. Med.*, 89, 365, 1955.
46. **Royce, P. C. and Sayers, G.,** Purification of hypothalamic corticotropin releasing factor, *Proc. Soc. Exp. Biol. Med.*, 103, 447, 1960.
47. **Dhariwal, A. P. S., Antunes-Rodrigues, J., Reeser, F., Chowers, I., and McCann, S. M.,** Purification of hypothalamic corticotrophin-releasing factor of ovine origin, *Proc. Soc. Exp. Biol. Med.*, 121, 8, 1966.
48. **McCann, S. M. and Porter, J. C.,** Hypothalamic pituitary stimulating and inhibiting hormones, *Physiol. Rev.*, 49, 240, 1969.
49. **Rivier, C. and Vale, W.,** Effects of corticotropin-releasing factor, neurohypophyseal peptides, and catecholamines on pituitary function, *Fed. Proc.*, 44, 189, 1985.
50. **Ono, N., Bedran de Castro, J. C., Khorram, O., and McCann, S. M.,** Role of arginine vasopressin in control of ACTH and LH release during stress, *Life Sci.*, 36, 1779, 1985.
51. **Ono, N., Samson, W. K., McDonald, J. K., Lumpkin, M. D., Bedran de Castro, J. C., and McCann, S. M.,** The effects of intravenous and intraventricular injection of antisera directed against corticotropin releasing factor (CRF) on the secretion of anterior pituitary hormones, *Proc. Natl. Acad. Sci. U.S.A.*, 82, 7787, 1985.
52. **Gibbs, D. M.,** Measurement of hypothalamic corticotropin-releasing factors in hypophyseal portal blood, *Fed. Proc.*, 44, 203, 1985.
53. **Gibbs, D. M., Vale, W.W., Rivier, J., and Yen, S. S. C.,** Oxytocin potentiates the ACTH-releasing activity of CRF but not vasopressin, *Life Sci.*, 34, 2245, 1984.
54. **Phillips, M. I.,** Functions of angiotensin in the central nervous system, *Annu. Rev. Physiol.*, 49, 413, 1987.
55. **Murkherjee, A., Kulkarni, P., McCann, S. M., and Negro-Vilar, A.,** Evidence for the presence and characterization of angiotensin II receptors in rat anterior pituitary membranes, *Endocrinology*, 110, 665, 1982.
56. **Ono, N. and McCann, S. M.,** unpublished data.
57. **Plotsky, P.,** Proceedings of Meeting on Stress held at NIH in 1986, in press.
58. **Piva, F., Motta, M., and Martini, L.,** Hypothalamic and pituitary function: long, short, and ultrashort feedback loops, in *Endocrinology*, Vol. 1, DeGroot, J., Cahill, G. F., Odell, W. D., Martin, L., Potts, J. T., Nelson, D. H., Steinberger, E., and Winegrad, A. J., Eds., Grune & Stratton, New York, 1979, 21.
59. **Hedge, G. A., Yates, M. B., Marcus, R., and Yates, F. E.,** Site of action of vasopressin in causing corticotropin release, *Endocrinology*, 79, 328, 1966.
60. **Ajika, K., Kalra, S. P., Fawcett, C. P., Krulich, L., and McCann, S. M.,** The effect of stress and Nembutal on plasma levels of gonadotropins and prolactin in ovariectomized rats, *Endocrinology*, 90, 705, 1972.
61. **Krulich, L., Hefco, E., Illner, P., and Read, C. B.,** The effects of acute stress on the secretion of LH, FSH, prolactin and GH in the normal male rat with comments on their statistical evaluation, *Neuroendocrinology*, 16, 293, 1974.
62. **Ono, N., Lumpkin, M. D., Samson, W. K., McDonald, J. K, and McCann, S. M.,** Intrahypothalamic action of corticotrophin-releasing factor (CRF) to inhibit growth hormone and LH release in the rat, *Life Sci.*, 35, 1117, 1984.
63. **Rivier, C. and Vale, W.,** Influence of CRF on reproductive functions in the rat, *Endocrinology*, 114, 914, 1984.
64. **Rivier, C. and Vale, W.,** CRF acts centrally to inhibit growth hormone secretion in the rat, *Endocrinology*, 114, 2409, 1984.
65. **Ono, N., Bedran de Castro, J., and McCann, S. M.,** Ultrashort-loop positive feedback of corticotropin-releasing factor (CRF) to enhance ACTH release in stress, *Proc. Natl. Acad. Sci. U.S.A.*, 82, 3528, 1985.
66. **Aguila, M. C. and McCann, S. M.,** The Influence of hpGRF, CRF, TRH, and LHRH on SRIF release from median eminence fragments, *Brain Res.*, 348, 180, 1985.
67. **Petraglia, E. T.,** Proceedings Capri Congress, in press, 1988.
68. **Aguila, M. C. and McCann, S. M.,** Stimulation of somatostatin release *in vitro* by synthetic human growth hormone-releasing factor by a nondopaminergic mechanism, *Endocrinology*, 117, 762, 1985.
69. **Aguila, M. C. and McCann, S. M.,** Evidence that GRF stimulates SRIF release via opioid mechanism *in vitro*, *Endocrinology*, 120, 341, 1987.
70. **Aguila, M. C., Khorram, O., and McCann, S. M.,** α-MSH discloses a stimulatory effect of β-endorphin on somatostatin release, *Brain Res.*, 417, 127, 1987.
71. **Rivier, C. and Vale, W.,** Stress-induced inhibition of reproductive functions: role of endogenous CRF, *Science*, 231, 607, 1986.
72. **Krieger, H. B. and Krieger, D. T.,** Chemical stimulation of the brain: effect on adrenal corticoid release, *Am. J. Physiol.*, 218, 16, 1970.

73. **Gillham, B., Lin, J. H., Torrellas, A., Beckford, U., Holmes, M. C., and Jones, M. T.**, in *Neuroendocrinology of Vasopressin Corticoliberin and Opiomelanocortins*, Baertschi, A. J. and Dreifuss, J. J., Eds., Academic Press, London, 1982, 249.
74. **Ganong, W. F.**, Evidence for a central noradrenergic system that inhibits ACTH secretion, in *Brain Endocrine Interactions, Median Eminence: Structure and Function*, Knigge, K., Scott, D. E., and Weindel, A., Eds., S. Karger, Basel, 1972, 254.
75. **Reese, L., Butler, P. W. P., Goshing, C., and Besser, G. M.**, Adrenergic blockade and the corticosterone and GH response to methylamphetamine, *Nature (London)*, 228, 565, 1970.
76. **Al-Damluji, Perry, L., Tomlin, S., Bouloux, P., Grossman, A., Rees, L. H., and Besser, G. M.**, Alpha-adrenergic stimulation of corticotropin secretion by a specific central mechanism in man, *Neuroendocrinology*, 45(1), 68, 1987.
77. **Hedge, G. A.**, Hypothalamic and pituitary effects of prostaglandins on ACTH secretion, *Prostaglandins II*, 293, 1976.
78. **Grossman, A., Moult, P. J. A., Dunnah, D., and Besser, M.**, Different opioid mechanisms are involved in the modulation of ACTH and gonadotrophin release in man, *Neuroendocrinology*, 42, 357, 1986.
79. **Grossman, A. and Besser, G. M.**, Opiates control ACTH through a noradrenergic mechanism, *Clin. Endocrinol.*, 17, 287, 1982.

Chapter 14

ELECTROPHYSIOLOGY OF CORTICOTROPIN-RELEASING FACTOR IN NERVOUS TISSUE

George Robert Siggins

TABLE OF CONTENTS

Considerable evidence, including data from physiological and immunocytochemical studies, indicates that central CRF immunoreactive neurons not only communicate through corticotropes to the rest of the organism, but may also communicate with neurons of the central nervous system (CNS). The demonstration of strong CRF immunoreactivity in rat,[1-4] sheep,[5,6] monkey,[7,8] and human[9,10] tissues with antisera raised against synthetic ovine and rat CRF suggests a highly conserved molecule likely to possess important functional attributes. Furthermore, release of endogenous CRF can be evoked from various brain regions (Chapter 8). These findings, combined with observation of potent effects of centrally injected CRF on autonomic[11] (see Chapter 20 and 21) and behavioral[12,13] (see Chapters 17 to 19) function in the rat, point to a role for CRF as an important neuromessenger in the CNS.

An important first test of a substance as a neuromessenger or transmitter (see Reference 14 for definitions of these terms) involves demonstration of effects on electrical activity in neurons. CRF recently has been administered to several neuronal preparations and cell lines while recording electrical activity. The results of these studies, reviewed below, have led to suggestions that the neuronal effects of CRF may mediate the stress-like response seen in behaving animals with intracerebral CRF injections. Furthermore, these electrophysiological data provide a better understanding of the cellular and molecular mechanisms involved in the hypothalamic regulation of ACTH release from pituitary corticotropes.

For these electrophysiological studies three types of model preparations have been used; (1) *in vivo* electroencephalographic (EEG) recording from awake, freely behaving rats; (2) *in vivo* extracellular recording of single unit activity in anesthetized rats; and (3) single unit recording, by both extracellular and intracellular methods, in brain slice preparations or cell lines *in vitro*. The following is a brief summary of CRF effects in these three preparations.

II. ELECTROENCEPHALOGRAPHIC (EEG) ACTIVITY

Ehlers (Chapter 16) provides an in-depth review of the effects of CRF on cortical and subcortical electrographic activity of freely moving rats. However, to summarize the early results of Ehlers, et al.,[15] CRF administered intracerebroventricularly (i.c.v.) in low doses (10 to 100 ng) elicited EEG activation associated with increased 6 to 8 Hz activity in the dorsal hippocampus and a desynchronization and decreased low frequency (less than 15 Hz) activity in cortex, after a delay of 15 to 20 min. These effects of low doses were not associated with notable increases in locomotor behavior. Higher doses (0.1 to 1 μg) produced a long-lasting EEG activation associated with increased locomotor and grooming behavior and, in some rats, after a delay of approximately 2 h, large interictal spikes spreading from amygdala to the dorsal hippocampus. Handling or loud sounds often produced amygdala afterdischarges associated with behavioral arrest and "wet dog shakes." At doses of 10 to 25 μg, CRF produced intense behavioral activation associated with interictal spikes and afterdischarges in amygdala. After a delay of 4 to 7 h, motor seizures developed in some rats in a manner akin to amygdala "kindling".[15]

The results of these and other EEG studies suggest a spectrum of CNS functions from CRF regulation of brain excitability to modulation of temporal lobe seizures, and may point to hippocampus and amygdala as important CRF receptive sites. Overall, CRF appears to increase limbic excitability, especially at the higher doses.

III. SINGLE UNIT STUDIES *IN VIVO*

A. LOCUS COERULEUS

One of several brainstem nuclei that appears to receive an innervation of CRF fibers is the locus coeruleus (LC). It is interesting that this nucleus has previously been implicated in behavioral activation and the mediation of stress responses.[16-18] In Chapter 15, Valentino reviews the effects of exogenous CRF on the spontaneous and evoked spike discharge activity of individual LC neurons recorded from rats anesthetized with halothane. CRF applied either i.c.v. or locally by pressure from three-barrel micropipettes increased spontaneous discharge rates in most LC cells.[19,20] However, spike activity evoked by sensory stimulation was reduced by CRF at the same time that spontaneous activity was increased.[20] These and other data suggest that CRF has two distinct actions in the LC.[20] Valentino et al.[19] suggest that endogenous CRF, possibly released into the ventricular system or from fibers reported to innervate the LC, may play a physiological role in the activity of this nucleus. The effective intraventricular dose of CRF in this study is comparable to that reported to produce behavioral and autonomic activation,[11-13] providing evidence that CRF has actions on the autonomic and central nervous system akin to those of the stress syndrome. LC neurons are known to become more active during those behavioral states involved in heightened vigilance or orientation to the external environment.[16-18] Valentino et al.[19] note that their results suggest that CRF activation of LC neurons could initiate or intensify those behavioral states that would be adaptive in stressful situations.

B. FOREBRAIN AND HYPOTHALAMUS

Another recent study involved testing the effects of iontophoretically applied CRF on forebrain and hypothalamic neurons in rats anesthetized with urethane.[21] When CRF was iontophoretically applied (with anionic currents of 5 to 25 nA from a pipette containing 0.1 m*M* CRF) to neurons of the cerebral cortex, about half of the cells showed increases in spontaneous discharge whereas 13% were depressed and 39% were unaffected by CRF. In the thalamus, 27% of the neurons tested showed excitatory effects, 41% were depressed, and 32%, no change. In the preoptic, anterior, and ventromedial hypothalamic areas, 59% of cells were activated, 5% depressed, and 36% unaffected.[21] Thus, the results of this study[21] suggest that, of those cells responding, the majority of hypothalamic and cortical neurons were excited by CRF, whereas thalamic and lateral septal neurons were more often depressed. These results suggest region-specific effects of CRF on neuronal excitability. Unfortunately,it is not yet clear if those cells tested would be likely targets for endogenous CRF-containing fibers. Furthermore, the somewhat high percentage of nonresponding cells could result from the use of iontophoresis rather than pressure application of CRF, as suggested by the data of Valentino et al.[19]

IV. *IN VITRO* PREPARATIONS

A. HIPPOCAMPAL SLICES

The first intracellular studies of CRF were performed on the hippocampal slice preparation,[22] because of the ease of intracellular recording and administration of known drug concentrations in this preparation. These studies have shown that superfusion of CRF concentrations above 0.25 μ*M* usually depolarizes both CA1 and CA3 pyramidal neurons, accompanied by elevations in spontaneous firing rate and a slight increase in input resistance[22] (Figure 1), that is, a decrease in ionic conductance. There was little effect of CRF on evoked synaptic activity. At lower concentrations (10 to 200 n*M*), CRF also reduced the magnitude and duration of the afterhyperpolarizations (AHPs) following spontaneous or current-evoked bursts of action potentials.[22] This effect also was obtained both from CA1 pyramidal neurons, where AHPs were produced by bursts of spikes evoked by injection of depolarizing current through the recording pipette, or from CA3 pyramids, where AHPs followed spontaneous bursts of spikes (Figure 1). These lower CRF concentrations had no apparent effect on membrane potential, although increased discharge activity occurred. Superfusion of the acidic, deamidated form of CRF (CRF-OH) had no affect on membrane potential or AHPs at concentrations of 1 μ*M*.

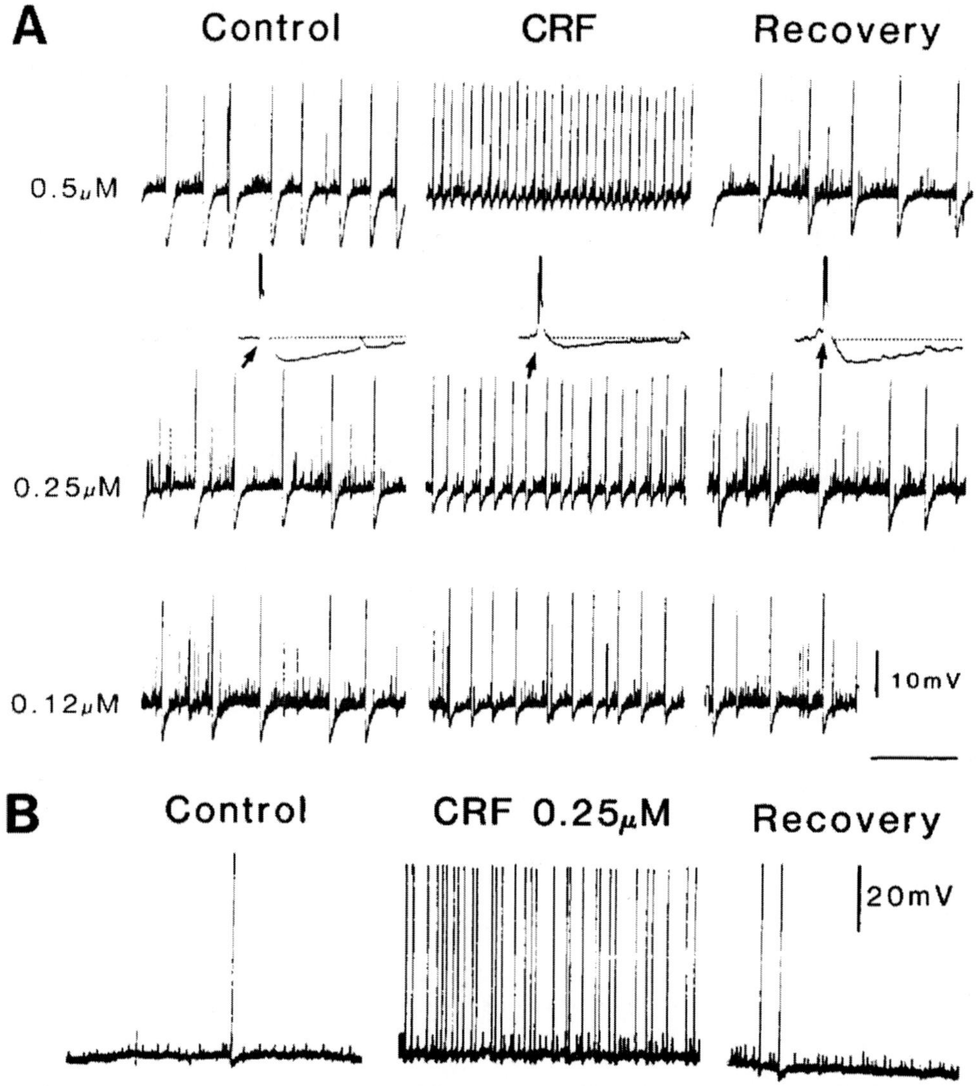

FIGURE 1. Effects of CRF on spontaneous spikes and transmembrane properties of pyramidal neurons. (A) Dose-dependent increases in firing and reductions of postburst afterhyperpolarizations by CRF in a CA3 neuron. Spontaneous large spikes followed by afterhyperpolarizations are actually bursts of multiple spikes that appear as single attenuated spikes because of the slow speed and rise time of the chart recorder. Arrows indicate representative bursts displayed at faster speed (insets). Dotted lines represented the baseline potentials in each condition. The spike firing rate increases at all CRF concentrations. (B) Oscillographs from a CA1 neuron: 0.25 μM CRF increased the spike discharge rate and slightly depolarized (about 3 mV) this cell. There is an apparent increase in the rate and size of activity subthreshold for somatic spike generation (excitatory postsynaptic potentials or dendritic spikes). Horizontal calibration bars are 20 s in (A), 0.8 s in the insets, and 2 s in (B). All CRF records were obtained within 2 to 5 min of control records; recovery records were taken at 10 to 40 min after CRF washout. (From Aldenhoff, J. B., Gruol, D. L., Rivier, J., Vale, W., and Siggins, G. R., *Science*, 221, 875, 1983. With permission.)

Since these AHPs are tetrodotoxin (TTX)-resistant and are blocked by calcium antagonists,[22-25] they probably arise from a calcium-dependent potassium conductance. The size of the "Ca^{++} spikes" recorded in the presence of TTX (to block the sodium component of the spike), are not reduced by CRF.[22] Therefore, CRF probably alters such a K^+ conductance directly rather than indirectly through alteration of calcium conductances (Figure 2). Indeed, it is possible that a CRF-induced reduction of a Ca^{++}-dependent K^+ conductance

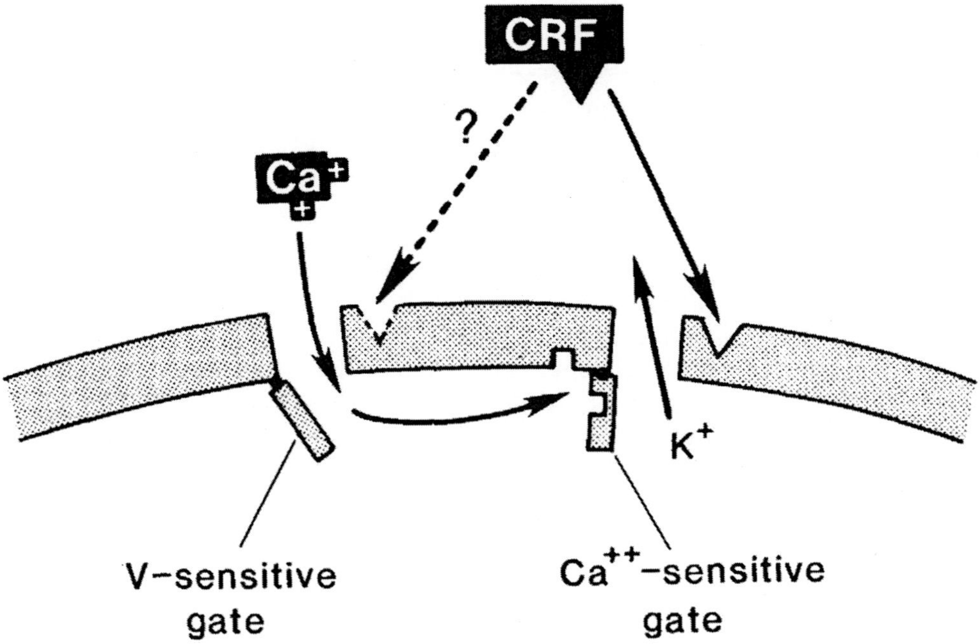

FIGURE 2. Schematic diagram of suggested locus of CRF action at the membrane of the hippocampal pyramidal cell. The CRF molecule, in binding with a CRF 'receptor' (triangular notches) on the neuronal membrane, could block either of two ion channels; (1) Ca^{++} channels, that are voltage sensitive channels opened by a depolarization such as during an action potential, or (2) the Ca^{++}-dependent K^+ channels that are opened by the influx of Ca^{++}. Data showing lack of a CRF effect on the Ca^{++} spike during TTX treatment[22] suggests that the second possibility is more likely. Since CRF inactivation of the Ca^{++}-dependent K^+ channels would retard repolarization of the membrane during an action potential, more Ca^{++} might then enter the cell during this state of depolarization, by virtue of the voltage-dependence of the Ca^{++} channel. (From Siggins, G. R., Gruol, D., Aldenhoff, J., and Pittman, Q., *Fed. Proc.*, 44, 237, 1985. With permission.)

would lead to influx of greater amounts of Ca^{++} during firing of spikes or bursts of spikes. Such an effect, if obtained also in corticotropes, could account for the enhanced release from these cells of ACTH, via a calcium-dependent process.[26]

It should be noted that the overall effect of CRF on hippocampal pyramidal neurons is similar to some of the actions of cholinergic muscarinic agonists on these cells.[14] Thus, the muscarinic agonists depolarize these neurons with increased discharge, decreased ionic conductance and a reduction of postburst AHPs. A portion of the muscarinic depolarization is attributable to inactivation of a voltage-dependent K^+ conductance (the M-current: I_M) persisting at membrane potentials slightly depolarized from resting levels.[27] The inactivation of this K^+ conductance would lead to further depolarization in cells already slightly depolarized and therefore to increased spiking probability and Ca^{++} entry during the spikes. The effect of CRF on the M-current in hippocampal neurons has not been tested, but it is possible that it may inactivate the M-current here like muscarinic agonists, and as CRF appears to do in human adenoma cells (see below). It is of interest, given the known antagonistic interaction of CRF and somatostatin,[29] that the M-current in pyramidal neurons recently has been shown to be augmented by somatostatin.[28]

The enhanced discharge activity of the hippocampal neurons elicited by CRF is consistent with CRF-induced excitations of neurons recorded in other brain regions *in vivo* (see above). A general reduction of AHPs or I_M by CRF could provide the mechanism for these excitations and, therefore the activation of the cortical and hippocampal EEG in awake animals. It might be argued that excitation of hippocampal neurons by CRF constitutes another electrophysiological sign of the adaptation of the organism to stress, especially given the role the hippocampus is likely to play in visceral, learning and emotional behaviors.[30,31] A few cell bodies and fibers immunoreactive for CRF have been found in the hippocampus.[2,3] However, to date it has been difficult to experimentally activate synaptic release of CRF in this area by stimulation of such cells or fibers, thus preventing satisfaction of several criteria[14,32] required for proof of CRF as a neurotransmitter.

B. COMPLEX OF THE SOLITARY TRACT (CST)

This group of brain stem nuclei, including the nucleus of the dorsal motor vagus and the various subnuclei of the tractus solitarius, are involved in regulation of the autonomic nervous system. These nuclei are invested with fibers immunoreactive for CRF[2] and would thus appear to be ideally organized to provide a central link in the autonomic response to stress (see Chapter 20). As in the hippocampus, somatostatin augments the M-current and inhibits spontaneous firing of CST neurons.[33] Because of the presence in these nuclei of CRF-containing fibers, a study of the effects of exogenous CRF on CST neurons has been initiated to determine the physiological role of this peptide (G. R. Siggins, J. Champagnat, T. Jacquin, and M. Saubié, in preparation).

To this end, extra- and intracellular recording was used in slices of brain stem prepared as for hippocampal slices and incubated *in vitro*.[34] Superfusion of CRF 0.2 to 0.5 μM increased the discharge rate of six of eight spontaneously firing neurons recorded extracellularly (Figure 3). In ten intracellular recordings, the same concentrations of CRF also increased spike discharge in spontaneously firing CST neurons, often (eight of ten neurons) in association with a slow depolarization of 3 to 8 mV. As in hippocampal neurons, the AHPs following current-evoked bursts of spikes were reduced in amplitude by CRF superfusion. Thus, the effect of CRF in this brain region appears to be roughly equivalent to that in the hippocampal pyramidal neurons. Again, because of the similarity between the effects of muscarinic agonists[14,35] and CRF, it would be instructive to examine the effects of CRF on the M-current in the CST.

C. HYPOTHALAMIC SLICE

Hypothalamic neurons would seem to be likely targets for CRF containing fibers in immunohistochemical studies.[1,2] Therefore, preliminary studies[36] have focused on the hypothalamic slice preparation,[37] because of the relative ease of identifying cell types, of recording stable neuronal activity and of applying known drug concentrations. The initial studies were of the basal arcuate area in more caudal slices, where early immunohistochemical studies[1,2] had suggested the presence of abundant CRF-containing fibers. These fibers were initially thought to be in contact with β-endorphin-containing cells known to be scattered throughout this area.[1,7] In preliminary studies using extracellular recording of nine neurons of the arcuate area, no cells were excited by CRF superfusion, four cells were depressed and five cells were unresponsive to 1 μM CRF.[36] The inconsistency of these responses probably arises from the fact that none of the recorded cells could be identified in terms of potential innervation by CRF fibers. Moreover, more recent immunohistochemical studies using staining of alternate 20 μm sections with either anti-β-endorphin or anti-CRF sera suggests that CRF-containing fibers previously seen in arcuate do not project to the β-endorphinergic cells but merely represent fibers of passage en route to the median eminence (Battenberg and Bloom, unpublished). If there is a correlation between numbers of CRF terminals and CRF receptors, these findings could help explain the high percentage of arcuate cells that did not respond to superfusion of CRF in 1 μM concentrations.

Other hypothalamic studies focused on the paraventricular nucleus (PVN), which contains abundant CRF-containing cell bodies, mostly in the medial dorsal parvocellular regions, and CRF-containing fibers projecting laterally through the magnocellular region and ventrally

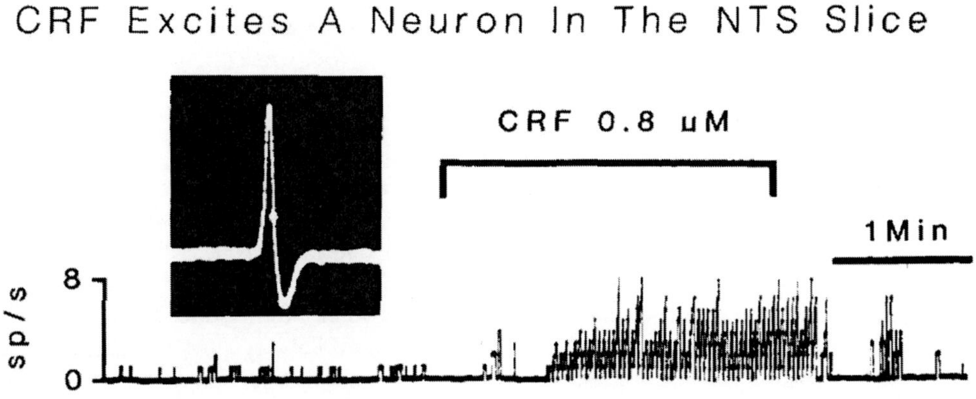

FIGURE 3. CRF superfusion elevates the spontaneous discharge rate of a CSF neuron *in vitro*. Extracellular recording with a glass micropipette filled with 3 *M* NaCl. Action potentials (see inset for examples) were gated with a window discriminator, converted to square pulses, integrated over 1-s intervals and plotted on a polygraph as firing rate (sp/s = spikes per s). The representative spike (inset) is 8 mV in amplitude. (From Siggins, Champagnat, and Saubié, unpublished.)

toward the median eminence, pituitary and more caudal structures[1,2] (see also Chapter 4). In mostly extracellular recordings from 9 neurons of the PVN in the rat hypothalamic slice, CRF superfusion (up to 1 μM) in PVN was more often effective than in the arcuate area.[36] However, inhibition still predominated: of seven neurons recorded extracellularly, five showed depression of spontaneous firing with threshold concentrations of about 0.1 μM or less; one cell was excited and one showed no effect of CRF at 0.5 μM. In two intracellularly recorded cells, one cell was inhibited by CRF with little change in membrane potential, whereas the other cell was depolarized with an increased firing rate. Most of these nine cells were recorded in the magnocellular region of PVN.[36]

The inhibitory effects of CRF in the PVN and arcuate nucleus are somewhat surprising,in the light of the predominant excitatory actions recorded elsewhere in brain (see above). These divergent effects might arise from inappropriate testing of neurons that do not receive an input of CRF-containing fibers. However, there is no *a priori* reason for every neuron to respond identically to CRF, even if they all receive CRF afferents. Ultrastructural immunohistochemistry of the CRF system will be required to resolve these questions.

However, an explanation for the inhibitory CRF effects in PVN might be forthcoming from results of anatomic and biochemical studies. Thus, van den Pol[38] has shown evidence for axon collaterals from PVN parvocellular neurons that contact magnocellular neurons. It is possible that many of these parvocellular neurons contain CRF, and that a population of the magnocellular neurons contain either vasopressin and/or oxytocin. Furthermore, Plotsky et al.[39] have shown that i.c.v. CRF diminishes the amount of vasopressin released into the portal blood of the pituitary stalk. Taken together with the preliminary results of CRF-induced inhibition of neurons in PVN, these findings may suggest a collateral inhibition of magnocellular vasopressinergic neurons by parvocellular CRF-containing cells. Experiments using pathway stimulation will be needed to test this idea and the possibility that CRF may act as a transmitter in this collateral pathway.

D. CULTURED HUMAN ADENOMA CELLS

A recent electrophysiological study[40] of the adrenocorticotropin-secreting adenoma, taken from patients with Cushing's disease and cultured *in vitro*, sheds considerable light on the mechanism(s) of CRF-evoked ACTH release. Dispersed cells from these tumors reveal, on intracellular recording, calcium-dependent spikes whose durations of depolarization appear markedly dependent upon K^+ conductances: agents like TEA that reduce some K^+ conductances greatly prolong the action potential duration. In standard intracellular recordings of voltage (so-called current-clamp mode), CRF alone at low concentrations (10 n*M*) increased spike discharge and, at ten times higher concentrations, caused slow depolarizations with: (1) increased firing, (2) increased input resistance, (3) increased duration and amplitude of the calcium-dependent spikes, and (4) decreased ionic conductance.[40] When applied together with angiotensin II (AII), CRF evoked depolarizations more pronounced than the sum of the effects of either releasing factor alone. CRF had no effect on the AHPs following single spikes; however, the effect of CRF on the long duration AHPs following burst of spikes apparently was not examined.

Use of the single-electrode voltage-clamp to study the ionic currents directly has revealed at least one possible ionic mechanism for the CRF-elicited excitability increase. These tumor cells appear to possess a slow, voltage-dependent K^+ current that persists at the low resting potentials of these cells and that is very similar to the M-current described for sympathetic ganglia[41] and hippocampal neurons.[27] CRF, and the combination of AII and CRF, appear to reduce this persistent K^+ current, thus causing a net inward current at normal resting potentials in these adenoma cells.[40] These results suggest that the effect of reducing the K^+ conductance would be to depolarize the adenoma cells, thus causing increased action potential discharge, and increasing the duration of each action potential. As the action potential currents appear to be carried mostly by calcium ions (the spikes are TTX resistant), the end result would be an increased calcium influx and enhanced ACTH release. However, these authors[40] caution that other mechanisms not studied (e.g., Ca^{++} conductances, second messengers) might also be involved in the ACTH release mechanism.

V. DISCUSSION

In spite of the relative paucity of electrophysiological studies of CRF reported to date, the findings provide several significant implications. First, it is clear that CRF has very potent, usually excitatory, actions on the electrical activity of central neurons. Since CRF fibers appear to project to neurons in many CNS areas, CRF may thus be considered a candidate for a neurotransmitter. However, it should be cautioned that at least two criteria for its role as a transmitter[14,32] have not yet been satisfied: (1) mimicry of the effects of exogenous CRF by stimulated release of endogenous CRF; and (2) comparable antagonism or potentiation of endogenous and exogenous CRF by pharmacological agents. Although the development of CRF antagonists[42] will aid satisfaction of the last criterion, pathway stimulation studies of CRF release likely will be hindered by the somewhat diffuse distribution of CRF-containing fibers.

Second, intracellular studies of hippocampal pyramidal neurons and human adenoma cells suggest at least two novel mechanisms for the excitatory action of CRF and may explain how CRF induces release of ACTH from corticotropes. Interestingly, the reduction of AHPs by block of a Ca^{++}-dependent K^+ conductance (Figure 2) has been described for the effects of norepinephrine,[43-45] histamine,[44] and acetylcholine,[14,46] also in hippocampal pyramidal cells. The end-effect of such a process and the reduction of other K^+ currents (e.g., I_M) may lead not only to the increased firing of spikes (due to the reduced inhibition afforded by the AHP), but also to a change in cellular Ca^{++} physiology, including increased intracellular Ca^{++}. The influence of intracellular Ca^{++} on other nonelectrical processes such as cell growth and motility, genome expression, enzyme activity (for example, adenylate cyclase) and exocytosis suggests that CRF might serve an important role in these processes as well. Of course, other as yet undetected effects of CRF on non-K^+ processes; e.g., directly on Ca^{++} channels or indirectly via second messengers,[47] could also be involved and need more study.

Finally, observations of excitatory actions of CRF in brain regions that may be involved in alerting behavior point to a role of CRF in the stress response of the extrahypothalamic brain as well as those areas concerned with ACTH and β-endorphin release. These observations, combined with histochemical findings of projections of CRF-containing fibers to these regions and those (e.g., n. tractus solitarius, spinal cord intermediolateral column) concerned with control of the autonomic nervous system (see Reference 2 and Chapter 4), may lead to a more unifying view of CRF function. Insofar as the "one transmitter — one function" concept can be validated (see Reference 14), CRF might function in a concerted way across all CNS areas so innervated, toward a common behavioral end: one involving arousal, alertness, recall of memory, hyperactivity, and increased sympathetic function. These behaviors might then have one common purpose: to prepare the organism for appropriate responses to novel or threatening situations.

VI. SUMMARY

Recent studies of effects of CRF on the electrical activity of central neurons indicate that CRF has predominantly excitatory actions in locus coeruleus, the solitary tract complex, hippocampus, cortex, and some regions of the hypothalamus. These brain areas are reported to contain immunoreactive CRF. Intracellular recordings in the hippocampal slice preparation and the solitary tract complex demonstrate that the excitations may arise from reduction of the afterhyperpolarizations (AHPs) following bursts of spikes. The postburst AHPs probably are produced by a Ca^{++}-dependent K^+ conductance. Since "Ca^{++} spikes" recorded in the presence of tetrodotoxin are not diminished by CRF, this peptide therefore would appear to be acting either at the level of the Ca^{++}-dependent K^+ conductance itself, or at the linkage between this conductance and Ca^{++} influx or Ca^{++} recognition sites. Furthermore, recent studies of ACTH-containing human adenoma cells in culture also show CRF-induced depolarizations and increased spiking correlated with a reduction of a voltage-dependent K^+ conductance similar to the M-current seen in neurons. All these excitatory effects are consistent with electroencephalographic recordings in awake animals, where intracerebroventricular CRF activates cortical and limbic areas and, at higher doses, evokes epileptiform activity in amygdala and hippocampus. However, predominantly inhibitory actions of CRF have been seen with extracellular single unit recordings in a few CNS areas such as lateral septum, thalamus, and the hypothalamic paraventricular nucleus. These findings, combined with those from immunohistochemical, biochemical and behavioral studies, suggest (1) a possible neuromessenger role for CRF in extrahypothalamic regions, (2) that inactivation of K^+ conductances may provide the mechanism for CRF-evoked ACTH release, and (3) a possible concerted function by CRF containing elements in the CNS in an integrated behavioral response to stress.

VII. ACKNOWLEDGMENTS

We thank Drs. Wylie Vale and Jean Rivier for providing the CRF and OH-CRF, and Nancy Callahan for typing the manuscript. Portions of these results were supported by NIH (DK 26741), AA-06420, AA-07456, the Alexander von Humboldt Stiftung, and the Fondation pour la Recherche Medicale.

REFERENCES

1. **Bloom, F. E., Battenberg, E. L. F., Rivier, J., and Vale, W.,** Corticotropin releasing factor (CRF): immunoreactive neurons and fibers in rat hypothalamus, *Regul. Pept.*, 4, 43, 1982.
2. **Swanson, L. W., Sawchenko, P. E., Rivier, J., and Vale, W. W.,** The organization of ovine corticotropin-releasing factor (CRF)-immunoreactive cells and fibers in the rat brain: an immunohistochemical study, *Neuroendocrinology*, 36, 165, 1983.
3. **Merchenthaler, I., Vigh, S., Petrusz, P., and Schally, A. V.,** Immunocytochemical localization of corticotropin-releasing factor (CRF) in the rat brain, *Am. J. Anat.*, 165, 385, 1982.
4. **Olschowka, J. A., O'Donohue, T. L., Mueller, G. P., and Jacobowitz, D. M.,** Hypothalamic and extrahypothalamic distribution of CRF-like immunoreactive neurons in the rat brain, *Neuroendocrinology*, 35, 305, 1982.
5. **Paull, W. K., Scholer, J., Arimura, A., Meyers, C. A., Chang, J. K., Chang, D., and Shimizu, M.,** Immunocytochemical localization of CRF in the ovine hypothalamus, *Peptides*, 3, 183, 1982.
6. **Palkovits, M., Brownstein, M. J., and Vale, W.,** Corticotropin releasing factor (CRF) immunoreactivity in hypothalamic and extrahypothalamic nuclei of sheep brain, *Neuroendocriology*, 37, 302, 1983.
7. **Battenberg, E. L. F., Bloom, F. E., Rivier, J., and Vale, W.,** Corticotropin releasing factor (CRF): immunoreactive neurons and fibers in rats and primate hypothalamus, *Soc. Neurosci. Abstr.*, 8, 110, 1982.
8. **Kawata, M., Hashimoto, K., Takahara, J., and Sano, Y.,** CRF immunoreactive nerve fibers in the circumventricular organs of the monkey, *Macaca fuscata, Cell Tissue Res.*, 232, 679, 1983.
9. **Pelletier, G., Desy, L., Cote, J., and Vaudry, H.,** Immunocytochemical localization of corticotropin-releasing factor-like immunoreactivity in the human hypothalamus, *Neurosci. Lett.*, 41, 259, 1983.
10. **Suda, T., Tomori, N., Tozawa, F., Demura, H., Shizume, K., Mouri, T., Miura, Y., and Sasano, N.,** Immunoreactive corticotropin and corticotropin-releasing factor in human hypothalamus, adrenal, lung cancer, and pheochromocytoma, *J. Clin. Endocrinol. Metab.*, 58, 919, 1984.
11. **Brown, M. R., Fisher, L. A., Spiess, J., Rivier, C., Rivier, J., and Vale, W.,** Corticotropin-releasing factor (CRF): actions on the sympathetic nervous system and metabolism, *Endocrinology*, 111, 928, 1982.
12. **Sutton, R. E., Koob, G. F., Le Moal, M., Rivier, J., and Vale, W.,** Corticotropin releasing factor produces behavioral activation in rats, *Nature (London)*, 297, 331, 1982.
13. **Britton, D. R., Koob, G. F., Rivier, J., and Vale, W.,** Intraventricular corticotropin releasing factor enhances behavioral effects of novelty, *Life Sci.*, 31, 363, 1982.
14. **Siggins, G. R. and Gruol, D. L.,** Synaptic mechanisms in the vertebrate central nervous system, in *Handbook of Physiology*, Volumne on Intrinsic Regulatory Systems of the Brain, Bloom, F. E., Ed., The American Physiological Society, Washington, D.C., 1986, 1.
15. **Ehlers, C. L., Henriksen, S. J., Wang, M., Rivier, J., Vale, W., and Bloom, F. E.,** Corticotropin releasing factor produces increases in brain excitability and convulsive seizures in rats, *Brain Res.*, 278, 332, 1983.
16. **Aston-Jones, G. and Bloom, F. E.,** Activity of norepinephrine containing locus coeruleus neurons in behaving rats anticipates fluctuations in the sleep-waking cycle, *J. Neurosci.*, 1, 976, 1981.
17. **Aston-Jones, G. and Bloom, F. E.,** Norepinephrine-containing locus coeruleus neurons in behaving rats exhibit pronounced responses to non-noxious environmental stimuli, *J. Neurosci.*, 1, 887, 1981.
18. **Foote, S., Aston-Jones, G., and Bloom, F. E.,** Impulse activity of locus coeruleus neurons in awake rats and squirrel monkeys is a function of sensory stimulation and arousal, *Proc. Natl. Acad. Sci. U.S.A.*, 77, 3033, 1980.
19. **Valentino, R. J., Foote, S. L., and Aston-Jones, G.,** Corticotropin-releasing factor activates noradrenergic neurons of the locus coeruleus, *Brain Res.*, 270, 363, 1983.
20. **Valentino, R. J. and Foote, S. L.,** Corticotropin-releasing factor disrupts sensory responses of brain noradrenergic neurons, *Neuroendocrinology*, 45, 28, 1987.
21. **Eberly, L. B., Dudley, C. A., and Moss, R. L.,** Iontophoretic mapping of corticotropin-releasing factor (CRF) sensitive neurons in the rat forebrain, *Peptides*, 4, 837, 1983.
22. **Aldenhoff, J. B., Gruol, D. L., Rivier, J., Vale, W., and Siggins, G. R.,** Corticotropin releasing factor decreases post-burst hyperpolarizations and excites hippocampal pyramidal neurons *in vitro*, *Science*, 221, 875, 1983.
23. **Alger, B. E. and Nicoll, R. A.,** Epileptiform burst after-hyperpolarization: calcium-dependent potassium potential in hippocampal CA1 cells, *Science*, 210, 1122, 1980.
24. **Hotson, J. R. and Prince, D. A.,** A calcium-activated hyperpolarization follows repetitive firing in hippocampal neurons, *J. Neurophysiol.*, 43, 409, 1980.
25. **Schwartzkroin, P. A. and Stafstrom, C. E.,** Effects of EGTA on the calcium-activated afterhyperpolarization in hippocampal CA3 pyramidal cells, *Science*, 210, 1125, 1980.
26. **Vale, W., Spiess, J., Rivier, C., and Rivier, J.,** Characterization of a 41-residue ovine hypothalamic peptide that stimulates secretion of corticotropin and β-endorphin, *Science*, 231, 1394, 1981.

27. **Halliwell, J. V. and Adams, P. R.,** Voltage-clamp analysis of muscarinic excitation in hippocampal neurons, *Brain Res.*, 250, 71, 1982.
28. **Moore, S.D., Madamba, S. G., Joels, M., and Siggins, G. R.,** Somatostatin augments the M-current in hippocampal neurons, *Science*, 239, 278, 1988.
29. **Reisine, T. and Axelrod, J.,** Prolonged somatostatin pretreatment desensitizes somatostatin's inhibition of receptor-mediated release of adrenocorticotropin hormone and sensitizes adenylate cyclase, *Endocrinology*, 113, 811, 1983.
30. **Isaacson, R. L. and Wickelgren, W. D.,** Hippocampal ablation and passive avoidance, *Science*, 138, 1104, 1962.
31. **O'Keefe, J. and Nadel, L.,** *The Hippocampus as a Cognitive Map*, Clarendon Press, Oxford, 1978.
32. **Werman, R.,** Criteria for identification of a central nervous system transmitter, *Comp. Biochem. Physiol.*, 18, 745, 1966.
33. **Jacquin, T., Chapagnat, J., Madamba, S., Denavit-Saubie, M., and Siggins, G. R.,** Somatostatin depresses excitability in neurons of the solitary tract complex through hyperpolarization and augmentation of I_M, a noninactivating voltage-dependent outward current blocked by muscarinic agonists, *Proc. Natl. Acad. Sci. U.S.A.*, 85, 948, 1988.
34. **Champagnat, J., Denavit-Saubie, M., and Siggins, G. R.,** Rhythmic neuronal activities in the nucleus of the tractus solitarius isolated in vitro, *Brain Res.*, 280, 155, 1983.
35. **Champagnat, J., Jacquin, T., and Richter, D. W.,** Voltage-dependent currents in neurons of the nucleus of the solitary tract of rat brainstem slices, *Pflugers Arch.*, 406, 372, 1986.
36. **Siggins, G. R., Gruol, D., Aldenhoff, J., and Pittman, Q.,** Electrophysiological actions of corticotropin-releasing factor in the central nervous system, *Fed. Proc.*, 44, 237, 1985.
37. **Pittman, Q. J., Hatton, J. D., and Bloom, F. E.,** Spontaneous activity in perfused hypothalamic slices: dependence on calcium content of perfusate, *Exp. Brain Res.*, 42, 49, 1981.
38. **van den Pol, A. N.,** The magnocellular and parvocellular paraventricular nucleus of rat: intrinsic organization, *J. Comp. Neurol.*, 206, 317, 1982.
39. **Plotsky, P. M., Bruhn, T. O., and Vale, W.,** Central modulation of corticotropin releasing factor-like immunoreactivity secretion by arginine vasopressin, *Soc. Neurosci. Abstr.*, 9, 391, 1983.
40. **Mollard, P., Vacher, P., Guerin, J., Rogawski, M. A. and Dufy, B.,** Electrical properties of cultured human adrenocorticotropin-secreting adenoma cells: effects of high K^+, corticotropin-releasing factor, and angiotensin II, *Endocrinology*, 121, 395, 1987.
41. **Brown, D. A. and Adams, P. R.,** Muscarinic suppression of a novel voltage-sensitive K^+ current in a vertebrate neurone, *Nature London*, 283, 673, 1980.
42. **Rivier, J., Rivier, C., and Vale, W.,** Synthetic competitive antagonists of corticotropin releasing factor: effect on ACTH secretion in the rat, *Science* 224, 889, 1984.
43. **Madison, D. V. and Nicoll, R. A.,** Noradrenaline blocks accomodation of pyramidal cell discharge in the hippocampus, *Nature*, 299, 636, 1982.
44. **Haas, H. L. and Konnerth, A.,** Histamine and noradrenaline decrease calcium-activated potassium conductance in hippocampal pyramidal cells, *Nature*, 302, 432, 1983.
45. **Gruol, D.L. and Siggins, G. R.,** Adrenergic agonists alter the intracellularly recorded spontaneous activity and postdischarge after-hyperpolarization of pyramidal neurons of the rat hippocampus in vitro, in *Catecholamines: Neuropharmacology and Central Nervous System — Theoretical Aspects*, Usdin, E., Carlsson, A., Dahlstrom, A., and Engel, J., Eds., Alan R. Liss, New York, 1984, 131.
46. **Cole, A. E. and Nicoll, R. A.,** Acetylcholine mediates a slow synaptic potential in hippocampal pyramidal cells, *Science*, 221, 1299, 1983.
47. **Guild, S. and Reisine, T.,** Molecular mechanisms of corticotropin-releasing factor stimulation of calcium mobilization and adrencorticotropin release from anterior pituitary tumor cells, *J. Pharmacol. Exp. Ther.*, 241, 125, 1987.

Chapter 15

EFFECTS OF CRF ON SPONTANEOUS AND SENSORY-EVOKED ACTIVITY OF LOCUS COERULEUS NEURONS

Rita J. Valentino

TABLE OF CONTENTS

I. INTRODUCTION

The isolation and characterization of corticotropin-releasing factor (CRF),[1-3] while relatively recent, has generated a vast literature describing aspects of CRF localization, behavioral effects, autonomic effects, endocrine effects, and cellular mechanisms for ACTH release. One reason for this productivity in CRF research is that CRF has provided investigators with a tool to study the stress response, which although important biologically, has not been well defined. CRF is the most useful tool investigators have to study stress because it is the primary neurohormone regulating ACTH release and this has been the hallmark of stress responses.[4,5] Investigations of CRF localization in the CNS,[6-10] CRF receptor localization,[11,12] and of the behavioral[13-16] and autonomic effects[17-19] of centrally administered CRF suggest that CRF plays more than a neuroendocrine role in stress. In addition, it may act as a neurotransmitter or neuromodulator in extrahypothalamic circuits to initiate an integrated response to various stimuli. For example, the coordinated release of CRF from cells in the hypothalamus, fibers innervating brainstem nuclei, and fibers innervating cortical structures, could elicit the multisystem responses generally associated with stress, including behavioral arousal, increased autonomic activity, and ACTH release. This hypothesis is supported by the findings that an analog of CRF which acts as an antagonist at CRF binding sites (α-helical CRF) and anti-CRF antibody block certain aspects of the stress response.[20,21] However, several key pieces of data are needed to confirm this hypothesis. For example, release of CRF in extrahypothalamic circuits must be demonstrated. Moreover, the effects of CRF on target neurons have not been sufficiently described, and little is known of cellular mechanisms by which CRF acts on neurons. A review of CRF effects of brain neurons by Siggins is included in Chapter 14. The following chapter describes the effects of CRF on one specific target, the noradrenergic nucleus locus coeruleus (LC). These effects are emphasized here because the anatomical and physiological characteristics of the LC make it a target by which many stress- or CRF-elicited effects may be integrated. The studies described being to test the hypothesis that CRF acts as a neurotransmitter or neuromodulator in the LC as one step in producing an integrated response to stressful stimuli.

II. THE LC AS A TARGET FOR CRF

The LC is a compact pontine nucleus which has widespread projections throughout the CNS providing noradrenergic input to cortical regions, the cerebellum, nuclei in the hypothalamus, nuclei in the brainstem involved in autonomic function, and to the spinal cord.[22-26] The neocortex and hippocampus receive their sole norepinephrine input from the LC.[24,25] CRF immunoreactive fibers have been visualized in both rat and primate LC.[6-8,10,27] The origin of the CRF-containing fibers is unknown. However, studies of LC afferents have demonstrated a restricted input originating primarily from the nucleus paragigantocellularis and the prepositus hypoglossal nucleus, with a minor input from the paraventricular nucleus of the hypothalamus.[28] The high density of CRF cell bodies in the hypothalamic paraventricular nucleus implicate this as a site of origin for LC CRF. However, CRF has also been localized in the paragigantocellularis which contains a number of different peptides. CRF localized in the LC may utilize the widespread noradrenergic projection system to exert effects throughout the CNS. Thus, the LC may serve as an amplification mechanism to convey certain information about stress via CRF release in the LC. To date, release of CRF in the LC has not been demonstrated; however, large increases in CRF content in LC have been observed after acute and chronic stress.[29] It is not clear whether these increased levels reflect enhanced processing due to increased turnover.

If CRF utilizes the LC to initiate aspects of the stress response its effects on LC neuronal activity should mimic those of stress. Several studies indicate that the LC is activated by

TABLE 1
CRF Effects on LC Spontaneous Discharge Rate[a]

	Anesthetized rats		Unanesthetized rats	
	3 min[b]	20 min	9 min	30 min
Ala[14]CRF (3.0 μg)	107 ± 1	104 ± 5	89 ± 5	101 ± 9
CRF (0.3 μg)	100 ± 3	100 ± 4	124 ± 27	130 ± 32
CRF (1.0 μg)	109 ± 2*	133 ± 22	132 ± 11*	186 ± 32*
CRF (3.0 μg)	147 ± 12**	150 ± 12*	178 ± 31*	284 ± 29**

Note: *, $p < 0.05$; **, $p < 0.005$.

[a] Rate is expressed as a percentage of the mean preinjection rate ± 1 SEM. Determinations are the means of at least five rats (one cell per rat and one treatment per cell). See References 43 and 48 for details.
[b] Times after injection during which rates were determined.

stressful stimuli. For example, shock, neutral stimuli that have been conditioned to shock, and hemorrhage all increase norepinephrine turnover in brain regions that receive their sole norepinephrine input from the LC.[30-34] Consistent with this, direct increases in LC discharge rate evoked by footshock, hypovolemia, and hypoxia have been measured using unit recordings in anesthetized rats.[35-37] Finally, although the LC is activated by many kinds of stimuli it has been reported that stimuli paired with noxious events are more effective in increasing LC discharge of unanesthetized cats than nonconditioned, nonnoxious stimuli.[38] Many of these stimuli that are condisered stressful also release CRF, at least in hypothalamic circuits.[39,40] Thus, certain stimuli that are considered stressful by their ability to release CRF and thereby ACTH, are among those that also increase LC activity. It is tempting to speculate that the mechanism for LC activation by these stimuli is through release of CRF from fibers within the LC, while other, nonstressful stimuli which increase LC discharge may do so by other mechanisms. As one step in testing this hypothesis the effects of centrally administered CRF on LC activity were characterized.

III. EFFECTS OF CRF ON LC SPONTANEOUS DISCHARGE RATE

A. ANESTHETIZED RATS

LC spontaneous discharge of anesthetized rats is relatively slow and stable irrespective of the anesthesia used, with discharge rates ranging from 0.5 to 5.0 Hz.[41] Intracerebroventricular (i.c.v.) injection of CRF increases LC discharge rates of halothane anesthetized rats in a dose-dependent manner with 3.0 μg producing an increase of about 50% (Table 1).[42-44] This increase in spontaneous activity begins approximately 3 min after injection and is maintained up to 20 min after injection (Table 1). Ala[14]CRF and CRF-OH, the free acid analog of CRF, are analogs of CRF that are relatively ineffective in releasing ACTH from anterior pituitary.[3] These peptides do not alter LC spontaneous discharge rates when injected in the same dose range as CRF (up to 3 μg). Thus, the structural requirements for LC activation by CRF are similar to the structural requirements for CRF at its receptor in the anterior pituitary.

Preliminary findings suggest that increased spontaneous discharge rate of LC neurons is a direct effect of CRF.[42] Similar increases in spontaneous discharge rate were observed when CRF was applied directly onto LC neurons by pressure application to a CRF-containing micropipette in 9 of 14 cells tested as are observed after i.c.v administration of CRF. In contrast, application of CRF to neurons in the cerebellum, or cells of the mesencephalic nucleus of the trigeminus did not increase spontaneous discharge. Interestingly, direct application of small amounts of CRF produced a marked excitation of a cell that was identified within the parabrachial nucleus. CRF fibers have been localized in the parabrachial nucleus and this nucleus is thought to be involved in certain autonomic functions. The potent effects of CRF on parabrachial neurons suggest that it may be another extrahypothalamic target for CRF. Although direct application of CRF to LC neurons by micropressure techniques increased spontaneous discharge of the majority of LC neurons tested, iontophoresis of CRF had no consistent effect. In order to verify direct actions of CRF it is necessary to test a greater population of neurons or to investigate effects of CRF on LC activity in *in vitro* preparations such as the LC slice preparation.

B. UNANESTHETIZED RATS

The studies of CRF neuronal effects in anesthetized rats demonstrate a dose-dependent, structurally specific, increase in LC discharge rate which may be a direct effect on LC neurons. However, anesthesia alters LC discharge such that it is usually slower and more stable than that recorded in unanesthetized animals. In unanesthetized animals LC spontaneous discharge is labile and associated with the state of arousal or behavioral vigilance.[45,46] For example, discharge rate is greatest when the animal is in an active waking state, attending to stimuli in the external environment. Discharge rates are relatively slower when animals are involved in vegetative behaviors such as grooming, and in slow wave sleep. Finally, LC cells are inactive during REM sleep. In both anesthetized and unanesthetized animals LC discarge can be evoked by phasic sensory stimuli.[46,47] However, potentially noxious stimuli such as tail or paw pressure, or footshock are required to elicit LC discharge of anesthetized rats while many modes of nonnoxious stimuli evoke LC discharge of unanesthetized rats. These characteristics of LC discharge in unanesthetized rats are consistent with the hypothesis that the LC activation is associated with states of arousal. LC activation produced by stress or by CRF administration may also be related to behavioral arousal associated with these conditions. To more fully understand the relationship between stress, LC discharge, and behavioral arousal, the effects of CRF on LC discharge of unanesthetized rats were characterized.

As in anesthetized rats, CRF increases LC spontaneous discharge rates recorded in both sling-restrained and freely moving unanesthetized rats.[48] No differences in spontaneous discharge rate or CRF efficacy were apparent in sling-restrained vs. freely moving rats. This was not surprising since the restrained animals were habituated to the sling prior to recording sessions. As in anesthetized rats, the activation of LC discharge by CRF is dose dependent and the inactive analog, Ala^{14}CRF (3.0 μg) is ineffective (Table 1). The important differences between CRF effects on LC discharge of anesthetized vs. unanesthetized rats lie in CRF potency and duration of action (Table 1). CRF is more potent in unanesthetized rats. Because complete dose-effect curves for CRF were not generated it is not known whether the efficacy of CRF differs in the two preparations. There are few comparisons of drug effects on LC activity in anesthetized vs. the unanesthetized state, but it is likely that dose-effect curves will differ in the two conditions. For example, in a study of morphine inhibition of LC activity the morphine dose-response curve determined in unanesthetized rats was shifted to the right 1 log unit from that determined in anesthetized rats even though the mean control discharge rate was similar in both preparations.[49] Thus, anesthesia may potentiate agents that inhibit LC discharge either directly, or by inhibiting excitatory inputs onto LC, and may antagonize the effects of agents like CRF, that increase LC discharge.

The doses of CRF that increase LC spontaneous discharge are in the range of doses that cause increased locomotor activity, neophobic behaviors in an open field paradigm, and increases in blood pressure and heart rate.[13,16,18] Thus, doses of i.c.v. CRF that activate LC neurons are similar to those that mimic behavioral and autonomic components of the stress response. These doses are somewhat higher than those that produce EEG activation.[50]

The long duration of action of CRF in unanesthetized rats is striking in that LC discharge is still increased 40 min after CRF injection and discharge rate is usually greater than that measured 9 min after injection (Table 1). In contrast, LC discharge of anesthetized rats is relatively constant 3 to 20 min after CRF injection. A long duration of action is observed for other effects of CRF including EEG activation, and certain behavioral effects which are not recorded until 1 h after injection.[13,16,50] The similar dose range and duration of action of CRF neuronal and behavioral effects is consistent with the hypothesis that LC activation by CRF is related to behavioral or autonomic aspects of the stress response. This is supported by the finding that increases in LC discharge by CRF are often accompanied by increased motor activity of unanesthetized rats.[48] This is particularly apparent in sling-restrained rats where struggling and escape attempts occur simultaneously with LC activation. Freely moving rats which usually sit quietly in a corner of the cage during recording sessions often show more exploratory movements as LC activity increases after CRF injection.

IV. EFFECTS OF CRF ON LC SENSORY-EVOKED DISCHARGE

A. ANESTHETIZED RATS

One characteristic of LC discharge that has been a basis for implicating the LC in arousal and behavioral vigilance is the response to phasic sensory stimuli.[46,47] In anesthetized rats LC discharge is elicited by potentially noxious stimuli such as pressure to the paw or tail, or footshock.[51] In contrast, LC discharge of unanesthetized rats may be evoked by a variety of nonnoxious phasic stimuli.[46,47] Stimulus novelty is an important determinant of the LC sensory response recorded from unanesthetized monkeys although this may not be true for rats.[46] The response to phasic stimuli is similar in both the anesthetized and unanesthetized state (Figure 1). Discharge is increased for 80 to 100 ms after the stimulus (onset = 10 to 20 ms) and a period during which relatively few spikes occur follows the activation (postactivation pause). This pattern of discharge is similar irrespective of the stimulus. However, the relationship between the type of stimulus and the intensity of the response has not been systematically studied. The response of LC neurons to sensory stimuli may be a signal to target neurons providing information about the environment.[41] Therefore, the effects of neurotransmitters and drugs on this response are useful in providing information about the interaction of the LC with environmental stimuli.

Behaviors produced by i.c.v. CRF in an open field paradigm have been interpreted to represent enhanced neophobia or an intensified response to novel stimuli.[13] These findings led to the prediction that similar doses of CRF would alter LC responses to sensory stimuli. This was investigated in both anesthetized and unanesthetized rats.[43,48] In anesthetized rats LC discharge was evoked by repeated footshock (80 to 100 trials) produced by sciatic nerve stimulation (5.0 ms duration, 1.3 mA, 0.1 Hz).[43] The control peristimulus time histogram (PSTH) in Figure 1 shows the typical pattern of LC discharge elicited by the stimulus. The initial 500 ms of the histogram represents unstimulated or tonic discharge rate which is followed by an excitatory component, or the evoked discharge, and finally a period of relative inhibition. CRF (1.0 and 3.0 μg, i.c.v.) alters this pattern of discharge such that unstimulated discharge is often increased, evoked activity decreased, and more discharges occur during the postactivation pause. The resultant effect of CRF on the pattern of LC discharge to sciatic nerve stimulation is that discharge is less temporally associated with the stimulus. The effect of CRF on different components of the sensory response depends on the dose of CRF (Table 2). For example, while there is a trend towards an increase in tonic discharge rate, this is significant for the 3.0 μg dose only. Discharge rates during the evoked component of the response are significantly decreased by 1.0 μg CRF but not by 3.0 μg CRF and all doses significantly increase the number of discharges during the postactivation pause. The net effect of 1.0 and 3.0 μg CRF on the LC response is a decrease in the ratio of evoked to tonic discharge rate (signal-to-noise).

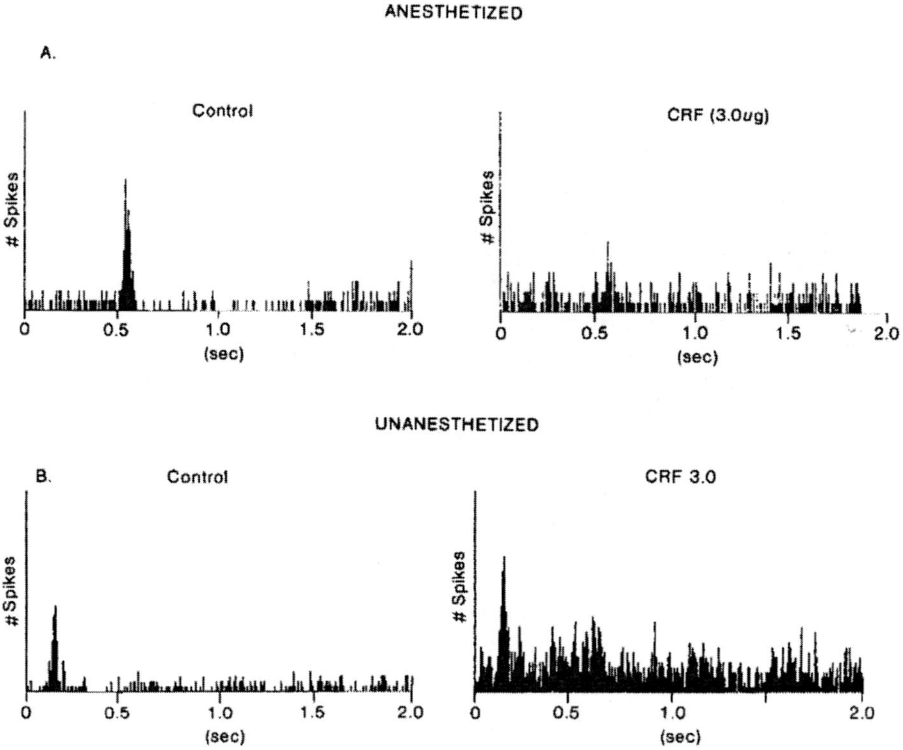

FIGURE 1. Effect of CRF on LC sensory-evoked response. A, shows PSTHs generated from an anesthetized rat during repeated trials of sciatic nerve stimulation (0.1 Hz). The stimulus is presented at 0.5 s. PSTHs show cumulative discharge for 80 presentations. Discharge is recorded for 0.5 sec prior to the stimulus and the following 1.5 s. Note the typical pattern of discharge in the control PSTH: discharge increases shortly after the stimulus and a period of inhibition follows, before a return to prestimulus discharge activity. In the CRF PSTH this pattern is disrupted and discharge is less clearly associated with the stimulus (see Reference 43, for details). B, shows PSTHs generated in an unanesthetized rat during repeated trials (50) of auditory stimulus presentation (20 ms, 0.033 Hz). In this case the stimuli are presented 100 ms after the start of the sweep. The pattern of discharge is similar to that recorded in anesthetized rats, i.e., excitation followed by inhibition. CRF increased background discharge to a greater percentage than evoked discharge (See Reference 48 for details). Bin length = 8 ms.

CRF attenuation of the LC sensory response may occur because CRF increases tonic LC discharge and sensory stimuli may not be able to increase discharge rate further, i.e., the result of a ceiling effect on LC discharge. Although this may be partly the mechanism for the disrupted sensory response in anesthetized rats, several findings suggest that a distinct mechanism is involved in CRF attenuation of the LC response to footshock. First, 1.0 μg CRF does not significantly increase tonic discharge rate (prestimulus discharge) recorded during a trial of sciatic nerve stimulation but does significantly decrease discharge rate in the evoked period (Table 2). Second, if one correlates the change in tonic LC discharge rates to a measure of disruption of the sensory response produced by CRF on the same neuron, the two effects are not highly correlated (Figure 2;[43]). The relationship between LC spontaneous discharge and stimulus-evoked discharge was further investigated using the excitatory muscarinic agonist, carbachol.[52] It was hypothesized that any agent that increases LC spontaneous discharge rate would attenuate the response to a sensory stimulus because of a ceiling effect on LC discharge. As predicted, carbachol increased LC spontaneous discharge rate and disrupted the response to sciatic nerve stimulation by decreasing the ratio of evoked to tonic discharge rate. Moreover, these two effects were highly correlated (Figure

TABLE 2
Effect of CRF on the LC Sensory Response

	Tonic		Evoked		Ratio	
	Control	Peptide	Control	Peptide	Control	Peptide
Anesthetized Rats[a]						
Ala[14]CRF (3.0 μg)	1.25 ± 0.4	1.13 ± 0.5	14 ± 1.8	14 ± 2.9	12.8 ± 3.5	13.5 ± 2.5
CRF (0.3 μg)	1.2 ± 0.4	1.17 ± 0.5	13 ± 2.1	12.3 ± 3.4	12.2 ± 2.0	12.7 ± 3.2
CRF (1.0 μg)	1.05 ± 0.55	1.12 ± 0.5	10.8 ± 1.7	8.4 ± 1.7**	12.2 ± 1.8	8.1 ± 1.2**
CRF (3.0 μg)	1.05 ± 0.5	1.61 ± 0.8*	13.8 ± 2.3	12.5 ± 3.3	14.4 ± 2.7	7.5 ± 1.2*
Unanesthetized Rats						
Ala[14]CRF (3.0 μg)	3.0 ± 0.9	2.2 ± 1.2	7 ± 1.4	5.8 ± 1.2	7.1 ± 4.8	6.1 ± 2.9
CRF (0.3 μg)	1.9 ± 0.6	1.7 ± 0.5	14.2 ± 4.0	12.1 ± 2.1	4.3 ± 1.2	10.5 ± 4.1
CRF (1.0 μg)	2.6 ± 0.5	5.8 ± 1.9	22.5 ± 8.0	26.7 ± 8.0	7.8 ± 1.0	5.3 ± 1.1*
CRF (3.0 μg)	4.5 ± 0.9	10.7 ± 2.5*	30.3 ± 5.7	38.5 ± 8.8	7.1 ± 0.7	4.0 ± 0.8**

Note: *, $p < 0.05$; **, $p < 0.005$.

[a] Discharge rate in spikes per second during an unstimulated component (tonic), and a stimulus-evoked component (evoked) of LC activity during blocks of repeated sensory stimulation. The ratio is that of evoked/tonic discharge rate. Details of the anesthetized study are given in Reference 43 and those of the study using unanesthetized rats are in Reference 48. Numbers represent the mean of single determinations in at least five animals ± 1 SEM.

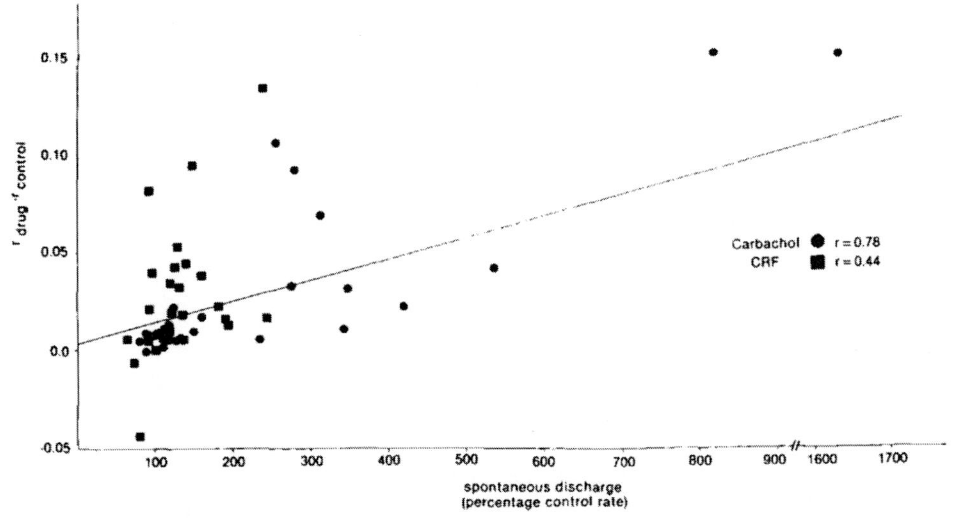

FIGURE 2. Relationship between changes in LC spontaneous discharge and the pattern of LC discharge to sciatic nerve stimulation in anesthetized rats. The abscissa indicates LC spontaneous rate after i.c.v. carbachol (●) or CRF (■) expressed as a percentage of the pre-injection rate. The ordinate is a measure of the disruption of the response to sciatic nerve stimulation based upon the difference in correlation coefficients determined for normalized pre- and postdrug histograms. Each point represents a single cell (one cell per rat) for which the effect of either 0.1 or 0.5 μg carbachol, or 0.3, 1.0, or 3.0 μg CRF was tested. Only one cell was tested per rat and only one rat per treatment. Both spontaneous discharge and effect on the sensory response were recorded and plotted for each individual cell. The correlation coefficients determined by linear regression were 0.78 for carbachol and 0.44 for CRF indicating that this relationship is much stronger for carbachol than for CRF. For details see References 43 and 52. (From Valentino, R. J., in *Mechanisms of Physical and Emotional Stress*, Chrousos, G. P., Loriaux, L., and Gold, P. W., Eds., Plenum Press, New York, in press. With permission.)

2;[52,53]). It appeared, as predicted, that when LC discharge is sufficiently increased, the sensory response becomes disrupted. In contrast, when compared to carbachol, doses of CRF that disrupt the LC sensory response produce much smaller increases in LC spontaneous discharge rates and the two effects are not as highly correlated (Figure 2;[43,53]). Taken together, these results suggest that when LC spontaneous discharge is sufficiently increased, evoked activity will be less intense probably because of a ceiling effect on LC discharge rate. However, this is not the sole mechanism by which CRF attenuates the response of LC cells of anesthetized rats to sciatic nerve stimulation.

Two, probably distinct, effects of CRF on LC neuronal activity of anesthetized rats have been described: (1) it increases spontaneous discharge rate; and (2) it attenuates the response to footshock. The possible significance of these effects with relation to stress is discussed below.

B. UNANESTHETIZED RATS

The unanesthetized rat is a better preparation in which to study CRF effects on LC sensory-evoked activity because LC neurons are responsive to multimodal nonnoxious stimuli in the unanesthetized state.[46,47] For example, the pattern of LC discharge to repeated auditory stimuli is similar to the pattern evoked by footshock recorded from anesthetized rats (Figure 1). However, the response may be less intense and the ratio of evoked to tonic discharge is somewhat lower (Table 2). This may be a result of differences in anesthetized vs. unanesthetized state or in the type of stimulus used to evoke LC discharge.

When repeated auditory stimuli (20 ms tone; 1 every 30 s) are used to evoke LC discharge the net effect of CRF on auditory evoked LC activity is similar to the effect of CRF on LC activity evoked by sciatic nerve stimulation in anesthetized rats. In both preparations CRF caused a dose-dependent decrease in the ratio of evoked to tonic discharge rate (signal-to-noise) and this is not produced by the inactive analog Ala^{14}CRF.[48] However, in contrast to its effects in anesthetized rats, CRF did not, at any dose, cause a significant decrease in discharge rate during the evoked component of the sensory response. Rather it appeared to increase tonic discharge to a greater extent than evoked discharge resulting in a net decrease in the signal-to-noise ratio. This may be the result of a ceiling effect on LC discharge such that tonic activity is greatly increased by CRF but evoked activity is already high and cannot be increased further. The attenuation of evoked LC discharge in studies using anesthetized rats and lack of attenuation of auditory-evoked discharge in unanesthetized rats may be due to several different variables between the two experiments. First, it may be more difficult to attenuate evoked activity in unanesthetized rats because anesthesia, which may mask excitatory inputs, is not present. Second, CRF may only attenuate the effect of noxious stimuli perhaps by releasing β-endorphin from pituitary cells. Finally, since the absolute discharge rate in unanesthetized rats during auditory stimulus presentation is greater in unanesthetized rats than rates during footshock presentation to anesthetized rats it may be more difficult for CRF to decrease these higher rates in unanesthetized animals.

The net effect of CRF on LC sensory-evoked discharge is similar in anesthetized and unanesthetized rats and this is a decrease in the ratio of evoked to tonic activity. In unanesthetized rats this occurs primarily because tonic activity is increased to a greater extent than evoked activity while in anesthetized rats this is due to both an increase in tonic discharge rate and an attenuation of the evoked rate. The functional significance of these effects is discussed below.

V. POSSIBLE CELLULAR MECHANISM FOR CRF EFFECTS ON LC NEURONS

The mechanism by which CRF alters LC activity is merely speculative since such studies are lacking. This speculation is based upon what is known of cellular effects of CRF on other neurons. To date the most extensive investigatons of CRF neuronal mechanisms have utilized the pyramidal cells of the hippocampal slice preparation[53,54] and these are reviewed in another chapter of this volume. Briefly, CRF increases firing rates of pyramidal neurons in the CA1 and CA3 region of the hippocampus and this is accompanied by depolarization and increased input resistance.[54,55] The magnitude and duration of the afterhyperpolarization seen after a burst of action potentials is decreased by low concentrations of CRF. Because this afterhyperpolarization is due to a calcium-dependent potassium conductance, CRF is either blocking influx of calcium or directly blocking the potassium conductance. Experiments demonstrating a lack of effect of CRF on calcium spikes suggest that the potassium conductance is directly affected by CRF.[54,55] A block of this channel would result in decreased afterhyperpolarization, slower repolarization, and more calcium entering the cell through voltage sensitive channels. A similar afterhyperpolarization produced by a calcium-dependent potassium conductance is present in LC neurons following a burst of action potentials and this is thought to be responsible for the postactivation pause observed in the LC sensory response.[56,57] A decreased potassium conductance by CRF would be predicted to increase LC tonic discharge rate and to attenuate the postactivation pause in the LC sensory response. This could result in the observed decrease in the signal-to-noise of the LC sensory response. The reported cellular mechanism of CRF on hippocampal pyramidal neurons provide a model which can be tested using intracellular studies on LC neurons.

VI. FUNCTIONAL SIGNIFICANCE OF CRF EFFECTS ON LC NEURONS

The finding that CRF increases LC spontaneous discharge as do stressful stimuli suggest that these particular stimuli produce their electrophysiological effects by releasing CRF from fibers localized within the LC. Some evidence for release is suggested by findings of increased CRF levels in LC of acute and chronically stressed rats.[29] This is not observed in many other CRF-containing brain regions. However, these results are not conclusive evidence of increased turnover. If stressful stimuli release sufficient amounts of CRF in the LC spontaneous discharge would be predicted to increase.

The importance of CRF activation of LC discharge is determined by the influence of LC discharge on its diverse target cells. Several investigations of the effect of LC stimulation or norepinephrine application on target cells in cortex, hippocampus, cerebellum, spinal cord, lateral geniculate nucleus, and olfactory bulb have yielded similar results with unified implications.[58-70] In general, spontaneous discharge of target cells is decreased or unaffected while synaptic activity, or the effect of other neurotransmitters, is enhanced. That is, the signal-to-noise of activity in target cells is increased when LC is activated or when these cells are exposed to norepinephrine. This has led to the hypothesis that increased LC activity biases its target neurons towards the receipt of synaptic input or towards signal processing.[41,71] This hypothesis is consistent with studies demonstrating that the LC activation is associated with, and may precede, behavioral indices of arousal.[45] Presumably, arousing stimuli activate the LC and the LC through its widespread projection system alters activity throughout the CNS in such a way that synaptic influence is enhanced. This would be expected to be adaptive for responding to such stimuli. The likely consequence of CRF release in the LC and activation of neuronal discharge would similarly be a bias of target cells toward information processing and an adaptive increase in behavioral arousal and/or autonomic activity as is observed in the stress response.

Behavioral indices of neophobia predict that CRF, and likewise stress, potentiate the behavioral response to novel stimuli.[13] Because of these findings it was initially hypothesized that CRF would enhance the response of LC neurons to phasic, sensory stimuli. In contrast, CRF was found not to enhance this response but rather to alter the pattern of the response such that the signal-to-noise is decreased, i.e., tonic activity is increased relative to evoked activity. It is not known whether LC activity of stressed animals would respond in a similar manner to phasic sensory stimuli as animals injected with CRF. The interpretation of a change in pattern of LC sensory response is difficult without knowing what this pattern of LC discharge to a phasic stimulus means to activity in target cells, or moreover, how this is translated to behavior or autonomic activity. It may be simplistic to assume that increased behavioral responsiveness of animals to stimuli should be associated with increased responsiveness of LC cells. The type of stimulus, its informative nature, its presentation, and its duration may all be complex determinants of the cellular response and its translation to behavior. What can be concluded from data presently available is that CRF increases the rate of tonic LC discharge to a greater extent than it increases evoked discharge and that this would result in persistent NE release in target areas. The behavioral consequence of this may be persistent arousal which is not further enhanced by presentation of neutral, phasic sensory stimuli.

Although the behavioral consequences of LC activation by CRF have been emphasized, this activation may have important consequences on autonomic activity and play a role in the well described autonomic components of the stress response. LC efferents project to regions of the brainstem involved in control of cardiovascular function[25,26,72-78] and some studies have demonstrated blood pressure changes during stimulation or pharmacologic manipulation of LC activity.[79-85] Moreover, LC discharge can vary depending on changes in blood volume.[36,37] Since i.c.v. CRF is known to increase blood pressure of unanesthetized rats[18] the hypothesis that increases in LC discharge were correlated to blood pressure increases by CRF was tested.[86] It was found that 3.0 μg of CRF, while increasing LC spontaneous discharge rate, had no effect on blood pressure of halothane-anesthetized rats.[86] Interestingly, anesthesia appears to impair the hypertensive effect of CRF.[87] The results demonstrate a separation between LC discharge and the blood pressure response to CRF at least in anesthetized rats. However, it is possible that further increases in LC activity are required to elicit blood pressure increases in anesthetized rats. These results do not preclude the possibility that LC activation is involved in other autonomic components of stress responses.

VII. PSYCHIATRIC IMPLICATIONS OF CRF EFFECTS ON LC NEURONS

Abnormal function of the hypothalamic-pituitary axis has been demonstrated in several psychiatric conditions including depression and anorexia nervosa.[88-90] In some patients with these disorders, CRF levels in CSF are increased and it is hypothesized that this is a result of enhaced CRF release. It is not clear whether enhanced CRF release is part of the etiology or a subsequent result of the disease. Nor is it certain that CRF release is enhanced in extrahypothalamic areas. In any case, the consequences of a chronic CRF excess on CNS neurons must be considered in the symptomology of the diseases. Exposure of LC neurons to excesses of CRF would be expected to result in continuous activation of these cells. LC cells in these patients may therefore have consistently higher rates of discharge and sensory stimuli may be less effective in eliciting further activity. These patients would be predicted to be persistently aroused and may not respond appropriately to external stimuli because of different baseline rates of LC discharge.

CRF within the LC may also play a role in the opiate withdrawal syndrome which may be considered to be stressful, as it involves increased activity of the autonomic nervous system and increased secretion of ACTH. Several lines of evidence have associated increased LC discharge with opiate withdrawal. For example, LC stimulation in opiate-naive monkeys elicits certain behaviors characteristic of opiate withdrawal and these are attenuated by clonidine which decreases LC discharge.[91] Supporting this, electrophysiological studies demonstrate activation of LC cells of morphine-dependent rats by iontophoretic application of the opiate antagonist, naloxone, to the LC.[92] This effect is thought to be an electrophysiological correlate of opiate withdrawal since naloxone has no effect on LC discharge of opiate-naive rats. In preliminary studies another opiate antagonist, naltrexone, was found to increase LC spontaneous discharge rates of anesthetized rats by 75% when injected i.c.v. after a dose of morphine (0.3 μg) that completely inhibited LC spontaneous activity.[49] Moreover, naltrexone disrupted the LC response to sciatic nerve stimulation in a manner similar to CRF. When naltrexone was administered after lower doses of morphine it only returned spontaneous rate and the pattern of sensory response to premorphine levels. The increase in spontaneous activity and disruption of the sensory response by naltrexone may indicate that an acute state of dependence has developed in these neurons after a high dose of morphine and these effects may be neuronal correlates of opiate withdrawal. It is of interest that these effects are similar to those produced by i.c.v. CRF since these neuronal correlates of withdrawal are not observed *in vitro*. Thus, opiate withdrawal in the LC requires input that is probably lesioned in *in vitro* preparations. Because of its localization within the LC and the similarity of its neuronal effects with those of naltrexone, CRF release by the opiate antagonist could be a mechanism for this withdrawal phenonomenon. Consistent with this hypothesis is the finding that CRF release is increased by opiate antagonists.[93]

VIII. SUMMARY

The localization of CRF immunoreactive fibers in rat and primate LC and the effects of CRF on LC neuronal activity implicate the LC as one extrahypothalamic target for CRF. Release of CRF from these fibers by stressful stimuli is predicted to activate the LC and alter LC responses to environmental stimuli. These effects may be amplified through the LC efferent system resulting in persistent release of norepinephrine onto target cells. From what is known about the influence of the LC on its target cells, the consequence of CRF release in the LC would be a bias towards enhanced synaptic influence in CNS regions involved in behavior, autonomic activity, and motor activity. The LC system may be operating at optimal levels during CRF release such that neutral, phasic stimuli will be less consequential on LC activity. It may be adaptive for LC activity to be maximal and not as discriminating to individual neutral, phasic stimuli when the organism is stressed. Alternatively, the results of chronic activation of this system are likely to be pathologic signs and symptoms perhaps resembling those seen in certain types of depression, such as persistent arousal and decreased responses to environmental stimuli. While the translation of LC discharge to specific behavioral, autonomic, or motor effects associated with stress or CRF administration has not been made, the influence of LC activity on its target cells implicates that it may set a tone such that the effects considered to be adaptive are more likely to occur.

REFERENCES

1. **Rivier, J., Spiess, J., and Vale, W.,** Characterization of rat hypothalamic corticotropin-releasing factor, *Proc. Natl. Acad. Sci. U.S.A.*, 80, 4851, 1983.
2. **Vale, W., Spiess, J., Rivier, C., and Rivier, J.,** Characterization of 41-residue ovine hypothalamic peptide that stimulates secretion of corticotropin and β-endorphin *Science*, 213, 1394, 1981.
3. **Vale, W., Rivier, C., Brown, M. R., Spiess, J., Koob, G., Swanson, L., Bilezikjian, L., Bloom, F., and Rivier, J.,** Chemical and biological characterization of corticotropin releasing factor, *Rec. Prog. Horm. Res.*, 39, 245, 1983.
4. **Rivier, C., Rivier, J., and Vale, W.,** Inhibition of adrenocorticotropic hormone secretion in the rat by immunoneutralization of corticotropin-releasing factor, *Science*, 218, 377, 1982.

5. **Rivier, C. and Vale, W.,** Modulation of stress-induced a ACTH release by corticotropin releasing factor, catecholamines and vasopressin, *Nature,* 305, 325, 1983.
6. **Cummings, S., Elde, R., Ells, J., and Lendall, A.,** Corticotropin-releasing factor immunoreactivity is widely distributed within the central nervous system of the rat: an immunohistochemical study, *J. Neurosci.,* 3, 1355, 1983.
7. **Merchenthaler, I., Hynes, M. A., Vigh, S., Schally, A. V., and Petrusz, P.,** Immunocytochemical localization of corticotropin-releasing factor (CRF) in rat brain, *Am. J. Anat.,* 165, 383, 1982.
8. **Olschowka, J. S., O'Donahue, T. L., Mueller, G. P., and Jacobowitz, D. M.,** The distribution of corticotropin-releasing factor-like immunoreactive neutrons in rat brain, *Peptides,* 3, 995, 1982.
9. **Palkovits, M., Brownstein, M. J., and Vale, W.,** Distribution of corticotropin releasing factor in rat brain, *Fed. Proc.,* 44, 215, 1985.
10. **Swanson, L. W., Sawchenko, P. E., Rivier, J., and Vale, W.,** Organization of ovine corticotropin-releasing factor immunoreactive cells and fibers in the rat brain: an immunohistochemical study, *Neuroendocrinology,* 36, 165, 1983.
11. **De Souza, E. B.,** Corticotropin-releasing factor receptors in the rat central nervous system: characterization and regional distribution, *J. Neurosci.,* 7, 88, 1987.
12. **De Souza, E. B., Insel, T. R., Perrin, M. H., Rivier, J., Vale, W. W., and Kuhar, M. J.,** Corticotropin-releasing factor receptors are widely distributed within the rat central nervous system: an autoradiographic study, *J. Neurosci.,* 5, 3189, 1985.
13. **Britton, D. R., Koob, G. F., Rivier, J., and Vale, W.,** Intraventricular corticotropin releasing factor enhances behavioral effects of novelty, *Life Sci.,* 31, 363, 1982.
14. **Kalin, N. H., Shelton, S. E., Karemer, G. W., and McKenny, W. T.,** Corticotropin-releasing factor administered intraventricularly to rhesus monkeys, *Peptides,* 4, 217, 1983.
15. **Kalin, N. H.,** Behavioral effects of ovine corticotropin-releasing factor administered to rhesus monkeys, *Fed. Proc.,* 44, 249, 1985.
16. **Sutton, R. E., Koob, G. F., LeMoal, M., Rivier, J., and Vale, W.,** Corticotropin-releasing factor produces behavioral activation in rats, *Nature,* 297, 331, 1982.
17. **Brown, M. R., Fisher, L. A., Spiess, J., Rivier, C., Rivier, J., and Vale, W.,** Corticotropin-releasing factor: actions on the sympathetic nervous system and metabolism, *Endocrinology,* 111, 928, 1982.
18. **Fisher, L. A., Rivier, J., Rivier, C., Spiess, J., Vale, W., and Brown, M. R.,** Corticotropin releasing factor (CRF): central effects on mean arterial pressure and heart rate in rats, *Endocrinology,* 110, 2222, 1982.
19. **Tache, Y., Goto, Y., Gunion, M., Vale, W., Rivier, J., and Brown, M.,** Inhibition of gastric acid secretion in rats by intracerebral injection of corticotropin-releasing factor (rCRF), *Science,* 222, 935, 1983.
20. **Rivier, J., Rivier, C., and Vale, W.,** Synthetic competitive antagonists of corticotropin-releasing factor: effect on ACTH secretion in the rat, *Science,* 224, 889, 1984.
21. **Brown, M. R., Fisher, L. A., Vale, W. W., and Rivier, J. E.,** Corticotropin-releasing factor: role in central nervous system regulation of the adrenal medulla, *Soc. Neurosci. Abstr.,* 10, 1117, 1984.
22. **Grzanna, R. and Molliver, M. E.,** The locus coeruleus in the rat: an immunohistochemical delineation, *Neuroscience,* 5, 21, 1980.
23. **Jones, B. E., Halaris, A. E., and Freeman, O. X.,** Innervation of forebrain regions by medullary noradrenaline neurons, a biochemical study in cats with central tegmental tract lesions, *Neurosci. Lett.,* 10, 251, 1978.
24. **Jones, B. E. and Moore, R. Y.,** Ascending projections of the locus coeruleus in the rat. II. Autoradiographic study, *Brain Res.,* 127, 23, 1977.
25. **Kobayashi, R. M., Palkovits, M., Kopin, I. J., and Jacobowitz, D. M.,** Biochemical mapping of noradrenergic nerves arising from the rat locus coeruleus, *Brain Res.,* 77, 269, 1974.
26. **Swanson, L. W. and Hartman, B. K.,** The central adrenergic system: an immunofluorescence study of the location of cell bodies and their efferent connections in the rat utilizing dopamine-β-hydroxylase as a marker, *J. Comp. Neurol.,* 163, 467, 1976.
27. **Valentino, R. J., Cha, C. I., and Foote, S. L.,** Anatomic and physiologic evidence for innervation of noradrenergic locus coeruleus by neuronal corticotropin-releasing factor, *Soc. Neurosci. Abstr.,* 12, 1003, 1986.
28. **Aston-Jones, G., Ennis, M., Pieribone, V. A., Nickell, W. T., and Shipley, M. T.,** The brain nucleus locus coeruleus: restricted afferent control of a broad efferent network, *Science,* 234, 734, 1986.
29. **Chappell, P. B., Smith, M. A., Kilts, C. D., Bissette, G., Ritchie J., Anderson, C., and Nemeroff, C. B.,** Alterations in corticotropin-releasing factor-like immunoreactivity in discrete rat brain regions after acute and chronic stress, *J. Neurosci.,* 6, 2908, 1986.
30. **Cassens, G., Roffman, M., Kuruc, A., Orsulak, P. J., and Schildkraut, J. J.,** Alterations in brain norepinephrine metabolism induced by environmental stimuli previously paired with inescapable shock, *Science,* 209, 1138, 1980.

31. **Cassens, G., Kuruc, A., Roffman, M., Orsulak, P. J., and Schildkraut, J. J.,** Alterations in brain norepinephrine metabolism and behavior induced by environmental stimuli previously paired with inescapable shock, *Behav. Brain Res.*, 2, 387, 1981.
32. **Korf, J., Aghajanian, G. K., and Roth, R. H.,** Increased turnover of norepinephrine in the rat cerebral cortex during stress: role of the locus coeruleus, *Neuropharmacology*, 12, 933, 1973.
33. **Solomon, R. A., McCormack, B. M., Lovitz, R. N., Swift, D. M., and Hegemann, M. T.,** Elevation of brain norepinephrine concentration after experimental subarachnoid hemorrhage, *Neurosurgery*, 19, 363, 1986.
34. **Thierry, A. M., Javoy, F., Glowinski, J., and Kety, S. S.,** Effects of stress on the metabolism of norepinephrine, dopamine and serotonin in the central nervous system of the rat: modification of norepinephrine turnover, *J. Pharmacol. Exp. Ther.*, 163, 163, 1968.
35. **Elam, M., Yao, T., Thoren, P., and Svensson, T. H.,** Hypercapnia and hypoxia: chemoreceptor-mediated control of locus coeruleus neurons and splanchnic, sympathetic nerves, *Brain Res.*, 222, 373, 1981.
36. **Elam, M., Yao, T., Svensson, T. H., and Thoren, P.,** Regulation of locus coeruleus neurons and splanchnic, sympathetic nerves by cardiovascular efferents, *Brain Res.*, 290, 281, 1984.
37. **Svensson, T. H. and Thoren, P.,** Brain noradrenergic neurons in the locus coeruleus: inhibition by blood volume load through vagal afferents, *Brain Res.*, 172, 173, 1979.
38. **Rasmussen, K., Morilak, D. A., and Jacobs, B. L.,** Single unit activity of locus coeruleus neurons in the freely moving cat. I. During naturalistic behaviors and in response to simple and complex stimuli, *Brain Res.*, 371, 324, 1986.
39. **Plotsky, P. M.,** Hypophyseotropic regulation of adrenohypophyseal adrenocorticotropin secretion, *Fed. Proc.*, 344, 207, 1985.
40. **Plotsky, P. M. and Vale, W.,** Hemorrhage-induced secretion of corticotropin-releasing factor immunoreactivity into the rat hypophyseal portal circulation and its inhibition by glucocorticoids, *Endocrinology*, 114, 164, 1984.
41. **Foote, S. L., Bloom, F. E., and Aston-Jones, G.,** Nucleus locus coeruleus: new evidence of anatomical and physiological specificity, *Physiol. Rev.*, 63, 844, 1983.
42. **Valentino, R. J., Foote, S. L., and Aston-Jones, G.,** Corticotropin-releasing factor activates noradrenergic neurons of the locus coeruleus, *Brain Res.*, 270, 363, 1983.
43. **Valentino, R. J. and Foote, S. L.,** Corticotropin-releasing factor disrupts sensory responses of brain noradrenergic neurons, *Neuroendocrinology*, 45, 28, 1987.
44. **Valentino, R. J. and Foote, S. L.,** Brain noradrenergic neurons, corticotropin-releasing factor, and stress, in *Neural and Endocrine Peptides and Receptors*, Moody, T., Ed., Plenum Press, New York, 1986, 101.
45. **Aston-Jones, G. and Bloom, F. E.,** Activity of norepinephrine-containing locus coeruleus neurons in behaving rats anticipates fluctuations in the sleep-waking cycle, *J. Neurosci.*, 1, 876, 1981.
46. **Foote, S. L., Aston-Jones, G., and Bloom, F. E.,** Impulse activity of locus coeruleus neurons in awake rats and monkeys is a function of sensory stimulation and arousal, *Proc. Natl. Acad. Sci. U.S.A.*, 77, 303, 1980.
47. **Aston-Jones, G. and Bloom, F. E.,** Norepinephrine-containing neurons in behaving rats exhibit pronounced responses to non-noxious environmental stimuli, *J. Neurosci.*, 1, 887, 1981.
48. **Valentino, R. J. and Foote, S. L.,** Corticotropin-releasing hormone increases tonic but not sensory-evoked activity of noradrenergic locus coeruleus neurons in unanesthetized rats, *J. Neurosci.*, in press.
49. **Valentino, R. J. and Wehby, R.,** Morphine effects on locus coeruleus neurons are dependent on the state of arousal and availability of external stimuli: studies in anesthetized and unanesthetized rats, *Soc. Neurosci. Abstr.*, in press.
50. **Ehlers, C. L., Henricksen, S. J., Wang, M., Rivier, J., Vale, W. W., and Bloom, F. E.,** Corticotropin-releasing factor produces increases in brain excitability and convulsive seizures in rats, *Brain Res.*, 278, 332, 1983.
51. **Cedarbaum, J. M. and Aghajanian, G. K.,** Activation of locus coeruleus neurons by peripheral stimuli: modulation by a collateral inhibitory mechanism, *Life Sci.*, 23, 1383, 1978.
52. **Valentino, R. J. and Aulisi, E. F.,** Carbachol-induced increases in locus coeruleus spontaneous activity are associated with an altered pattern of response to sensory stimuli, *Neurosci. Lett.*, 74, 297, 1987.
53. **Valentino, R. J.,** CRH effects on central noradrenergic neurons: Relationship to stress, in *Mechanisms of Physical and Emotional Stress*, Chrousos, G. P., Loriaux, L., and Gold, P. W., Eds., Plenum Press, New York, in press.
54. **Aldenhoff, J. B., Gruol, D. L., Rivier, J., Vale, W., and Siggins, G. R.,** Corticotropin-releasing factor decreases postburst hyperpolarizations and excites hippocampal neurons, *Science*, 221, 875, 1983.
55. **Siggins, G. R., Gruol, D. L., Aldenhoff, J., and Pittman, Q.,** Electrophysiological actions of corticotropin-releasing factor in the central nervous system, *Fed. Proc.*, 44, 237, 1985.
56. **Andrade, R. and Aghajanian, G. K.,** Locus coeruleus activity in vitro: intrinsic regulation by a calcium-dependent potassium conductance but not α_2-adrenoreceptors, *J. Neurosci.*, 4, 161, 1984.

57. **Williams, J. T., North, R. A., Shefner, S. A., Nishi, S., and Egan, T. M.,** Membrane properties of rat locus coeruleus neurons, *Neuroscience,* 13, 137, 1984.
58. **Foote, S. L., Freedman, R., and Oliver, A. P.,** Effects of putative neurotransmitter on neuronal activity in monkey auditory cortex, *Brain Res.,* 86, 229, 1975.
59. **Freedman, R., Hoffer, B. J., Woodward, D. J., and Riro, D.,** Interaction of norepinephrine with cerebellar activity evoked by mossy and climbing fibers, *Exp. Neurol.,* 55, 269, 1977.
60. **Jarh, C. E. and Nicoll, R. A.,** Noradrenergic modulation of dendrodendritic inhibition in the olfactory bulb, *Nature,* 297, 227, 1982.
61. **Kayama, Y., Negi, T., Sugitani, M., and Iwama, K.,** Effects of locus coeruleus stimulation on neuronal activities of dorsal lateral geniculate nucleus and perigeniculate reticular nucleus of rat, *Neuroscience,* 7, 655, 1982.
62. **Moises, H. C., Waterhouse, B. D., and Woodward, B. J.,** Locus coeruleus stimulation potentiates Purkinje cell responses to afferent input: the climbing fiber system, *Brain Res.,* 222, 43, 1981.
63. **Moises, H. C., Woodward, D. J., Hoffer, B. J., and Freedman, R.,** Interactions of norepinephrine with Purkinje cell responses to putative amino acid neurotransmitters applied by microiontophoresis, *Exp. Neurol.,* 64, 493, 1979.
64. **Rogawski, M. A. and Aghajanian, G. K.,** Activation of lateral geniculate neurons by norepinephrine: mediation by an alpha-adrenergic receptor, *Brain Res.,* 182, 345, 190.
65. **Segal, M. and Bloom, F. E.,** The action of norepinephrine in the rat hippocampus. I. Iontophoretic studies, *Brain Res.,* 72, 79, 1974.
66. **Segal, M. and Bloom, F. E.,** The action of norepinephrine in the rat hippocampus. II. Activation of the input pathway, *Brain Res.,* 72, 99, 1974.
67. **Segal, M. and Bloom, F. E.,** The action of norepinephrine in the rat hippocampus. III. Hippocampal cellular responses to locus coeruleus stimulation in the awake rat, *Brain Res.,* 107, 499, 1976.
68. **Waterhouse, B. D., Moises, H. C., and Woodward, D. J.,** Noradrenergic modulation of somatosensory cortical neuronal responses to iontophoretically applied putative neurotransmitters, *Exp. Neurol.,* 69, 30, 1980.
69. **Waterhouse, B. D., Moises, H. C., Yeh, H. H., and Woodward, D. J.,** Norepinephrine enhancement of inhibitory synaptic mechanisms in cerebellum and cerebral cortex: mediation by β-adrenergic receptors, *J. Pharmacol. Exp. Ther.,* 221, 495, 1982.
70. **White, S. R. and Neuman, R. S.,** Facilitation of spinal motorneuron excitability by 5-hydroxytryptamine and noradrenaline, *Brain Res.,* 188, 119, 1980.
71. **Woodward, D. J., Moises, H. C., Waterhouse, B. D., Hoffer, B. J., and Freedman, R.,** Modulatory actions of norepinephrine in the central nervous system, *Fed. Proc.,* 38, 2109, 1979.
72. **Conard, L. C. A., Leonard, C. M., and Pfaff, D. W.,** Connections of the median and dorsal raphe nuclei in the rat: an autoradiographic degeneration study, *J. Comp. Neurol.,* 156, 179, 1974.
73. **Gunn, C. G., Sevelius, G., Puiggari, J., and Myers, F. K.,** Vagal cardioinhibitory mechanisms in the hindbrain of the dog and cat, *Am. J. Physiol.,* 214, 258, 1968.
74. **Jacobowitz, D. and Kostzewa, R.,** Selective action of 6-hydroxydopa on noradrenergic terminals: mapping of preterminal axons of the brain, *Life Sci.,* 10, 1329, 1971.
75. **Loizou, L. A.,** Projections of the nucleus locus coeruleus in the albino rat, *Brain Res.,* 15, 563, 1969.
76. **McBride, R. L. and Suttin, J.,** Projections of the nucleus locus coeruleus and adjacent pontine tegmentum in the cat, *J. Comp. Neurol.,* 165, 265, 1976.
77. **Olson, L. and Fuxe, F.,** Further mapping out of central noradrenaline neuron systems: projections of the subcoeruleus area, *Brain Res.,* 43, 289, 1972.
78. **Pickel, V. M., Joh, T. H., and Reis, D. J.,** Serotonergic innervation of noradrenergic neurons in nucleus locus coeruleus: demonstration by immunohistochemical localization of the transmitter specific enzyme tyrosine hydroxylase and tryptophan hydroxylase, *Brain Res.,* 131, 197, 1977.
79. **Chida, K., Kawamura, H., and Hatano, M.,** Participation of the nucleus locus coeruleus in DOCA-salt hypertensive rats, *Brain Res.,* 273, 53, 1983.
80. **Gurtu, C. G., Pant, K. K., Sinha, J. N., and Bhargava, K. P.,** An investigation into the mechanism of cardiovascular responses elicited by electrical stimulation of locus coeruleus and subcoeruleus in the cat, *Brain Res.,* 301, 59, 1984.
81. **Kawamura, H., Gunn, C. G., and Frohlich, E. D.,** Cardiovascular alteration by nucleus locus coeruleus in spontaneously hypertensive rat, *Brain Res.,* 140, 137, 1978.
82. **Przuntek, H. and Philippu, A.,** Reduced pressor responses to stimulation of the locus coeruleus after lesion of the posterior hypothalamus, *Naunyn-Schmiedebergs Arch. Pharmacol.,* 276, 119, 1973.
83. **Ward, D. G and Gunn, C. G.,** Locus coeruleus complex: elicitation of a pressor response and brainstem region necessary for its occurrence, *Brain Res.,* 107, 401, 1976.
84. **Pant, K. K., Gurtu, S., Sharma, D. K., Sinha, J. N., and Bhargava, K. P.,** Cardiovascular effects of microinjection of morphine into the nucleus locus coeruleus of the cat, *Jpn. J. Pharmacol.,* 33, 253, 1983.

85. **Perlman, R. and Guideri, G.**, Cardiovascular changes produced by the injection of aconitine at the area of the locus coeruleus in unanesthetized rats, *Arch. Int. Pharmacodyn. Ther.*, 268, 202, 1984.
86. **Valentino, R. J., Martin, D. L., and Suzuki, M.**, Dissociation of locus coeruleus activity and blood pressure: effects of clonidine and corticotropin-releasing factor, *Neuropharmacology*, 25, 603, 1986.
87. **Kurosawa, M., Sato, A., Swenson, R. S., and Takahasi, Y.**, Sympatho-adrenal medullary functions in response to intracerebroventricularly injected corticotropin-releasing factor in anesthetized rats, *Brain Res.*, 367, 250, 1986.
88. **Gold, P. W., Loriaux, D. L., Roy, A., Kling, M. A., Calabrese, J. R., Kellner, C. H., Nieman, L. K., Post, R. M., Pickar, D., Gallucci, W., Avgerinos, P., Paul, S., Oldfield, E. H., Cutler, G. B., Jr., and Chrousos, G. P.**, Responses to corticotropin-releasing hormone in the hypercorticolism of depression and Cushing disease, *N. Engl. J. Med.*, 314, 1329, 1986.
89. **Kay, W. H., Gwirtsman, H. E., George, D. T., Ebert, M. H., Jimerson, D. C., Tomai, T. P., Chrousos, G. P., and Gold, P. W.**, Elevated cerebrospinal fluid levels of immunoreactive corticotropin-releasing hormone in anorexia nervosa: relation to state of nutrition, adrenal function and intensity of depression, *J. Clin. Endocrinol. Metab.*, 64, 203, 1987.
90. **Nemeroff, C. B., Widerlov, E., Bissette, G., Wallerus, H., Karlsson, I., Eklund, K., Kilts, C. D., Loosen, P. T., and Vale, W.**, Elevated concentrations of CSF corticotropin-releasing factor-like immunoreactivity in depressed patients, *Science*, 226, 1342, 1984.
91. **Redmond, D. E., Jr. and Huang, Y. H.**, The primate locus coeruleus and effects of clonidine on opiate withdrawal, *J. Clin. Psychol.*, 43, 25, 1982.
92. **Aghajanian, G. K.**, Tolerance of locus coeruleus neurones to morphine and suppression of withdrawal response by clonidine, *Nature*, 276, 186, 1978.
93. **Plotsky, P. M.**, Opioid inhibition of immunoreactive corticotropin-releasing factor secretion into the hypophysial-portal circulation of rats, *Regul. Pept.*, 16, 235, 1986.

Chapter 16

CRF EFFECTS ON EEG ACTIVITY: IMPLICATIONS FOR THE MODULATION OF NORMAL AND ABNORMAL BRAIN STATES

Cindy L. Ehlers

TABLE OF CONTENTS

I. INTRODUCTION

Over the last several decades a myriad of neurophysiological, biochemical, and pharmacological evidence has amassed to support various neural or neurohumoral mechanisms suspected to underlie the regulation of states of sleep and wakefulness (see Koella[1] for review). In an attempt to explain how several different classes of drugs may produce qualitatively different states of waking, sleep, sedation, and/or anesthesia, Winters et al.[2,3] proposed a model whereby most drugs could be seen as acting within a continua depending on the dosage administered (see Figure 1). Based on neurophysiological findings,[4,5] it was suggested that anesthetics and CNS excitants induce an initial activation characterized by increased motor activity. Some CNS depressants then were seen as inducing hypnosis, followed by deep anesthesia and ultimately death. Other drugs produce activation followed by dissociative states associated with reduced motor activity, and in some cases, myoclonus and seizure activity. The brain area where these drugs were envisioned to induce these changes in brain states was thought to be the ascending reticular activating system of Moruzzi and Magoun.[6] This system had been demonstrated to be reactive to changes in the spontaneous sleep-wake cycle[7] as well as to be sensitive to states induced by excitant and depressant drugs.[8,9]

Evidence that a hypothalamic site may participate in the control of brain states has also been advanced. Several studies in both rats[10] and diurnal primates[11] have demonstrated that when lesions are made in the region of the suprachiasmatic nucleus, rhythms in rest and activity are shifted or severely disturbed. These studies have also been used to suggest that a hypothalamic "clock" or "clocks" may control the onset and/or duration of the sleep-wake cycle in mammals.[11] How such a hypothalamic clock or pacemaker may transduce event timing into the behavioral state transitions of the sleep-wake cycle is unclear; however, a humoral messenger either neural or endocrine seems likely.

A relationship between circadian or diurnal rhythms in the sleep-wake cycle and several well-known pituitary hormones has been recognized. For instance, it has been demonstrated that in humans,[12,13] baboons,[14] and rats[15] growth hormone (GH) secretion is augmented during sleep, whereas studies of ACTH and corticosteroids suggest that values of these hormones are generally higher during arousal[15] and peak in the early morning[16,17] under normal lighting conditions. It is still unclear, however, whether these hormones serve any regulatory function over the sleep-wake cycle or are just passively related in time. The recent discovery of the hypothalamic-releasing factors has introduced new probes with which to further elucidate this relationship.

The discovery of a central nervous system (CNS) activating role for the hypothalamic neuropeptide corticotropin-releasing factor (CRF)[18] has raised the possibility that this peptide may play a pivotal role in the regulation of brain activation states (also see Chapter 17). This chapter will attempt to review the data available which have linked this neuropeptide with: (1) the regulation of sleep-wake cycles in normal and depressed subjects, (2) the induction of abnormal excitatory brain activity and seizures, and (3) interactions with other CNS activating or sedating drugs, in an attempt to develop a preliminary hypothesis on how endogenous neuromodulatory compounds may modulate brain states.

II. CRF AND THE SLEEP-WAKE CYCLE

Since ancient times, both Eastern and Western philosophers have postulated that the states of sleep are regulated by circulating factors in the blood.[19] However, it wasn't until the early 20th century that Legendre and Pieron (see Reference 19) suggested that a "hypnotoxin", which was present in the blood and cerebrospinal fluid of sleep-deprived dogs, could induce what appears to be a state of sleep when injected into the cerebral ventricles

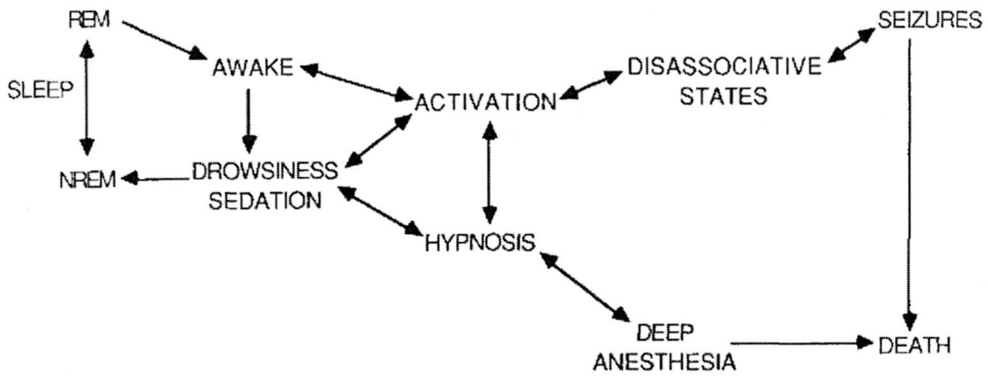

FIGURE 1. A continuum of activation and sedation states devised to describe the relationship between the normal sleep-wake cycle and the effects of psychoactive drugs.[1,2]

of normal dogs. Since that time there have been numerous reports of sleep-promoting substances (see Inoue et al.[19] for review). Many of these substances have been determined to be peptides, such as delta sleep-inducing peptide, factor S, arginine vasotocin, and vasoactive intestinal polypeptide. It has been sugested that true sleep-promoting factors should meet the criteria of (1) being produced endogenously according to the body necessity and (2) primarily playing an essential role in triggering and maintaining the state of sleep, either rapid eye movement (REM) sleep, or non-REM sleep (NREM) or both.[20] To date none of the above-mentioned compounds have convincingly met these criteria.[19] However, recent data from studies which evaluated the effects of the hypothalamic peptides CRF and growth hormone-releasing factor (GRF) have provided new evidence to suggest that these substances may eventually meet these criteria.

Unfortunately, following the initial examination of almost every new neurotransmitter candidate, reports appear that the substance produces an effect on animal behavior or EEG activity following intracerebroventricular or intracerebral administration. The amounts administered may be larger than those contained in the entire animals nervous system and may produce behaviors such as bizarre postures or abnormal movements which have little obvious physiological significance. However, use of the cerebrospinal fluid as a route of administration can provide a "first attempt screen" at identifying neuropeptides actions if it is not viewed as a physiological event, but rather as a means of gaining access to several CNS areas following a single injection.

With this caveat in mind, we have found that administration of CRF and GRF into the lateral cerebral ventricles (i.c.v.) can produce potent effects on sleep and activity in rats. CRF at extremely low doses (0.0015 to 0.015 nmol) has been found to produce significant increases in wakefulness, and reductions in slow wave sleep (SWS).[21] Where rats treated with saline would normally display SWS for 55% of a scored epoch, rats given low doses of CRF only slept 3 to 6% of the test time. In addition, these doses of i.c.v. CRF are also below the threshold for producing behavioral activation (see Reference 18). Spectral analysis of the electroencephalogram (EEG) also confirmed that CRF could produce significant changes in sleep as quantified by the frequency content of the EEG.[21,22] As seen in Figure 2, CRF at the 0.0015 nmol dose produced a spectral pattern characterized by decreases in EEG slow waves in the 1 to 6 HZ range, and increased activity in the 32 to 64 Hz range. In parallel experiments the effects of i.c.v. injections of GRF were also studied in rats.[21,23] GRF (2.0 nmol) was found to double the amount of time that the animals spent in slow-wave sleep during the recording epoch, as well as to reduce the time to sleep onset by a half. These sleep promoting effects of GRF at even lower doses have recently been confirmed in studies in rats as well as in rabbits.[24] At the 2.0 nmol dose of GRF significant increases

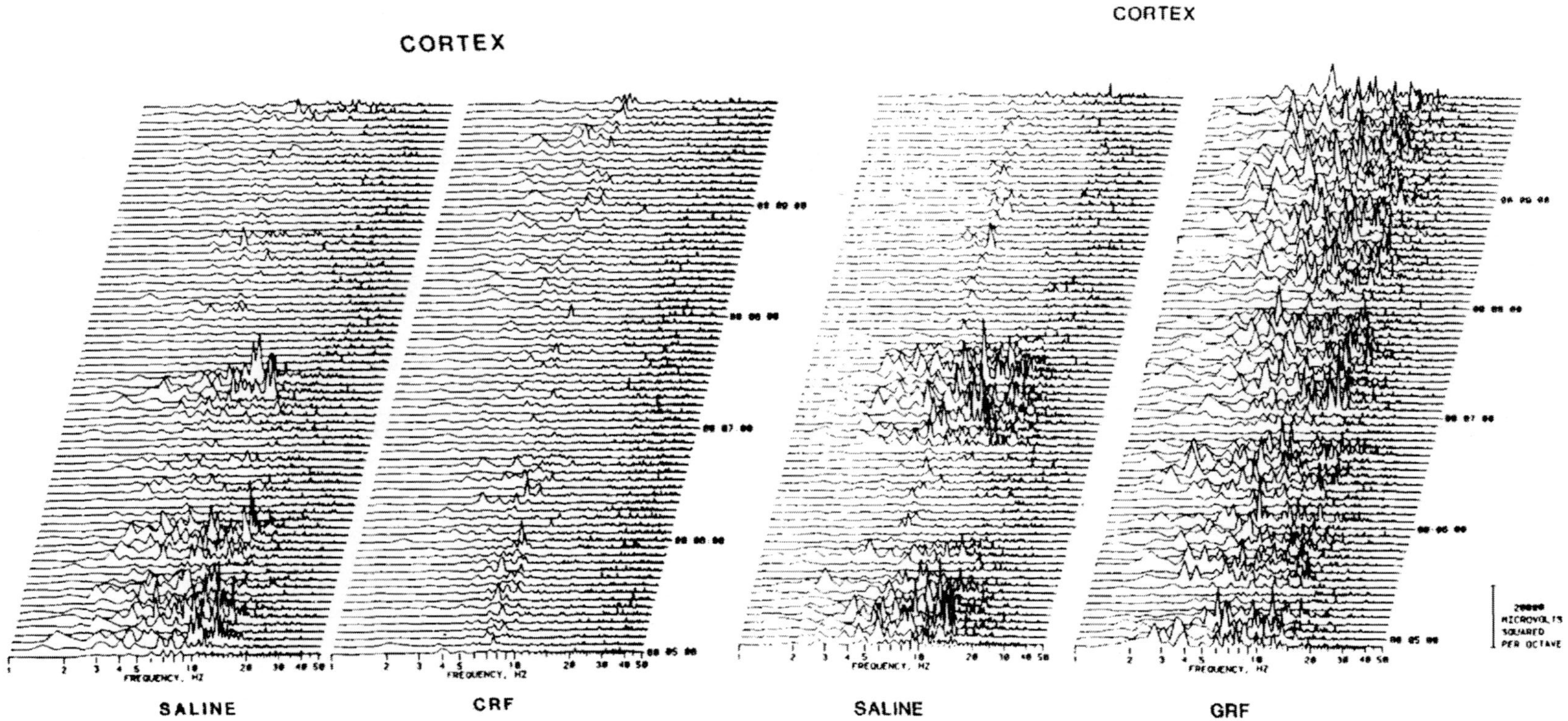

FIGURE 2. The effects of i.c.v. saline injections and injections of CRF (0.0015 nmol) and GRF (2.0 nmol) on two individual rats EEG spectra. In each graph frequency (Hz) is represented on the x-axis, power density (μV^2/octave) on the y-axis, and time (min) on the z-axis. Recordings are from frontoparietal cortex. Note that CRF produces a continuous waking state whereas GRF produces slow wave sleep.

in the delta wave frequency band (1 to 2 Hz) and decreases in high frequency (32 to 64 Hz) EEG activity could also be quantified by spectral analysis as seen in Figure 2. Thus, these studies have provided evidence to suggest that the central physiological actions of the two peptides may, in rats, be oppositional, with CRF producing activation and wakefulness and GRF promoting slow-wave sleep. Whether these neurohormones, CRF and GRF, actually participate in a significant manner in the normal regulation of the sleep-wake cycle under physiological conditions cannot yet be determined. However, it is of interest that Kimura and Kawakami[15] have reported that mean GH values measured in rat serum peak 30 min after sleep onset and were higher overall during sleep than arousal; whereas, corticosterone values were highest in the arousal state and decreased abruptly after 20 min of sleep. These secretory patterns of anterior pituitary hormones reported by Kimura and Kawakami[15] in the rat are also quite similar to those observed during the sleep-wake cycle in man.[12]

One model which may help to integrate these findings of hypothalamic peptide modulation of sleep is the two-process model of sleep regulation proposed by Borbely.[25,26] The two processes, or dynamics, in the original model are process S, a sleep-dependent process, and process C, a sleep-independent circadian process. Process S is thought to be a measure of sleep intensity which presumably builds up during the day and is released in an exponential function over the night beginning around sleep onset. The time course of process S was derived from spectral analysis studies of slow-wave activity in the human EEG, demonstrating an exponential decay during sleep. Thus process S is thought to be linked to the mechanisms which also regulate the amount or intensity of slow-wave sleep. Process C, the circadian oscillator, is reflected by a rhythmic variation of sleep propensity and it imposes a stable period upon the sleep-wake cycle. During sleep, REM activity is assumed to approximate the level of process C.

Although the two-process model has stimulated much thought and debate,[27] it does not offer any insights into the physiological mechanisms that may "drive" the described dynamic processes. As a first approximation, we have suggested that the neurohormones CRF and GRF may actively participate in the two processes.[28] GRF would be envisioned as being involved in process S, as GRF has been shown to modify slow wave activity and since GH release seems to be intimately associated with sleep onset rather than being entrained to a circadian pacemaker. CRF would be seen as modulating process C. Decreases in the level of CRF would theoretically produce increased sleep propensity. The circadian rhythm of CRF might also represent a marker of the level of process C over the 24-h cycle.

A. HYPOTHALAMIC PEPTIDES AND THE TWO PROCESS MODEL OF SLEEP IN DEPRESSION

Several lines of evidence suggest that a constellation of EEG monitored sleep abnormalities exists in the vast majority of depressed patients. Disturbances of sleep during depressive episodes can include: sleep maintenance difficulties, a delay in sleep onset, and a reduced amount of slow-wave sleep.[29] REM sleep alterations during depressive episodes include a shortened first NREM period (which is usually referred to as a decrease in REM latency) and an increase in REM density during the first few hours of sleep, especially in the first REM period. The majority of these sleep abnormalities, whether they represent changes in REM or NREM sleep, are most pronounced during the first 100 to 120 min following sleep onset.

More recently, it has been confirmed that these sleep abnormalities are also associated with consistent changes in neuroendocrine secretion. Studies in depressed subjects have established that the hypothalamic-pituitary-adrenal axis (HPA) displays definite abnormalities, as increases in cortisol secretion rates[30] a flattening of the circadian rhythm in cortisol[31] and elevated cortisol nadir[32] have been reported and replicated in numerous investigations. The failure to suppress plasma cortisol in the dexamethasone suppression test (DST) in many depressed patients has also supported the notion of a specific HPA abnormality in depression.[33] A hyposecretion of adrenocorticotrophic hormone (ACTH) in response to CRF administration[34] and an increase in the cerebrospinal fluid levels of CRF[35] have also been observed in clinical studies (also see Chapter 23). Both findings are consistent with the hypothesis that CRF is hypersecreted in the state of depression. Recently, Gold et al.[36] have also argued that CRF abnormalities may be even more specific for depression, rather than reflecting a general stress or anxiety response.

Disturbances in the secretion of growth hormone have also been uncovered in depressed patients. Jarrett et al.[37] have demonstrated that the secretion of growth hormone in the first half of the night is decreased in depressed patients as compared to control subjects. These findings have been confirmed by other investigators, including Holsboer et al.[38] and Berger et al.,[39] who have also reported a reduction in sleep-associated growth hormone release in depressed patients. Additionally, Mendelwicz et al.[40] have demonstrated that when growth hormone (GH) is measured over the entire 24-h period in depressed patients, an increase in GH during the day is found in comparison to normals. Mendelson et al.[41] suggested earlier that this release of GH before sleep onset may, in fact, act as an inhibitor to the release of GH after sleep onset. Thus, decreased nighttime GH secretion associated with increased daytime GH secretion may be consistent with almost all of the available empirical data.

If a "dysregulation" of the hypothalamic peptides CRF and GRF is a prominent factor in the initiation or maintenance of a depressive sleep state, this might occur through a mechanism that produces a hypersecretion of CRF in the face of lowered levels of GRF. How this change in peptide dynamics would ensue is not clear; however, a cascade of events may occur. For instance, in a "high-risk" individual, significant stressors may induce a state of CRF hypersecretion. This hypersecretion, if sustained, may be responsible for the higher cortisol nadir observed during the night and the delay in sleep onset seen in depressed individuals. This delay in sleep onset would be associated with decreased GH release after sleep onset and would induce reductions in delta sleep and increased awakenings. In fact, it has been shown by Holsboer et al.[38] that the administration of CRF in repeated pulses produces a blunting in growth hormone release, perhaps by action at the level of the pituitary. Thus, if GRF release is inhibited directly or indirectly by CRF, then growth hormone release might occur before sleep onset, or much later in the night, as opposed to immediately after sleep onset, as is usually seen in normal subjects.

According to our hypothesis, these presumed changes in peptide dynamics in depression should also be reflected as a disturbance in the patients sleep patterns. These changes in sleep which occur in depressed patients might also be explained by the use of the two-process model. It has been previously suggested that a deficiency in process S accumulation is the "main dynamic" responsible for the sleep EEG alterations observed in depressed patients.[42]

If process S is deficient before or during depressive episodes then we would propose that lowered levels of GRF would be seen associated with a weakened or impaired slow-wave (delta) sleep generator. When the S system is somewhat weakened, less of stages 3 and 4 sleep would also be seen as well as a lighter overall sleep pattern. The growth hormone increases seen during the day in depressed patients may reflect the increased level of process S during the day in terms of "leakage" and, therefore, may lead to a decrease in process S during sleep time. The increased levels of CRF over the 24-h period would be envisioned as producing not only an elevated and a flattened appearance of the C process, but might also result in increased REM activity and density seen during the night in depressed patients. The CRF/GRF ratio may also be a marker of the strength of the S process.

A summary of this model is presented in Table 1. Although further studies are needed to confirm or refute this hypothesis, the advantage of the "extended" two-process model are several-fold: (1) it provides additional experimental tools to test the hypothesis, so that

TABLE 1
CRF/GRF and the Two-Process Model of Sleep Depression

	Process S A sleep dependent process that builds up during the day and is released at night	Process C A sleep-independent circadian process
Sleep stage relationship	Slow-wave sleep	REM sleep
Neuroendocrine relationship	GRF and/or CRF ratio	CRF-HPA axis
Alteration in depression	Deficient	Increased
Physiological consequences in depression	GH hyposecretion after sleep onset, decreased delta density, delayed sleep onset, lighter overall sleep pattern	Cortisol hypersecretion over 24-h period, increased REM activity and density

neuropeptide challenge strategies on sleep EEG may be employed; (2) it facilitates the examination of interactions among various biological rhythms; and (3) the application of the model provides an opportunity to make specific predictions concerning the level of dissociation among selected biological rhythms in the depressive state and in the state of clinical recovery.

III. CRF AND EPILEPTIFORM DISCHARGES

The potential role of CRF in the development of epileptiform discharges has been recently reviewed[43] and thus it will only be summarized briefly in this chapter. Hypothalamic peptides have long been implicated by indirect evidence to epilepsy by data which linked electrolyte and endocrine imbalances to seizures frequency and/or seizure thresholds in epileptic patients and in animal models (see References 44 and 45). In addition, there is increasing evidence that hypothalamic neurohormones may also act centrally as neurotransmitters or neuromodulators and in this capacity alter brain excitability. For instance, Kasting et al.[46] have suggested that arginine vasopressin (AVP) may be involved in the etiology of febrile convulsions through its action as a central nervous system (CNS) neuromodulator, independent of its ability to produce antidiuresis.

One line of research which has been fruitful in the exploration of the etiology of human seizure disorders is the development of animal models of the various seizure types which respond to anticonvulsants in a clinically relevant manner (see Reference 47). If a neurochemical can be found to initiate or significantly alter seizures which occur in these animal models, then a bridge to human epilepsy may be, at least theoretically, constructed. In the case of corticotropin-releasing factor (CRF), preliminary data suggest that such a connection may exist. In these studies CRF was administered to rats who had been previously implanted with electrodes in the dorsal hippocampus (DHPC), amygdala (AMYG), and the cerebral cortex (CTX). Intracerebroventricular administration of CRF as opposed to saline was found to produce, after a delay of 10 to 20 min, EEG activation associated with increased locomotion, grooming, and rearing as well as a grossly exaggerated startle response to environmental stimuli. After a delay of 1 to 3 h, paroxysmal EEG activity was also observed. Epileptiform activity began with the appearance of large spikes in the amygdala and later spread to dorsal hippocampus and cerebral cortex as seen in Figure 3. Following the spread of EEG seizure activity to cortical areas, behavioral signs of seizure activity occurred. These behavioral seizures, which developed over 3 to 7 h, began with myoclonic movement of the jaw, and with subsequent seizures predictively progressed to rearing, forelimb clonus, and falling to the floor associated with apparent loss of consciousness. Seizures which led to loss of consciousness were followed by EEG flattening, postictal depression and explosive

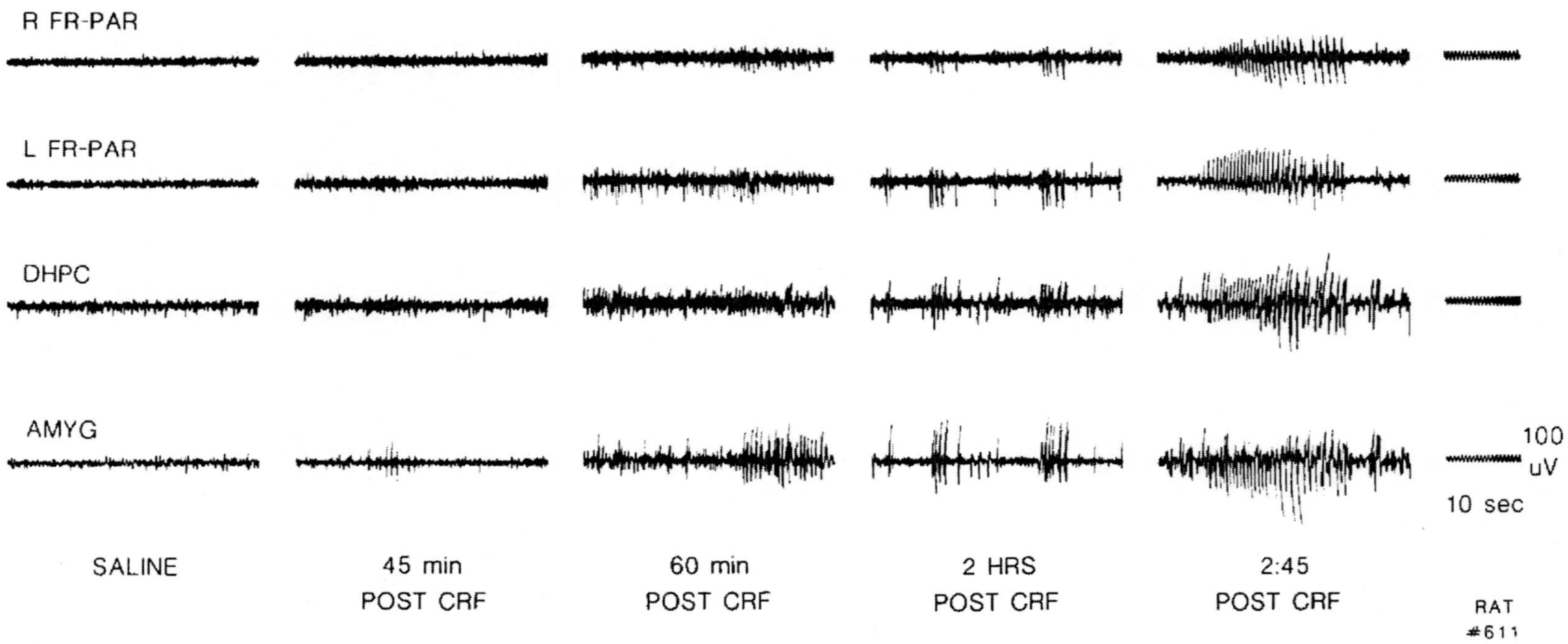

FIGURE 3. Examples of spontaneous EEG records obtained one hour following i.c.v. injection of saline and 45 min, 60 min, 2 h, and 2:45 min post-i.c.v CRF. Recordings are from right frontoparietal cortex (RFR-PAR), left frontoparietal cortex (LFR-PAR), dorsal hippocampus (DHPC), and amygdala (AMYG). Note that both spikes and spontaneous afterdischarges appear to originate in the amygdala and with time spread to other brain areas.

motor behavior. It is of interest that these electrographic and behavioral seizures, which occurred spontaneously following a single i.c.v. injection of CRF, appeared to develop in a manner indistinguishable from seizures which occur following repeated electrical ("kindling") stimulation of the amygdala.[48] In a separate study, it was also confirmed that the i.c.v. administration of CRF produced seizures which resemble those seen following electrical "kindling" of the amygdala.[49] "Kindling" is thought to be a model of temporal lobe epilepsy (partial complex seizures). These data represent the first description of an endogenous neurochemical (CRF) found to be capable of spontaneously producing "kindling" following a single i.c.v. dose. We have also found that "kindling" produced by i.c.v. CRF and the electrical kindling first described, in name, by Goddard et al.[50] share some common features. Goddard et al.[50] have demonstrated that electrical kindling of convulsions in rats can be induced by applications of current to a number of brain sites; however, the amygdala required the fewest stimulations. We have found that when CRF is administered to rats via the cerebral spinal fluid, electrographic seizures are elicited first in the amygdala and later in other brain sites. The behavioral concomitants of CRF induced seizures also appear to be identical in development to those produced by electrical kindling of the amygdala. Anatomical support for these EEG and behavioral observations has also been provided by immunohistochemical studies which have shown that CRF pathways can be localized in the central nucleus of the amygdala.[51,52]

Intracerebroventricular administration of CRF to rats has also been demonstrated to significantly reduce the number of stimulations necessary to later produce electrical kindling of the amygdala.[49] However, electrical kindling of the amygdala appears to produce a decrease in seizure susceptibility to subsequent i.c.v. administration of CRF, suggesting that electrical and CRF kindling do not entirely share the same mechanisms.[50] Tolerance has also been reported to CRF kindling.[49]

However, it should also be kept in mind that the EEG and behavioral effects observed following central administration of peptides, particularly neurohormones, may actually be produced by "downstream" effects of these substances. For instance, CRF may act centrally to produce its effect on EEG and behavior or it may act in concert with circulating ACTH, β-endorphin, and corticosteroids to produce the full repertoire of observed EEG and behavioral effects. A number of studies, in animal models of epilepsy, have shown that ACTH and corticosteroids can increase susceptibility to seizures.[53-56] In humans, ACTH and corticosteroids have also been reported to produce convulsant effects in epileptics with certain seizure types and in some nonepileptic patients.[57-59] Therefore, it is possible that the release of ACTH and corticosteroids, known to occur following central administration of CRF, could act to potentiate the seizure inducing effects of CRF observed in these studies.

Neuropeptides may also initiate or modify epileptic phenomena by interacting with other neurotransmitter systems such as the monoamines, amino acids, or modified amino acids. For instance, in some cases it seems that a peptidergic neuron may release both an amine and a peptide, suggesting that pairs of transmitters may act on target cells.[60] Some neuropeptide actions may also be indirect in nature in that they cause such effects as "disinhibition". In the hippocampus the opioid-evoked naloxone-reversible excitations have been attributed to a suppression of the discharge of inhibitory GABA neurons.[61,62] Opioid-induced reductions of EPSP activity in hippocampal pyramidal cells have also been suggested to be the result of an antagonism of the depolarizing effects of glutamate.[60] However, recent studies on the effects of CRF on hippocampal pyramidal cells have suggested a direct action of this peptide on neuronal responses.[63] Thus, many possible mechanistic models may be needed to describe the actions of the various classes of neuropeptides on epileptiform phenomena.

Thus far it cannot be acknowledged with confidence that any one neuropeptide is the "cause of epilepsy". However, the demonstration that CRF has the potential to elicit

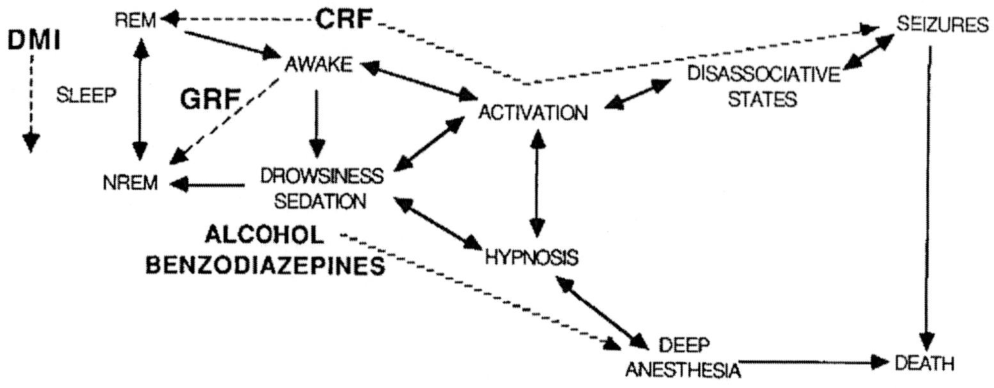

FIGURE 4. The suggested locus of action of CRF, GRF, DMI, and ETHANOL in the continuum of states first described by Winters et al.[1,2]

spontaneous seizure activity in rats similar to "kindling," a model of partial complex seizures, suggests that these substances should be given serious consideration by epileptologists.

IV. INTERACTIONS OF CRF WITH PSYCHOTROPIC DRUGS

A. ALCOHOL

If one were to evaluate, where, within the continua of brain states first described by Winters[2,3] that CRF was active, we would suggest that it would modulate the activating upper arm as indicated in the model presented in Figure 4. This is because CRF has been shown to produce, depending on the dose, activation, dissociative states, and eventually seizures. Depressant drugs such as alcohol and benzodiazepines, on the other hand, have been thought to act on the "sedating lower arm" also as shown in Figure 4. Alcohol, for instance, has been demonstrated, depending again on the dose, to produce activation, drowsiness, hypnosis, anesthesia, and ultimately coma and death (see Reference 64). Thus, one would expect that CRF may antagonize the effects of the depressant drugs. In fact, it has been reported that CRF can reduce single dose ethanol induced sleeping time in rats[65] and that chlordiazepoxide can block CRF potentiation of acoustic startle.[66]

The effects of CRF on chronic ethanol administration, however, appears to be a more complicated scenario. Chronic ethanol ingestion has been previously demonstrated to produce an activating effect on the hypothalamic-pituitary-adrenal (HPA) axis in both clinical and animal studies.[67] In fact, in certain chronic alcoholics, hypercortisolemia can be severe enough so that several features of "pseudo-Cushing's syndrome" are present.[68-72] Some subjects may even display nonsuppression of plasma cortisol in response to the synthetic glucocorticoid, dexamethasone.[73-78] In animal studies dose-related increases in corticosteroids have also been reported following acute ethanol exposure[78-82] which may persist during chronic administration.[79,81]

Recent studies by Rivier et al.[83] have revealed that the increases in ACTH secretion seen following acute ethanol administration in rats is highly dependent on CRF production, as immunoneutralization of endogenous corticotropin-releasing factor (CRF) was found to totally abolish the ethanol-induced release of ACTH. Chronic exposure to ethanol vapors has also been shown to cause a decrease in hypothalamic content of CRF.[83]

The mechanism whereby chronic ethanol exposure induces this release of CRF has not been elucidated. It has been suggested that ethanol administration may increase the activity of the HPA axis, and thus behavior, by acting as a nonspecific stressor. However, several studies have provided evidence that ethanol may increase cortisol levels through a different

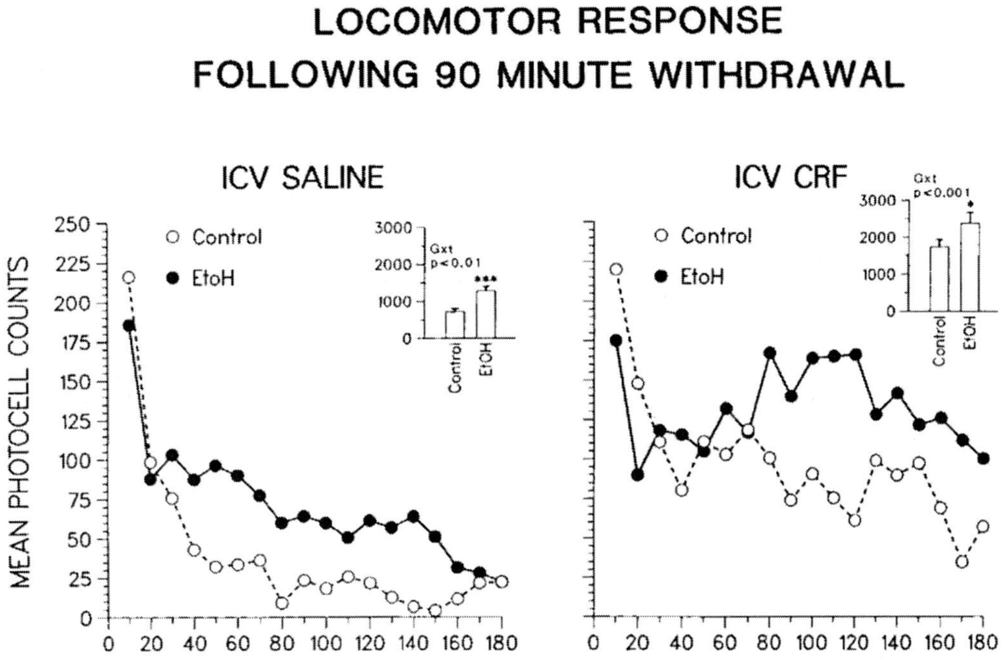

FIGURE 5. Comparison of locomotor response to i.c.v. saline or CRF in control rats (open circles) and rats withdrawing from chronic ethanol treatment (closed circles). Note that chronic ethanol can significantly enhance the locomotor response to CRF.

mechanism. For instance, during acute stress, cortisol levels are increased but the normal diurnal variation is intact. Even during chronic stress, response to dexamethasone is normal, no Cushingoid features appear, and ultimately glucocorticoid levels return to normal regardless of the duration of the stressful stimuli.[67] Whereas in alcoholism, glucocorticoid levels can remain high, Cushingoid features can appear,[68-72] and nonsuppression to dexamethasone administration may be finding.[73-78] In addition, Rivier et al.[83] have demonstrated in rats that while anti-CRF serum can completely abolish ethanol-induced ACTH-secretion it can only partially block ACTH response to ether stress.[84] Taken together, these studies suggest that at least some aspects of stress-induced hypothalamic-pituitary-adrenal activation occurs through different mechanisms than those produced by chronic ethanol exposure.

We surmised that the presumed release of CRF following ethanol exposure may also have behavioral consequence separate from its stimulation of the pituitary-adrenal axis. Therefore, we studied the effects of chronic exposure (21 d) to ethanol vapors on the locomotor activating effects of CRF in male Wistar rats. Responses to CRF were tested during chronic exposure, 1 $^1/_2$ h following removal of ethanol vapors, and 2 weeks after withdrawal of ethanol. A greater sensitivity to the locomotor-activating effects of CRF was found in ethanol treated rats as compared to their controls during ethanol exposure and 90 min following removal of ethanol vapors (as seen in Figure 5) but not 2 weeks following withdrawal.[85] These results suggest that chronic ethanol exposure may enhance behavioral responsiveness to exogenously administered CRF under these conditions. One process which may account for this increased behavioral response following ethanol exposure is the ability of CRF to produce behavioral "kindling" effects or sensitization to other CNS-activating stimuli as previously described. Thus, ethanol-induced CRF release may be producing a heightened behavioral sensitivity to subsequent CRF exposure through a type of positive feedback mechanism. CRF may be acting only centrally to induce this sensitization effect or it may act in concert with circulating adrenocorticotrophic hormone (ACTH), β-endorphin and/or corticosteroids to produce the observed increase in locomotion. However, although β-endorphin and ACTH can produce an increase in activity,[86,87] the time course and nature of the response to these peptides differ from the behaviors we have observed. As following CRF, the rats' responses appear to be more similar to the exaggeration of normal activation produced by introduction into a familiar environment.[88] In addition, the synthetic glucocorticoid dexamethasone has been previously found to be devoid of behaviorally activating effects under these conditions.[88] Thus, it is most probable that the observed increase in CRF-induced locomotion is centrally mediated.

In our study the greatest difference in CRF response between ethanol-treated and control rats was seen in the acutely withdrawing rats. These data are consistent with those provided by Mendelson and Stein[89] who showed that human alcoholics actually had their greatest cortisol levels during ethanol withdrawal. In fact, one hypothesis which may explain many of the symptoms of ethanol withdrawal is the possibility that the CRF release, which may occur during chronic ethanol exposure, may greatly sensitize the system to subsequent, even larger, surges of CRF which occur during withdrawal.

We have observed that following 2 weeks of withdrawal from ethanol vapors the CRF-induced locomotor response of the ethanol treated rats was not significantly different from controls. This finding is in agreement with clinical studies where cortisol response to exogenous dexamethasone has been reported to be normalized in alcoholics by 2 to 4 weeks following withdrawal.[75,76,78]

Our results indicate that chronic ethanol exposure can produce an increased behavioral sensitivity to the locomotor-activating effects of central administered CRF in rats. This increased behavioral sensitivity to CRF also appears to be further enhanced during acute ethanol withdrawal, but locomotor responses were found to be normalized after 2 weeks of withdrawal from ethanol vapors. Thus, these studies suggest that brain mechanisms involving CRF should be considered as a possible explanation for some of the behavioral effects seen in chronic ethanol exposure and/or withdrawal syndromes.

B. ANTIDEPRESSANTS

If CRF or GRF play a role in the triggering or maintenance of a depressive episode then one might suspect that antidepressant drugs should modify responses to CRF, perhaps through alterations in the patients sleep and activity cycles as represented in Figure 4. In preliminary experiments, we have evaluated whether chronic administration of lithium (Li) or desmethylimipramine (DMI), could modify the behavioral effects of central administration of CRF and/or GRF in rats.[90,91] In these studies 75 male Wistar rats were exposed to one of four treatment conditions: 18 d of DMI (10 mg/kg) or water injections, or 21 d of lithium (30 mmol/kg) or control diets. CRF (0.15 nmol) and GRF (2.0 nmol) were administered intracerebroventricularly and locomotion was measured using computer-controlled photocell cages. As seen in Figure 6, CRF was found to produce increases in locomotion in all four treatment groups. DMI, but not lithium, was found to significantly modify the pattern of locomotion seen following CRF administration.

We have previously reported that chronic DMI administration, while not affecting mean locomotion in rats, can increase the variance and frequency of locomotor rhythms in rats tested during the active (nocturnal) phase of their circadian cycle.[92] We have further postulated that depression may represent a state where the CNS is of low variance and "desensitized" to neurochemical input. DMI treatment may then induce its antidepressant effects by "resensitizing" the brain to peptide inputs and thus reestablishing the normal variance of biological rhythms. How this resensitization following DMI may be enacted is not clear; however, it is possible that CRF receptors are down-regulated in depression, perhaps in response to CRF hypersecretion. DMI could then potentially act by reversing this down-regulation of CRF receptors.

In contrast, growth hormone-releasing factor was found to produce decreases in locomotion in all four treatment groups in our study. GRF was also observed to modify the pattern of locomotor response to both LI and DMI, with DMI producing an enhancement of GRF induced decreases in locomotion and LI producing less response to GRF. Thus these two peptides CRF and GRF appear to have different interactions with antidepressant treatments.

V. CONCLUSIONS

The findings that central administration of CRF can produce behavioral activation, lightening of EEG monitored sleep, and at higher doses seizure kindling, suggests that it may play an important role in a brain system important for the regulation of brain activation states. To the extent that disorders such as alcoholism and depression are viewed as syndromes of a dysregulation of activation states, then CRF would be inferred to be a chemical mediator. Although the data supporting peptidergic mediation of affective and drug seeking states are still somewhat limited, the models presented in this chapter should provide a tentative theoretical framework for further in depth studies.

ACKNOWLEDGMENTS

The authors would like to thank R. I. Chaplin, A. Lopez, T. K. Reed, and M. Wang for technical assistance in these studies. The development of the extension of the S-process hypothesis was made in collaboration with D. J. Kupfer. Nancy Callahan and Susan Lopez helped with typing and editing the manuscript.

Preparation of this chapter was supported by Grants 00098, 06059, and 06420 from NIAAA, MH 24652, and a grant from the John and Catherine MacArthur Foundation.

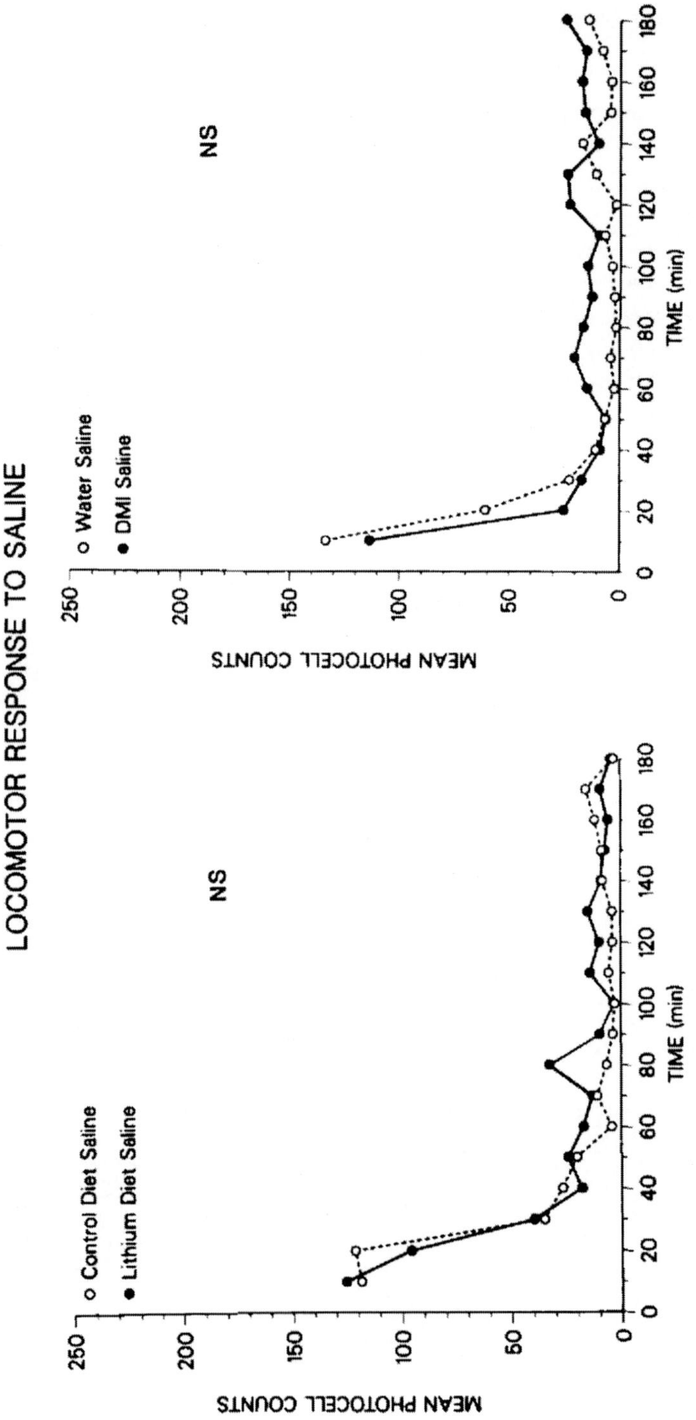
LOCOMOTOR RESPONSE TO SALINE
○ Control Diet Saline
● Lithium Diet Saline
NS
MEAN PHOTOCELL COUNTS
TIME (min)
0
20
40
60
80
100
120
140
160
180
0
50
100
150
200
250
○ Water Saline
● DMI Saline
NS
MEAN PHOTOCELL COUNTS
TIME (min)

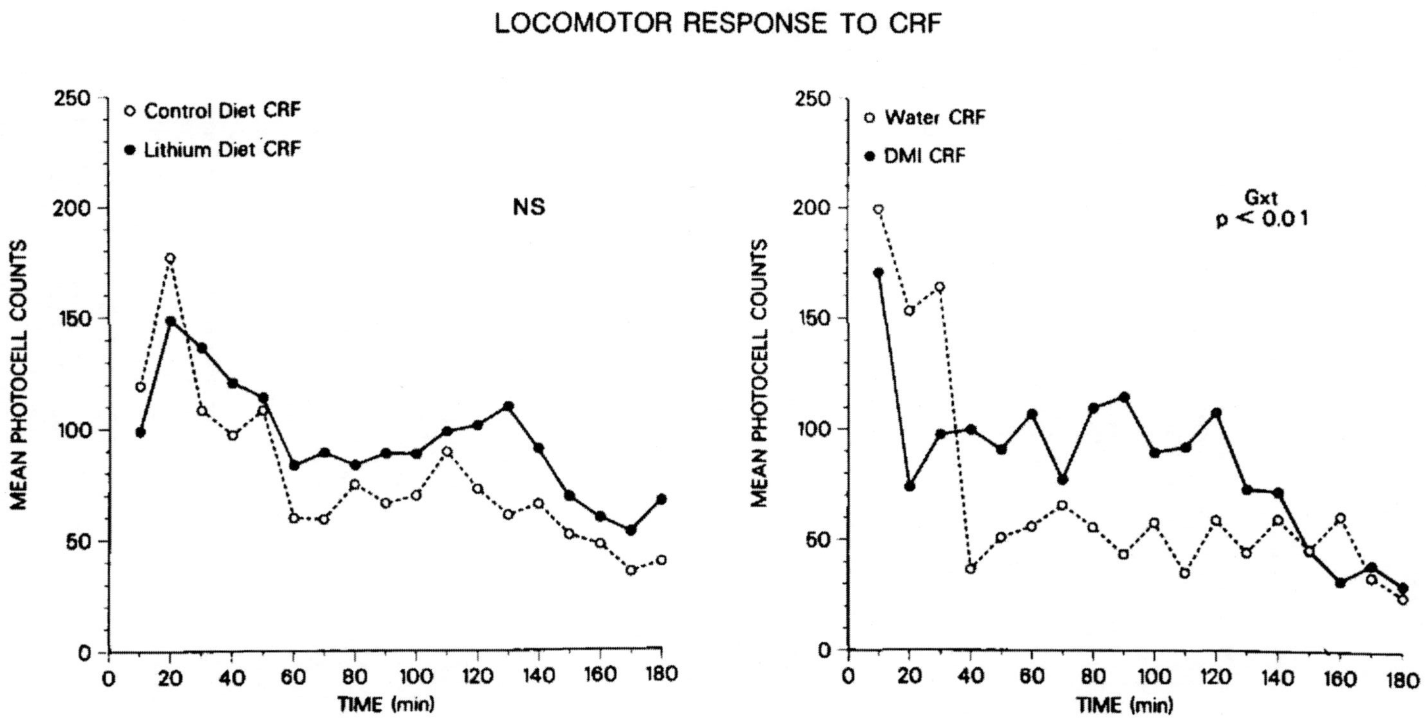

FIGURE 6. Comparison of locomotor response to i.c.v. injections of saline (upper graphs) following control vs. lithium diet (upper left) and DMI vs. water injections. Comparisons of locomotor response to i.c.v. injections of CRF following control vs. lithium diet (lower left). Note DMI treatment significantly modified the locomotor response to i.c.v. CRF injections.

REFERENCES

1. **Koella, W. P.,** The organization and regulation of sleep, *Experientia,* 40, 309, 1984.
2. **Winters, W. D., Ferrar-Allado, T., Guzman-Flores, C., and Alcaraz, M.,** The cataleptic state induced by ketamine: a review of the neuropharmacology of anesthesia, *Neuropharmacology,* 11, 303, 1972.
3. **Winters, W. D. and Kott, K.,** Continuum of sedation activation and hypnosis or hallucinosis: a comparison of low doses effects of pentobarbital, diazepam or gamma-hydroxybutyrate in the cat, *Neuropharmacology,* 18, 877, 1979.
4. **Winters, W. D., Mori, K., Spooner, C. E., and Bauer, R. O.,** The neurophysiology of anesthesia, *Anesthesiology,* 218, 65, 1967.
5. **Winters, W. D. and Wallach, M. B.,** Drug induced states of CNS excitation: a theory of hallucinosis, in *Psychotomimetric Drugs,* Effrod, D. H., Ed., Raven Press, New York, 1970, 193.
6. **Moruzzi, G. and Magoun, H. W.,** Brain stem reticular formation and activation of the EEG, *Electroencephalogr. Clin. Neurophysiol.,* 1, 455, 1949.
7. **Winters, W. D., Mori, K., Wallach, M. B., Marcus, R. J., and Spooner, C. E.,** Reticular multiple unit activity during a progression of states induced by CNS excitants, *Electroencephalogr. Clin. Neurophysiol.,* 27, 514, 1969.
8. **Mori, K., Winters, W. D., and Spooner, C. E.,** Comparison of reticular and cochlear multiple unit activity with auditory evoked responses during various stages induced by anesthetic agents, *Electroencephalogr. Clin. Neurophysiol.,* 24, 242, 1968.
9. **Winters, W. D., Mori, K., Wallach, M. B., Marcus, R. J., and Spooner, C. E.,** Reticular multiple unit activity during a progression of states induced by CNS excitants, *Electroencephalogr. Clin. Neurophysiol.,* 27, 514, 1969.
10. **Moore, R. Y.,** Central neural control of circadian rhythms, in *Frontiers in Neuroendocrinology,* Ganong, W. F. and Martini, E., Eds., Raven Press, New York, 1978, 185.
11. **Fuller, C. A., Lydic, R., Sulzman, F. M., Albers, H. E., Tepper, B., and Moore-Ede, M. C.,** Circadian rhythm of body temperature persists after suprachiasmatic lesions in the squirrel monkey, *Am. J. Physiol.,* 241, 385, 1981.
12. **Parker, D. C., Sassin, J. F., Mace, J. W., Gotlin, R. W., and Rossman, L. G.,** Human growth hormone release during sleep-electroencephalographic correlation, *J. Clin. Endocrinol. Metab.,* 29, 871, 1969.
13. **Takahashi, Y., Kipnis, D. M., and Daughaday, W. H.,** Growth hormone secretion during sleep, *J. Clin. Invest.,* 47, 2079, 1968.
14. **Parker, D. C., Morishima, M., Koerker, D. J., Gale, C. C., and Goodner, C. J.,** Pilot study of growth hormone release in sleep of the chair-adapted baboon. Potential as a model of human sleep release, *Endocrinology,* 91, 1462, 1972.
15. **Kimura, F. and Kawakami, M.,** Serum hormone levels during sleep and wakefulness in the immature female rat, *Neuroendocrinology,* 33, 276, 1981.
16. **Ehlers, C. L. and Killam, E. K.,** Circadian periodicities in brain activity and urinary excretion in the epileptic baboon, *Am. J. Physiol.,* 239, 404, 1979.
17. **Weitzman, E. D., Fukushima, D., Nogeire, C., Roffwarg, H., Gallagher, T. F., and Hellman, L.,** Twenty-four hour pattern of the episodic secretion of cortisol in normal subjects, *J. Clin. Endocrinol. Metab.,* 33, 14, 1971.
18. **Sutton, R. E., Koob, G. F., Le Moal, M., Rivier, J., and Vale, W.,** Corticotropin releasing factor produced behavioral activation in rats, *Nature (London),* 297, 331, 1982.
19. **Inoue, S., Vchizono, K., and Nagasaki, H.,** Endogenous sleep-promoting factors, *TINS,* 0218, 1982.
20. **Tobler, I. and Borbely, A. A.,** *Waking Sleeping,* 4, 139, 1980.
21. **Ehlers, C. L., Reed, T. K., and Henriksen, S. J.,** Effects of corticotropin-releasing factor on sleep and activity in rats, *Neuroendocrinology,* 42, 467, 1986.
22. **Ehlers, C. L.,** EEG stability following corticotropin releasing factor in rats, *Psychoneuroendocrinology,* 11, 121, 1986.
23. **Wehrenberg, W. B. and Ehlers, C. L.,** Effects of growth hormone releasing factor in the brain, *Science,* 232, 1271, 1986.
24. **Obal, F., Jr., Johannsen, L., Cady, A. B., and Krueger, J. M.,** Growth hormone-releasing factor increases both non-REM sleep and REM sleep in rabbits, *Soc. Neurosci. Abstr.,* 13, 76.4, 1987.
25. **Borbely, A. A.,** A two-process model of sleep regulation, *Hum. Neurobiol.,* 1, 195, 1982.
26. **Daan, S., Beersma, D. G. M., and Borbely, A. A.,** Timing of human sleep: recovery process gated by a circadian pacemaker, *Am. J. Physiol.,* 246, R161, 1984.
27. **Borbely, A. A.,** Models of sleep regulation, meeting held at Neurosciences Institute, February 21—22, 1985.
28. **Ehlers, C. L. and Kupfer, D. J.,** Hypothalamic peptide modulation of EEG sleep in depression: a further application of the S-process hypothesis, *Biol. Psychiatry,* 22, 513, 1987.

29. **Kupfer, D. J., Ulrich, R. F., Coble, P. A., Jarrett, D. B., Grochocinski, V. J., Doman, J., Matthews, G., and Borbely, A. A.,** Application of automated REM and slow wave sleep analysis. I. Normal and depressed subjects, *Psychiatry Res.*, 13, 325, 1984.
30. **Linkowski, P., Van Cauter, E., Leclercq, R., Desmedt, D., Brasseur, M., Goldstein, J., Copinschi, G., and Mendelwicz, J.,** ACTH, cortisol and GH 24-hour profiles in major depressive illness, *Acta Psychiatry Belgium*, 85, 615, 1985.
31. **Sachar, E. I.,** Neuroendocrine dysfunction in depressive illness, *Annu. Rev. Med.*, 27, 389, 1976.
32. **Jarrett, D. B., Coble, P. A., and Kupfer, D. J.,** Reduced cortisol latency in depressive illness, *Arch. Gen. Psychiatry*, 40, 506, 1983.
33. **Carroll, B. I.,** The dexamethasone suppression test for melancholia, *Br. J. Psychiatry*, 140, 292, 1982.
34. **Gold, P. W., Chrousos, G., Kellner, C., Post, R., Roy, A., Augerinos, P., Schulte, H., Oldfield, E., and Loriaux, L.,** Psychiatric implications of basic and clinical studies with corticotropin-releasing factor, *Am. J. Psychiatry*, 141, 619, 1984.
35. **Nemeroff, C. B., Widerlov, E., Bissette, G., Walleus, H., Karlsson, I., Eklund, K., Kilts, C. D., Loosen, P. T., and Vale, W. J.,** Elevated concentrations of CSF corticotropin-releasing factor-like immunoreactivity in depressed patients, *Science*, 226, 1342, 1984.
36. **Gold, P. W., Loriaux, D. L., Roy, A., Kling, M. A., Calabrese, J. R., Keller, C. H., Nieman, L. K., Post, R. M., Pickar, D., Galluci, W., Augerinos, P., Paul, S., Oldfield, E. H., Cutler, G. B., and Chrousos, G. P.,** Responses to corticotropin-releasing hormone in the hypercortisolism of depression and Cushing's disease: pathophysiologic and diagnostic implications, *N. Engl. J. Med.*, 314, 1329, 1986.
37. **Jarrett, D. B., Greenhouse, J. B., Coble, P. A., and Kupfer, D. J.,** Sleep EEG and neuroendocrine secretion in depression, in *Biological Psychiatry 1985. Proc. 4th World Congress of Biological Psychiatry*, Shagass, C., Josiassen, R. C., Bridger, W. H., Weiss, K. J., Stoff, D., and Simpson, G. M., Eds., Elsevier, New York, 1986, 966.
38. **Holsboer, F., von Bardeleben, U., Benkert, O., Herdt, T., Hiller, W., Nehring, K., Steigee, A., and Stein, A.,** Human corticotropin-releasing factor induced modulation of sleep architecture hormone secretion and penile tumescence, in *Biological Psychiatry 1985. Proc. 4th World Congress of Biological Psychiatry*, Shagass, C., Josiassen, R. C., Bridger, W. H., Weiss, K. J., Stoff, D., and Simpson, G. M., Eds., Elsevier, New York, 1986, 150.
39. **Berger, M., Zulley, J., Hochli, D., Reimann, D., and Von Zerssen, D.,** The activity of different neuroendocrine axes during sleep and sleep deprivation in patients with a major depressive disorder, in *Biological Psychiatry 1985. Proc. 4th World Congress of Biological Psychiatry*, Shagass, C., Josiassen, R. C., Bridger, W. H., Weiss, K. J., Stoff, D., and Simpson, G. M., Eds., Elsevier, New York, 1986, 156.
40. **Mendelwicz, J., Linkowski, P., Kerkhofs, M., Desmedt, D., Goldstein, J., Chopinschi, G., and Van Cauter, E.,** Diurnal hypersecretion of growth hormone in depression, *J. Clin. Endocrinol. Metab.*, 60, 505, 1985.
41. **Mendelson, W. B., Jacobs, L. S., and Gillin, J. C.,** Negative feedback suppression of sleep-related growth hormone secretion, *J. Clin. Endocrinol. Metab.*, 56, 486, 1983.
42. **Borbely, A. A. and Wirz-Justice, A.,** Sleep, sleep deprivation and depression: a hypothesis derived from a model of sleep regulation, *Hum. Neurobiol.*, 1, 205, 1982.
43. **Ehlers, C. L.,** Role of selected neuropeptides in the development of epileptiform discharges, in *Neurotransmitters Seizures, and Epilepsy II*, Fariello, R. G. et al., Eds., Raven Press, New York, 1984, 295.
44. **Millchap, J. G.,** Systemic electrolyte and neuroendocrine mechanisms, in *Basic Mechanisms of Epilepsies*, Jasper, H. H., Ward, A. A., and Pope, A., Eds., Little Brown, Boston, 1969, 709.
45. **Woodbury, D. M.,** Relation between the adrenal cortex and the central nervous system, *Pharmacol. Rev.*, 10(2), 275, 1958.
46. **Kasting, N. W., Veale, W. L., Cooper, K. E., and Lederis, K.,** Vasopressin may mediate febrile convulsion, *Brain Res.*, 213, 327, 1981.
47. **Purpura, D. P., Penry, J. K., Tower, D. B., Woodbury, D. M., and Walter, R. D.,** *Experimental Models of Epilepsy*, Raven Press, New York, 1972.
48. **Ehlers, C. L., Henriksen, S. J., Wang, M., Rivier, J., Vale, W., and Bloom, F. E.,** Corticotropin releasing factor produces increases in brain excitability and convulsive seizures in rats, *Brain Res.*, 278, 332, 1983.
49. **Weiss, S. R. B., Post, R. M., Gold, P. W., Chrousos, G., Sullivan, T. L., Walker, D., and Pert, A.,** CRF-induced seizures and behavior: interaction with amygdala kindling, *Brain Res.*, 372, 345, 1986.
50. **Goddard, G. V., McIntyre, D. C., and Leech, C. K.,** A permanent change in brain function resulting from daily electrical stimulation, *Exp. Neurol.*, 25, 295, 1969.
51. **Olschowka, J. A., O'Donohue, T. L., Mueller, G. P., and Jacobowitz, D. M.,** Hypothalamic and extrahypothalamic distribution of GRF-like immunoreactive neurons in the rat brain, *Neuroendocrinology*, 35, 305, 1982.

52. **Paull, W. K., Scholer, J., Arimura, A., Meyers, C. A., Chang, J. K., Chang, D., and Shimizu, M.,** Immunocytochemical localization of CRF in the ovine hypothalamus, *Peptides,* 3, 183, 1982.
53. **Ehlers, C. L. and Killam, E. E.,** The influence of cortisone on EEG and seizure activity in the baboon, *Papio papio, Electroencephalogr. Clin. Neurophysiol.,* 47, 404, 1979.
54. **Feldman, S.,** The interaction of neural and endocrine factors regulating hypothalamic activity, in *Brain Pituitary-Adrenal Interrelationships,* Bradish, A. and Redgate, E. S., Eds., S. Karger, Basel, 1973, 224.
55. **Glaser, G. H., Kornfeld, D. S., and Knight, R. P.,** Intravenous hydrocortisone, corticotropin and the electroencephalogram, *AMA Arch. Neurol. Psychiatry,* 73, 338, 1955.
56. **Torda, C. and Wolff, H. G.,** Effects of various concentrations of adrenocorticotrophic hormone on electrical activity of brain and on sensitivity to convulsion-induced agents, *Am. J. Physiol.,* 168, 406, 1952.
57. **Dorfman, A., Apter, N. S., Smull, K., Bergenstal, D. M., and Richter, R. B.,** Status epilepticus coincident with use of pituitary adrenocorticotrophic hormone: report of 3 cases, *JAMA,* 146, 1, 1951.
58. **Gastaut, H.,** ACTH, adrenocortical hormones and juvenile epilepsy, *Epilepsia,* 2, 343, 1961.
59. **Stephen, E. H. M. and Noad, K. B.,** Status epilepticus occurring during cortisone therapy, *Med. J. Aust.,* 38, 334, 1951.
60. **Siggins, G. R. and Zieglgansberger, W.,** Morphine and opioid peptide reduce inhibitory synaptic potentials in hippocampal pyramidal cells in vitro without alteration of membrane potential, *Proc. Natl. Acad. Sci. U.S.A.,* 78(8), 5235, 1981.
61. **Alger, B. E. and Nicoll, R. A.,** Spontaneous inhibitory post-synaptic potentials, in hippocampus: mechanism for tonic inhibition, *Brain Res.,* 200, 195, 1980.
62. **Nicoll, R. A., Schenker, C., and Leeman, S. E.,** Substance P as a transmitter candidate, *Annu. Rev. Neurosci.,* 3, 227, 1980.
63. **Aldenhoff, J. B., Gruol, D. G., Rivier, J., Vale, W., and Siggins, G. R.,** Corticotropin releasing factor decreases post burst hyperpolarizations and excites hippocampal neurons, *Science,* 221 (4613), 875, 1983.
64. **Wallgren, H. and Barry, H., III,** *Actions of Alcohol,* Elsevier, Amsterdam, 1970.
65. **Wenger, J. R., Alwerud, E. C., Rivier, J., Vale, W., and Koob, G. F.,** Central administration of corticotropin-releasing factor antagonizes the ethanol-induced impairment of the righting reflex in rats, *Neurosci. Abstr.,* 8, 368, 1982.
66. **Swerdlow, N. R., Geyer, M. A., Vale, W., and Koob, G. F.,** Corticotropin-releasing factor potentiates acoustic startle in rats: blockade by chlordiazepoxide, *Psychopharmacology,* 88, 147, 1986.
67. **Van Thiel, D.,** Adrenal response to ethanol: a stress response, in *Stress and Alcohol Use,* Pohorecky, L. A. and Brick, J., Eds., Elsevier, Amsterdam, 1983, 23.
68. **Jenkins, R. M. and Page, McB.,** A typical case of alcohol-induced Cushingoid syndrome, *Br. Med. J.,* 282, 1117, 1981.
69. **Smals, A. G., Kloppenborg, P. W., Njo, K. T., Knoben, J. M., and Ruland, C. M.,** Alcohol-induced Cushingoid Syndrome, *J. R. Call. Physicians London,* 12, 36, 1977.
70. **Rees, L. H., Besser, G. M., Jeffcoate, W. J., Goldie, D. J., and Marks, V.,** Alcohol-induced pseudo-Cushing's syndrome, *Lancet,* 412, 726, 1977.
71. **Lambers, S. W. J., Klijn, J. G. M., and De Jong, J. C., Birkenhager,** Hormone secretion in alcohol-induced pseudo-Cushing's syndrome, *JAMA,* 242, 1640, 1978.
72. **Proto, G., Barberi, M., and Bertolissi, F.,** Pseudo-Cushing's syndrome: an example of alcohol-induced central disorder in corticotropin-releasing factor — ACTH release?, *Drug Alcohol Dependence,* 16, 111, 1985.
73. **Oxenkrug, G. F.,** Dexamethasone test in alcoholics, *Lancet,* ii, 795, 1978.
74. **Swartz, C. M. and Dunner, F. J.,** Dexamethasone suppression testing of alcoholics, *Arch. Gen. Psychiatry,* 39, 1309, 1982.
75. **Newsom, G. and Murray, N.,** Reversal of dexamethasone suppression test: nonsuppression in alcohol abusers, *Am. J. Psychiatry,* 140, 353, 1983.
76. **Kroll, P., Palmer, C., and Greden, J. F.,** The dexamethasone suppression testing in patients with alcoholism, *Biol. Psych.,* 18, 441, 1983.
77. **Willenbring, M. L., Morely, J. E., Niewoehner, C. B., Heilman, R. O., Carlson, C. H., and Shafter, R. B.,** Adrenocortical hyperactivity in newly administered alcoholics: prevalence, course and associated variables, *Psychoneuroendocrinology,* 914, 415, 1984.
78. **Porto, J. A., Monteiro, M. G., Laranjeira, R. R., Jorge, M. R., and Masur, J.,** Reversal of abnormal dexamethasone suppression test in alcoholics abstinent for 4 weeks, *Biol. Psychiatr.,* 20, 1156, 1985.
79. **Ellis, F. W.,** Effect of ethanol on plasma corticosterone levels, *J. Pharmacol. Exp. Ther.,* 153, 121, 1966.
80. **Kakihana, R. and Moore, J. A.,** Circadian rhythm of corticosterone in mice: the effect of chronic consumption of alcohol, *Psychopharmacology,* 46, 301, 1976.
81. **Tabakoff, B., Jaffe, R. C., and Ritzmann, R. F.,** Corticosterone concentrations in mice during ethanol drinking and withdrawal, *J. Pharm. Pharmacol.,* 30, 371, 1978.
82. **Knych, E. T. and Prohaska, J. R.,** Effect of chronic intoxication and naloxone on ethanol-induced increase in plasma corticosterone, *Life Sci.,* 28, 1987, 1981.

83. **Rivier, C., Bruhn, T., and Vale, W.,** Effect of ethanol on the hypothalamic-pituitary-adrenal axis in the rat: role of corticotropin-releasing factor (CRF), *J. Pharmacol. Exp. Ther.*, 229, 1, 127, 1984.
84. **Rivier, C., Rivier, J., and Vale, W.,** Inhibition of adrenocorticotrophic hormone secretion in the rat by immunoneutralization of corticotropin-releasing factor (CRF), *Science*, 218, 377, 1982.
85. **Ehlers, C. L. and Chaplin, R. I.,** Chronic ethanol exposure potentiates the locomotor activating effects of corticotropin-releasing factor in rats, *Regul. Pept.*, 19, 345, 1987.
86. **Dunn, A. J., Gren, E. J., and Isaacson, R. L.,** Intracerebral ACTH mediates novelty-induced grooming in the rat, *Science*, 203, 281, 1979.
87. **De Wied, D.,** Behavioral effects of neuropeptides related to ACTH, MSH, and b-LPH, *ACTH and Related Peptides, Structure, Regulation and Action*, Krieger, D. and Ganong, W., Eds., Ann. New York Acad. Sci., 1977, 263.
88. **Britton, K. T., Lee, G., Dana, R., Risch, S. C., and Koob, G. F.,** Activating and "anxiogenic" effects of corticotropin releasing factor are not inhibited by blockade of the pituitary-adrenal system with dexamethasone, *Life Sci.*, 39, 1281, 1986.
89. **Mendelson, J. H. and Stein, S.,** Serum cortisol levels in alcoholic and nonalcoholic subjects during experimentally induced ethanol intoxication, *Psychosom. Med.*, 28, 616, 1966.
90. **Ehlers, C. L., Chaplin, R. I., and Koob, G. F.,** Antidepressants modulate the CNS effects of corticotropin releasing factor in rats, *Med. Sci. Res.*, 15, 719, 1987.
91. **Ehlers, C. L., Lopez, A., and Chaplin, R. I.,** Chronic lithium and desmethylimipramine modulation of locomotor response to growth hormone releasing factor, *Psychopharmacol. Bull.*, 22, 991, 1986.
92. **Ehlers, C. L., Russo, P. V., Mandell, A. J., and Bloom, F. E.,** Architecture of rat nocturnal locomotion: a predictive descriptor of the effects of antidepressant and antimanic treatments, *Psychopharmacol. Bull.*, 19(4), 692, 1983.

Chapter 17

BEHAVIORAL EFFECTS OF CORTICOTROPIN-RELEASING FACTOR

George F. Koob and Karen T. Britton

TABLE OF CONTENTS

I. INTRODUCTION

The discovery of a central nervous system activating role for corticotropin-releasing factor (CRF) at the cellular, electroencephalographic, and behavior level has raised the possibility of a role for this peptide not only in mobilizing physiological function in the classical hypothalamic pituitary axis[1] but also the possibility of a role for CRF as a primary mediator of the behavioral state of stress and the behavioral responses to stress. The non-specific endocrine response originally described by Selye[2,3] may be paralleled by a nonspecific behavioral (emotional) response under control of this peptide neurotransmitter in the central nervous system.[4,5]

Stress is a hypothetical construct which is a ubiquitous concept in physiology, psychology, and medicine that often eludes precise definition. Selye conceptualized stress as "a nonspecific response to any demand upon the body (usually, but not always, noxious), or anything which causes an alteration of homeostatic processes".[6] A more modern version of this definition taken from Burchfield[7] has emphasized the concept of psychological homeostasis or the maintenance of a normal mood state at rest. All emotions then are changes from this state. An extension of this hypothesis is that without emotional changes accompanying a stressor, the stress response is minimal.[8] Thus, commonly accepted physiological "stressors" (heat, exercise, hunger) do not elicit a "stress response" when they are presented in a way that eliminates their psychological (emotional) effects (fear, conflict, uncertainty, frustration).

The importance of psychological variables in the stress response emphasizes the need for a neurobiological substrate to process the interaction of sensory stimuli with the ultimate psychoendocrine stress response. It is presumably through central nervous system pathways of the limbic system to the hypothalamus that the stress response is triggered by psychological factors. There is, however an alternative means by which behavioral or physiological responses to stress or anxiety might be mediated by neurohormones classically involved in stress, i.e., via a direct neurotropic action in the central nervous system.

The sequencing and subsequent synthesis of CRF[1] has provided a direct opportunity to study the central control of the HPA axis, and numerous studies support the hypothesis that CRF is the critical mediator of stress-induced changes in the pituitary adrenocortical response to stress. Although CRF was originally isolated and characterized on the basis of its ability to induce ACTH release, it is now known to exert a broad spectrum of action in the central nervous system. CRF appears to act within the brain as a mediator of certain autonomic nervous system, visceral, and behavioral actions independent of its action on the pituitary. Many of these effects are not abolished by hypophysectomy or adrenalectomy. These actions, in concert with the wide anatomical distribution of CRF in brain regions outside the areas hypothalamic-pituitary-adrenal axis,[9] provide a reasonable basis for hypothesizing that CRF may simultaneously activate and coordinate neuroendocrine, autonomic circulatory, metabolic, and behavioral responses to stressful stimuli.

II. CRF AND CENTRAL NERVOUS SYSTEM AROUSAL

CRF activates the central nervous system when administered directly into the central nervous system. Intracerebroventricular (i.c.v.) injection of CRF produces elevation of plasma epinephrine, norephinephrine, and glucose.[10,11] These effects are reproducible in hypophysectomized animals but are abolished by ganglionic blocking agents[12] suggesting an involvement of the sympathetic autonomic system. CRF injected i.c.v. also produces a profound dose-dependent activation of the electroencephalogram (EEG).[13] Doses of 0.015 to 0.15 nmol produced a long-lasting activation of EEG (see Chapter 16). At the cellular level, CRF produces increases in the firing frequencies of cells within the locus coeruleus[14]

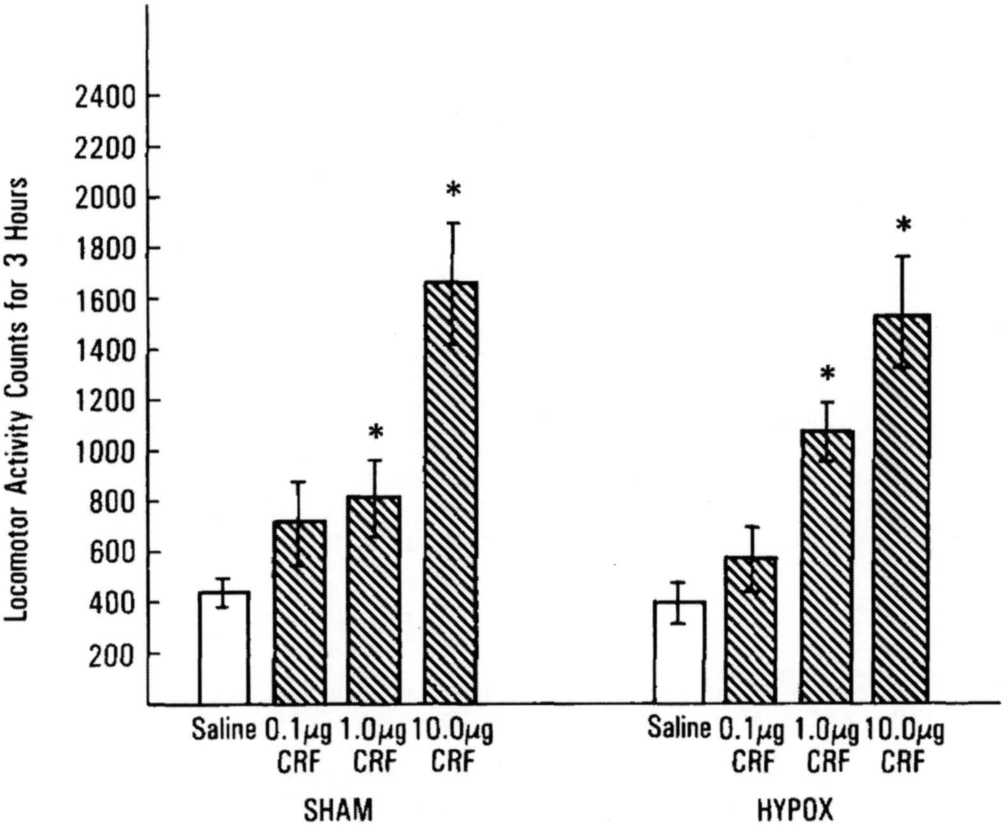

FIGURE 1. Effects of CRF injected i.c.v. on locomotor activity in photocell cages in hypophysectomized rats treated chronically with rat growth hormone and in sham operated animals. Results represent the total activity counts over 3 h (mean ± SEM) n = 9 in each group. *, Significantly different from saline injection. $p < 0.05$ main effects ANOVA. (From Eaves, M., Britton, K. T., Rivier, J., Vale, W., and Koob, G. F., *Peptides*, 6, 923, 1985. With permission.)

a system thought to be of importance in the mechanisms by which the brain is able to attend selectively to certain novel external events.

The autonomic and electrophysiological activation produced by central administration of CRF is paralleled by a dose-dependent locomotor activation.[15,16] These effects appear to be independent of direct mediation by the pituitary-adrenal system since they were observed in hypophysectomized and dexamethasone-treated rats (see Figure 1).[17-19] Given that this activation is not seen with systemic administration, these observations suggested that CRF exerted its effects within the central nervous system.

Although other peptides such as the endorphins and ACTH have been shown to produce increases in spontaneous behavioral activity, the locomotor activation caused by CRF is not antagonized by the opiate antagonist naloxone, or by low doses of the dopamine receptor antagonist alpha flupenthixol.[16] Nor is this activation reversed by 6-hydroxydopamine lesions of the region of the nucleus accumbens, lesions that reverse the locomotor-stimulated effects of indirect sympathomimetics.[20] ACTH, by itself, injected centrally failed to increase locomotor activity but instead produced an increase in grooming behavior as has been observed by others.[21]

To delineate further the site of action for this activating effect, experiments were conducted examining lateral ventricular and cisterna magna injections of CRF combined with obstruction of the cerebral aqueduct. Rats injected intracerebroventricularly with corticotro-

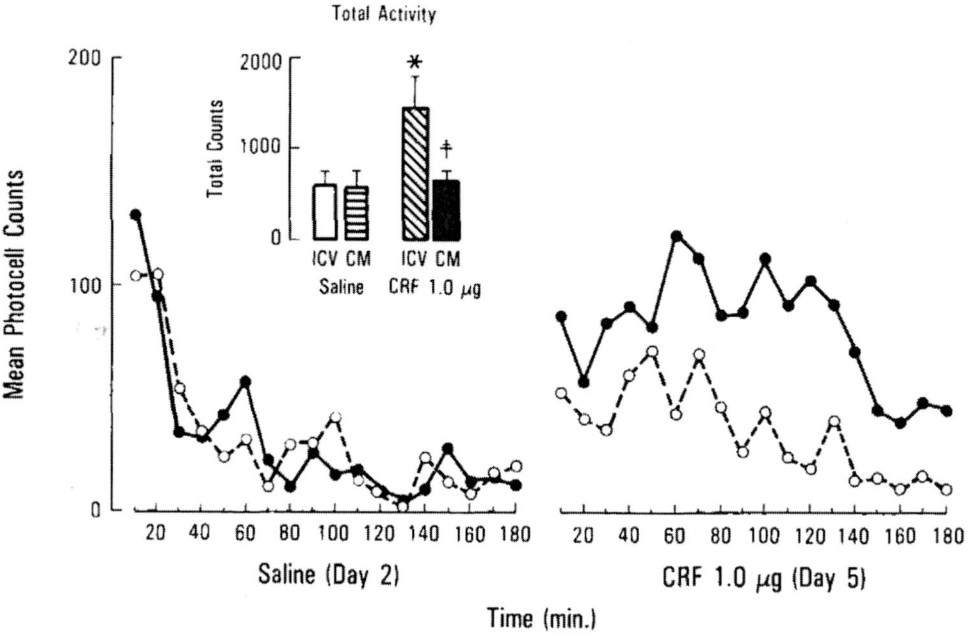

FIGURE 2. Effects on locomotor activity of CRF injected either into the lateral ventricle (i.c.v.) or the cisterna magna (CM). The values represent the mean photocell counts of the *eight rats* tested in each group, over 18 10-min intervals. Rats were tested on day 2 after saline treatment. On day 5, all the rats were injected with 10 μl of "Nivea Cream" into the cerebral aqueduct, and left undisturbed for 1 h before the 90 min habituation period. The right-hand figure show the locomotor activity after 1.0 μg CRF injected into the lateral ventricle and the cisterna magna in the plugged rats. The inset represents the mean total counts for the i.c.v. and CM groups under each condition. $^*p < 0.05$ in comparison to i.c.v. saline-treated rats, paired t-test; $^{\ddagger}p < 0.05$ in comparison to i.c.v. CRF-treated rats, Student's t-test. (From Tazi, A., Swerdlow, N. R., Le Moal, M., Rivier, J., Vale, W., and Koob, G. F., *Life Sci.*, 41, 41, 1987. With permission.)

pin-releasing factor at the level of the lateral ventricle or cisterna magna showed a dose-dependent increase in locomotor activity. The increase in locomotor activity from injections of CRF into the cisterna magna was blocked by a cold cream plug in the cerebral aqueduct.[22] An identical plug failed to block the increase in locomotor activity produced by CRF injected into the lateral ventricle (see Figure 2).

In a subsequent study, rats were surgically implanted with cannulae aimed at either the lateral ventricle, frontal cortex (FC), nucleus accumbens (NAC), amygdala centralis (AC), substantia innominata/lateral preoptic area (SI/LPO) or pedunculopontine nucleus (PPN). Injection of 0.5 μg rat CRF into discrete brain regions produced increases in locomotor activity that were inversely related to the distance of the injection site from the SI/LPO: activation was least intense when produced by CRF injection into the FC and PPN, and most intense when produced by CRF injection into the SI/LPO. Only SI/LPO injection sites yielded activation levels that exceeded those produced by i.c.v. CRF injection.[22] These results suggest some degree of neuroanatomical specificity for the neural substrates of CRF activation.

There is a significant amount of empirical and theoretical evidence which suggests that performance of learned behavior can vary with the level of activation or "arousal" of the animal. Correlations have been obtained between independent measures of "arousal" such as exploratory behavior, electroencephalogram and galvanic skin response and the speed of acquisition of a response.[23] Indeed, it has been hypothesized that both animals and man strive to maintain intermediate amounts of arousal potential, and it follows that optimal learning will correspond to these optimal levels of arousal. More recent treatments of this theoretical issue have centered on "arousal" as the intervening variable in the incentive-motivation required for all learning.[24] Others have argued that both stress and arousal can be considered as representing "state" phenomena reflecting diffuse central nervous system connections. In this conceptual framework activity in the pituitary adrenal system was considered a reliable, and sensitive measure of arousal and reflects the aroused "state" of the animal.

A central nervous system (CNS) component stress-arousal continuum is proposed that may be a CNS extension of the pituitary adrenal system. The hypothesis for the work to be reviewed in this chapter is that central nervous system corticotropin-releasing factor forms a basic arousal system that mediates behavioral responses to stressors. Thus the previously described homology of stress and arousal[25] may be reflected in a specific neurochemical substrate in the CNS.

Performance in a variety of test situations varies with the level of activation or "arousal" of the animal. Given that CRF has "arousal" properties it follows that a reflection of this activation would be the performance of an animal during acquisition of a learned response. Preliminary results provided support for this hypothesis. CRF administered prior to the daily acquisition of an appetitively motivated visual discrimination task significantly improved acquisition.[5] CRF-injected rats made significantly fewer errors and much more rapidly acquired the visual discrimination reaching 70 to 80% correct by days 3 and 4 whereas the control animals only reached this level by days 5 and 6.

III. CRF, STRESS, AND PERFORMANCE

Perhaps of more importance for the conceptualization of CRF as a peptide involved in the organism's behavioral response to stress were the experiments showing that CRF can potentiate the effects of exposure to a novel, presumably aversive, environment. When mildly food-deprived rats were exposed to an open field with food in the center, rats injected i.c.v. with CRF showed a marked decrease in exploratory and ingestive behavior and remained close to the corners of the field,[26] in contrast to an untreated animal which made increasingly frequent forays into the center of the field to consume food.[27] Similar results were observed with nondeprived rats in an open field without food (see Figure 3).[15] Rats tested in a novel open field following i.c.v. injection of doses of CRF (0.0015 to 0.15 nmol) showed responses that are consistent with an increased "emotionality" or increased sensitivity to the stressful aspects of the situation. Here rats showed *decreases* in locomotion and rearing (see Figure 3). In this open field test, a typical saline-injected rat rapidly circled the outer squares of the open field during the first 3 to 4 min of the 5-min test. During the last 1 to 2 min of the test these saline-injected animals then made some forays into the center of the open field, usually accompanied by rearing on their hind legs. Typically a rat injected with 0.15 nmol of CRF and placed 60 min later in the open field moved hesitantly to the outer squares and then either circled the open field remaining close to the floor or remained in one of the corners grooming or hesitantly moving forwards and backwards.[15] These behavioral changes were consistent with heightened "emotionality" or increased sensitivity to the presumably stressful aspects of the test.

These findings led to the suggestion that CRF might be related in some way to the expression of human anxiety or fear. Since it is not possible to identify animals with anxiety, CRF was tested in a series of tests that are sensitive to antianxiety (anxiolytic) compounds such as the benzodiazepines. In an operant conflict test, CRF produced a significant decrease in punished responding, an effect opposite to that observed with benzodiazepines, and this "anxiogenic" effect was reversed by concomitant treatment with a benzodiazepine (see Figure 4).[28] However, this increased sensitivity to aversive events was not paralleled by an increased sensitivity to pain. CRF also decreases food intake and muscimol-, norepine-

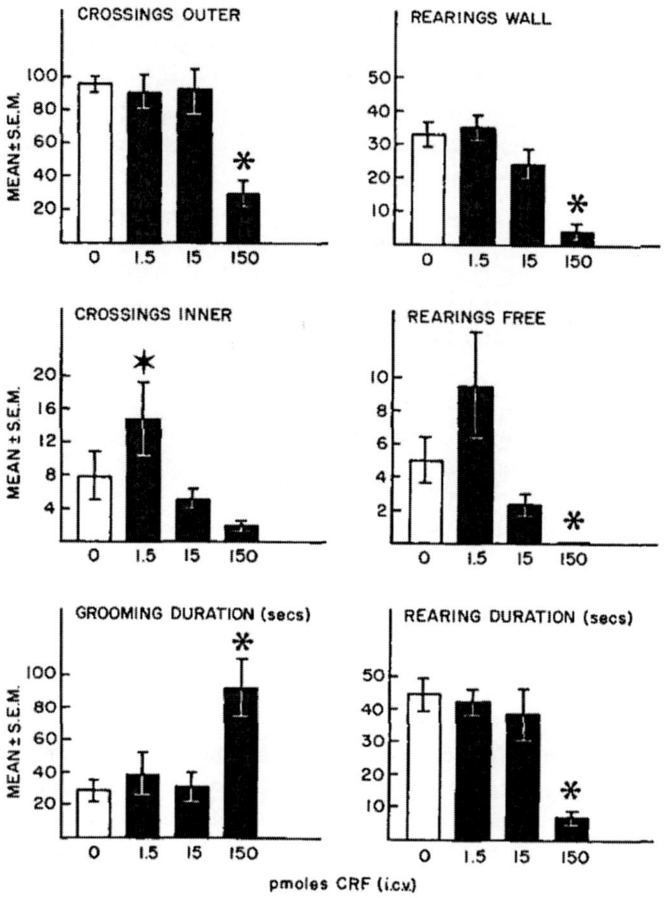

FIGURE 3. Effects of CRF injected i.c.v. on behavior of rats in an open field. Values represent mean ± SEM of the total number of observations during the 5 min test. 1.5, 15, and 150 pmol of CRF correspond to 0.01, 0.1, and 1.0 μg/rat. *Significantly different from saline injected rats, Newman-Keuls test $p < 0.05$ following ANOVA. (Adapted from Sutton, R. E., Koob, G. F., Le Moal, M., Rivier, J., and Vale, W., *Nature*, 297, 331, 1982.)

phrine-, dynorphin-, and insulin-induced feeding, effects attributed to a stress-related suppression of food intake.[29,30] These effects appear to be independent of the pituitary adrenal axis in that dexamethasone treatment also failed to alter the suppression in operant behavior produced by CRF (see Figure 5).[19]

Other more recent work has extended and elaborated this stress-like effect of CRF. CRF injected i.c.v. into mice allowed to explore a novel environment (multicompartment chamber) produced a decreased mean time per contact with novel stimuli.[31] Interestingly, this effect was reversed by naloxone, but it should be noted that previous studies have shown that naloxone can produce the opposite result on its own, i.e., increased exploratory behavior.[31]

In a recent study in our laboratory CRF injected i.c.v. in rats produced a further suppression during the conditioned stimulus (CS) presentation in a conditioned suppression task.[42] The observed decrease in the suppression rate is of significance because it shows that the decrease in performance was more dramatic during the CS period that signaled impending shock.

Another paradigm differentially sensitive to anxiogenic and anxiolytic compounds is the acoustic startle test. The acoustic startle reflex is an easily quantified muscular contraction

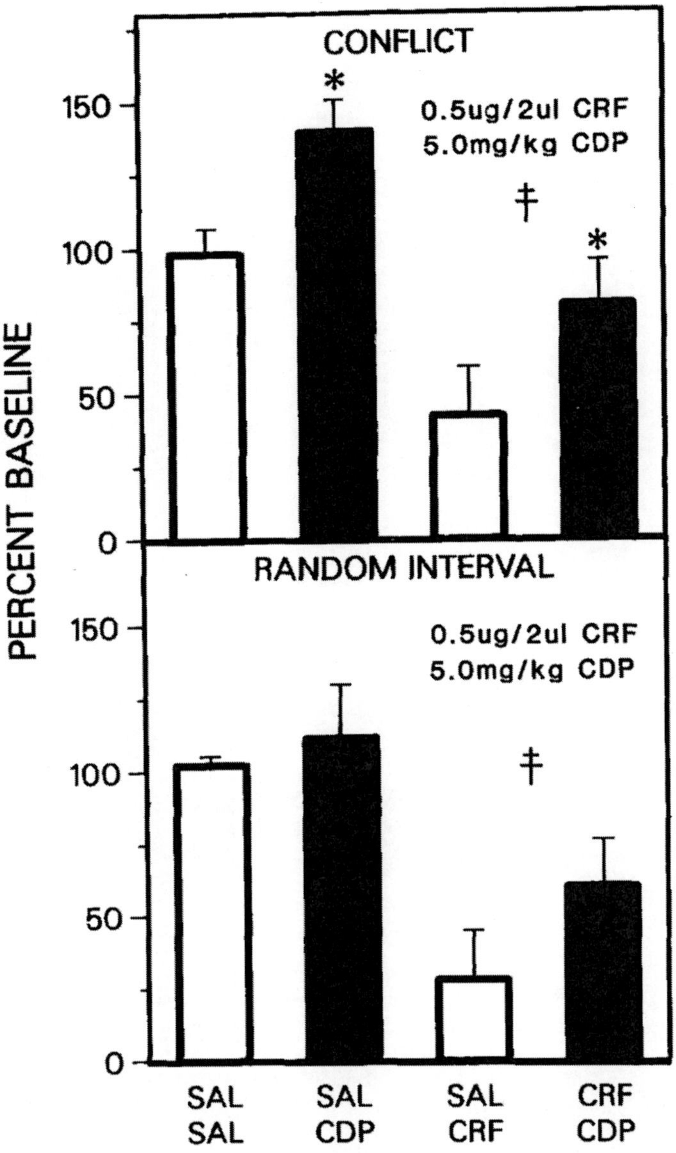

FIGURE 4. The interaction of 0.5 μg corticotropin-releasing hormone injected i.c.v. and of 5.0 mg/kg chlordiazepoxide (CDP) injected intraperitoneally on responding during the random interval and conflict components of an operant conflict test. Results are expressed as percent of baseline from the previous day. Asterisk denotes significant difference from saline, main effect CDP; double cross significant different from saline, main effect CRF, $p < 0.05$, two-way analysis of variance. (From Thatcher-Britton, K., Morgan, J., Rivier, J., Vale, W., and Koob, G. F., *Psychopharmacology*, 86, 170, 1985. With permission.)

in response to an intense acoustic stimulus. CRF (1 μg i.c.v.) significantly potentiated acoustic startle amplitude.[33] Pretreatment with the benzodiazepine chlordiazepoxide in doses that did not by themselves lower startle baseline attenuated this effect. These results also supported the hypothesis that CRF might act to potentiate behavioral responses normally expressed during states of enhanced fear or anxiety.

While the results discussed above with appetitively motivated tasks are consistent with

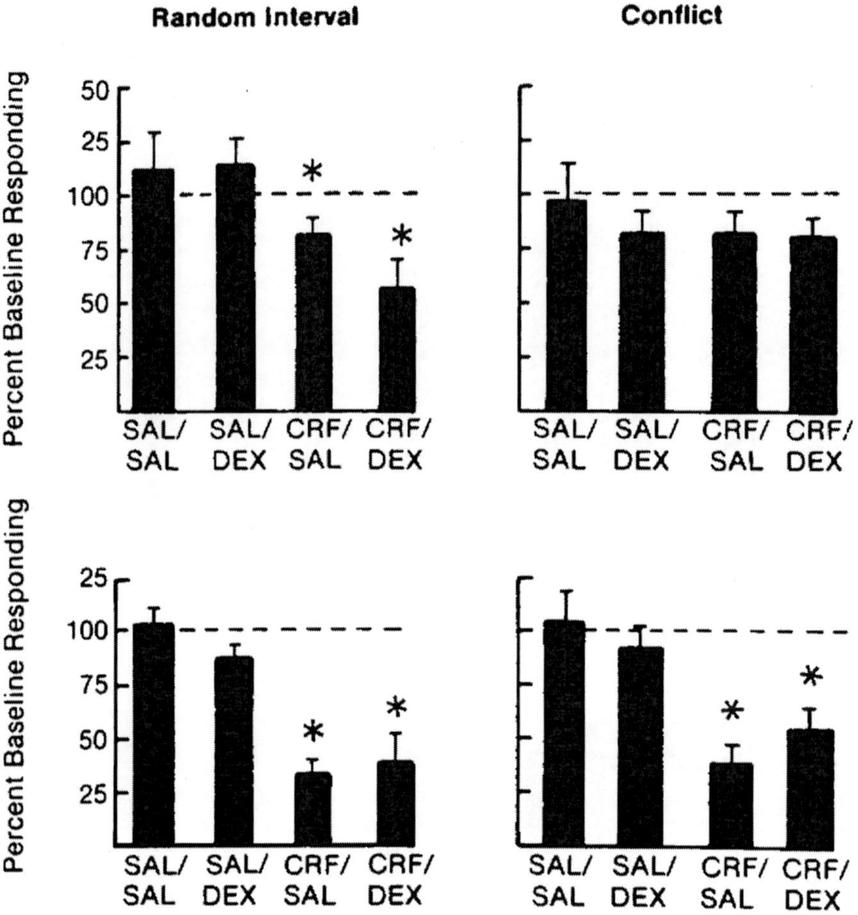

FIGURE 5. The interaction of 0.5 μg (top panel) and 1.0 μg (lower panel) of CRF and dexamethasone (100 μg/kg) on responding during the random interval and conflict components of an operant conflict test. Results are expressed at percent of baseline (mean ± SEM) from the previous day. *, Denotes significant difference from saline control, $p < 0.05$, analysis of variance. n = 5 to 6 per group. (From Britton, K. T., Lee, G., Dana, R., Risch, S. C., and Koob, G. F., *Life Sci.*, 39, 1281, 1986. With permission.)

an improvement in learning due to an arousing action, one would presumably observe an inverted U dose-response curve where lower doses should have weak effects and higher doses should interfere with performance. Alternatively, if the unconditioned stimulus is aversive such as in an avoidance situation, lower doses of CRF might become effective and moderate doses might disrupt performance. Preliminary results support this hypothesis. CRF at doses as low as 0.01 μg i.c.v. significantly improved performance (Koob et al., unpublished results).

Perhaps even more interesting, however, was the observation that the rats receiving 0.1 and 1.0 μg of CRF and trained in these avoidance tasks showed extreme "emotionality" for long periods after testing. When returned to their home cages these rats assumed boxing positions, vocalized, and fought with other rats for hours following testing. This unusual behavior was never observed in rats receiving these doses without avoidance experience.

This unusual interaction of CRF with the stress produced by aversive events suggested a systematic investigation using a stress-induced fighting paradigm might exaggerate this phenomenon. This fighting test has been previously shown to be sensitive to environmental parameters such as stress level[34] and to be modulated by peptides.[35] Rats exposed in weight-

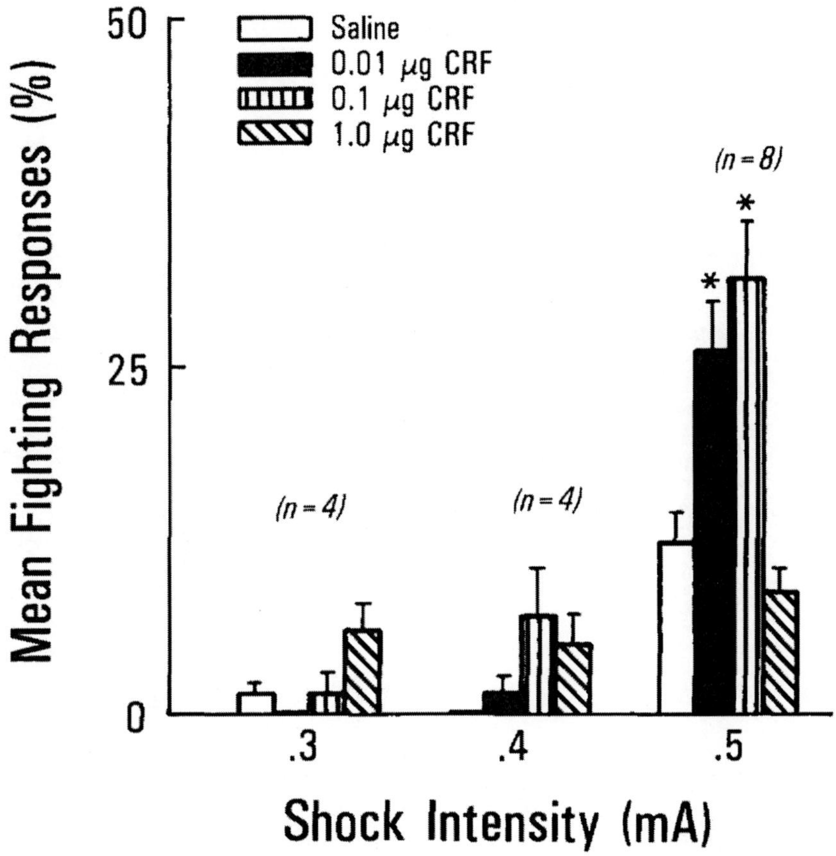

FIGURE 6. Effects of CRF on fighting responses observed under three different shock levels (0.3, 0.4, and 0.5 mA) in animals treated 30 min before testing with 0, 0.01, 0.1, or 1.0 μg CRF. CRF was injected i.c.v. to both animals fo the same pair. The data represent the mean over five consecutive daily 10-min sessions. The numbers between brackets represent the number of pairs tested. *$p < 0.05$ in comparison to saline-treated rats, Newman-Keuls test, following a significant analysis of variance. (From Tazi, A., Dantzer, R., Le Moal, M., Rivier, J., Vale, W., and Koob, G. F., *Regul. Pept.*, 18, 37, 1987. With permission.)

matched pairs to *mild* inescapable electric footshock during daily consecutive sessions show upright postures (boxing positions) or fighting (upright postures followed by physical contacts) that develop within a few sessions. These behaviors depend on the shock intensity,[35] and the frequency of fighting responses is increased gradually with the shock intensity, probably as a consequence of the gradual increase in physiological changes accompanying the increased stress.

CRF injected i.c.v. in the two rats of the same pair increased the frequency of these responses in a dose-dependent, and shock-dependent manner (see Figure 6).[38] At the lower shock intensities (0.3 and 0.4 mA), there was no fighting between the control rats. However, "boxing positions" were observed at these shock levels in control rats and "boxing" was increased significantly by 0.1 and 1.0 μg of CRF. CRF doses of 0.01 and 0.1 μg significantly facilitated fighting at the 0.5 mA shock level. CRF at the highest dose and highest shock level totally disrupted the behavior of the animals.[38]

IV. BEHAVIORAL EFFECTS OF A CRF ANTAGONIST

Until recently this work has succeeded only in characterizing a behavioral action of CRF administered exogenously in amounts, at first glance, of dubious physiological relevance (it

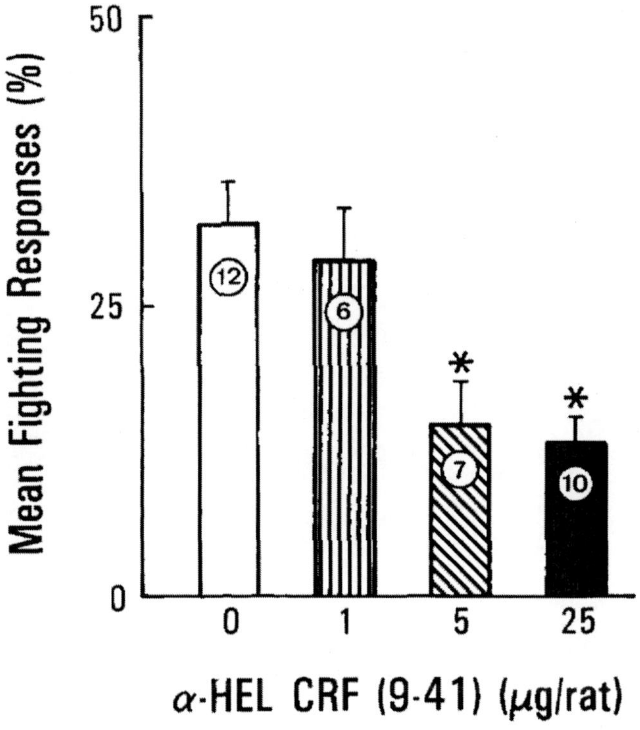

FIGURE 7. Effects of 0, 1, 5, and 25.0 μg/rat of α-HEL CRF(9-41) on fighting responses. Pairs of weight-matched rats were tested under 0.6 mA shock intensity. The CRF antagonist was injected i.c.v. 5 min before testing. No significant change was observed in the number of pairs showing the "boxing position" at any dose (data not shown). The data represent the means over three consecutive daily 10-min sessions. Numbers inside the columns represent the number of pairs in each group. *, $p < 0.05$ in comparison to saline-treated rats, Newman-Keuls test, following a significant analysis of variance. (From Tazi, A., Dantzer, R., Le Moal, M., Rivier, J., Vale, W., and Koob, G. F., *Regul. Pept.*, 18, 37, 1987. With permission.)

is estimated that the total hypothalamic content of CRF is 600 to 700 pg or 0.1 pmol).[36] Nevertheless, given that CRF is a large molecule (41 amino acids) and that the site of action for these effects remains largely unexplored, it is still possible that at some distant point from the ventricular injection, particularly at the lower doses, these effects represent the hypothesized exaggeration of normal function.

In addition recent work has demonstrated behavioral actions of intraventricularly administered CRF antagonist, α-helical 9-41. This CRF antagonist injected i.c.v. partially reversed the attenuation of feeding induced by stress.[37] In recent studies in our laboratory α-helical CRF injected i.c.v. attenuated stress-induced fighting.[38]

A slightly higher shock level (0.6 mA) produced a higher fighting frequency in control animals and this fighting was reversed by 5 and 25 μg/rat of αHEL CRF(9-41), a CRF antagonist, see Figure 7. These rats had not received exogenously administered CRF. This CRF antagonist has been shown to decrease both *in vitro* and *in vivo* baseline release of ACTH, as well as ether-induced ACTH secretion.[39] At the behavioral level, αHEL CRF(9-41) has been shown to block the activating and "anxiogenic" actions of CRF in the rat.[40] These observations, while pointing to the involvement of CRF receptors in such effects, did not specify a behavioral role for endogenous CRF released during stress. The present results, however, suggest that under certain conditions of high arousal and stress endogenous CRF systems may play a role in mediating behavioral responses.

In a recent study, the effects on the acquisition of conditioned suppression of the CRF antagonist α-helical CRF(9-41) were examined. Conditioned suppression is a task where an animal learns to suppress responding to a previously neutral stimulus that has been paired with a *mild* shock. The suppression that develops has been hypothesized to be a measure of the behavioral state of "fear" and is less sensitive to benzodiazepines than the conflict procedure.[41]

Food-deprived rats were trained to respond on a random interval-90 (RI-90) schedule for food reinforcement. After a stable level of responding had been attained, all the subjects were implanted with a guide cannula aimed at the lateral ventricle, and assigned to one of four experimental groups, equated for baseline response rates. After a postoperative recovery period, the effects of α-helical CRF on the acquisition of a conditioned emotional response were examined, by infusing the antagonist (0, 1, 5, and 25 μg/5 μl) into the lateral ventricle, 30 min before each test session. During these test sessions, four pairings of a conditioned stimulus (CS) (light) and a 0.5 s, 2.1 mA (biphasic, direct, constant current) footshock were presented, while the animals were responding for food reinforcement. The control animals showed a significant decrease in response rate during the presence of the CS in the nine test sessions, demonstrating their acquisition of a conditioned emotional response. However, this response suppression was significantly attenuated by α-helical CRF, at all of the doses studied.[42]

These results provide further evidence for the importance of CRF in the behavioral response to stress. However, further experiments are necessary to determine the nature of the behavioral impairment. One possibility is that α-helical CRF produces an "anxiolytic-like" effect, qualitatively similar to benzodiazepines. Alternatively, endogenous CRF may be specifically involved in the acquisition of conditioned fear. Particularly intriguing is that in experiments to date, the CRF antagonist has not produced a reliable release of punished responding in the conflict test. The possibility that CRF interacts on different behavioral mechanisms than benzodiazepines, perhaps more related to central noradrenergic function is an intriguing hypothesis for future study.

V. CONCLUSIONS

These results describing neuronal activation, sympathetic activation, EEG arousal, general behavioral activation, and stress-enhancing actions of CRF all suggest a possible role for CRF as a fundamental activating system. The functional significance of this system may have developed as a means for an organism to mobilize not only the pituitary adrenal system, but also the central nervous system in response to environmental challenge. Indeed, results in our laboratory suggest that treatment with CRF can improve performance and this is dose and task related. Aversive situations appear much more sensitive to exogenous CRF and preliminary results suggest that these aversive states may be sensitive to administration of a weak CRF antagonist. Clearly, a hypothetical central nervous system activation system definitively linked to the pituitary adrenal system, that can improve behavioral performance at low levels of output but attenuate behavioral performance at high levels of output would be of certain survival value. It is not difficult either to imagine a possible role for such a system in clinical disorders such as anxiety, affective disorders, and other psychopathology.

ACKNOWLEDGMENT

Preparation of this chapter was supported in part by NIH Grant AM26741 and a merit review grant from the Veterans Administration (KTB).

This is publication number 5331BCR from the Research Institute of Scripps Clinic, La Jolla, California.

REFERENCES

1. **Vale, W., Spiess, J., Rivier, C., and Rivier, J.,** Characterization of a 41 residue ovine hypothalamic peptide that stimulates the secretion of corticotropin and beta-endorphin, *Science,* 213, 1394, 1981.
2. **Selye, H.,** A syndrome produced by diverse noxious agents, *Nature,* 32, 138, 1936.
3. **Selye, H.,** The general adaptation syndrome and the diseases of adaptation, *J. Clin. Endocrinol.,* 6, 117, 1946.
4. **Koob, G. F.,** Stress corticotropin releasing factor and behavior, in *Perspectives on Behavioral Medicine: Neuroendocrine Control and Behavior,* Vol. 2, Williams, R. B., Ed., Academic Press, New York, 1985, 39.
5. **Koob, G. F. and Bloom, F. E.,** Corticotropin releasing factor and behavior, *Fed. Proc.,* 44, 259, 1985.
6. **Selye, H.,** *Selye's Guide to Stress Research,* D. van Nostrand Reinhold, New York, 1980, 5.
7. **Burchfield, S.,** The stress responses: a new perspective, *Psychosom. Med.,* 41, 661, 1979.
8. **Mason, J. W.,** A re-evaluation of the concept of "non-specificity" in stress specificity in stress theory, *J. Psychiatr. Res.,* 8, 323, 1971.
9. **Swanson, L. W., Sawchenko, P. E., Rivier, J., and Vale, W.,** The organization of ovine corticotropin releasing factor (CRF) immunoreactive cells and fibres in the rat brain: immunohistochemical study, *Neuroendocrinology,* 36, 165, 1983.
10. **Brown, M. R., Fisher, L. A., Rivier, J., Spiess, J., Rivier, C., and Vale, W.,** Corticotropin-releasing factor: Effects on the sympathetic nervous system and oxygen consumption, *Life Sci.,* 30, 207, 1982.
11. **Brown, M. R., Fisher, L. A., Spiess, J., Rivier, J., Rivier, C., and Vale, W.,** Comparison of the biologic actions of corticotropin releasing factor and sauvagine, *Regul. Pept.,* 4, 107, 1982.
12. **Brown, M. R. and Fisher, L. A.,** Central nervous system effects of corticotropin releasing factor in the dog, *Brain Res.,* 280, 75, 1983.
13. **Ehlers, C. L., Hendricksen, S. J., Wang, M., Rivier, J., Vale, W. W., and Bloom, F. E.,** Corticotropin releasing factor produces increases in brain excitability and convulsive seizures in the rat, *Brain Res.,* 278, 332, 1983.
14. **Valentino, R. J., Foote, S. L., and Aston-Jones, G.,** Corticotropin-releasing factor activates noradrenergic neurons of the locus coeruleus, *Brain Res.,* 270, 363, 1983.
15. **Sutton, R. E., Koob, G. F., Le Moal, M., Rivier, J., and Vale, W.,** Corticotropin releasing factor produces behavioral activation in rats, *Nature,* 297, 331, 1982.
16. **Koob, G. F., Swerdlow, N., Seelingson, M., Eaves, M., Sutton, R., Rivier, J., and Vale, W.,** CRF-induced locomotor activation is antagonized by alpha flupenthixol but no naloxone, *Neuroendocrinology,* 39, 459, 1984.
17. **Eaves, M., Britton, K. T., Rivier, J., Vale, W., and Koob, G. F.,** Effects of corticotropin releasing factor on locomotor activity in hypophysectomized rats, *Peptides,* 6, 923, 1985.
18. **Britton, D. R., Varela, M., Garcia, A., and Rivier, J.,** Dexamethasone suppresses pituitary-adrenal but not behavioral effects of centrally administered CRF, *Life Sci.,* 38, 211, 1986.
19. **Britton, K. T., Lee, G., Dana, R., Risch, S. C., and Koob, G. F.,** Activating and "anxiogenic" effects of CRF are not inhibited by blockade of the pituitary-adrenal system with dexamethasone, *Life Sci.,* 39, 1281, 1986.
20. **Swerdlow, N. R. and Koob, G. F.,** Separate neural substrates of the locomotor-activity properties of amphetamine, heroin, caffeine, and corticotropin releasing factor (CRF) in the rat, *Pharmacol. Biochem. Behav.,* 23, 303, 1985.
21. **Gispen, W. H., Weigant, V. M., Greven, H. H., and De Wied, D.,** The induction of excessive grooming in the rat by intraventricular application of peptides derived from ACTH: structure activity, *Life Sci.,* 17, 645, 1975.
22. **Tazi, A., Swerdlow, N. R., Le Moal, M., Rivier, J., Vale, W., and Koob, G. F.,** Behavioral activation of CRF: evidence for the involvement of the vertral forebrain, *Life Sci.,* 41, 41, 1987.
23. **Berlyne, D. E.,** *Conflict Arousal and Curiosity,* McGraw-Hill, New York, 1960.
24. **Killeen, P. R., Hanson, S. J., and Osbourne, S. R.,** Arousal: its genesis and manifestation as response rate, *Psychol. Rev.,* 85, 571, 1978.
25. **Hennessy, J. W. and Levine, S.,** A psychoendocrine hypothesis, in *Progress in Psychobiology and Physiological Psychology,* Sprague, J. M. and Epstein, A. N., Eds., Academic Press, New York, 1979, 8.

26. **Britton, D. R., Koob, G. F., Rivier, J., and Vale, W.,** Intraventricular corticotropin-releasing factor enhances behavioral effects of novelty, *Life Sci.*, 31, 363, 1982.
27. **Britton, D. R. and Britton, K. T.,** A sensitive open field measure of analytic drug activity, *Pharmacol. Biochem. Behav.*, 15, 577, 1981.
28. **Thatcher-Britton, K., Morgan, J., Rivier, J., Vale, W., and Koob, G. F.,** Chlordiazepoxide attenuates CRF-induced responses suppression in the conflict test, *Psychopharmacology*, 86, 170, 1985.
29. **Levine, A. S., Rogers, B., Kneip, J., Grace, M., and Morley, J. E.,** Effect of centrally administered corticotropin releasing factor (CRF) on multiple feeding paradigms, *Neuropharmacology*, 22, 337, 1983.
30. **Morley, J. E. and Levine, A. S.,** Corticotropin-releasing factor elicits naloxone sensitive stress-like alterations in exploratory behavior in mice, *Regul. Pept.*, 16, 83, 1986.
32. **Arnsten, A. F. T. and Segal, D. S.,** Naloxone alters locomotion and interaction with environmental stimuli, *Life Sci.*, 25, 1035, 1979.
33. **Swerdlow, N. R., Geyer, M. A., Vale, W. W., and Koob, G. F.,** Corticotropin releasing factor potentiates acoustic startle in rats: blockade by chlordiazepoxide, *Psychopharmacology*, 88, 142, 1986.
34. **Tazi, A., Dantzer, R., Mormede, R., and Le Moal, M.,** Effects of post-trial injection of beta-endorphin on shock-induced fighting are dependent on baseline of fighting, *Behav. Neural Biol.*, 43, 322, 1985.
35. **Tazi, A., Dantzer, R., Mormede, R., and LeMoal, M.,** Effects of post-trial administration of naloxone and beta-endorphin on shock-induced fighting in rats, *Behav. Neural Biol.*, 39, 192, 1983.
36. **Fischman, A. J. and Moldow, R. L.,** Extra-hypothalamic distribution of CRF-like immunoreactivity in the rat brain, *Peptides*, 3, 149, 1982.
37. **Krahn, D. D., Gosnell, B. A., Grace, M., and Levine, A. S.,** CRF antagonist partially reverses CRF- and stress-induced effects on feeding, *Brain Res. Bull.*, 17, 285, 1986.
38. **Tazi, A., Dantzer, R., Le Moal, M., Rivier, J., Vale, W., and Koob, G. F.,** Corticotropin releasing factor antagonist blocks stress-induced fighting in rats, *Regul. Pept.*, 18, 37, 1987.
39. **Rivier, J., Rivier, C., and Vale, W.,** Synthetic competitive antagonists of corticotropin-releasing factor: effect on ACTH secretion in the rat, *Science*, 224, 889, 1984.
40. **Thatcher-Britton, K., Lee, G., Vale, W., Rivier, J., and Koob, G. F.,** Corticotropin releasing factor antagonists blocks activating and "anxiogenic" actions of CRF in the rat, *Brain Res.*, 369, 303, 1986.
41. **Dantzer, R.,** Behavioral analysis of anxiolytic drug action, in *Experimental Approaches in Neuropharmacology*, Greenshaw, A. J. and Dourish, C. T., Eds., Humana Press, Clifton, NJ, 1987.
42. **Cole, B. J., Britton, K. T., and Koob, G. F.,** Central administration of alpha helical corticotropin releasing factor attenuates the acquisition of a conditioned emotional response, *Neurosci. Abstr.*, 13, 427, 1987.

Chapter 18

CORTICOTROPIN-RELEASING FACTOR AND INGESTIVE BEHAVIORS

John E. Morley and Allen S. Levine

TABLE OF CONTENTS

I. INTRODUCTION

Psychological and physical stress in animals and humans has been shown to alter feeding behavior. Both the type and the duration of the stressor appear to play a role in determining the effect on food intake. In laboratory animals, some stressors, e.g., mild tail pinching, will lead to overeating, while other stressors, e.g., immobilization stress or exposure to a novel environment, result in anorexia.[1] In a study conducted in humans, we found that 44% increased eating and 48% decreased eating when stressed.[2] Pathological stress overeating can also be precipitated by stress, as seen in bulimia, in which the eating binge tends to be precipitated by a traumatic event occurring during a period of voluntary dieting.[3] Anorexia nervosa may be considered a pathological form of stress-induced undereating.

Much evidence has accumulated suggesting that, at least in animals, stress-induced eating involves activation of the endogenous opioid system.[4] The opioid antagonist, naloxone, inhibits tail pinch-induced feeding[5,6] as it does in the defeated mouse model of stress-induced eating.[7] The pattern of eating observed in the defeated mouse model is similar to that observed after morphine injection.[7] Dynorphin levels in the rat brain are altered during tail-pinch[8] and tail-pinch has been shown to produce naloxone-reversible analgesia when writhing is used as the nociceptive stimulus.[9]

In this chapter, we will develop the thesis that stress-induced anorexia is mediated through corticotropin-releasing factor (CRF).

II. CRF AND FEEDING

In 1982, Britton et al.[10] and Morley and Levine[11] simultaneously reported that CRF, when administered centrally, is a potent anorectic agent (Figure 1). Increased grooming is recognized to occur under certain stressful situations in the wild and has been considered a displacement behavior. Not only did CRF decrease feeding, but it also increased grooming.[11] As CRF releases both ACTH (which increases grooming) and beta-endorphin (which decreases feeding when administered peripherally) from the pituitary, we examined the effect of CRF on these behaviors in hypophysectomized rats. CRF continued to decrease feeding and increase grooming in hypophysectomized animals, suggesting that its effect on these behaviors was independent of its ability to release ACTH and/or beta-endorphin. CRF also decreased fluid ingestion following an 18-h water deprivation period, suggesting relative nonspecificity of CRF on ingestive behaviors.

In a second study, we investigated the ability of CRF to modulate feeding induced by a number of pharmacological agents.[12] Norepinephrine increases feeding by activating an alpha-2 receptor mechanism,[13] dynorphin increases feeding by activating kappa opioid receptors in the central nervous system,[14] and muscimol induces feeding by acting as a gamma-amino, butyric agonist.[15] CRF decreased the ability of these agents to stimulate feeding. Insulin increases feeding secondary to the hypoglycemia it produces. It also causes activation of the hypothalamic-pituitary-adrenal axis. Insulin-induced feeding is attenuated by CRF, but it requires a higher dose than that necessary to attenuate the other pharmacological inducers of feeding. This is not surprising in view of the fact that insulin would be expected to have increased endogenous CRF. The ability of CRF to suppress feeding induced by a variety of pharmacological manipulations supports the concept that CRF is a potent suppressor of ingestive behavior.

Sauvagine is a 40 amino acid peptide isolated from the skin of the frog, ***Phyllomedusa sauvagei,*** which has structural similarities to CRF and releases ACTH from the pituitary. Like CRF, sauvagine inhibits spontaneous and starvation-induced feeding, as well as feeding produced by the administration of ethylketacyclazocine.[16] Sauvagine produced a larger and longer-lasting suppressive effect than did CRF. Similar results have been reported by Britton

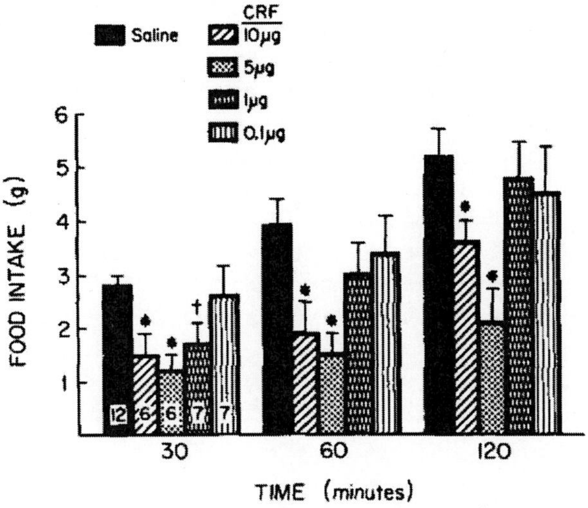

FIGURE 1. The effect of CRF on food intake following a 30-h deprivation period. *$p < 0.01$, **$p < 0.05$. (From Morley, J. E. and Levine, A. S., *Life Sci.*, 31, 1459, 1982. With permission.)

et al.[17] Both sauvagine and CRF produced a conditioned aversion when paired with a novel saccharin taste.[16]

III. CRF, THE ADRENALS, AND FEEDING

CRF produces the release of corticosterone from the adrenal cortex secondary to the release of ACTH from the pituitary and also activates the adrenal medulla to release epinephrine. Epinephrine has been demonstrated to inhibit feeding when administered peripherally.[18] Corticosteroids can increase or decrease feeding, depending on the dose and circumstances of administration.[19,20] The presence of corticosterone is also essential to allow norepinephrine to increase feeding.[21]

Gosnell et al.[22] found that adrenalectomy markedly attenuated the suppressive effect of CRF on food intake. As hypophysectomy failed to attenuate the CRF effect on feeding, this suggested that either some circulating corticosterone was necessary for the expression of the CRF effect on feeding, or, alternatively, that the adrenal medullary secretions were necessary for the full expression of the CRF effect on feeding. Corticosterone replacement in adrenalectomized animals failed to restore the CRF effect. Adrenal demedullation, on the other hand, markedly attenuated the effect of central CRF on food ingestion. This suggests that epinephrine or other adrenal medullary secretions plays an important role in mediating or modulating the effects of CRF on feeding.

Vagotomy attenuates the effect of norepinephrine on feeding[23] and has also been shown to alter the effects of a number of peripherally administered gastrointestinal hormones (namely, cholecystokinin, glucagon, somatostatin, and thyrotropin-releasing hormone) on feeding.[24-27] Slowing of gastric emptying has been implicated as a mechanism by which satiety is produced.[28] Sauvagine slows gastric emptying.[29] Vagotomy, however, does not alter the ability of CRF to inhibit feeding, suggesting that the effects of CRF on gastric emptying play a relatively minor role in the overall effects of CRF on feeding.[22]

IV. LOCALIZATION OF THE EFFECT OF CRF ON FEEDING

The paraventricular nucleus (PVN) of the hypothalamus represents an area within which many of the putative neurotransmitter modulators of feeding have been shown to exert their effects. Lesions of the PVN lead to hyperphagia,[23] suggesting that the PVN has an inhibitory effect on feeding. In view of the presence of both CRF cell bodies and CRF receptors in the PVN and the potent inhibitory action of CRF on feeding, it appeared that the PVN was a reasonable area in which CRF may produce its action.

Krahn et al.[30] administered CRF by microinjection into five different sites within the central nervous system. These included the PVN, ventromedial hypothalamus, striatum, lateral hypothalamus, and globus pallidus. Only within the PVN was CRF demonstrated to decrease feeding. CRF administered into the PVN also blocked norepinephrine-induced feeding. These data are compatible with the concept that the PVN is the site within the central nervous system at which CRF produces its inhibitory effect on feeding.

V. DOES CRF PLAY A PHYSIOLOGICAL ROLE IN THE REGULATION OF FEEDING?

Rivier et al.[31,32] synthesized alpha-helical CRF(9-41), a peptide which competitively antagonized the ACTH-releasing effects of CRF and partially reversed the stress-induced inhibition of luteinizing hormone release in the rat. Krahn et al.[33] showed that alpha-helical CRF(9-41) would block the inhibitory effect of CRF on feeding. This effect appeared relatively specific as it did not alter the inhibitory effect of calcitonin on feeding. They then produced anorexia by placing 24-h food-deprived rats in a restraint cage. Using this paradigm, they found that 50 μg of alpha-helical CRF(9-41) reversed the stress-induced anorexia. This finding supports a role for CRF in the physiological regulation of feeding.

Shibaski et al.[34] confirmed that alpha-helical CRF(9-41) reversed the anorexic effect of restraint stress. They also found that ODT8-SS, a somatostatin analog, reversed the restraint stress anorexia and prevented CRF anorexia and the CRF increase in epinephrine. Previous studies on somatostatin have reported that it may have either inhibitory[35] or stimulatory[36] effects on feeding. The inhibitory effect of somatostatin appears to be mediated through the lateral hypothalamus and, perhaps, to involve a vagally dependent mechanism.[37] The studies by Shibaski et al.[34] would suggest that the enhancing effects of somatostatin on feeding involve the inhibition of the CRF stimulation of the adrenomedullary discharge, perhaps within the PVN.

Overall, the two studies cited above would suggest that CRF plays a physiological role in the production of stress-induced anorexia. Further evidence in support of this concept comes from the demonstration that food deprivation has been shown to alter CRF levels within the PVN.[38]

VI. INTERACTIONS OF CRF AND OTHER NEUROTRANSMITTERS IN THE REGULATION OF FEEDING

In 1980, Morley[39] suggested that the regulation of feeding was dependent on the interaction of a number of neurotransmitters within the central nervous system. It was suggested that these neurotransmitters were in a delicate balance, with small changes in any one of them resulting in an increase or decrease in feeding until the system came back into balance. In addition, it seemed clear that a number of feeding systems existed, allowing a fail-safe mechanism which would prevent death by starvation should one of these neurotransmitters fail. Today, at least two neuropeptide feeding systems are recognized. These are the Neuropeptide Y system which predominantly is involved in carbohydrate intake[40] and the opioid system which is responsible for the regulation of the intake of fatty or highly palatable foods.[41] It is now clear that not only can specific neurotransmitters exert their own specific effects on feeding, but that anatomy lends specificity to the effects of certain neurotransmitters. For example, GABA enhances feeding when administered into the ventromedial hypothalamus, while inhibiting it when injected into the lateral hypothalamus.[42]

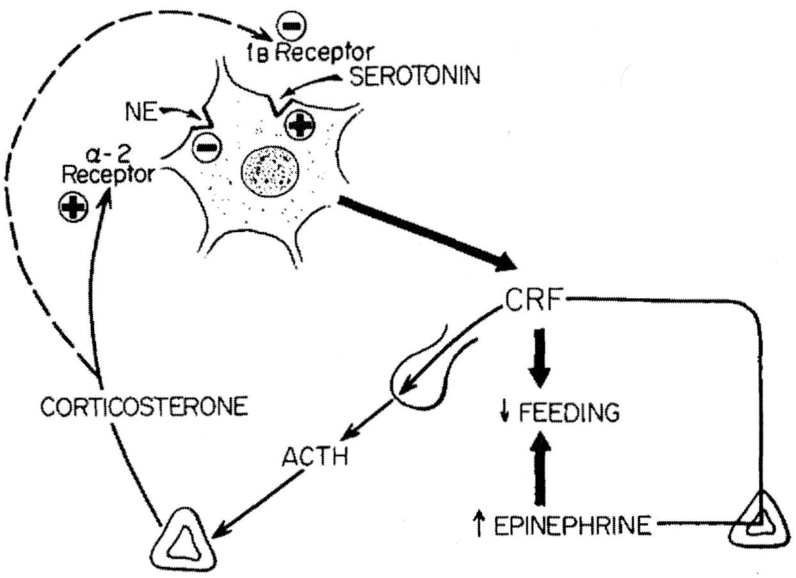

FIGURE 2. Interaction of CRF with norepinephrine (NE) and serotonin in the regulation of feeding. Corticosterone provides a classical negative feedback system, enhancing the inhibitory effect of norepinephrine and attenuating the stimulatory effect of serotonin on CRF secretion.

Previous studies have suggested that norepinephrine enhances feeding by activating alpha-2 receptors within the PVN.[43] This effect is dependent on the presence of circulating corticosterone.[21] Adrenalectomy results in a region-specific decline in alpha-2 adrenergic receptor binding within the PVN.[44] This effect appears to involve the Type I or corticosterone receptors (rather than the Type II or glucocorticoid receptors) and to produce activational effects, such as the increased feeding and exploratory behavior that occurs at the initiation of the waking period. This norepinephrine effect on feeding, which requires the permissive action of corticosterone, appears to be mediated through the inhibition of release of a feeding inhibitor.

The evidence that CRF is this inhibitor can be summarized as follows: (1) CRF inhibits norepinephrine-stimulated feeding; (2) adrenalectomy increases CRF within the PVN (45) and attenuates norepinephrine-induced feeding; and (3) norepinephrine inhibits CRF release from the hypothalamus.[46] In addition, brain norepinephrine-depleting lesions produce the same selectively enhanced behavioral responses to novelty as does centrally administered CRF;[47] and CRF reduced norepinephrine turnover within the PVN.[48]

Stressful procedures, such as restraint stress, result in an increase in serotonin turnover.[49,50] Serotonin inhibits norepinephrine-induced feeding within the PVN.[51] Serotonin has been shown to promote the release of CRF from the hypothalamus.[52] Glucocorticoids reduce serotonin-1 receptor density,[53] providing a negative feedback control system.

Figure 2 represents a schematic representation of the interactions of other neurotransmitters within CRF. This schema suggests that CRF inhibits feeding by both central mechanisms and the peripheral release of epinephrine. CRF is depicted as the mediator of norepinephrine and serotonin effects on feeding.

VII. CRF AND THE PATHOPHYSIOLOGY OF FEEDING DISORDERS

Activation of the hypothalamic-pituitary-adrenal axis is well recognized to occur in anorexia nervosa.[54] In view of the potent anorectic effects of CRF, it seemed reasonable to postulate that it plays a role in the pathophysiology of anorexia nervosa. Confirmation of this hypothesis has come from the recent reports from Japan[55] and the NIMH[56] that CRF levels are elevated in the cerebrospinal fluid of patients with anorexia nervosa.

Anorexia and weight loss occur in approximately two thirds of patients with major depression. Similarly, these patients have a hyperactive hypothalamic-pituitary-adrenal axis. A number of groups[57-59] have shown that depressed patients also have elevated CRF levels in the cerebrospinal fluid. CRF may, therefore, play a role in the anorexia and weight loss associated with depression. Besides the effects on appetite, animal studies have suggested that CRF also may produce psychomotor agitation, decreased sex drive, and alterations in cognition, all of which are included in the DSM-III diagnosis of major depressive disorders.

VIII. CONCLUSION

CRF is a potent anorectic agent in animals. It produces its effect within the PVN and appears to mediate the effects of serotonin and norepinephrine on feeding. It is possible that CRF plays a role in the pathophysiology of anorexia nervosa and it may play a role in the anorexia and weight loss associated with major depressive disorders.

REFERENCES

1. **Morley, J. E., Levine, A. S., and Willenbring, M. L.,** Stress-induced feeding disorders, in *Psychopharmacology of Eating Disorders: Theoretical and Clinical Advances,* Blundell, J. E. and Carruba, M. O., Eds., Raven Press, New York, 1986, 71.
2. **Willenbring, M. L., Levine, A. S., and Morley, J. E.,** Stress-induced eating: a pilot study, *Int. J. Eating Disorders* 5, 855, 1986.
3. **Pyle, R. L., Mitchell, J. E., and Eckert, E. D.,** Bulimia: a report of 34 cases, *J. Clin. Psychiatry,* 42, 60, 1981.
4. **Morley, J. E., Levine, A. S., and Rowland, N.,** Stress-induced eating, *Life Sci.,* 32, 2169, 1983.
5. **Morley, J. E. and Levine, A. S.,** Stress-induced eating is mediated through endogenous opiates, *Science,* 209, 1259, 1980.
6. **Lowy, M. T., Maickel, R. P., and Yim, G. K. W.,** Naloxone reduction of stress-related feeding, *Life Sci.,* 26, 2113, 1980.
7. **Teskey, G. C., Kavaliers, M., and Hirst, M.,** Social conflict activates opioid analgesic and ingestive behaviors in male mice, *Life Sci.,* 35, 303, 1984.
8. **Morley, J. E., Elson, M. K., Levine, A. S., and Shafer, R. B.,** The effects of stress on central nervous system concentrations of the opioid peptide, dynorphin, *Peptides,* 3, 901, 1982.
9. **Levine, A. S. and Morley, J. E.,** Tail pinch induced eating: is it the tail or the pinch?, *Physiol. Behav.,* 28, 565, 1982.
10. **Britton, D. R., Koob, G., Rivier, J., and Vale, W.,** Intraventricular corticotropin-releasing factor enhances behavioral effects of novelty, *Life Sci.,* 31, 363, 1982.
11. **Morley, J. E. and Levine, A. S.,** Corticotropin-releasing factor, grooming and ingestive behavior, *Life Sci.,* 31, 1459, 1982.
12. **Levine, A. S., Rogers, B., Kneip, J., Grace, M., and Morley, J. E.,** Effect of centrally administered corticotropin-releasing factor (CRF) on multiple feeding paradigms, *Neuropharmacology,* 22, 337, 1983.
13. **Goldman, C. K., Marino, L., and Leibowitz, S. F.,** Postsynaptic alpha-2 noradrenergic receptors mediate feeding induced by paraventricular nucleus injection of norepinephrine and clonidine, *Eur. J. Pharmacol.,* 115, 11, 1985.
14. **Morley, J. E. and Levine, A. S.,** Involvement of dynorphin and the kappa opioid receptor in feeding, *Peptides,* 4, 797, 1983.
15. **Morley, J. E., Levine, A. S., and Kneip, J.,** Muscimol-induced feeding: a model to study the hypothalamic regulation of appetite, *Life Sci.,* 29, 1213, 1981.
16. **Gosnell, B. A., Morley, J. E., and Levine, A. S.,** A comparison of the effects of corticotropin-releasing factor and sauvagine on food intake, *Pharmacol. Biochem. Behav.,* 19, 771, 1983.
17. **Britton, D. R., Hoffman, D. K., Lederis, K., and Rivier, J.,** A comparison of the behavioral effects of CRF, sauvagine and urotensin, *Brain Res.,* 304, 210, 1984.

18. **Hinton, V., Esquierra, M., Farhoody, N., Granger, J., and Geary, N.,** Epinephrine inhibits feeding nonspecifically in the rat, *Physiol. Behav.*, 40, 109, 1987.
19. **Hollifield, G.,** Glucocorticoid-induced obesity: a model and a challenge, *Am. J. Clin. Nutr.*, 21, 1471, 1968.
20. **Lowy, M. T. and Yim, G. K. W.,** Selective reduction by dexamethasone of stress-related hyperphagias, *Life Sci.*, 27, 2553, 1980.
21. **Bhakthavatsalam, P. and Leibowitz, S. F.,** Alpha-2-noradrenergic feeding rhythm in the paraventricular nucleus: relation to corticosterone, *Am. J. Physiol.*, 250, R83, 1986.
22. **Gosnell, B. A., Morley, J. E., and Levine, A. S.,** Adrenal modulation of the inhibitory effect of corticotropin releasing factor on feeding, *Peptides*, 4, 807, 1983.
23. **Leibowitz, S. F.,** Neurochemical systems of the hypothalamus: control of feeding and drinking behavior and water electrolyte excretion, in *Handbook of the Hypothalamus*, Vol. 3, Morgane, P. J. and Panksepp, J., Eds., Raven Press, New York, 1980, 299.
24. **Smith, G. P., Jerome, C., and Norgren, R.,** Afferent axons in abdominal vagus mediate the satiety effect of cholecystokinin in rats, *Am. J. Physiol.*, 249, R635, 1985.
25. **Geary, N. and Smith, G. P.,** Selective hepatic vagotomy blocks pancreatic glucagon's satiety effect, *Physiol. Behav.*, 31, 391, 1983.
26. **Morley, J. E., Levine, A. S., Kneip, J., and Grace, M.,** The effect of vagotomy on the satiety effects of neuropeptides and naloxone, *Life Sci.*, 30, 1943, 1982.
27. **Levine, A. S. and Morley, J. E.,** Peripherally administered somatostatin reduces feeding by a vagal mediated mechanism, *Pharmacol. Biochem. Behav.*, 16, 897, 1982.
28. **McHugh, P. R.,** The control of gastric emptying, *J. Autonom. Nerv. Syst.*, 9, 221, 1983.
29. **Broccardo, M., Improta, G., and Melchiorri, P.,** Effect of sauvagine on gastric emptying in conscious rats, *Eur. J. Pharmacol.*, 85, 111, 1982.
30. **Krahn, D. D., Gosnell, B. A., Levine, A. S., and Morley, J. E.,** Localization of the effects of corticotropin-releasing factor on feeding, *Proc. Soc. Neurosci. Abstr.*, 10, 302, 1984.
31. **Rivier, J., Rivier, C., and Vale, W.,** Synthetic competitive antagonists of corticotropin releasing factor: effect on ACTH secretion in the rat, *Science*, 224, 889, 1984.
32. **Rivier, C., Rivier, J., and Vale, W.,** Stress-induced inhibition of reproductive function: role of endogenous corticotropin-releasing factor, *Science*, 231, 607, 1986.
33. **Krahn, D. D., Gosnell, B. A., Grace, M., and Levine, A. S.,** CRF antagonist partially reverses CRF- and stress-induced effects on feeding, *Brain Res. Bull.*, 17, 285, 1986.
34. **Shibaski, T., Masuda, A., Imaki, T., and Yamauchi, N.,** CRF and somatostatin in stress-induced suppression of food intake in rats, *Endocrinology*, 120, 85A, 1987.
35. **Vijayan, E. and McCann, S. M.,** Suppression of feeding and drinking activity in rats following intraventricular injection of thyrotropin-releasing hormone (TRH), *Endocrinology*, 100, 1727, 1977.
36. **Aponte, G., Leung, P., Gross, D., and Yamada, T.,** Effect of somatostatin on food intake in rats, *Life Sci.*, 35, 741, 1984.
37. **Lin, M. T., Chen, J. J., and Ho, L. T.,** Hypothalamic involvement in the hyprglycemia and satiety actions of somatostatin in rats, *Neuroendocrinology*, 45, 62, 1987.
38. **Suemaru, S., Hashimoto, K., Hatton, J., Inoue, H., Kageyama, J., and Ota, Z.,** Starvation-induced changes in rat brain corticotropin-releasing factor (CRF) and pituitary adrenal cortical responses, *Life Sci.*, 39, 1161, 1986.
39. **Morley, J. E.,** The neuroendocrine control of appetite: the role of the endogenous opiates, cholecystokinin, TRH, gamma-amino butyric acid and the diazepam receptor, *Life Sci.*, 27, 355, 1980.
40. **Morley, J. E., Levine, A. S., Gosnell, B. A., Kneip, J., and Grace, M.,** Studies on the effect of neuropeptide Y on ingestive behaviors in the rat, *Am. J. Physiol.*, 252, R599, 1987.
41. **Morley, J. E., Levine, A. S., Yim, G. K., and Lowy, M. T.,** Opioid modulation of appetite, *Neurosci. Biobehav. Rev.*, 7, 281, 1983.
42. **Kelly, J., Alherd, G. F., Newberg, A., and Grossman, S. P.,** GABA stimulation of blockade of the hypothalamus and midbrain: effects on feeding and locomotor activity, *Pharmacol. Biochem. Behav.*, 7, 523, 1977.
43. **Leibowitz, S. F.,** Brain monoamines and peptides: role in the control of eating behavior, *Fed. Proc.*, 45, 14, 1986.
44. **Jhanwar-Uniyal, M. and Leibowitz, S. W.,** Impact of circulating corticosterone on alpha-1- and alpha-2-noradrenergic receptors in discrete brain areas, *Brain Res.*, 368, 404, 1986.
45. **Sawchenko, P. E., Swanson, L. W, and Vale, W. W.,** Co-expression of corticotropin-releasing factor and vasopressin immunoreactivity in parvocellular neurosecretory neurons of the adrenalectomized rat, *Proc. Natl. Acad. Sci. U.S.A.*, 81, 1883, 1984.
46. **Rivier, C. L. and Plotsky, P. M.,** Mediation by corticotropin releasing factor (CRF) of adenohypophyseal hormone secretion, *Annu. Rev. Physiol.*, 48, 475, 1986.

47. **Britton, D. R., Ksir, C., Britton, K. T., Young, D., and Koob, G. F.,** Brain norepinephrine depleting lesions selectively enhance behavioral responsiveness to novelty, *Physiol. Behav.*, 33, 473, 1984.
48. **Andersson, K., Agnati, L. F., Fuxe, K., Eneroth, P., Harfstand, A., and Benefenati, F.,** Corticotropin-releasing factor increases noradrenaline turnover in the median eminence and reduces norepinephrine turnover in the paraventricular region of the hypophysectomized male rat, *Acta Physiol. Scand.*, 120, 621, 1984.
49. **Thierry, A. M., Fedkete, M., and Glowinski, J.,** Effects of stress on the metabolism of noradrenaline, dopamine, and serotonin (5-HT) in the central nervous system of the rat. II. Modifications of serotonin metabolism, *Eur. J. Pharmacol.*, 4, 384, 1968.
50. **Curzon, G., Joseph, M. H., and Knott, P. J.,** Effects of immobilization and food deprivation on rat brain tryptophan metabolism, *J. Neurochem.*, 19, 1967, 1972.
51. **Weiss, G. F., Papadakos, P., Knudson, K., and Leibowitz, S. F.,** Medial hypothalamic serotonin: effects on deprivation and norepinephrine-induced eating, *Pharmacol. Biochem. Behav.*, 25, 1223, 1986.
52. **Buckingham, J. C.,** Corticotropin-releasing factor, *Pharmacol. Rev.*, 31, 253, 1980.
53. **McEwen, B. S.,** Glucocorticoid-biogenic amine interactions in relation to mood and behavior, *Biochem. Pharmacol.*, 36, 1755, 1987.
54. **Beaumont, P. J. V.,** The endocrinology of anorexia nervosa, *Med. J. Aust.*, 2, 225, 1983.
55. **Hotta, M., Shibasaki, T., Masuda, A., Imaki, T., Demura, H., Ling, N., and Shizume, M.,** The responses of plasma adrenocorticotropin-releasing hormone (CRH) and cerebrospinal fluid immunoreactive CRH in anorexia nervosa patients, *J. Clin. Endocrinol. Metab.*, 62, 319, 1986.
56. **Kaye, W. H., Gwirtsman, H. E., George, D. T., Ebert, M. H., Jimesson, D. C., Tomai, T. P., Chrousos, G. P., and Gold, P. W.,** Elevated cerebrospinal fluid levels of immunoreactive corticotropin-releasing hormone in anorexia nervosa: relation to state of nutrition, adrenal function and intensity of depression, *J. Clin. Endocrinol. Metab.*, 64, 203, 1987.
57. **Nemeroff, C. B., Widerlow, E., and Bissette, G.,** Elevated concentrations of CSF corticotropin-releasing factor-like immunoreactivity in depressed patients, *Science*, 226, 1342, 1984.
58. **Davis, K., Davis, B., and Mohs, R.,** CSF corticotropin-releasing factor in neuropsychiatric diseases, New Research Abstracts, 137th Annual Meeting of American Psychiatric Association, Washington, D.C., 1984.
59. **Roy, A., Pickar, D., Paul, S., Doran, A., Chrousos, G. P., and Gold, P. W.,** CSF corticotropin-releasing hormone in depressed patients and normal control subjects, *Am. J. Psychiatry*, 144, 641, 1987.

Chapter 19

BEHAVIORAL AND ENDOCRINE STUDIES OF CORTICOTROPIN-RELEASING HORMONE IN PRIMATES

Ned H. Kalin

TABLE OF CONTENTS

I. INTRODUCTION

This chapter will summarize studies performed in our laboratory exploring how brain corticotropin-releasing hormone (CRH) systems mediate the stress hormone response in rhesus monkeys. We have used rhesus monkeys for our studies because they are neuroendocrinologically similar to humans[1-4] and well-established models of stress and psychopathology exist in this species.[5] Thus, an understanding of CRH systems in rhesus monkeys may be particularly relevant to humans.

Our laboratory has taken two approaches to explore the function of the primate's brain CRH systems. The first has been to study cerebrospinal fluid (CSF) concentrations of CRH as an indicator of brain CRH activity, since under various circumstances CSF peptide concentrations reflect CNS activity of peptidergic neuronal systems.[6,7] The second approach has been to study the effects of centrally administered CRH on endocrine, autonomic, and behavioral systems.

II. STUDIES OF CRH IN CEREBROSPINAL FLUID

In clinical neuropsychiatric research, CSF is frequently sampled as the most direct indicator of brain function and assessed for concentrations of peptides and other neurotransmitters. Studies in humans have already demonstrated alterations in CSF concentrations of CRH in patients with depression,[8] anorexia nervosa,[9] amyotrophic lateral sclerosis,[10] and Alzheimer's disease.[11] These findings are very interesting; however, their meaning in relationship to altered function of brain CRH systems is unknown. This is because the factors that regulate CSF CRH concentrations are not clearly understood. Our aim was to elucidate the factors regulating CSF CRH concentrations and establish whether concentrations of CRH in CSF reflect activity of hypothalamic and/or extrahypothalamic CRH neuronal systems.

A. MEASUREMENT AND CHARACTERIZATION OF CRH-IMMUNOREACTIVITY IN PRIMATE CSF

CSF-CRH immunoreactivity (IR) was measured in unextracted CSF with a specific antiserum directed against the N-terminal portion of the intact peptide (IgG Corp., Nashville, TN) The ED_{10} sensitivity of the assay was 0.3 pg/tube. Unextracted CSF was used, since we demonstrated that unextracted CSF, extracted CSF, and a rat hypothalamic extract all showed displacement of the CRH trace parallel to the synthetic human CRH (hCRH) standard.[12] Figure 1 illustrates the HPLC analysis of the CRH immunoreactivity (CRH-IR) measured in CSF. As can be seen, two peaks of immunoreactivity exist, which comigrate with synthetic hCRH[1-41] and hCRH [Met(0)^{21,38}]. Our finding is consistent with other reports of oxidized forms of CRH in hypothalamic[13] and CSF[14] extracts.

B. DIURNAL VARIATIONS IN CSF-CRH-IR CONCENTRATIONS

The first question we asked was whether CSF concentrations of CRH follow a diurnal variation. We were interested in this because the pituitary release of ACTH follows a diurnal rhythm and CRH neurons located in the paraventricular nucleus (PVN) play a major role in regulating ACTH secretion.[15,16] We hypothesized that if the source of CSF CRH is from CRH-releasing neurons originating in the PVN, then diurnal changes in CSF CRH concentrations should be apparent and should parallel the diurnal variation observed in activity of the peripheral pituitary-adrenal system. To perform this study we used a system for continuously sampling CSF from the cisterna magna in unanesthetized, partially restrained adult male rhesus monkeys. CSF was collected at a rate of 1 ml/h over a 48-h period. We measured concentrations of cortisol in CSF rather than in plasma, as an indicator of peripheral pituitary-adrenal activity. This is a valid approach, since studies from our laboratory and others have

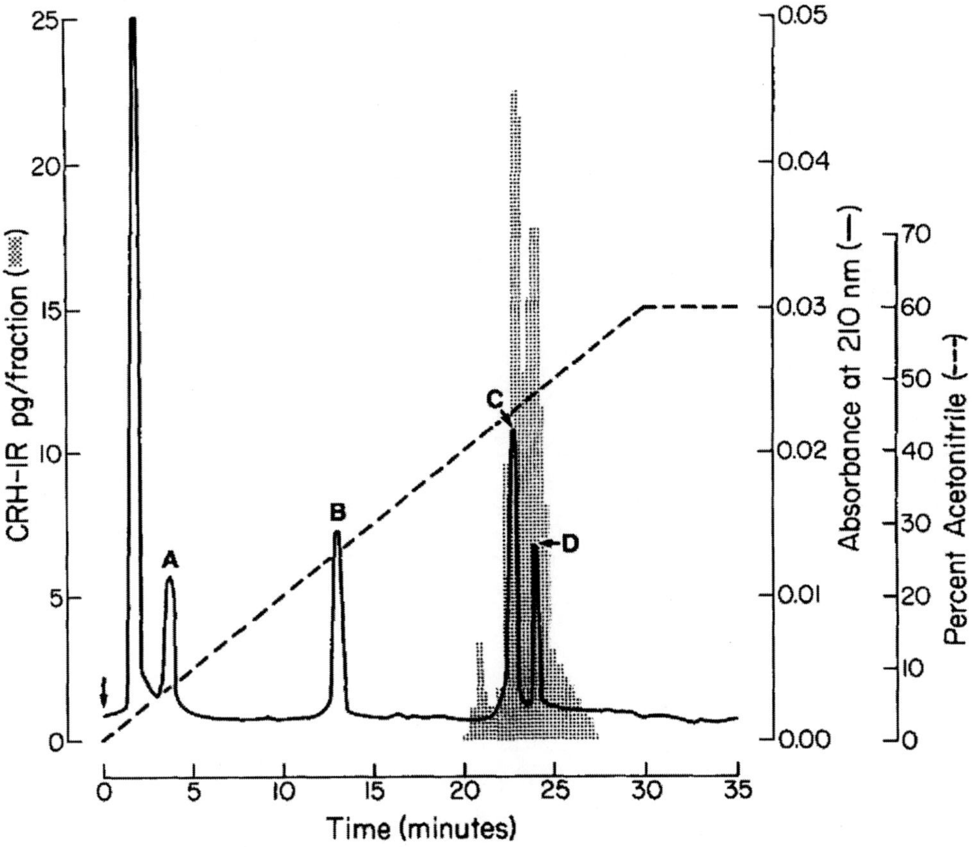

FIGURE 1. HPLC of extracted CSF. Samples of extracted monkey CSF and synthetic standard peptides (A) hCRH[38-41], (B) hCRH[28-41], (C) hCRH[MET(0)^{21,38}], and (D) hCRH[1-41] (Bachem Inc., Torrance, CA; Peninsula Labs., Belmont, CA) were applied in two successive runs to an Altex C_3 (No. 244454) end-capped (5 μm column. The initial solvent was 0.1% TFA in 100% distilled water; the eluting solvent was 0.1% TFA in 100% acetonitrile. Samples were run at a flow rate of 1 ml/min under gradient conditions shown by the broken line. Fractions of 250 μl were collected and assayed for CRH-IR. (From Kalin, N. H., Shelton, S. E., Barksdale, C. M., and Brownfield, M. S., *Brain Res.*, 426, 385, 1987.)

demonstrated that changes in CSF cortisol concentrations parallel changes in plasma cortisol levels.[12]

As can be seen in Figure 2, significant time-related variations occurred for concentrations of both CSF CRH-IR and cortisol. However, the changes in CSF cortisol did not parallel those in CSF CRH-IR. The peak concentration of CRH-IR, 19.1 ± 3.04 pg/ml (mean ± SEM), occurred at 1700 h and was nearly twice the nadir concentration of 10.0 ± 1.47, which occurred at 0600 h. The peak cortisol concentration was observed at 1000 h and the nadir was at 2100 h.

We confirmed the results of this experiment by sampling femoral blood and cisternal CSF at 0600 and 1800 h in a different group of eight adult male monkeys that were anesthetized with ketamine-HCl. From each monkey we collected samples at the different time points 1 week apart. A counterbalanced design was used to control for order of sampling. As before, we found that CSF-CRH-IR concentrations varied significantly, with the greatest concentration at 1800 h (Figure 3). Compared with CRH, plasma concentrations of ACTH and cortisol and CSF cortisol followed an inverse pattern, being greater at 0600 h than at 1800 h ($p < 0.01$).

These studies are described in greater detail in a recent publication[12] and clearly dem-

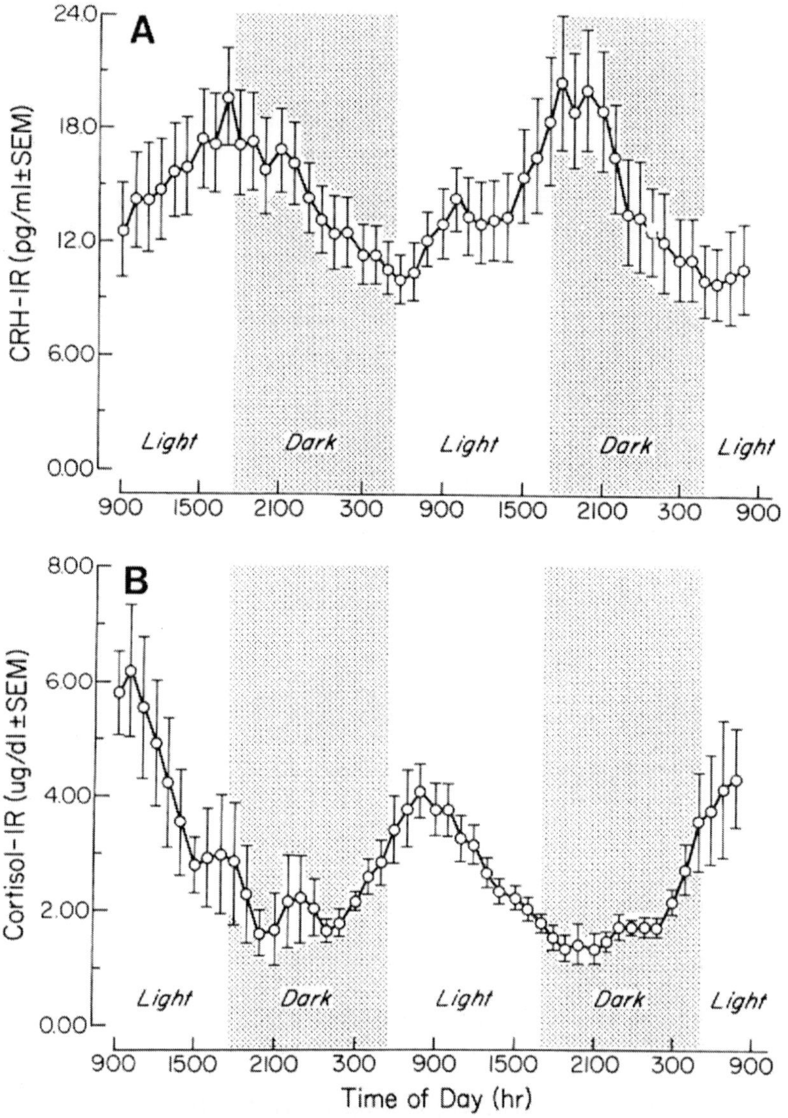

FIGURE 2. Significant changes in CSF concentrations (mean ± S.E.M.) of (A) CRH-IR (n = 9) (F = 11.016; d.f. = 23,184; $p < 10^{-6}$) and (B) cortisol (n = 6) (F = 11.008; d.f. = 23,115; $p < 10^{-6}$) over 48 h of continuous sampling. (From Kalin, N. H., Shelton, S. E., Barksdale, C. M., and Brownfield, M. S., *Brain Res.*, 426, 385, 1987.)

onstrate that time of day is an important variable in determining concentrations of CRH in CSF. The finding that diurnal changes in CSF-CRH-IR concentrations are not paralleled by changes in pituitary-adrenal activity is consistent with results recently reported by Garrick et al.[17] and suggests that under basal conditions, CSF CRH does not originate from PVN neurons that release CRH into the median eminence.

C. EFFECTS ON CSF-CRH CONCENTRATIONS OF MANIPULATIONS THAT ENHANCE PITUITARY ACTH RELEASE

The aim of the next series of studies was to establish whether manipulations that increase the release of pituitary ACTH via enhanced activation of PVN CRH-releasing neurons are associated with increased CSF concentrations of CRH-IR. To test this we used three ap-

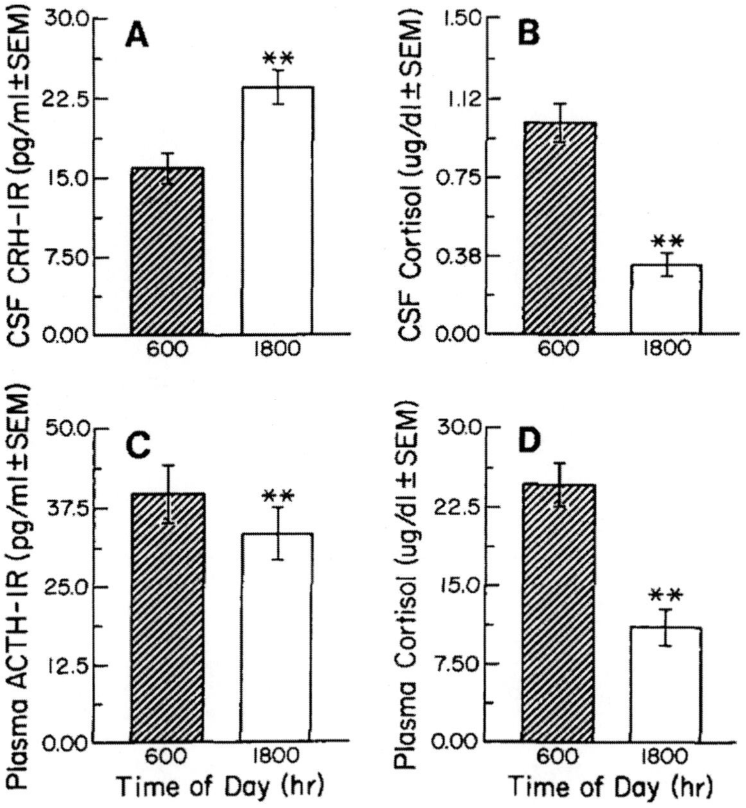

FIGURE 3. Concentrations of (A) CSF-CRH-IR, (B) CSF cortisol, (C) plasma ACTH-IR, and (D) plasma cortisol at 0600 and 1800 h, illustrating diurnal variation (** $p < 0.01$). Change in CSF-CRH-IR concentration is opposite in direction to changes in cortisol and plasma ACTH-IR concentrations. (From Kalin, N. H., Shelton, S. E., Barksdale, C. M., and Brownfield, M. S., *Brain Res.*, 426, 385, 1987.)

proaches (cholinergic stimulation, reduction of glucocorticoid negative feedback, and exposure to stress) that promote the release of CRH into the median eminence, presumably through different mechanisms. In the first study,[12] we administered the cholinesterase inhibitor physostigmine 66 μg/kg intramuscularly (i.m.) to nine adult male monkeys. Plasma and cisternal CSF were sampled either 60, 120, or 180 min after physostigmine administration, always at the same time of day, while the monkeys were anesthetized with ketamine. Baseline samples, without physostigmine administration, were also obtained at the same time of day. As can be seen in Figure 4, physostigmine did not affect CSF-CRH-IR concentrations even though a threefold increase in plasma ACTH concentrations occurred.

In the next study we administered the 11-β-hydroxylase inhibitor metyrapone, which reduces production of cortisol, subsequently promoting ACTH release by diminished negative feedback. On the 1st day of the study, two chair-restrained monkeys received normal saline intravenously, and on the 2nd day they were infused with metyrapone 10 mg/ml at a rate of 9 ml/h for 24 h. Blood and CSF samples were collected repeatedly throughout the 48 h of study. Figure 5 shows that metyrapone markedly decreased plasma cortisol and increased plasma ACTH concentrations. In one animal ACTH levels increased approximately 30-fold, and in the other the increase was approximately 12-fold, yet no major changes in concentrations of CSF-CRH-IR were observed.

The next experiment assessed the effects of environmental stress on CSF-CRH-IR concentrations. Blood and cisternal CSF were sampled from 18 adult monkeys before and after

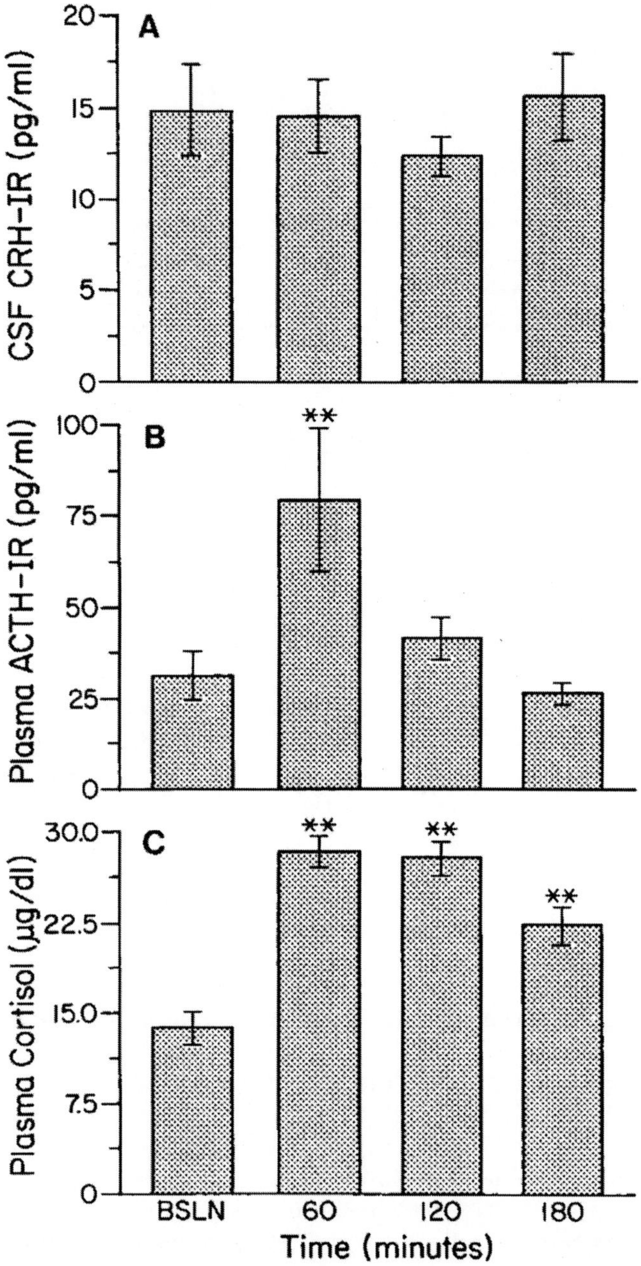

FIGURE 4. Effects of physostigmine 66 µ/kg administered to nine monkeys. (A) CSF-CRH-IR concentration was unaffected; (B) plasma ACTH concentration increased sharply 1 h after physostigmine administration but returned to baseline after 2 to 3 h; (C) plasma cortisol concentration was still elevated 3 h after physostigmine administration (**$p < .01$). (From Kalin, N. H., Shelton, S. E., Barksdale, C. M., and Brownfield, M. S., *Brain Res.*, 426, 385, 1987.)

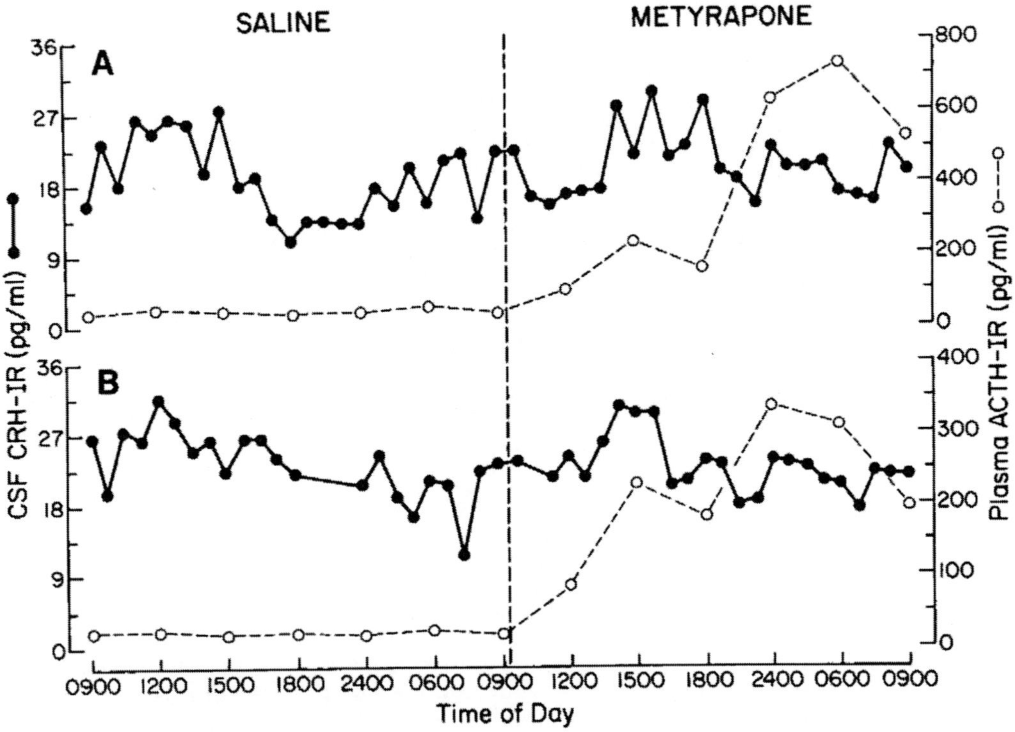

FIGURE 5. Effects of metyrapone administered over 24 h to two rhesus monkeys (A and B). There was no significant effect on CSF-CHR-IR concentrations, but a marked increase in plasma ACTH concentrations. (From Kalin, N. H., Shelton, S. E., Barksdale, C. M., and Brownfield, M. S., *Brain Res.*, 426, 385, 1987.)

the stress of being placed in a novel confinement cage. To determine the time course of stress effects, our sampling points were established (baseline, 15 min of stress, 30 min of stress, and 30 min after the stress). Assessments were made at weekly intervals at each of these time points in each animal. As can be seen in Figure 6, the stressor induced significant increases in plasma concentration of ACTH after 15 and 30 min of stress ($F = 11.69$; d.f. $= 3,51$ $p < 0.00005$) but did not affect CSF-CRH-IR concentrations.

This series of studies demonstrates that marked increases in pituitary ACTH secretion occurring acutely and lasting 24 h are not accompanied by similar changes in concentrations of CRH in CSF. Taken together with the data from the diurnal studies, these results suggest that the site of origin of CSF-CRH-IR is not the same as that which controls pituitary ACTH release. This is not the case for all hypothalamic releasing hormones. For example, other studies in monkeys show that when increases in CSF levels of gonadotropin-releasing hormone are observed they are immediately followed by increases in plasma concentrations of luteinizing hormone.[18]

D. SUMMARY OF CSF-CRH STUDIES

Our work has important implications for clinical studies, in its strong suggestion that CSF levels of CRH-IR do not reflect hypothalamic CRH systems but more likely reflect function of extrahypothalamic CRH neurons. In preliminary studies, increased levels of CSF-CRH-IR were found in patients with major depression[8] and anorexia nervosa.[9] Our work suggests that increased CSF-CRH concentrations in these illnesses are secondary to increased function of extrahypothalamic CRH neurons. Further support for the concept that CSF-CRH reflects activity of extrahypothalamic systems comes from findings that brains of Alzheimer's patients have a decreased CRH-IR content in cerebral cortex[19,20] and the demonstration of decreased CSF-CRH-IR concentrations in Alzheimer's patients.[1,11]

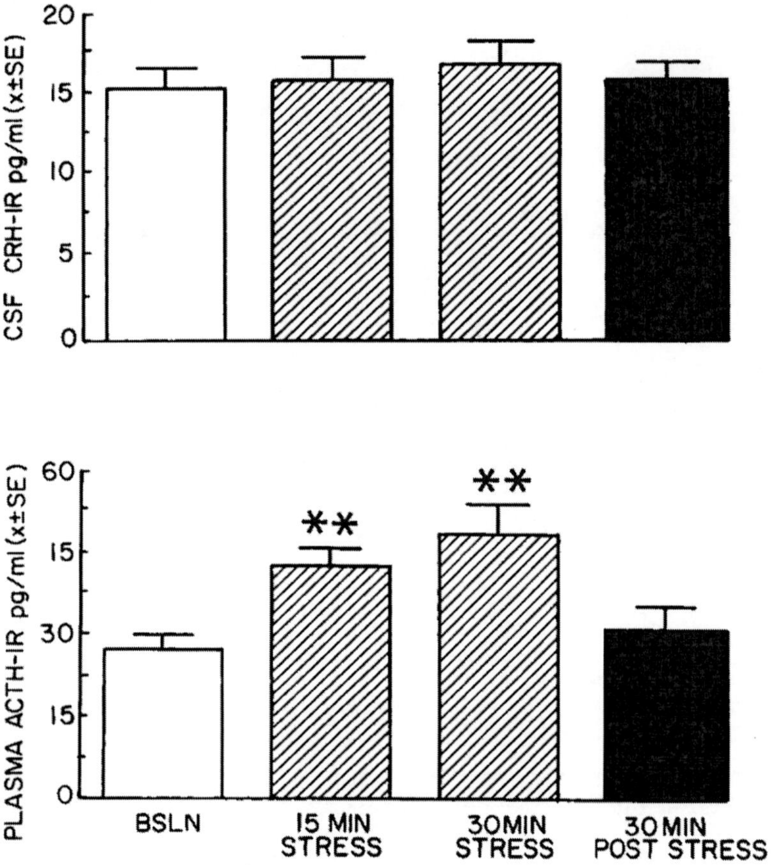

FIGURE 6. Time course of effects of exposure to a novel confinement cage on concentrations of (A) CSF-CRH-IR and (B) plasma ACTH-IR (**$p < 0.01$).

III. EFFECTS OF CRH ON BEHAVIORAL, ENDOCRINE, AND AUTONOMIC SYSTEMS

The behavioral effects of CRH administered intracerebroventricularly (i.c.v.) have been explored primarily in rodents. In these experiments i.c.v. administration of CRH causes activation of locus coeruleus neurons,[21] increases in plasma concentrations of epinephrine and norepinephrine,[22,23] autonomic activation,[22] and increased arousal assessed by electroencephalography[24] and behavioral observation.[25-28] The effects of CRH are modulated by environmental influences,[25,29] and in some experiments, CRH has been demonstrated to further enhance stress-induced behavior.[13,31] However, the ability of exogenous CRH to enhance behaviors activated by stress appears to be specific only to certain stressors, as we demonstrated that in rats, i.c.v. CRH does not enhance the behavioral effects of exposure to novelty[32] but does increase foot shock-induced freezing behavior (unpublished data).

A. EFFECTS OF INTRACEREBROVENTRICULAR CRH ADMINISTERED TO PARTIALLY RESTRAINED AND SINGLE-CAGED MONKEYS

Little is known about the effects of i.c.v. CRH in primates. Because of our interest in how CRH systems mediate the stress response in primates, we investigated these effects. In our first study,[29] we administered ovine CRH i.c.v. (0, 20, and 180 μg) to four adult male monkeys that had indwelling lateral ventricular cannulas and were adapted to chair restraint. The order of dose administration was counterbalanced and doses were administered

FIGURE 7. Adult male home-caged rhesus monkey before i.c.v. CRH administration. (From Kalin, *Fed. Proc.*, 44 (Part 2), 249, 1985.)

at least 24 h apart. At these doses, significant increases in plasma ACTH and cortisol concentrations were observed as well as a slight increase in mean arterial pressure. Compared with vehicle, CRH increased behavioral arousal. While each animal responded with its own individual pattern of behaviors, increases in vocalization, environmental exploration, head-shaking, and struggling occurred. None of these behavioral changes were statistically significant because of the small sample size and amount of individual variability.

Three of the same monkeys received i.c.v. CRH in their home cages.[29] Here the behavioral response was quite different. Vocalizations increased after both 20 and 180 μg doses. However, occasionally after 20 μg and always after 180 μg, the monkeys displayed a combination of huddling and lying-down behaviors (see Figures 7 and 8). These effects were not simply sedative, since the animals responded normally when approached by the person performing the behavioral observations. It is also unlikely that they were due to hypotension, because in another experiment we found that these doses of CRH administered i.c.v. to anesthetized monkeys resulted in small increases in blood pressure.

B. EFFECTS OF INTRACEREBROVENTRICULAR CRH ON INFANTS DISTRESS INDUCED BY BRIEF SEPARATION

We administered CRH to infant monkeys undergoing separation from their mothers because we wanted to test the effects of this peptide in stressful situations. We expected

FIGURE 8. The monkey in Figure 7 shortly after i.c.v. administration of 20 μg ovine CRH. (From Kalin, N. H., *Fed. Proc.*, 44 (Part 2), 249, 1985.)

that CRH would enhance behavioral and physiological changes induced by the stressor. We selected brief separation because it is a naturally occurring stressor that causes activation of the pituitary-adrenal system and species-specific distress vocalizations.[30-36] In addition, from earlier work with this model, we already knew the effects of anxiolytic[35] and anxiogenic (unpublished data) compounds on separation-induced distress. We expected that CRH would have effects similar to anxiogenic compounds.

When diazepam was administered to infants immediately before they were separated from their mothers for 1 h, this antianxiety compound decreased distress vocalizations, attenuated ACTH and cortisol secretion, and increased levels of activity. The effects of diazepam on activity and HPA hormones were mediated specifically by benzodiazepine receptors, since they were blocked by prior administration of the benzodiazepine antagonist Ro 15-1788. The effects of diazepam on reducing distress vocalizations were not reversed by Ro 15-1788, suggesting that this effect of diazepam may be mediated by nonbenzodiazepine systems.[35]

In another study, we administered the anxiogenic compound β-CCE, a β-carboline known to be an inverse agonist when interacting with the benzodiazepine receptor. The effects of β-CCE (500 μg/kg administered i.m.) were opposite those of diazepam. β-CCE significantly increases separation-induced vocalizations at the same time it decreased activity levels (unpublished data).

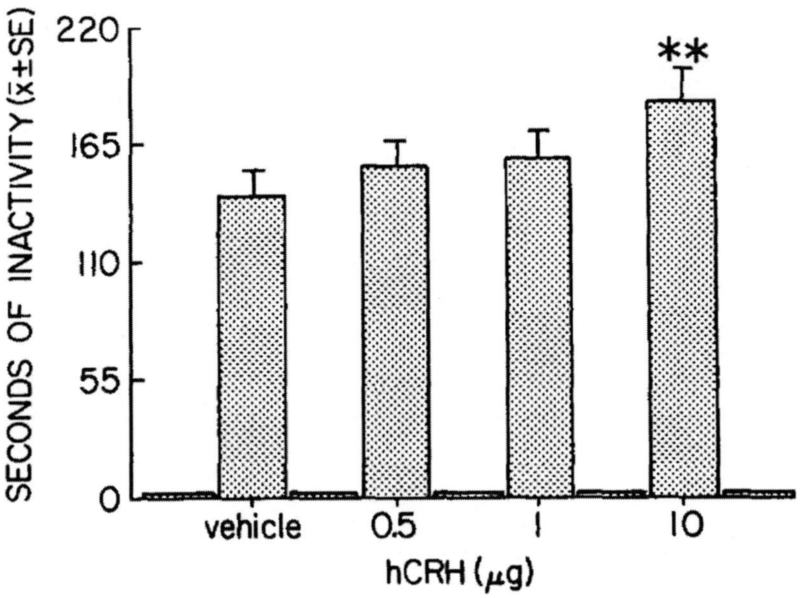

FIGURE 9. Effects of i.c.v. hCRH on levels of inactivity in 11 rhesus monkey infants separated from their mothers for 1 h (** $p < 0.01$).

We expected that the effects of CRH on separated infants would be similar to those of β-CCE, increasing distress vocalizations and decreasing activity levels. We administered human CRH (0, 0.5, 1.0, and 10.0 μg) i.c.v. to 11 infant monkeys. The CRH used in this study was obtained from Wylie Vale and Jean Rivier of the Salk Institute and was administered in normal saline solution. Using a design counterbalanced for order of dose administration, each animal received an i.c.v. infusion at weekly intervals. Immediately after peptide administration the infant was separated from its mother for approximately 1 h. Behavioral observations were collected during two 5-min observation sessions separated by a 30-min interval. At the end of the separation period, femoral venous blood was sampled and heart rate, mean arterial pressure, and rectal temperature were measured.

No significant effects of CRH on distress vocalizations were observed. However, CRH significantly reduced activity levels. This is demonstrated in Figure 9 as an increase in inactivity ($F = 6.452$; d.f. $= 3,30$; $p < 0.002$). The dose of CRH that decreased activity, 10 μg, also significantly decreased body temperature, marginally reduced mean arterial pressure, had no effect on heart rate, and significantly increased plasma concentrations of ACTH and cortisol (see Figure 10).

C. SUMMARY OF CRH ADMINISTRATION STUDIES

Our studies with adult monkeys demonstrated that, as in rats, the behavioral effects of CRH are modulated by environmental input. When monkeys were chair restrained, CRH produced behavioral activation. However, when animals were in a more familiar environment, CRH caused behavioral inhibition. This finding was unexpected because rats receiving CRH in a familiar environment respond with increased locomotion.[25] In addition, lying down and huddling behavior was observed. Whether the differences between rats and monkeys in their responses to CRH can be explained by species differences or are due to methodologic factors remains to be determined.

We used the mother-infant separation paradigm to investigate the effects of CRH in a stressful situation. A dose-response study established that 10 μg was the lowest dose that had reliable effects. With this dose we showed that CRH reduced activity without significantly affecting distress vocalizations and increased plasma concentrations of ACTH and cortisol.

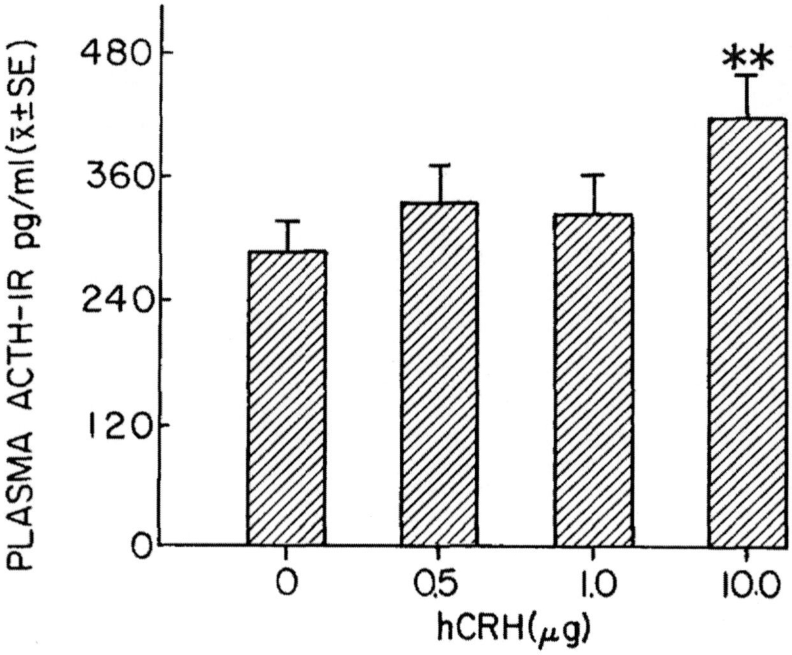

FIGURE 10. Effects of i.c.v. hCRH on separation-induced increases in plasma concentrations of ACTH in 11 rhesus monkey infants (** $p < 0.01$).

It is worth noting that at these doses no lying down or huddling behaviors were seen. Nor were the infants sedated, as they were able to respond appropriately when approached by an investigator. The effects of CRH on activity were similar to those of the anxiogenic compound β-CCE and opposite to those of diazepam. However, unlike β-CCE, CRH did not increase the frequency of distress vocalizations.

It is of interest that in the separated infants, i.c.v. CRH reduced body temperature. In rabbits, i.c.v. CRH has been reported to reduce elevations in rectal temperature induced by the administration of leukocytic pyrogen, but is without effect on basal body temperature.[37]

IV. DISCUSSION AND CONCLUSIONS

Our CSF studies have established that in primates CSF-CRH-IR concentrations follow a diurnal variation and that the origin of CSF-CRH is probably from extrahypothalamic brain regions. These findings aid in the interpretation of clinical studies reporting altered CSF-CRH levels in neuropsychiatric patients. Based on our work, we believe that the increased CSF-CRH levels reported in depressed patients reflect dysfunction of extrahypothalamic CRH neurons. Our studies also suggest that a shift in the CSF-CRH diurnal rhythm could account for the increased levels observed in clinical studies in which only a single CSF sample is obtained.

To directly explore the role of CRH systems in mediating behavior associated with stress, we administered i.c.v. CRH to adult and infant monkeys under a variety of conditions. We found that the effects of CRH depend on the environment in which the animal was tested. CRH-induced behavioral inhibition, without sedation, occurred when adult animals were observed in their home cages and when infants were separated from their mothers. This behavioral response was associated with increased secretion of ACTH and cortisol. These findings are of interest because behavioral inhibition or "freezing" occurs in many species when an animal is presented with danger and becomes frightened.[38,39]

We were intrigued by our observation that higher doses of CRH produced huddling and lying-down behavior when the adult monkeys were treated in their home cages. These behaviors are characteristically seen in despairing monkeys experiencing a significant loss.[40,41] It is interesting that the behaviors induced by i.c.v. CRH were also accompanied by activation of the pituitary-adrenal system, since some rhesus monkeys undergoing separation[42] and about half of depressed humans[43] also exhibit HPA overactivity.

As already discussed, our data suggest that the increased CRH levels seen in some depressed patients[8] reflect increased release of CRH from extrahypothalamic neurons. These data, along with data from our behavioral studies of CRH, support the idea that enhanced function of extrahypothalamic CRH systems mediates some of the behavioral aspects of the depressive syndrome.

ACKNOWLEDGMENTS

This chapter was supported in part by U.S. Veterans Administration Medical Research and by the National Institute of Diabetes and Digestive and Kidney Diseases (DK35641). The author would like to thank Steven E. Shelton and Charles M. Barksdale for their technical assistance, Wylie Vale and Jean Rivier for their generous gift of hCRH, and Carol Steinhart for her editorial assistance.

REFERENCES

1. **Hawking, F.**, Circadian rhythms in monkeys, dogs, and other animals, *J. Interdiscip. Cycle Res.*, 2, 153, 1971.
2. **Kalin, N. H., Cohen, R. M., Kraemer, G. W., Risch, S. C., Shelton, S., Cohen, M., McKinney, W. T., and Murphy, D. L.**, The dexamethasone suppression test as a measure of hypothalamic-pituitary feedback sensitivity and its relationship to behavioral arousal, *Neuroendocrinology*, 32, 92, 1981.
3. **Leshner, A. I., Toivola, P. T. K., and Terasau, E.**, Circadian variations in cortisol concentrations in the plasma of female rhesus monkeys, *J. Endocrinol.*, 78, 155, 1978.
4. **Spies, H. G., Norman, R. L., and Buhl, A. E.**, Twenty-four-hour patterns in serum prolactin and cortisol after partial and complete isolation of the hypothalamic-pituitary unit in rhesus monkeys, *Endocrinology*, 105, 1361, 1979.
5. **McKinney, W. T.**, Biobehavioral models of depression in monkeys, in *Animal Models in Psychiatry and Neurology*, Hanin, I. and Usdin, E., Eds., Pergamon Press, New York, 1977, 117.
6. **Kaji, H., Chihara, K., Minamitani, N., Kodama, H., Yanaihara, N., and Fujita, T.**, Release of vasoactive intestinal polypeptide into the cerebrospinal fluid of the fourth ventricle of the rat: involvement of cholinergic mechanism, *Brain Res.*, 269, 303, 1983.
7. **O'Donohue, T. L., Charlton, C. G., Thoa, N. B., Helke, C. J., Moody, T. W., Pert, A., Williams, A., Miller, R. L., and Jacobowitz, D. M.**, Release of alpha-melanocyte stimulating hormone into rat and human cerebrospinal fluid in vivo and from rat hypothalamus slices *in vitro*, *Peptides*, 2, 95, 1981.
8. **Nemeroff, C. B., Widerlöv, E., Bissette, G., Walleus, H., Karlsson, I., Eklund, K., Kilts, C. C., Loosen, P. T., and Vale, W.**, Elevated concentrations of CSF corticotropin-releasing factor-like immunoreactivity in depressed patients, *Science*, 226, 1342, 1984.
9. **Kaye, W. H., Gwirtsman, H. E., George, D. T., Ebert, M. H., Jimerson, D. C., Tomai, T. P., Chrousos, G. P., and Gold, P. W.**, Elevated cerebrospinal fluid levels of immunoreactive corticotropin-releasing hormone in anorexia nervosa: relation to state of nutrition, adrenal function, and intensity of depression, *J. Clin. Endocrinol. Metab.*, 64, 203, 1984.
10. **Klimek, A., Cieslak, D., Szulc-Kuberska, J., and Stepien, H.**, Reduced lumbar cerebrospinal fluid corticotropin releasing factor (CRF) levels in amyotrophic lateral sclerosis, *Acta Neurol. Scand.*, 74, 72, 1986.
11. **Mouradian, M. M., Farah, J. M., Mohr, E., Fabbrini, G., O'Donohue, T. L., and Chase, T. N.**, Spinal fluid CRF reduction in Alzheimer's disease, *Neuropeptides*, 8, 393, 1986.
12. **Kalin, N. H., Shelton, S. E., Barksdale, C. M., and Brownfield, M. S.**, A diurnal rhythm in cerebrospinal fluid corticotropin-releasing hormone different from the rhythm of pituitary-adrenal activity, *Brain Res.*, 426, 385, 1987.

13. **Rivier, J., Spiess, J., and Vale, W.,** Characterization of rat hypothalamic corticotropin-releasing factor, *Proc. Natl. Acad. Sci. U.S.A.*, 80, 4851, 1983.
14. **Tomori, N., Suda, T., Tozawa, F., Demura, H., Shizume, K., and Mouri, T.,** Immunoreactive corticotropin-releasing factor concentrations in cerebrospinal fluid from patients with hypothalamic pituitary-adrenal disorders, *J. Clin. Endocrinol. Metab.*, 57, 1305, 1983.
15. **Bruhn, T. O., Plotsky, P. M., and Vale, W. W.,** Effect of paraventricular lesions on corticotropin-releasing factor (CRF)-like immunoreactivity in the stalk-median eminence: studies on the adrenocorticotropin response to ether stress and exogenous CRF, *Endocrinology*, 114, 57, 1984.
16. **Vale, W., Rivier, C., Brown, M., Plotsky, P., Smith, M., Bilezikjian, L., Bruhn, T., Perrin, M., Spiess, J., and Rivier, J.,** Corticotropin-releasing factor, in *Secretory Tumors of the Pituitary Gland, Progress in Endocrine Research and Therapy*, Vol. 1, Black, P. McL. et al., Eds., Raven Press, New York, 1984, 213.
17. **Garrick, N. A., Hill, J. L., Szele, F. G., Tomai, T., Gold, P. W., and Murphy, D. L.,** Corticotropin-releasing factor: a marked circadian rhythm in primate cerebrospinal fluid peaks in the evening and is inversely related to the cortisol circadian rhythm, *Endocrinology*, 121, 1329, 1987.
18. **Van Vugt, D. A., Diefenbach, W. D., Alston, E., and Ferin, M.,** Gonadotropin-releasing hormone pulses in third ventricular cerebrospinal fluid of ovariectomized rhesus monkeys: correlation with luteinizing hormone pulses, *Endocrinology*, 117, 1550, 1985.
19. **Bissette, G., Reynolds, G., Kilts, C. D., Widerlöv, E., and Nemeroff, C. B.,** Corticotropin-releasing factor-like immunoreactivity in senile dementia of the Alzheimer type, *JAMA*, 254, 3067, 1985.
20. **De Souza, E. B., Whitehouse, P. J., Kuhar, M. J., Price, D. L., and Vale, W. W.,** Reciprocal changes in corticotropin-releasing factor (CRF)-like immunoreactivity and CRF receptors in cerebral cortex of Alzheimer's disease, *Nature*, 319, 593, 1986.
21. **Valentino, R. J., Foote, S. L., and Aston-Jones, G.,** Corticotropin-releasing factor activates noradrenergic neurons of the locus coeruleus, *Brain Res.*, 270, 363, 1983.
22. **Brown, M., Fisher, L., Rivier, J., Spiess, J., Rivier, C., and Vale, W.,** Corticotropin-releasing factor: effects on the sympathetic nervous system and oxygen consumption, *Life Sci.*, 30, 207, 1982.
23. **Brown, M. R., Fisher, L. A., Webb, V., Vale, W., and Rivier, J.,** Corticotropin-releasing factor receptor antagonist: a physiologic regulator of adrenal epinephrine secretion, *Brain Res.*, 328, 355, 1985.
24. **Ehlers, C., Henriksen, S., Wang, M., Rivier, J., Vale, W., and Bloom, F.,** Corticotropin releasing factor produces increases in brain excitability and convulsive seizures in rats, *Brain Res.*, 278, 332, 1983.
25. **Britton, D. R., Koob, G. F., Rivier, J., and Vale, W.,** Intraventricular corticotropin-releasing factor enhances behavioral effects of novelty, *Life Sci.*, 31, 363, 1982.
26. **Morley, J. E. and Levine, A. S.,** Corticotropin releasing factor, grooming and ingestive behavior, *Life Sci.*, 31, 1459, 1982.
27. **Sherman, J. and Kalin, N.,** ICV-CRH potently affects behavior without altering nociceptive responding, *Life Sci.*, 39, 433, 1985.
28. **Sutton, R. E., Koob, G. F., LeMoal, M., Rivier, J., and Vale, W. W.,** Corticotropin-releasing factor produces behavioral activation in rats, *Nature*, 297, 331, 1982.
29. **Kalin, N. H., Shelton, S. E., Kraemer, G. W., and McKinney, W. T.,** Corticotropin-releasing factor administered intraventricularly to rhesus monkeys, *Peptides*, 4, 217, 1983.
30. **Seay, B., Hansen, E., and Harlow, H.,** Mother-infant separation in monkeys, *J. Child Psychol. Psychiatry*, 3, 123, 1962.
31. **Suomi, S. J., Collins, M. L., Harlow, H. F., and Ruppenthal, G. C.,** Effects of maternal and peer separations on rhesus monkeys, *J. Child Psychol. Psychiatry*, 17, 101, 1976.
32. **Kalin, N. H. and Carnes, M.,** Biological correlates of attachment bond disruption in human and non-human primates, *Prog. Neuropsychopharmacol. Biol. Psychiatry*, 8, 459, 1984.
33. **Goodall, J.,** Life and death at Gombe, *Natl. Geogr.*, 155, 592, 1979.
34. **Gunnar, M. R., Gonzalez, C. A., Goodlin, B. L., and Levine, S.,** Behavioral and pituitary-adrenal responses during a prolonged separation period in infant rhesus macaques, *Psychoneuroendocrinology*, 6, 65, 1981.
35. **Kalin, N. H., Shelton, S. E., and Barksdale, C. M.,** Separation distress in infant rhesus monkeys: effects of diazepam and Ro 15-1788, *Brain Res.*, 408, 192, 1987.
36. **Kalin, N. H., Shelton, S. E., and Barksdale, C. M.,** Opiate modulation of separation-induced distress in nonhuman primates, *Brain Res.*, in press.
37. **Bernardini, G. L., Richards, D. B., and Lipton, J. M.,** Antipyretic effect of centrally administered CRF, *Peptides*, 5, 57, 1984.
38. **Bolles, R. C.,** Species-specific defense reactions and avoidance learning, *Physiol. Rev.*, 77, 32, 1970.
39. **Bouton, M. E. and Bolles, R. C.,** Conditioned fear assessed by freezing and by the suppression of three different baselines, *Anim. Learning Behav.*, 8, 429, 1980.
40. **Harlow, H. F. and Suomi, S. J.,** Induced depression in monkeys, *Behav. Biol.*, 12, 273, 1974.
41. **Kaufman, I. C. and Rosenblum, L. A.,** The reaction to separation in infant monkeys: anaclitic depression and conservation-withdrawal, *Psychosom. Med.*, 29, 648, 1967.

42. **Kalin, N. H., Weiler, S. J., McKinney, W. T., Kraemer, G. W., and Shelton, S. E.,** Where is the 'lesion' in patients who fail to suppress plasma cortisol concentrations with dexamethasone?, *Psychopharmacol. Bull.*, 18, 219, 1982.
43. **Kalin, N. H. and Dawson, G. W.,** Neuroendocrine dysfunction in depression: hypothalamic-anterior pituitary systems, *Trends Neurosci.*, 9, 261, 1986.

Chapter 20

REGULATION OF THE AUTONOMIC NERVOUS SYSTEM BY CORTICOTROPIN-RELEASING FACTOR

Marvin R. Brown and Laurel A. Fisher

TABLE OF CONTENTS

I. INTRODUCTION

Corticotropin-releasing factor (CRF) is a 41 amino acid-containing peptide originally isolated and characterized from ovine hypothalamus on the basis of its ability to stimulate the secretion of ACTH from the anterior pituitary gland.[1] Subsequently, CRF has been isolated and characterized from the hypothalami of rats and humans.[1] CRF is structurally related to urotensin I, a peptide isolated from the urophysis of teleost fish, and sauvagine, a peptide isolated from frog skin.[2,3] CRF is now accepted as being a physiologic regulator of pituitary ACTH secretion.[1]

II. EFFECTS OF CRF ON THE SYMPATHETIC NERVOUS SYSTEM AND ADRENAL MEDULLA

Intracerebroventricular (i.c.v.) administration of CRF to the rat or dog elicits dose-related elevations of plasma epinephrine and norepinephrine concentrations.[4-6] Ovine and rat CRFs are similarly potent in elevating plasma catecholamine levels.[4-6] That CRF acts in the central nervous system (CNS) to elevate plasma concentrations of epinephrine and norepinephrine is supported by two lines of evidence. First, CRF administered i.c.v. cannot be detected in peripheral plasma by radioimmunoassay.[8] Second, the elevation of plasma catecholamine concentrations following i.c.v. administration of CRF is not altered by systemic administration of CRF antisera or the CRF receptor antagonist, CRF(9-41).[7] If elevated plasma concentrations of catecholamines following i.c.v. CRF administration resulted from leakage of CRF from the brain it would be anticipated that CRF concentrations in plasma would increase and that peripheral administration of a neutralizing antibody or CRF receptor antagonist would prevent this biological response to CRF. Thus, these findings support the contention that CRF given i.c.v. acts within the CNS to stimulate sympathetic outflow and thus increase plasma concentrations of catecholamines.

The site of origin of epinephrine secretion following i.c.v. administration of CRF is the adrenal medulla; CRF does not increase plasma concentrations of epinephrine in adrenalectomized rats.[9] The origin of norepinephrine that appears in plasma following any particular stimulus is uncertain because of the diffuse distribution of noradrenergic nerve terminals. The tissue sites of origin of norepinephrine release following CRF treatment were explored by assessing the accumulation of tissue dopamine following inhibition of the enzyme dopamine β-hydroxylase. This method provides an indirect index of norepinephrine turnover; an increase in tissue dopamine levels may be equated to augmented noradrenergic activity in that site. CRF given i.c.v. increased dopamine accumulation in the kidney, but in no other organ tested.[7] Decreases in dopamine concentrations were observed in interscapular brown fat and pancreas. These data suggest that CRF elicits an increase in noradrenergic activity to certain organs, e.g., kidney, while producing a reduction of noradrenergic activity in others, e.g., brown fat and whole pancreas. Consistent with these observations, CRF given i.c.v. increases adrenal sympathetic and renal nerve activity[10] and elevates plasma renin activity (Figure 1). The proportion of the increase of norepinephrine concentrations in plasma following CRF administration that is of renal origin is not known at this time. Of interest is the question of whether CRF-induced norepinephrine release from the kidney represents spillover from the synaptic cleft, or whether the kidney might serve as a primary source of norepinephrine for use as a circulating hormone.

III. EFFECTS OF CRF ON THE PARASYMPATHETIC NERVOUS SYSTEM

CRF is suspected to act within the CNS to influence the parasympathetic nervous system on the basis of the effects of this peptide on heart rate (HR) and gastric acid secretion.

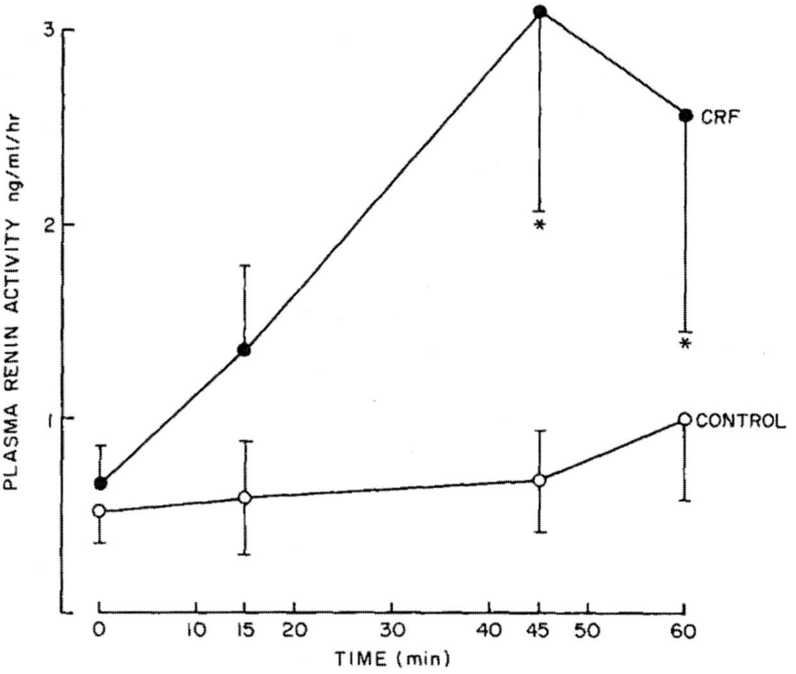

FIGURE 1. Effects of CRF (2.2 nmol) given i.c.v. on plasma renin activity. CRF was administered immediately following the zero time point. *$p < 0.05$ compared to zero value. Plasma renin activity was performed by Dr. Jean Sealey, Cornell Medical Center, New York.

Methyl atropine administration has no effect on CRF-induced tachycardia, suggesting that cardiac vagal tone is largely withdrawn in CRF-treated rats.[11] By virtue of its CNS action to inhibit cardiac vagal activity, CRF reduces the gain and range of baroreflex regulation of heart rate.[12] Thus, CRF given i.c.v. produces simultaneous elevations of arterial pressure and heart rate and also blunts the reflex bradycardia in response to phenylephrine-induced increases of arterial pressure.[7] That CRF acts within the CNS to influence gastric parasympathetic outflow in the rat is inferred from the observation that subdiaphragmatic vagotomy prevents the inhibitory effects of i.c.v. CRF treatment on gastric acid secretion.[13]

IV. CNS SITE OF ACTION OF CRF

CRF (200 pmol) has been microinjected into 48 different brain sites followed by measurement of plasma concentrations of catecholamines.[14] No consistent changes of plasma epinephrine levels were observed. Numerous sites were identified where CRF microinjection resulted in elevation of plasma norepinephrine levels. The most robust responses occurred in the dorsal hypothalamus, zona incerta, ventromedial hypothalamus, and lateral hypothalamus. However, administration of the same dose of CRF into the third ventricle produced responses similar to those observed in these tissue sites. Thus, a periventricular site of action of CRF cannot be excluded. These data lead to the conclusions that CRF may have multiple CNS sites of action and that this peptide acts at different brain sites to stimulate peripheral sympathetic efferents vs. the adrenal medulla.

Most recently, it was observed that CRF microinjection into the paraventricular nucleus of the hypothalamus or rostral nucleus of the solitary tract results in an increase of arterial pressure (AP) and HR greater than that occurring following ventricular administration of the peptide (M. Brown, unpublished data). However, lesions of the anteroventral third ventricle (AV3V) region did not alter CRF-induced changes in plasma catecholamine levels or cardiovascular function (Figures 2 and 3).

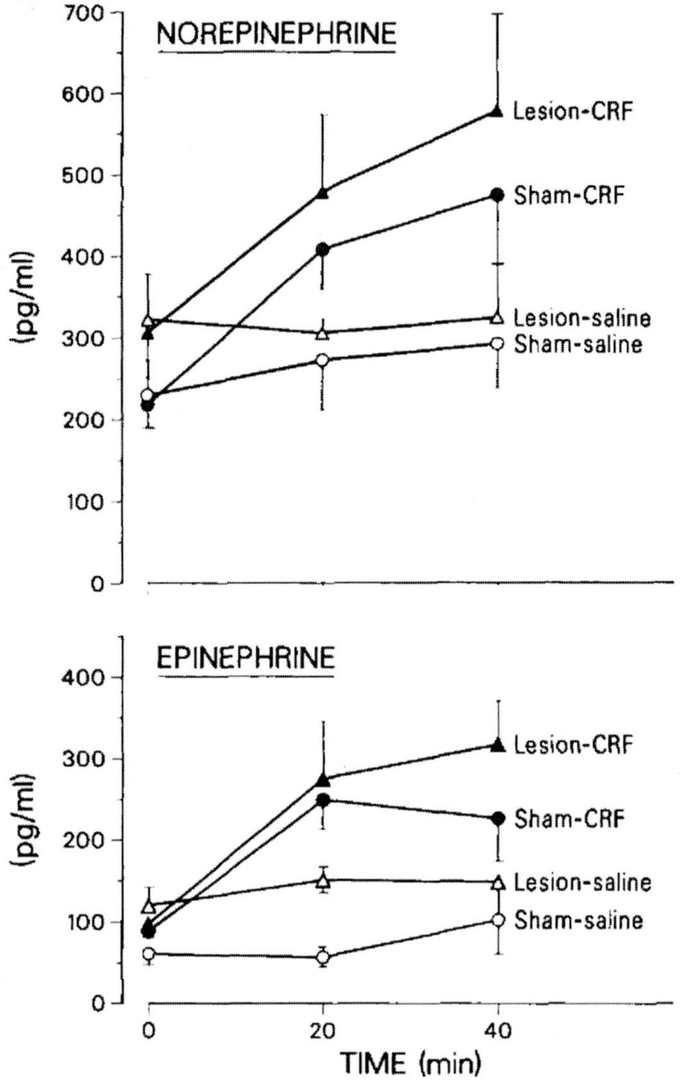

FIGURE 2. Effects of CRF (2.2 nmol) given i.c.v. on plasma concentrations of norepinephrine and epinephrine in sham and anteroventral third ventricle (AV3V) region lesioned rats. CRF was administered immediately following the zero time point. Experiments were performed in collaboration with Dr. W. Lind, The Salk Institute, La Jolla, CA.

V. CNS EFFECTS OF CRF ON PHYSIOLOGIC SYSTEMS

A. CARDIOVASCULAR EFFECTS

CRF injected i.c.v. increases AP and HR in the dog and rat.[6,15,16] In contrast, intravenous administration of CRF lowers AP.[6,16] Recent studies in the rat demonstrate that CRF given i.c.v. produces an increase in cardiac output and stroke volume, and a decrease in total peripheral vascular resistance.[17] Thus, in spite of relatively high plasma catecholamine concentrations, CRF-induced cardiovascular changes represent those of increased flow and decreased resistance favoring maximum tissue perfusion. These results suggest that the elevation in AP following i.c.v. administration of CRF is largely due to an increase in cardiac output. CRF-induced elevations of AP and HR are not prevented by hypophysectomy

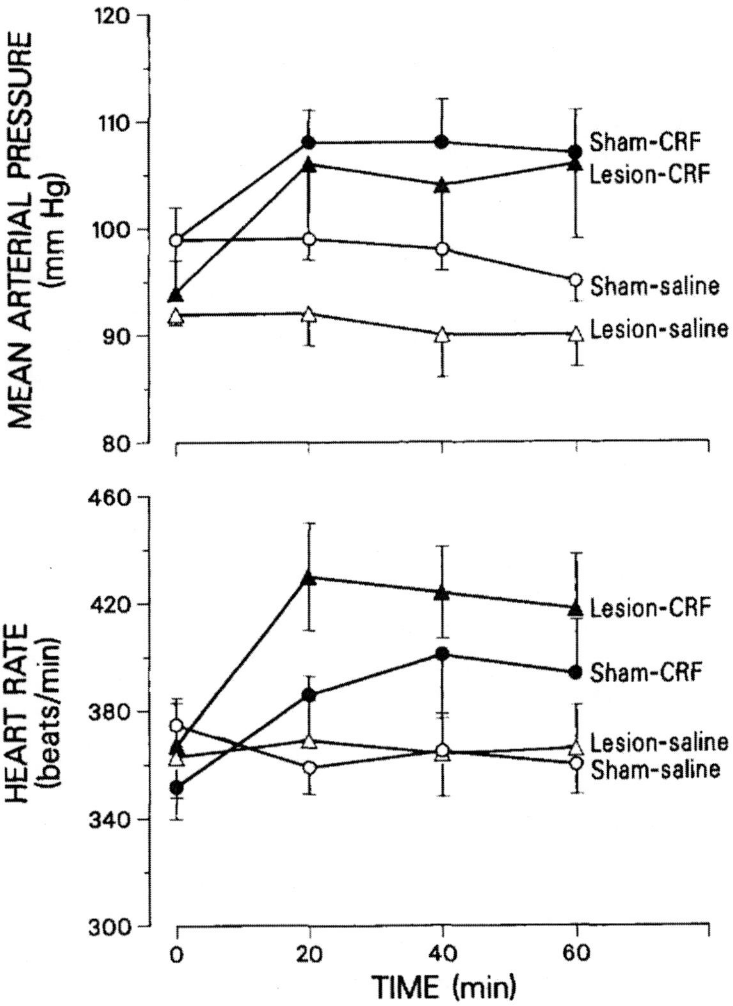

FIGURE 3. Effects of CRF (2.2 nmol) given i.c.v. on mean arterial pressure and heart rate in sham and anteroventral third ventricle (AV3V) region lesioned rats. CRF was administered immediately following the zero time point. Experiments were performed in collaboration with Dr. W. Lind, The Salk Institute, La Jolla, CA.

or adrenalectomy, or by peripheral administration of a vasopressin receptor antagonist or the angiotensin converting enzyme inhibitor, captopril.[16,17] CRF-induced elevations of AP and HR are prevented by administration of the ganglionic blocker, chlorisondamine.[15,16,18,19] As mentioned above, CRF-induced tachycardia is mediated mainly by cardiac parasympathetic withdrawal, although high levels of circulating catecholamines may contribute to this effect as well.

In summary, CRF acts in the central nervous system to influence sympathetic and parasympathetic nervous efferents so as to increase venous return and elevate heart rate, thereby producing an increase in cardiac output. The CRF-induced increase in arterial pressure is secondary to an increase in cardiac output rather than an increase in peripheral vascular resistance; in fact, CRF given i.c.v. decreases total peripheral vascular resistance, perhaps reflecting hindquarter vasodilation associated with increased locomotor activity.

B. METABOLIC EFFECTS

CRF given i.c.v. produces an increase in plasma glucose concentration that is not prevented by hypophysectomy or adrenalectomy, but is prevented by ganglionic blockade with chlorisondamine or by systemic administration of somatostatin.[4] CRF-induced hyperglycemia is accompanied by an elevation of plasma glucagon concentrations and a relative lowering of plasma insulin concentrations.[4] Thus, CRF-induced hyperglycemia is mediated by autonomic nervous-dependent elevations of plasma glucagon levels and suppression of insulin secretion with a resultant increase of hepatic glucose production.

C. GASTROINTESTINAL EFFECTS

CRF given i.c.v. to rats and dogs inhibits basal as well as pentagastrin-, 2-deoxyglucose-, TRF- and meal-stimulated gastric acid secretion.[13,20] Adrenalectomy, cervical cord transection, ganglionic blockade, and opiate and vasopressin antagonists are reported to completely or partially prevent CRF-induced inhibition of gastric acid secretion.[13,20] Vagotomy prevents CRF-induced inhibition of gastric acid secretion in the rat, but not in the dog.[13,20] These studies suggest a complex of neural efferent mechanisms activated by CRF capable of inhibiting gastric acid secretion.

VI. PHYSIOLOGIC ROLE OF CRF IN REGULATING THE AUTONOMIC NERVOUS SYSTEM

A physiologic role of CRF in the regulation of the autonomic nervous system (ANS) is supported by data obtained in experiments utilizing CRF receptor antagonists.[21,22] The CRF receptor antagonists, alpha-helical-CRF(9-41) or alpha-helical-CRF(10-41), given i.c.v., prevent CRF-induced elevations of plasma catecholamine concentrations.

The CRF receptor antagonist alpha-helical CRF(9-41) acts within the brain to attenuate adrenal epinephrine secretion following insulin-induced hypoglycemia, 30% hemorrhage, and exposure to ether vapor (Figure 4).[21,22] This CRF antagonist, given i.c.v., does not, however, alter basal AP or HR in spontaneously hypertensive rats (SHR, at 6 or 12 weeks of age), and does not influence the cardiovascular responses to administration of sodium nitroprusside or following electrical stimulation of the central nucleus of the amygdala or anterior hypothalamus in normotensive rats.[22] Intracerebroventricular injection of alpha-helical CRF (9-41), however, significantly attenuates the elevation of both epinephrine and norepinephrine concentrations in plasma following electrical stimulation of the anterior hypothalamus (monophasic pulses, 0.5 ms duration at 50 Hz at a current of 30 μA delivered over a 30-s period) (Brown, unpublished data).

The question of why the CRF antagonists inhibit stress-induced elevations of plasma epinephrine levels but not norepinephrine levels is not resolved at this time. Since exogenous CRF administration elevates both epinephrine and norepinephrine concentrations in plasma, several possibilities exist that may explain this finding. First, the CRF receptor antagonists could possess different binding affinities to brain CRF receptors involved in the regulation of adrenal epinephrine release vs. noradrenergic nerve release of norepinephrine. Second, the CRF antagonist may not gain access to the site of action of endogenous CRF that is involved in the regulation of norepinephrine release. Third, endogenous CRF may not participate in the regulation of peripheral norepinephrine release. Fourth, endogenous CRF may participate in the regulation of norepinephrine release, but if CRF receptors are blocked, other redundant pathways may co-opt this function. These results do suggest, as previously reported, that adrenal medullary epinephrine secretion and peripheral noradrenergic activity are regulated independently.[23,24]

Studies were performed using SHRs in an attempt to identify an animal model in which CRF may participate in a pathophysiologic process. We have found that the SHR exhibits an exaggerated elevation of plasma concentrations of epinephrine and glucose and blunted AP and HR responses following CNS administration of CRF.[25] Efforts have continued to

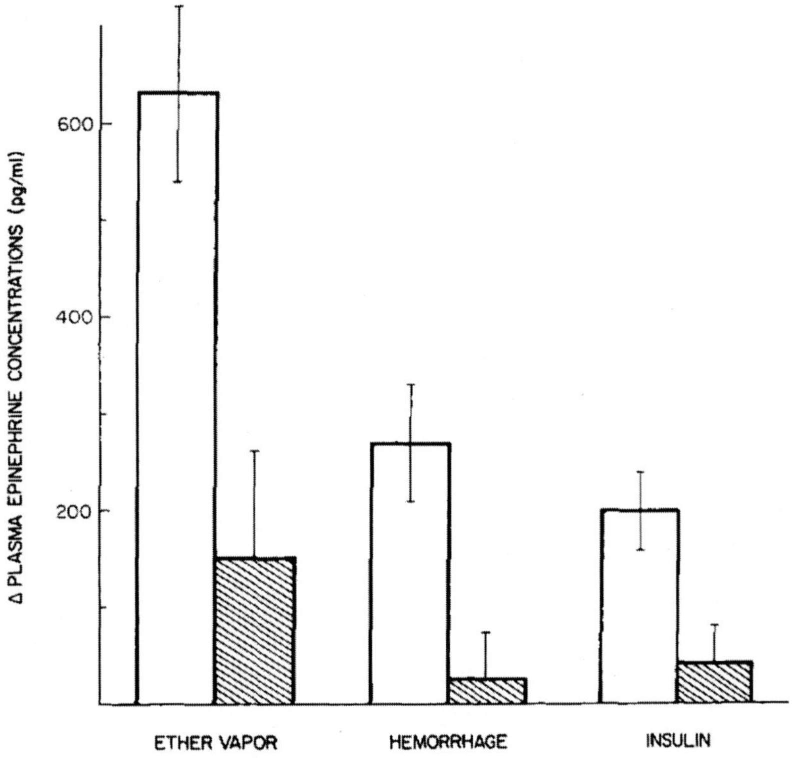

FIGURE 4. Effects of alpha-helical CRF(9-41) given i.c.v. on plasma concentrations of epinephrine in rats exposed to ether vapor, 30% hemorrhage or insulin treatment (1 U). Open bars represent each treatment without alpha-helical CRF(9-41). Hatched bars represent each treatment plus alpha-helical CRF(9-41) (10 to 100 μg given i.c.v. 15 min prior to treatment).

determine if CRF-containing pathways and effector systems are involved in the pathophysiology of these animals. We (Brown and Vale, unpublished data) and others have found decreased concentrations of CRF in some brain regions.[26] In collaboration with Dr. Richard Hauger, UCSD, we have also investigated the binding affinity and number of CRF receptors in SHR and Wistar-Kyoto (WKY) rat brains.[25] The results of this study fail to show any difference in receptor number or affinity for CRF among SHR, WKY and Sprague-Dawley rat brains.

VII. CONCLUSION

In conclusion, CRF, in addition to the regulation of pituitary ACTH secretion, may be involved in the physiologic regulation of the autonomic nervous system. CRF acts within the CNS to modify sympathetic, parasympathetic and adrenomedullary efferent activity which in turn results in metabolic, gastrointestinal, and cardiovascular changes. These neuroendocrine-induced changes in metabolic substrate availability, gastrointestinal activity and cardiovascular function may represent a generalized adaptive response to various types of stress that is coordinated by a single peptide, CRF.

REFERENCES

1. **Vale, W., Rivier, A., Brown, M. R., Spiess, J., Koob, G., Swanson, L., Bilezikjian, L., Bloom, F., and Rivier, J.,** Chemical and biological characterization of corticotropin-releasing factor, *Rec. Prog. Horm. Res.*, 39, 245, 1983.
2. **Lederis, K., Leffer, A., McMaster, D., and Moore, G.,** Complete amino acid sequence of urotensin-I, a hypotensive and corticotropin-releasing neuropeptide from *Catostomus*, *Science*, 218, 162, 1982.
3. **Montecucchi, P. C., Anastasi, A., de Castiglione, R., and Erspamer, V.,** Isolation and amino acid composition of sauvagine, *Int. J. Pept. Protein Res.*, 16, 191, 1980.
4. **Brown, M. R., Fisher, L. A., Spiess, J., Rivier, C., Rivier, J., and Vale, W.,** Corticotropin-releasing factor: actions on the sympathetic nervous system and metabolism, *Endocrinology*, 111, 928, 1982.
5. **Brown, M. R., Fisher, L. A., Rivier, J., Spiess, J., Rivier, C., and Vale, W.,** Corticotropin-releasing factor: effects on the sympathetic nervous system and oxygen consumption, *Life Sci.*, 30, 207, 1982.
6. **Brown, M. R. and Fisher, L. A.,** Central nervous system effects of corticotropin releasing factor in the dog, *Brain Res.*, 280, 75, 1983.
7. **Brown, M. R. and Fisher, L. A.,** Corticotropin-releasing factor: effects on the autonomic nervous system and visceral systems, *Fed. Proc.*, 44, 243, 1985.
8. **Lenz, H. J., Fisher, L. A., Vale, W. W., and Brown, M. R.,** Corticotropin-releasing factor, sauvagine, and urotensin I: effects on blood flow, *Am. J. Physiol.*, 249 *(Regul. Integr. Comp. Physiol.*, 18), R85, 1985.
9. **Brown, M. R. and Fisher, L. A.,** unpublished data, 1982.
10. **Kurosawa, M., Sato, A., Swanson, R. S., and Takahashi, Y.,** Sympathoadrenal medullary function in response to intracerebroventricularly injected corticotropin-releasing factor in anesthetized rats, *Brain Res.*, 367, 250, 1986.
11. **Fisher, L. A. and Brown, M. R.,** unpublished data, 1983.
12. **Fisher, L. A.,** Corticotropin-releasing factor: effects on baroreflex control of heart rate, *Soc. Neurosci. Abstr.*, 11, 195, 1985.
13. **Tache, Y., Goto, Y., Gunion, M. W., Vale, W., Rivier, J., and Brown, M.,** Inhibition of gastric acid secretion in rats by intracerebral injection of corticotropin-releasing factor, *Science*, 222, 935, 1983.
14. **Brown, M.,** Corticotropin releasing factor: central nervous system sites of action, *Brain Res.*, 399, 10, 1986.
15. **Fisher, L. A., Rivier, J., Rivier, C., Spiess, J., Vale, W., and Brown, M. R.,** Corticotropin-releasing factor (CRF): central effects on mean arterial pressure and heart rate in rats, *Endocrinology*, 110, 2222, 1982.
16. **Fisher, L. A., Jessen, G., and Brown, M. R.,** Corticotropin-releasing factor (CRF): mechanism to elevate mean arterial pressure and heart rate, *Regul. Pept.*, 5, 153, 1983.
17. **Brown, M. R. and Fisher, L. A.,** CRF: integration of the neuroendocrine and autonomic nervous system responses to stress, in *Endocrinology*, Labrie, F. and Proulx, L., Eds., Elsevier, Amsterdam, 1984, 597.
18. **Fisher, L. A. and Brown, M. R.,** Corticotropin-releasing factor and angiotensin II: comparison of CNS actions to influence neuroendocrine and cardiovascular function, *Brain Res.*, 296, 41, 1984.
19. **Fisher, L. A., and Brown, M. R.,** Corticotropin-releasing factor: central nervous system effects on the sympathetic nervous system and cardiovascular regulation, in *Current Topics in Neuroendocrinology*, Vol. 3, Ganten, D. and Pfaff, D., Eds., Springer-Verlag, Heidelberg, 1983, 87.
20. **Lenz, H. J., Hester, S. E., and Brown, M. R.,** Corticotropin-releasing factor, mechanisms to inhibit gastric acid secretion in conscious dogs, *J. Clin. Invest.*, 75, 889, 1985.
21. **Brown, M. R., Fisher, L. A., Webb, V., Vale, W. W., and Rivier, J. E.,** Corticotropin-releasing factor: a physiologic regulator of adrenal epinephrine secretion, *Brain Res.*, 328, 355, 1985.
22. **Brown, M. R., Gray, T. S., and Fisher, L. A.,** Corticotropin-releasing factor receptor antagonist: effects on the autonomic nervous system and cardiovascular function, *Regul. Pept.*, 16, 321, 1986.
23. **Brown, M. R. and Fisher, L. A.,** Brain peptide regulation of adrenal epinephrine secretion, *Am. J. Physiol.*, 247 *(Endocrinol. Metab.* 10), E41, 1984.
24. **Young, J. B., Rosa, R. M., and Landsberg, L.,** Dissociation of sympathetic nervous system and adrenal medullary responses, *Am. J. Physiol.*, 247 *(Endocrinol. Metab.* 10), E35, 1984.
25. **Brown, M. R., Hauger, R., and Fisher, L. A.,** Autonomic and cardiovascular effects of corticotropin-releasing factor in the spontaneously hypertensive rat, *Brain Res.*, 441, 33, 1988.
26. **Hattori, T., Hushimoto, K., and Ota, Z.,** Brain corticotropin releasing factor in the spontaneously hypertensive rat, *Hypertension*, 8, 1027, 1986.

Chapter 21

CRF: CENTRAL NERVOUS SYSTEM ACTION TO INFLUENCE GASTROINTESTINAL FUNCTION AND ROLE IN THE GASTROINTESTINAL RESPONSE TO STRESS

Yvette Taché, Mark M. Gunion, and Robert Stephens

TABLE OF CONTENTS

I. INTRODUCTION

Pioneer clinical observations have pointed out the stomach's link to the brain as early as the beginning of the last century. Beaumont noted in his fistulous subject that "fear, anger, or whatever depresses or disturbs the nervous system" was accompanied by suppression of gastric secretion and by a marked delay in digestion and emptying of the stomach.[1] In the 1930s, Cannon[2] and Selye[3] brought experimental evidence that the fight or flight response is associated with a diminished gastric acidity and motor activity and that various stressors induced the formation of gastric ulcers. Despite the early recognition that emotional, chemical, or physical stimuli influence gastrointestinal function, neurohumoral mechanisms underlying such alterations have not been fully elucidated. In 1981, Vale et al.[4] characterized from ovine hypothalami the structure of the corticotropin-releasing factor (CRF) and established its major regulatory role in modulating the endocrine response to stress.[5-7] These findings added to the presence of CRF-like immunoreactivity in brain areas involved in the regulation of visceral function,[8] gave new impetus to evaluate the possible role of neuronal CRF in mediating the gastrointestinal response to stress.

This chapter reviews the central nervous system actions of CRF to influence gastric secretion and gastrointestinal motor activity. The implications of these findings in the understanding of the mechanisms by which stress alters gastrointestinal function will also be discussed.

II. ALTERATION OF GASTRIC SECRETION BY CENTRAL INJECTION OF CRF

A. INHIBITORY EFFECT ON ACID SECRETION

In rats, cerebrospinal fluid (CSF) injection of CRF (0.1 to 2.3 nmol) and the amphibian counterpart, sauvagine (0.002 to 0.2 nmol), inhibits gastric acid secretion stimulated by intravenous pentagastrin, or vagus-dependent stimuli such as central thyrotropin-releasing factor (TRH), pylorus ligation, gastric distention and 2-deoxy-D-glucose but does not modify the gastric response to histamine.[9,10] The inhibition of gastric acid output induced by a maximal effective dose of CRF or sauvagine results from decreases in both, the concentration of acid and the volume of secretion.[9,10] Ovine and rat CRF show similar potency[11] (Figure 1) whereas sauvagine appears several times more potent that CRF.[10] Based on ED_{50}, intracerebroventricular injection of CRF is a more potent inhibitor of acid secretion than β-endorphin and neurotensin but less potent than bombesin, calcitonin, and calcitonin gene-related peptide.[12] Similar results have been obtained in conscious dogs. Injection into the third ventricle of CRF (0.2 to 6 nmol/kg) induces a long-lasting inhibition of pentagastrin-, meal-, and 2-deoxy-D-glucose- but not histamine-stimulated gastric acid secretion.[13-14] The ED_{50} to inhibit pentagastrin-stimulated gastric secretion is 2 nmol/kg.[13] Urotensin-I, a CRF-like peptide isolated from the teleost fish, and sauvagine appear twice as potent as CRF.[14]

The mechanisms through which CSF injection of CRF inhibits gastric acid secretion have been investigated. Peripheral injection of CRF and sauvagine also inhibits gastric acid secretion in the rat and the dog.[10,15-18] However, the inhibitory effect observed upon CSF injection is mediated through the central nervous system (CNS) and not by leakage of the peptides into the peripheral blood. This was demonstrated in the dog by the absence of CRF-, sauvagine-, and urotensin-immunoreactivity in the circulation following injection of the peptides into the third ventricle.[14,19] Additional studies in rats have attempted to identify the central site(s) at which CRF acts to inhibit gastric secretion. Responsive sites, investigated by direct microinjection of CRF into the brain parenchyma, have been localized in the ventromedial hypothalamus, the paraventricular nucleus and to a lower degree the lateral

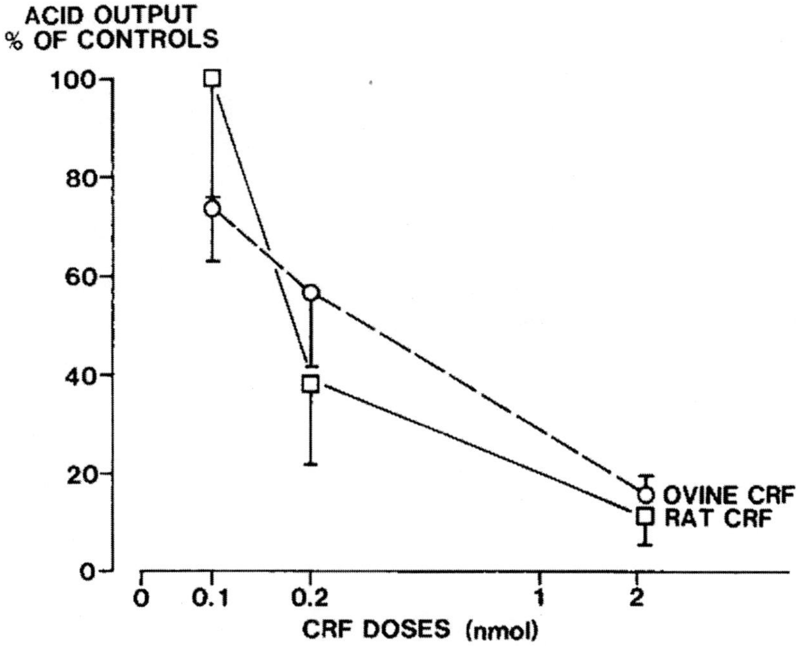

FIGURE 1. Dose-dependent inhibitory effect of ovine and rat CRF injected intracisternally on gastric acid secretion in pylorus-ligated rats. CRF or saline was injected under light ether anesthesia just prior to pylorus ligation. Conscious rats were sacrificed 2 h later and gastric secretion collected.

hypothalamus whereas the caudate putamen, and the dorsomedial frontal cortex were found to be inactive.[9,20] Interestingly, the inhibition of gastric acid output induced by intrahypothalamic injection of CRF is associated with an increase in the volume of gastric secretion (Figure 2).[20] Intrathecal injection of CRF also suppresses dose dependently gastric acid secretion in rats suggesting the existence of inhibitory sites in the spinal cord.[21]

Central injection of CRF has been well established to stimulate the sympathoadrenal activity as assessed by the marked elevation in plasma concentration of norepinephrine and epinephrine in rats and dogs.[22,23] Sauvagine appears five to ten times more potent than CRF to increase plasma levels of catecholamine in rats.[24] The CNS action of CRF on the parasympathetic nervous system activity has not been directly assessed, although pharmacologic evidence suggests a decrease in the parasympathetic outflow to the heart.[25] In the rat, acute adrenalectomy, vagotomy, spinal cord transection at the fifth cervical level, and α_2-adrenergic blockade using yohimbine prevented the inhibitory effect of intracisternal injection of CRF on acid secretion in the rat.[9,21] By contrast, naloxone, indomethacin, and hypophysectomy did not alter CRF or sauvagine action.[9-11] Taken together, these results suggest that the inhibitory effect of centrally injected CRF in the rat is independent of prostaglandin synthetic pathways, opioid receptor interaction, and pituitary hormone secretion but is mediated by the activation of the sympathetic nervous system with concurrent inhibition of vagal outflow to the stomach. In the dog, CRF inhibitory effect on acid secretion results mostly from the activation of the sympathetic nervous system to the stomach[19,23] since intracerebroventricular CRF-induced inhibition of pentagastrin-stimulated acid secretion was not altered by vagotomy or adrenalectomy but was completely reversed by ganglionic blockade with hexamethonium.[13,14,26] In addition, CRF action appears to involve interaction with peripheral vasopressin receptors. This was demonstrated by the facts that injection of CRF into the CSF increases plasma levels of vasopressin in the dog, unlike in the rat.[19,25] Moreover, intravenous injection of vasopressin inhibits gastric acid secretion and a vasopressin antagonist given intravenously partly reverses CRF inhibitory effect.[13]

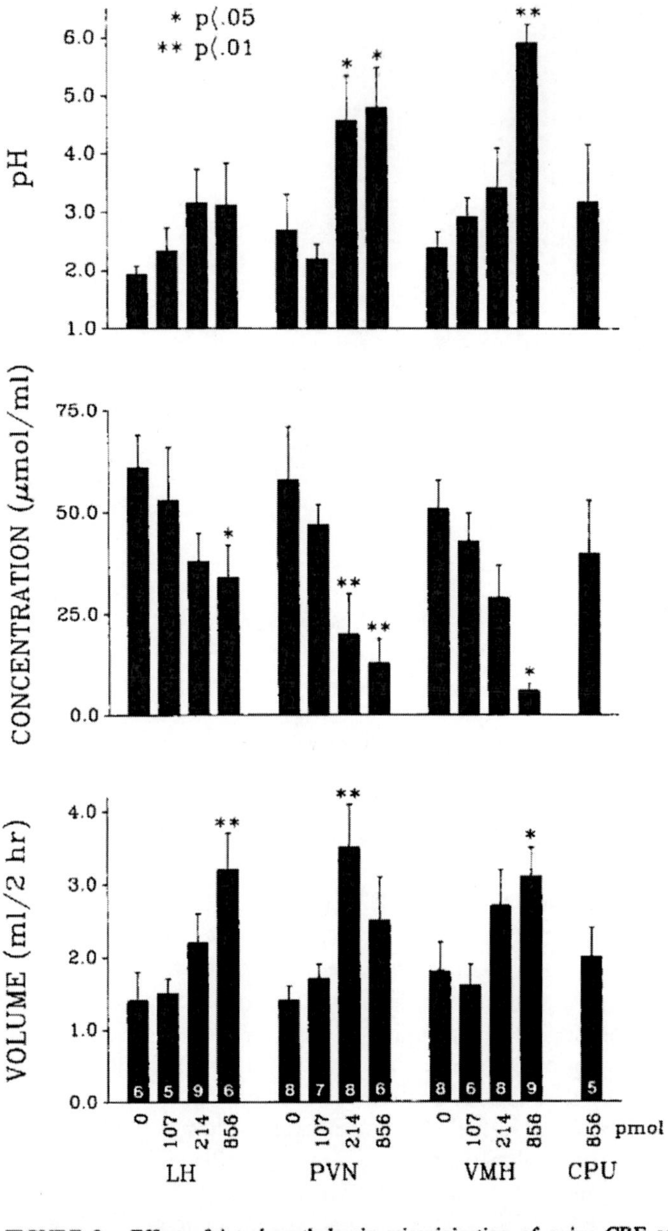

FIGURE 2. Effect of intrahypothalamic microinjection of ovine CRF on gastric secretion in pylorus-ligated rats. Rats under ether anesthesia were microinjected bilaterally (1 μl/site) with saline or CRF into the lateral hypothalamus (LH), ventromedial hypothalamus (VMH), or paraventricular nucleus (PVN) just prior to pylorus ligation. Conscious rats were sacrificed 2 h later and gastric secretion collected.

B. EFFECTS ON GASTRIN RELEASE

In pylorus-ligated rats, plasma levels of gastrin are enhanced by intracisternal injection of CRF.[9] Whether such an increase reflects a direct stimulation of G cells in the gastric mucosa or a regulatory feedback mechanisms in response to the inhibition of acid secretion has not been established. In the dog, under conditions of constant intragastric pH, injection of CRF into the CSF does not modify basal and meal-stimulated gastrin secretion.[13,14] These results clearly demonstrate that the inhibitory effect of CRF on acid secretion in both rats and dogs is not secondary to the suppression of the hormone stimulating acid secretion.[27]

C. STIMULATORY EFFECT ON BICARBONATE SECRETION

Intrahypothalamic injection of CRF in rats is associated with an increase in gastric secretory volume whereas the concentration of acid is decreased (Figure 2).[20] Microinjection into the hypothalamus of other peptides inhibiting acid secretion such as bombesin, calcitonin-gene related peptide, or calcitonin does not increase gastric secretory volume as does CRF.[28] Further analysis of the nonparietal gastric secretion demonstrated that microinjection of CRF into the ventromedial hypothalamus markedly stimulated both the concentration and the output of bicarbonate, and potassium in conscious pylorus-ligated rats. The stimulatory effect on gastric bicarbonate secretion is unrelated to the inhibition of acid secretion.[29]

D. EFFECTS ON BLOOD FLOW

In conscious dog, third ventricle injection of sauvagine or urotensin induced an immediate and sustained rise in superior mesenteric blood flow. By contrast, CRF injected into the third ventricle did not modify superior mesenteric or left gastric artery blood flow.[19,26] The effect of the central injection of CRF or related peptides on gastric mucosal blood flow has not yet been investigated either in rats or dogs.

III. ALTERATION OF GASTROINTESTINAL TRANSIT BY CENTRAL INJECTION OF CRF

A. INHIBITORY EFFECT ON GASTRIC EMPTYING

CRF injected into the cisterna magna or the lateral ventricle inhibits gastric emptying of a liquid, nonnutrient meal in conscious rats.[30-32] A CRF dose of 0.2 nmol inhibits by 93% gastric emptying measured 10 min after ingestion of a saline meal.[30] The CRF-related peptide, sauvagine, injected into CSF appears equipotent as CRF to delay gastric emptying whereas other hypothalamic releasing factors such as growth hormone-releasing factor, GRF(1-40), had no effect and TRH stimulated gastric emptying.[30,33,34] The inhibitory effect of CRF and sauvagine is dose dependent, rapid in onset and long acting (Figure 3).[30,33] In the dogs, third ventricle injection of CRF (1 nmol/kg) delays gastric emptying of a peptone meal.[26] By contrast, other study reported that lateral ventricle injection of CRF at doses (0.1 to 0.4 nmol/kg) not associated with behavioral changes, did not modify gastric emptying of a saline meal.[35]

CRF administered intravenously and into the CSF is equipotent to inhibit gastric emptying in rats and dogs (Figure 3).[30,36] However, intravenous injection of anti-CRF serum completely prevented intravenous but not intracisternal injection of CRF-induced inhibition of gastric emptying in the rat.[30] These data indicate that the effect of CRF injected into the CSF represents a CNS-mediated effect through brain sites which remain to be established. Central injection of CRF or sauvagine increases β-endorphin release.[37] Opiates are known to delay gastric emptying, an effect which is reversible by naloxone.[38] In rats, the inhibition of gastric emptying induced by CSF injection of CRF or sauvagine is not modified by naloxone or acute adrenalectomy but reversed by vagotomy. These results indicate that CRF action is unlikely to be secondary to changes in opioid peptides and adrenal-derived substances but may be mediated by modulation of the parasympathetic nervous system.[30,33]

B. STIMULATORY EFFECT ON LARGE INTESTINAL TRANSIT

Recent studies demonstrate that intracerebroventricular injection of CRF in rats inhibits also small bowel intestinal transit in a dose-dependent manner. However, the most pronounced effect on intestinal tract is a stimulation of the large intestinal transit and fecal pellet output.[31,32] Brain sites and peripheral mechanism of action involved in CRF-induced alteration of intestinal transit are still unknown.

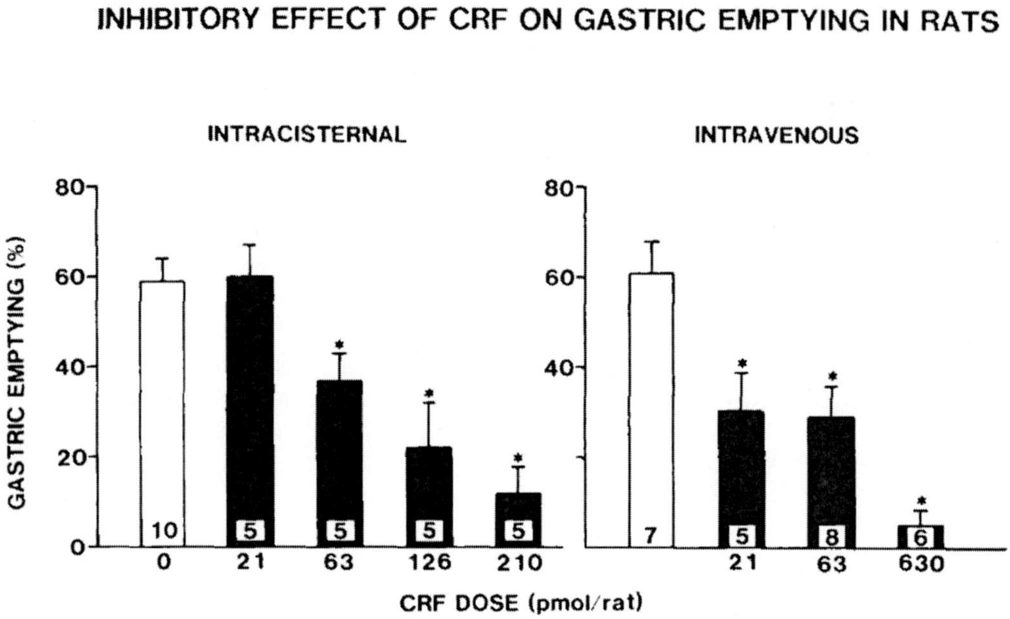

FIGURE 3. Dose-dependent inhibitory effect of rat CRF injected intracisternally or intravenously on gastric emptying of a liquid meal in conscious rats. Five minutes after injection of peptide or saline, a methylcellulose phenol red solution was given per gavage and gastric emptying was measured 20 min later.

IV. ALTERATIONS OF GASTROINTESTINAL CONTRACTILITY BY CENTRAL INJECTION OF CRF

A. INHIBITION OF GASTRIC CONTRACTILITY

Intercisternal injection of CRF suppresses gastric contractility stimulated by intracisternal TRH and 2-deoxy-D-glucose but not by peripheral administration of carbachol in anesthetized and fasted rats with strain-gauges implanted into the distal part of the corpus of the stomach.[39] In fasted dogs, gastrointestinal motility is organized into migrating motor complexes (MMC) which occur cyclically (90 to 120 min) in the stomach and are propagated to the ileum. Injection of CRF into the lateral brain ventricle immediately abolishes the cyclic activity fronts of the gastric antrum which is replaced by irregular contractions of small amplitude.[40] The disruption of the gastric cyclic contractions lasts for several hours.[40] The effect of CRF injected into the CSF on gastric contractility is centrally mediated based on the diminished activity or the inactivity of intravenous injection of the peptide in rats and dogs.[39,40] The inhibitory effect of intracerebroventricular injection of CRF on gastric contractility in the dog is not secondary to the activation of the pituitary adrenal axis[40-42] but involved vagal dependent pathways and the blockade of the cyclic release of motilin.[41,42]

B. ALTERATION OF JEJUNAL CONTRACTILITY

Intracerebroventricular injection of CRF at a low dose which abolishes the cyclic activity fronts of the dog antrum did not alter the MMC pattern in the proximal jejunum.[40,42] Higher doses of CRF delay for 4 to 5 h the occurrence of MMC in the proximal jejunum.[40]

V. PHYSIOLOGICAL ROLE OF CENTRAL CRF IN STRESS-INDUCED ALTERATION OF GASTROINTESTINAL FUNCTION

Previous studies demonstrate that CRF plays a major modulatory role in eliciting the endocrine, autonomic, and behavioral responses to stress.[5,7,43-46] Several lines of anatomical and pharmacological evidence also support the possibility that central CRF may be involved in mediating the alterations of gastrointestinal function induced by some stressors. CRF cell bodies and/or fibers and specific binding sites have been identified in brain structures such as the anterior cingulate cortex, hypothalamus, amygdala, locus coeruleus and dorsal motor nucleus complex[8,47,48] that are known to regulate gastric secretion,[30] gastrointestinal contractility,[49] and gastric ulceration.[50] Injection of CRF into the CSF elicits a pattern of gastrointestinal secretory and motor dysfunction similar to that elicited by certain models of stress. Restraint stress in rats inhibits gastric secretion, delays gastric emptying, and small intestinal transit, and enhances large intestinal transit and fecal pellet output.[31,51-55] Acoustic stress in dogs has been shown to delay the occurrence of gastric MMC without modifying jejunal MMC.[41] Moreover, the CRF antagonist, α-helical CRF(9-41), injected into the CSF, completely prevents surgical stress-induced hypogastric secretion[59] and restraint-induced increase in large intestinal transit and associated increase in fecal excretion.[31] Taken together these neuroanatomic and pharmacologic data support the view that neuronal CRF may be directly involved in the inhibition of gastric acid secretion and enhanced propulsive activity of the large intestine elicited by restraint or surgical stress. By contrast, CRF does not appear to mediate the ulcerogenic response in the cold restraint stress model. Injection of the peptide into the CSF does not cause gastric erosions[56] and when injected prior to exposure to cold restraint stress, it prevents the formation of gastric pathology.[57,58] Moreover the CRF antagonist does not inhibit the development of gastric ulceration induced by cold restraint stress.[58]

VI. CONCLUSIONS

Convergent data in the rat and the dog indicate that CRF or related peptides such as sauvagine and urotensin I acts into the brain to induce inhibition of vagally mediated stimulation of gastric acid secretion and contractility, delay of gastric emptying, prevention of gastric ulceration induced by cold restraint stress, and stimulation of gastric bicarbonate secretion and propulsive activity of the large intestine. CRF action on gastrointestinal function appears mostly expressed by altering the autonomic nervous system activity. The gastrointestinal secretory and motor responses to central injection of CRF reproduced the effects of elicited by certain stressors. Moreover, preliminary data indicate that the CRF antagonist, α-helical CRF(9-41), blocked the gastric secretory and large intestine responses to restraint and surgical stress. Collectively these pharamacological studies and the presence of CRF like-immunoreactivity in brain structures regulating the autonomic nervous system are consistent with a role for endogenous neuronal CRF in the mechanisms by which certain stressors alters gastrointestinal secretory and motor function. By contrast, CRF does not appear to be involved in mediating the formation of gastric ulceration elicited by cold restraint stress exposure.

ACKNOWLEDGMENT

The author's work cited here was supported by the National Institute of Diabetes & Digestive & Kidney Diseases, Grants AM 33061 and AM 17238 and the David H. Murdock Foundation for Advanced Brain Studies. Dr. J. Rivier (Salk Institute, La Jolla, CA) is greatly acknowledged for providing rat and ovine CRF and α-helical CRF(9-41).

REFERENCES

1. **Beaumont, W.,** *Experiments and Observations on the Gastric Juice and the Physiology of Digestion,* Osler, W., Ed., Dover Publications, New York, 1833, 1.
2. **Cannon, W. B.,** *Bodily Changes in Pain, Hunger, Fear and Rage,* Appleton, New York, 1929, 1.
3. **Selye, H.,** A syndrome produced by diverse nocuous agent, *Nature,* 138, 32, 1936.
4. **Vale, W., Spiess, J., Rivier, C., and Rivier, J.,** Characterization of a 41-residue ovine hypothalamic peptide that stimulates secretion of corticotropin and β-endorphin, *Science,* 213, 1394, 1981.
5. **Rivier, J., Rivier, C., and Vale, W.,** Synthetic competitive antagonists of corticotropin-releasing factor: effect on ACTH secretion in the rat, *Science,* 224, 889, 1984.
6. **Rivier, C., Rivier, J., and Vale, W.,** Stress-induced inhibition of reproductive functions: role of endogenous corticotropin-releasing factor, *Science,* 231, 607, 1986.
7. **Rivier, C. and Vale, W.,** Involvement of corticotropin-releasing factor and somatostatin in stress-induced inhibition of growth hormone secretion in the rat, *Endocrinology,* 117, 2478, 1985.
8. **Swanson, L. W., Sawchenko, P. E., Rivier, J., and Vale, W. W.,** Organization of ovine corticotropin-releasing factor immunoreactive cells and fibers in the rat brain: an immunohistochemical study, *Neuroendocrinology,* 36, 165, 1983.
9. **Taché, Y., Goto, Y., Gunion, M. W., Vale, W., Rivier, J., and Brown, M.,** Inhibition of gastric acid secretion in rats by intracerebral injection of corticotropin-releasing factor, *Science,* 222, 935, 1983.
10. **Improta, G. and Broccardo, M.,** Sauvagine: effects on gastric acid secretion in rats, *Peptides,* 9, 843, 1988.
11. **Taché, Y. and Gunion, M.,** Corticotropin-releasing factor: central action to influence gastric secretion, *Fed. Proc.,* 44, 255, 1985.
12. **Lenz, H. J., Forquignon, I., Druge, G., Friemel, E., Rivier, J. E., Raedler, A., and Greten, H.,** Central nervous system effects of hypothalamic peptides on gastric acid and proximal duodenal bicarbonate secretions in rats, *Gastroenterology,* 92, 1501 (Abstr.), 1987.
13. **Lenz, H. J., Hester, S. E., and Brown, M. R.,** Corticotropin-releasing factor mechanisms to inhibit gastric acid secretion in conscious dogs, *J. Clin. Invest.,* 75, 889, 1985.
14. **Lenz, H. J., Hester, S. E., Klapdor, R., Hestor, S. E., Webb, V. J., Galyean, R. F., Rivier, J. E., and Walsh, J. H.,** Inhibition of gastric acid secretion by brain peptides in the dog: role of the autonomic nervous system and gastrin, *Gastroenterology,* 91, 905, 1986.
15. **Pappas, T. N., Hamel, D., Debas, H. T., and Taché, Y.,** Corticotropin-releasing factor (CRF) inhibits meal-stimulated acid but not gastrin release, *Gastroenterology,* 88, 1530 (Abstr.), 1985.
16. **Taché, Y., Goto, Y., Gunion, M., Rivier, J., and Debas, H.,** Inhibition of gastric acid secretion in rats and in dogs by corticotropin-releasing factor, *Gastroenterology,* 86, 281, 1984.
17. **Konturek, S. J., Bilski, J., Pawlik, W., Thor, P., Czarnobilski, K., Szoke, B., and Schally, A. V.,** Gastrointestinal secretory, motor and circulatory effects of corticotropin releasing factor (CRF), *Life Sci.,* 37, 1231, 1985.
18. **Todisco, A., Park, J., Lezoche, E., Debas, H., Taché, Y., and Yamada, T.,** Peripheral acid inhibitory action of corticotropin releasing factor in rats: mediation by non-gastric mechanisms, *Gastroenterology,* 92, 919, 1987.
19. **Lenz, H. J., Fisher, L. A., Vale, W. W., and Brown, M. R.,** Corticotropin-releasing factor, sauvagine, and urotensin I: effects on blood flow, *Am. J. Physiol.,* 249, R85, 1985.
20. **Gunion, M. W. and Taché, Y.,** Intrahypothalamic microinfusion of corticotropin-releasing factor inhibits gastric acid secretion but increases secretion volume in rats, *Brain Res.,* 411, 156, 1987.
21. **Taché, Y., Hamel, D., and Gunion, M.,** Inhibition of gastric acid secretion in rats by intracisternal or intrathecal injection of rat corticotropin releasing factor (rCRF), *Dig. Dis. Sci.,* 29, 86S (Abstr.), 1984.
22. **Brown, M. R., Fisher, L. A., Rivier, J., Spiess, J., Rivier, C., and Vale, W.,** Corticotropin-releasing factor: effects on the sympathetic nervous system and oxygen consumption, *Life Sci.,* 30, 207, 1982.
23. **Brown, M.,** Corticotropin releasing factor: Central nervous system sites of action, *Brain Res.,* 399, 10, 1986.
24. **Brown, M. R., Fisher, L. A., Spiess, J., Rivier, J., Rivier, C., and Vale, W.,** Comparison of the biological actions of corticotropin-releasing factor and sauvagine, *Regul. Pept.,* 4, 107, 1982.
25. **Fisher, L. A. and Brown, M. R.,** Corticotropin-releasing factor and angiotensin II: comparison of CNS actions to influence neuroendocrine and cardiovascular function, *Brain Res.,* 296, 41, 1984.
26. **Lenz, H. J.,** Brain regulation of gastric secretion, emptying and blood flow by neuropeptides in conscious dogs, *Gastroenterology,* 92, 1500, 1987.
27. **Walsh, J. H.,** Gastrointestinal Peptide Hormones, in *Gastrointestinal Disease,* 2nd ed., Sleisenger, M. H. and Fordtran, J. S., Eds., W. B. Saunders, Philadelphia, 1983, 54.
28. **Taché, Y.,** Central regulation of gastric acid secretion, in *Physiology of the Gastrointestinal Tract,* 2nd ed., Johnson, L. R., Christensen, J., Jackson, M., Jacobson, E. D., and Walsh, J. H., Eds., Raven Press, New York, 1987, 911.

29. **Gunion, M. W., Taché, Y., and Kauffman, G. L.,** Intrahypothalamic corticotropin-releasing factor (CRF) increases gastric bicarbonate content, *Gastroenterology,* 88, 1407 (Abstr.), 1985.
30. **Taché, Y., Maeda-Hagiwara, M., and Turkelson, C. M.,** Central nervous system action of corticotropin-releasing factor to inhibit gastric emptying in rats, *Am. J. Physiol.,* 253, G241, 1987.
31. **Williams, C. L., Peterson, J. M., Villar, R. G., and Burks, T. F.,** Corticotropin-releasing factor (CRF) directly mediates colonic responses to stress, *Am. J. Physiol.,* 253, G582, 1987.
32. **Lenz, H. J., Burlage, M., Raedler, A., and Greten, H.,** Central nervous system effects of corticotropin-releasing factor on gastric emptying and intestinal transit in rats, *Gastroenterology,* 92, 1500 (Abstr.), 1987.
33. **Broccardo, M., Improta, G., and Melchiorri, P.,** Effect of sauvagine on gastric emptying in conscious rats, *Eur. J. Pharmacol.,* 85, 111, 1982.
34. **Maeda-Hagiwara, M. and Taché, Y.,** Central nervous system action of TRH to stimulate gastric emptying in rats, *Regul. Pept.,* 17, 199, 1987.
35. **Pappas, T. N., Debas, H. T., and Taché, Y.,** Corticotropin-releasing factor inhibits gastric emptying in dogs: studies on its mechanism of action, *Peptides,* 8, 1011, 1987.
36. **Pappas, T. N., Debas, H. T., and Taché, Y.,** Corticotropin-releasing factor inhibits gastric emptying in dogs, *Regul. Pept.,* 11, 193, 1985.
37. **Rivier, C. L. and Plotsky, P. M.,** Mediation by corticotropin releasing factor (CRF) of adenohypophysial hormone secretion, *Annu. Rev. Physiol.,* 48, 475, 1986.
38. **Burks, T. F., Galligan, J. J., Porreca, F., and Barber, W. D.,** Regulation of gastric emptying, *Fed. Proc.,* 44, 2897, 1985.
39. **Garrick, T., Veiseh, A., Weiner, H., and Taché, Y.,** Corticotropin releasing hormone (CRH) acts centrally to suppress gastric contractility, *Gastroenterology,* 92 (abstr.), 1400, 1987.
40. **Bueno, L. and Fioramonti, J.,** Effects of corticotropin-releasing factor, corticotropin and cortisol on gastrointestinal motility in dogs, *Peptides,* 7, 73, 1986.
41. **Gue, M., Fioramonti, J., Frexinos, J., Alvinerie, M., and Bueno, L.,** Influence of acoustic stress by noise on gastrointestinal motility in dogs, *Dig. Dis. Sci.,* 32, 1411, 1987.
42. **Bueno, L., Fargeas, M. J., Gue, M., Peeters, T. L., Bormans, V., and Fioramonti, J.,** Effects of corticotropin-releasing factor on plasma motilin and somatostatin levels and gastrointestinal motility in dogs, *Gastroenterology,* 91, 884, 1986.
43. **Krahn, D. D., Gosnell, B. A., Grace, M., and Levine, A. S.,** CRF antagonist partially reverses CRF- and stress-induced effects on feeding, *Brain Res. Bull.,* 17, 285, 1986.
44. **Brown, M. R., Gray, T. S., and Fisher, L. A.,** Corticotropin-releasing factor receptor antagonist: effects on the autonomic nervous system and cardiovascular function, *Regul. Pept.,* 16, 321, 1986.
45. **Brown, M. R., Fisher, L. A., Webb, V., Vale, W. W., and Rivier, J. E.,** Corticotropin-releasing factor: a physiologic regulator of adrenal epinephrine secretion, *Brain Res.,* 328, 355, 1985.
46. **Tazi, A., Dantzer, R., Le Moal, M., Rivier, J., Vale, W., and Koob, G. F.,** Corticotropin-releasing factor antagonist blocks stress-induced fighting in rats, *Regul. Pept.,* 18, 37, 1987.
47. **Skofitsch, G., Insel, T. R., and Jacobowitz, D. M.,** Binding sites for corticotropin releasing factor in sensory areas of the rat hindbrain and spinal cord, *Brain Res. Bull.,* 15, 519, 1985.
48. **De Souza, E. B., Insel, T. R., Perrin, M. H., Rivier, J., Vale, W. W., and Kuhar, M. J.,** Corticotropin-releasing factor receptors are widely distributed within the central nervous system: an autoradiography study, *J. Neurosci.,* 5, 3189, 1985.
49. **Roman, C. and Gonella, J.,** Extrinsic control of digestive tract motility, in *Physiology of the Gastrointestinal Tract,* 2nd ed., Johnson, L. R., Ed., Raven Press, New York, 1987, 507.
50. **Taché, Y.,** The peptidergic brain-gut axis: influence on gastric ulcer formation, *Chronobiol. Int.,* 4, 11, 1987.
51. **Menguy, R.,** Effects of restraint stress on gastric secretion in the rat, *Dig. Dis.,* 5, 911, 1960.
52. **Schwille, P. O., Engelhardt, W., Wolf, U., and Hanisch, E.,** Infusion of somatostatin antiserum in the rat — failure to raise stress induced low gastric acid secretion, *Horm. Metab. Res.,* 13, 710, 1981.
53. **Hayase, M. and Takeuchi, K.,** Gastric acid secretion and lesion formation in rats under water-immersion stress, *Dig. Dis. Sci.,* 31, 166, 1986.
54. **Paré, W. P. and Livingston, A., Jr.,** Shock predictability and gastric secretion in the chronic gastric fistula rat, *Physiol. Behav.,* 11, 521, 1973.
55. **Koo, M. W. L., Ogle, C. W., and Cho, C. H.,** The effect of cold-restraint stress on gastric emptying in rats, *Pharmacol. Biochem. Behav.,* 23, 969, 1985.
56. **Goto, Y. and Taché, Y.,** Gastric erosions induced by intracisternal thyrotropin-releasing hormone (TRH) in rats, *Peptides,* 6, 153, 1985.
57. **Gunion, M. W. and Taché, Y.,** Gastric mucosal damage is inhibited by intraventromedial hypothalamic corticotropin releasing factor, *Soc. Neurosci. Abstr.,* 12, 644, 1986.
58. **Krahn, D. D., Wright, B., Billington, C. J., and Levine, A. S.,** Exogenous corticotropin-releasing factor inhibits stress-induced gastric ulceration, *Soc. Neurosci. Abstr.,* 12, 1063, 1986.
59. **Stephens, R. L., Jr., Yang, H., Rivier, J., and Taché, Y.,** Intracisternal injection of CRF antagonist blocks surgical stress-induced inhibition of gastric secretion in the rat, *Peptides,* 9, 1067, 1988.

Chapter 22

CLINICAL STUDIES WITH OVINE AND HUMAN CRF IN PSYCHIATRIC PATIENTS AND NORMAL CONTROLS

Ulrich von Bardeleben and Florian Holsboer

TABLE OF CONTENTS

I. INTRODUCTION

In response to enhanced physical or psychological demands the organism releases pro-opiomelanocortin-derived peptides into the peripheral circulation which activate biosynthesis of corticosteroids from the adrenocortex.[1] In addition, following a stressful experience the adrenal medulla secretes epinephrine and the nerve endings release norepinephrine. These hormones act in a concerted fashion to adapt the body to stress. The major central controller of corticotropin cell function remained unidentified until Spiess et al.[2] characterized the corticotropin-releasing hormone (CRH) from ovine hypothalami and after the primary structure of the human CRH was derived from the nucleotid sequence of cloned DNA.[3] Since enhanced activation of the pituitary adrenocortex system represents the most robust neuroendocrine abnormality of the depressed state, psychoneuroendocrinologists rapidly studied the hormonal and behavioral effects of this novel neuropeptide. These studies were focused upon the question — Is CRH of possible relevance to components of the overall symptom complex of affective disorders?

The present overview summarizes our neuroendocrine studies employing ovine (o)- and human (h)-CRH in patients with depression, alcoholism, and panic disorder and in normal controls. These investigations were designed in order to analyze pathophysiology of hypothalamic-pituitary-adrenocortical aberrancies in several affective disorders and to generate hypotheses about a possible role of CRH in mediating depressive behavior.

II. NEUROENDOCRINE RESPONSE TO OVINE AND HUMAN CRH

A. STIMULATION RESPONSE AFTER OVINE CRH IN DEPRESSION

In a first attempt to study the neuroendocrine effects of ovine (o)-CRH we administered 100 μg of the neuropeptide intravenously to eight unmedicated patients with severe endogenous depression.[4] We observed that o-CRH is a potent stimulator of the pituitary adrenal axis in this disorder and compared the responses of ACTH and the adrenal glucocorticosteroids corticosterone, 11-deoxycortisol, cortisol and cortisone with those following synthetic (1-24) ACTH stimulation and a dexamethasone suppression test (DST). From a comparative evaluation of these three pituitary adrenal function tests no convincing correlations among hormonal responses emerged, which suggested to us that hypersecretion of ACTH and adrenocortical steroids in depression reflects a central dysfunction rather than an altered responsiveness of pituitary or adrenal cells.

Next we studied 16 patients with a major endogenous depression who were classified according to their DST status.[5] Ten depressed patients who were refractory to the ACTH and cortisol suppressing effect of dexamethasone had ACTH and cortisol responses to o-CRF which were indistinguishable from six DST-suppressive patients. This finding was further supporting the view that hypersensitivity of the pituitary and/or the adrenocortex toward their specific stimulators is not the prime cause of increased pituitary adrenal function in depressed patients with dexamethasone resistant cortisol levels. A limbic hypothalamic overactivity was considered to be the relevant mechanism underlying hypercortisolism associated with depression.

In a subsequent study we investigated eight patients suffering from a major endogenous depression who first received 100 μg o-CRH and thereafter a DST.[6] All patients showed inadequate suppression of cortisol following dexamethasone. After clinical remission and normalization of the DST they were reinvestigated with an o-CRH stimulation test. We observed that their mean ACTH release was indistinguishable at both sessions. Cortisol and corticosterone output after o-CRH tended to be higher during depression than after recovery. From these data we concluded that increased pituitary ACTH reserve or adrenocortical steroid reserve would not likely be responsible for dexamethasone nonsuppression in depressives.

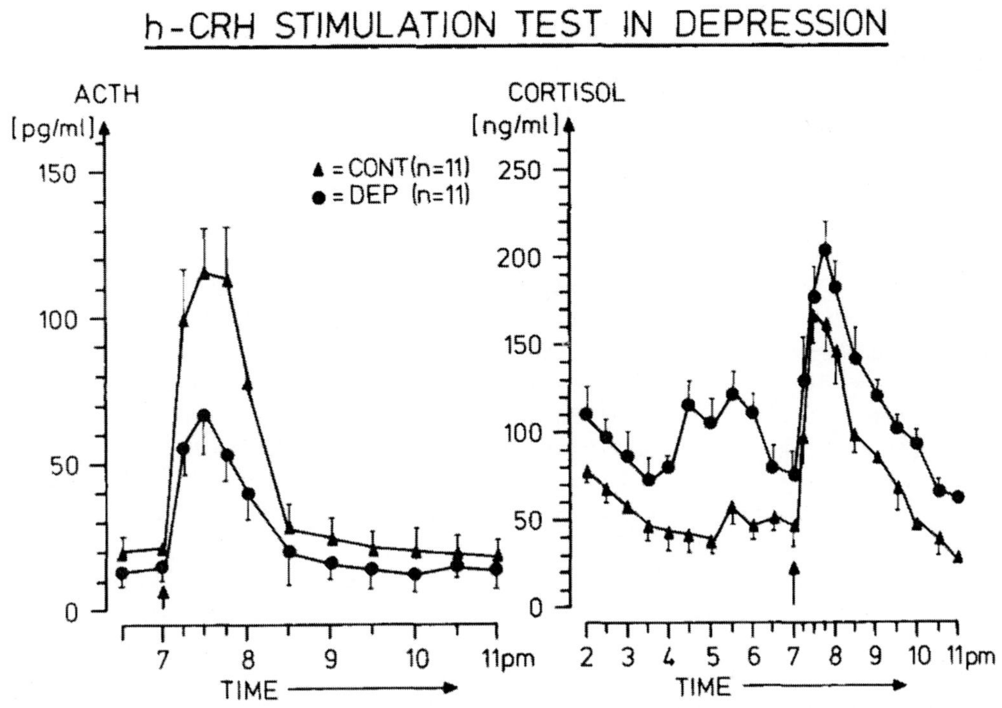

FIGURE 1. ACTH (left) and cortisol (right) secretion (mean ± SEM) at baseline and following (arrows) 100 μg h-CRH administered as an intravenous bolus at 7:00 P.M. to 11 patients with major endogenous depression and 11 controls. (From Holsboer, F., von Bardeleben, U., Gerken, A., Stalla, G. K., and Müller, D. A., *Biol. Psychiatry*, 20, 276, 1985. With permission.)

Taken together, these studies provided good preliminary evidence that neuroendocrine studies with o-CRH would generate data, which strongly support that pituitary-adrenocortical dysfunction in depression is primarily driven by a central disturbance.

B. STIMULATION RESPONSE AFTER HUMAN CRH IN DEPRESSION

Soon after the homologous human (h)-CRH became available for clinical studies we changed our original study protocol and injected 100 μg h-CRH intravenously as a bolus after recording an afternoon cortisol profile from 2:00 P.M. until 7:00 P.M.[7,8] Following administration of h-CRH at 7:00 P.M. we drew blood as indicated in the accompanying figure (Figure 1) until 11:00 P.M. A major advantage of this changed design was that h-CRH is a more physiologic challenge as it mimics spontaneous ACTH pulses and is devoid of antigenic properties.[9] Furthermore, pituitary-adrenocortical challenges during the evening hours, when the endogenous activation is almost absent, provide more vigorous pituitary-adrenocortical responses to exogenous stimuli.

As illustrated in Figure 1, h-CRH administration prompted a marked increase of ACTH and cortisol. ACTH secretion following a bolus of 100 μg h-CRH was significantly blunted in 11 depressives when compared with a matched control population. This difference is also documented by calculated areas under time course curves (AUC), where the net ACTH output for depressives was 3.0 ± 2.6 pg/ml/min × 10^3 and for controls 6.2 ± 3.4 pg/ml/min × 10^3 ($t = 2.4$, $df = 20$, $p < 0.05$; $U = 21$, $p < 0.01$). Mean cortisol secretion between 2:00 and 7:00 P.M. was significantly higher in depressives than in controls (95.3 ± 25.8 ng/ml vs. 56.6 ± 14.6 ng/ml, $t = 4.3$, $df = 20$, $p < 0.001$). However, net cortisol release (expressed as AUC) following h-CRH resulted in indistinguishable cortisol responses (11.5 ± 4.2 ng/ml/min × 10^3 vs. 11.8 ± 4.6 ng/ml/min × 10^3, p = n.s.). The high mean baseline cortisol secretory activity between 2:00 and 7:00 P.M. was found to be associated with blunted ACTH output after h-CRH, because a significantly negative correlation between these two measures emerged ($r = -0.48, p < 0.05$). Such an inverse correlationship between the individual baseline cortisol value at 7:00 P.M. and the corresponding areas under the time course curves of ACTH concentrations was absent. Our observation of blunted ACTH and normal cortisol response to h-CRH in patients with depression is compatible with the following mechanism. Hypersecretion of CRH from the nucleus paraventricularis into hypophyseal portal vessels increases transcription of the POMC gene resulting in a variety of POMC gene products. The major adrenogenic factor is ACTH which stimulates adrenal corticosteroid biosynthesis and gives rise to elevated plasma glucocorticosteroid concentrations. The enhanced output of those glucocorticosteroids which are hydroxylated at the carbon-C-11-position results in inhibitory action at several levels of the feedback system. Thus, elevated glucocorticosteroid levels acting via feedback inhibition and endogenous CRH hyperproduction resulting probably in down-regulation of CRH receptors at corticotrophic pituitary cells may reduce the capacity to release adequate amounts of ACTH after specific stimulation with exogenous h-CRH (see also Section III.B). Persistent LHPA overactivity leads primarily to elevated levels of ACTH which is a trophic hormone, inducing a mild functional hyperplasia of the adrenocortex. Such a phenomenon is also present in endocrine diseases associated with ACTH excess, e.g., congenital adrenal hyperplasia, and would explain why in depressives less ACTH after h-CRH is sufficient to produce cortisol output being indistinguishable from that in controls.

In this regard it should be emphasized that also other final products of POMC processing are most likely involved in adrenocortical hyperplasia.[10] Several lines of evidence now converge to assume that the N-terminal of POMC (N-POMC 1-49) is an active adrenocortical mitogen, while its parent peptide Pro-y-MSH (N-POMC 1-76) is inactive. Thus, the main N-terminal POMC peptide fragment secreted from the anterior pituitary is in the inactive form, which needs further processing before it exerts its adrenogenic capacity. The partial purification of an adrenal growth factor (AGF) was recently accomplished from cultured bovine anterior pituitary cells and it was further demonstrated that ACTH can block mitogenic activity of this AGF whose exact nature remains to be elucidated.[11]

These new developments show quite clearly that the entirety of POMC products released from the pituitary along with other extrapituitary and neural factors have to be taken into consideration if one attempts to speculate meaningfully about the origin of enhanced responsiveness of adrenocortical cells after specific stimulation. These aspects are particularly relevant for the work by Amsterdam et al.[12] who first reported enhanced release of cortisol after administration of synthetic ACTH to hypercortisolemic depressives and later on showed that adrenocortical hyperresponsiveness normalizes in parallel with successful treatment.[13] Similar reports were provided by our group[14] and others.[15]

The current studies employing h-CRH provided the first experimental evidence supporting the hypothesis that LHPA hyperactivity in depression originates from pathology in the CNS. Our findings and conclusions are in agreement with parallel studies by Gold et al.[16] who employed the long-acting ovine analog of h-CRH. Interestingly, this group found that responses to h-CRH were virtually identical in depressed patients and in controls[17] while another recent study by Lesch et al.[18] applying also the homologous h-CRH confirmed our original results. This underscores the urgent need for exact endocrine characterisation of study samples over an extended baseline period, because in the absence of elevated cortisol levels in the afternoon prior to h-CRH or o-CRH stimulation, the likelihood to find blunted ACTH following CRH is diminished. Such an endocrine evaluation at baseline would be helpful to eliminate sources of laboratory data variance across centers, which have plagued neuroendocrine research over years and which derived from the misconception that samples which are labelled as identical after verbal psychopathology assessments would represent biological homogeneity.

C. DIFFERENTIAL MINERALOCORTICOID AND GLUCOCORTICOID RELEASE FOLLOWING h-CRH IN DEPRESSION

Another area of interest which could be further studied with h-CRH challenge tests is disturbed mineralocorticoid homeostasis in depression. Hullin et al.[19] reported that *in vitro* production of aldosterone is altered if adrenocortical cell cultures were exposed to serum of patients during depression when contrasted to effects by serum collected during the manic states. Our group has further demonstrated that aldosterone surges following dexamethasone are blunted in depression.[20] Despite increasing evidence that aldosterone is regulated in part by ACTH-independent processes, h-CRH proved to be a powerful stimulus also for the mineralocorticoids deoxycorticosterone and aldosterone. We recently reported a comparative evaluation of ACTH, cortisol and aldosterone responses to h-CRH in depressives and matched controls.[21] We also used corticosterone as an additional reference hormone, because we detected that this steroid, which is pharmacologically a glucocorticoid, but is synthesized via the mineralocorticoid pathway responds most sensitively among all tested corticosteroids to an ovine- or human-CRH challenge.

We studied ten psychiatric inpatients (five men, five women) suffering from a major depressive episode according to DSM-III. Their mean age was 48.8 ± 7.0 years and all had been free from any medication for at least 14 d. The patient sample was contrasted with a sex- and age-matched control sample. Particular care was taken to exclude all factors which could invalidate evaluations of a mineralocorticoid status. Therefore, we examined electrolytes in plasma and 24-h urine collections, confirmed present normotension and absence of past history or family history of hypertension, investigated hormonal status and excluded present or previous abuse of nicotine or caffeine-containing beverages. The net ACTH release in depressed patients was significantly blunted (2.7 ± 1.1 pg/ml/min $\times 10^3$ vs. 6.0 ± 2.1 pg/ml/min $\times 10^3$; $t = 4.29$, d.f. $= 18$, $p < 0.01$). Both glucocorticoids, cortisol, and corticosterone failed to provide markedly different release patterns after h-CRH injection (see Figure 2). However, aldosterone responses are significantly lower in depressed patients than in control subjects (1.7 ± 0.4 pg/ml/min $\times 10^3$ vs. 2.7 ± 0.5 pg/ml/min $\times 10^3$ pg/ml/min; $t = 4.91$, d.f. $= 18$, $p < 0.01$). At baseline, plasma cortisol levels in the patients with depression were markedly higher than in the control subjects (71.9 ± 21.7 vs. 50.1 ± 14.4 ng/ml; $t = 2.7$, d.f. $= 18$, $p < 0.05$).

Our findings of a significantly blunted aldosterone and ACTH response in depressed patients amplify the impression that mineralocorticoid secretion is regulated by factors which are at least in part different from those which control glucocorticoid regulation. Whether diminished aldosterone output after h-CRH injection directly results from decreased circulating ACTH concentrations or is an indirect result of disturbed LHPA feedback circuitry remains unresolved. However, the recent characterization of different corticosteroid receptor types may provide a better understanding of these effects (for review see Reference 22). In addition, the probability of variable POMC gene processing in patients with depression leading to adrenogenic factors which act differentially at glomerulosa and fasciculata cells warrants further investigation. If corroborated in larger study samples under various clinical conditions, including drug status, the finding of altered mineralocorticosteroid function may provide a basis to understand vulnerability of patients with affective disorders to suffer from kidney pathology, particularly if treated with lithium.

D. STIMULATION RESPONSE AFTER h-CRH IN ALCOHOLISM

In patients with alcoholism several abnormalities of LHPA function are well documented. Sometimes also other stigmata of Cushing's syndrome can appear among alcoholics and therefore the term "pseudo-Cushing's syndrome" has been coined. Neuroendocrine studies of alcoholism are relevant for the psychiatrist because these patients constitute a major portion of their clientele which has several genetic and behavioral traits in common with

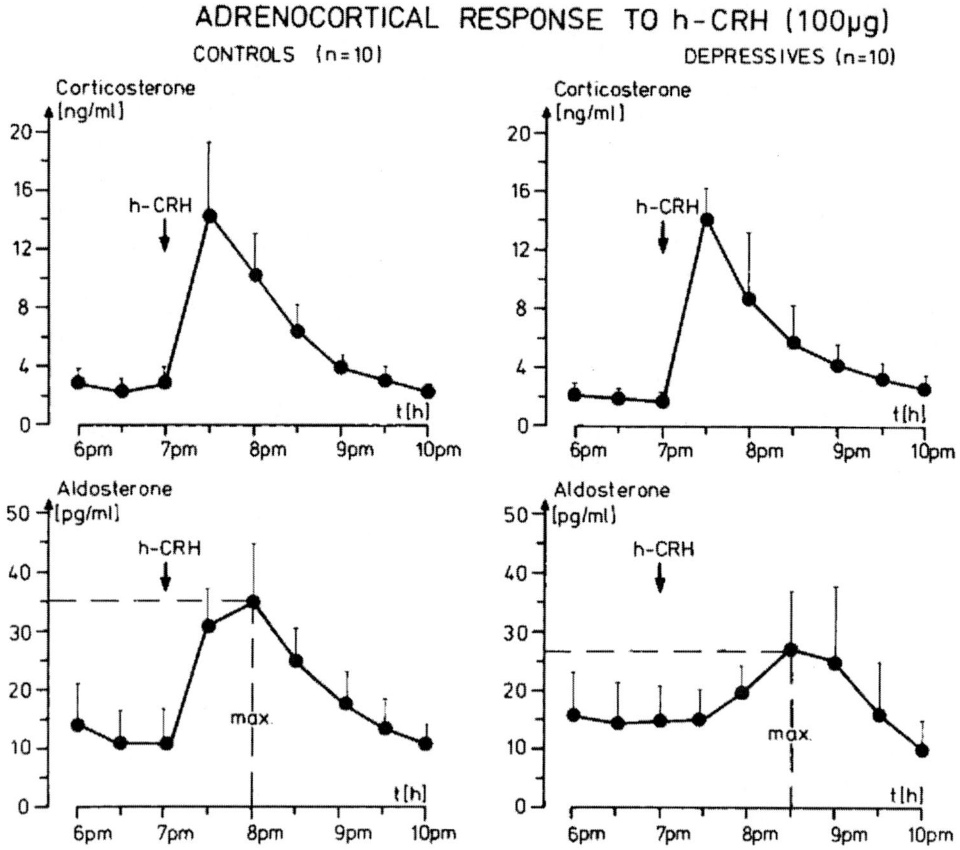

FIGURE 2. Corticosterone (above) and aldosterone (below) secretion (mean ± SEM) in ten patients with major endogenous depression and ten matched controls. (From Holsboer, F., Doerr, H. G., and Sippell, W. G., *Psychoneuroendocrinology,* 7, 155, 1982. With permission.)

uni- or bipolar depressives. As a result, family studies of alcoholic probands have yielded consistently higher rates of both alcoholism and depression than would be expected in the general population.[23]

To date, it remained unclear whether the hypercortisolism in these patients results from an abnormality at or above a hypothalamic or at a pituitary-adrenocortical site. To further elucidate alcohol-induced alteration of LHPA physiology we performed a standard h-CRH test in eight unmedicated patients (age 20 to 48 years) who were acutely withdrawn from excessive alcohol consumption.[24] The hormonal response patterns were compared with eight matched healthy controls (aged 24 to 48 years). As illustrated in Figure 3, ACTH responses were significantly lower (mean area under time course curves: 1.6 ± 1.0 pg/ml/min $\times 10^3$; in controls: 5.5 ± 3.0 pg/ml/min $\times 10^3$, $t = 3.0$, d.f. $= 14$, $p < 0.01$). Also the cortisol output was less marked in patients after short-term abstention from alcohol (6.8 ± 4.3 ng/ml/min $\times 10^3$ vs. 11.8 ± 4.2 ng/ml/min $\times 10^3$, $t = 2.3$, $p < 0.05$). Mean cortisol secretion before stimulation was higher among the patients than among controls (105.4 ± 48 ng/ml vs. 55.6 ± 24.7 ng/ml, $t = 3.0$, $p < 0.01$). The mechanism for this restrained ACTH release remains unclear. Overexposure to ethanol over a prolonged period of time and subsequent withdrawal of this drug may precipitate numerous neurochemical changes in the CNS. In fact, little is known for certain about the underpinning of alcohol withdrawal syndrome. Several of the symptoms, such as sweating, tachycardia, hypertension, or tremor and the therapeutic efficacy of clonidine suggest that α-adrenergic autoreceptors probably

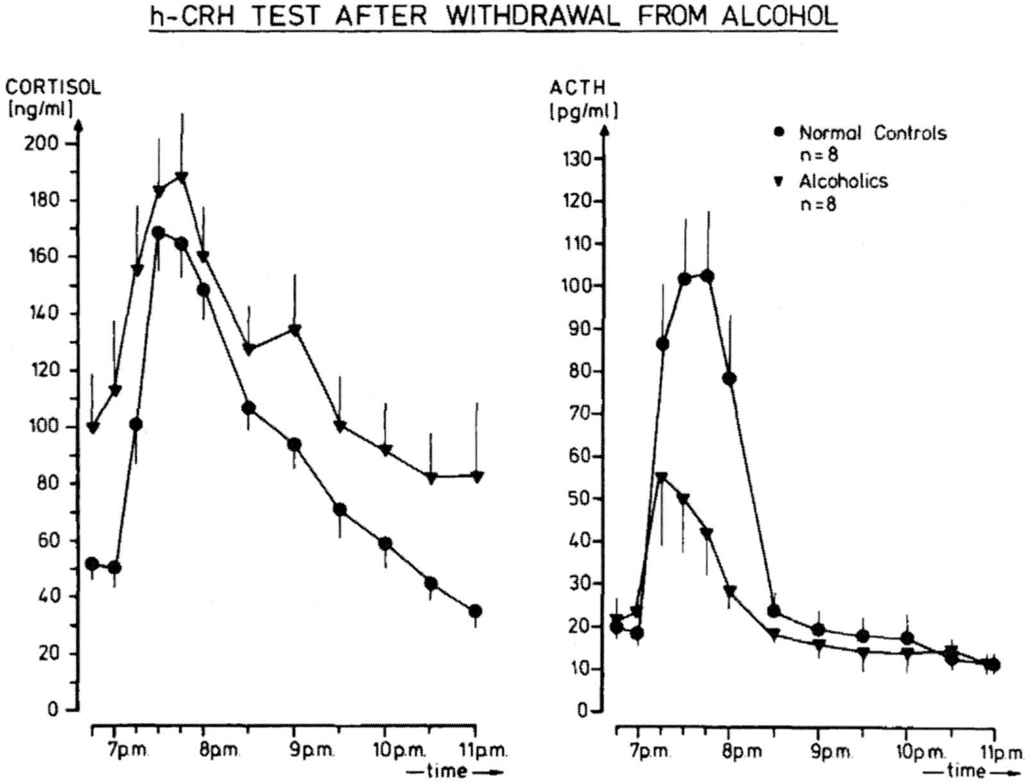

FIGURE 3. Cortisol (left and ACTH (right) secretion (mean ± SEM) after 100 μg h-CRH administered as intravenous bolus to eight patients acutely withdrawn from alcohol abuse and eight normal controls. (From Heuser, I., von Bardeleben, U., Boll, E., and Holsboer, F., *Biol. Psychiatry*, 24, 316, 1988. With permission.)

located at the locus coeruleus (LC) are involved. In alcohol withdrawal, the elevation of norepinephrine in lumbar CSF is thought to stem from central cerebral nuclei, especially the LC.[25] The finding of blunted ACTH after exogenous CRH may be secondary to central overactivity of CRH-secreting neurons maintaining increased plasma cortisol levels which, in turn, attenuate ACTH response following exogenous h-CRH. Importantly, one extrahypothalamic site where CRH has been visualized is the noradrenergic LC. The association between blunted ACTH response to exogenous h-CRH, and the reported elevation of the sympathetic nervous system (increased norepinephrine and 3-methoxy-4-hydroxyphenylglycol in lumbar CSF; withdrawal symptoms, etc.) can be understood as indicative for a functional link between CRH release, LC activity, and norepinephrine turnover. After ethanol withdrawal we surmise that decreased postsynaptic α-2-receptor sensitivity results in enhanced endogenous CRH secretory activity and increased norepinephrine release. The LC, through its widespread efferent system, may amplify CRH-mediated effects in an attempt to restore homeostatic alterations secondary to alcohol withdrawal.

Several additional mechanisms underlying blunted ACTH responses after h-CRH in alcoholism have to be considered. Rivier et al.[26] recently reported that pretreatment of cultured anterior pituitary cells with 0.2% ethanol for 24 h resulted in decreased basal and o-CRH stimulated ACTH secretion and suggested that chronic ethanol administration reduces pituitary responsiveness. Also, other studies indicate a reduction of hypothalamic CRH contents, resulting in decreased rate of POMC gene transcriptions and subsequent lowering of POMC mRNA and circulating POMC gene products. The effects seen after cessation of drug intake may point to an excessive rebound phenomenon which leads to stepwise re-

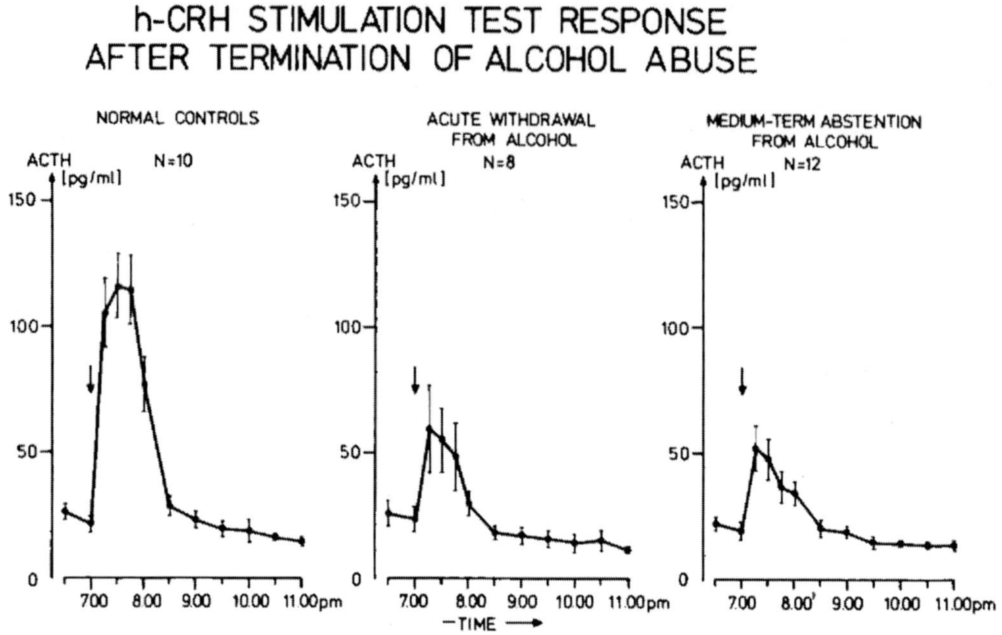

FIGURE 4. ACTH secretion (mean ± SEM) in normal controls (left) after acute abstention with associated withdrawal symptoms (center) and after medium-term abstention (right). (From von Bardeleben, U., Stalla, G. K., and Holsboer, F., *Psychoneuroendocrinology*, in press. With permission.)

storation of homeostasis. Several findings support this view. For example, lumbar CSF concentrations of 3-methoxy-4-hydroxyphenylglycol and norepinephrine normalize in dry alcoholics. However, as illustrated in Figure 4, ACTH blunting after h-CRH persists after termination of clinical withdrawal symptoms, which indicates that the effects seen during withdrawal are not only due to the acute physical stress induced by abrupt cessation of alcohol abuse.[27] For clarification to which extent the LHPA axis is capable to normalize, serial h-CRH tests after long-term abstention are needed. Studies of this kind may shed some light upon the contribution of the behaviorally active LHPA axis as a risk factor for relapse into alcohol drinking behavior, in similarity as persistent alteration of the LHPA axis in depression is indicative for a risk to relapse into an affective episode.

E. STIMULATION RESPONSE AFTER h-CRH IN PATIENTS WITH PANIC DISORDER

From recent classificatory efforts considering anxiety states two new definitions emerged, panic disorder and generalized anxiety disorder. The interest in these concepts for psychiatrists stems from the observation that patients who experienced clinical panic attacks (unexpected occurrence of marked apprehension or fear, dyspnea, palpitations, choking, chest pain, sweating, cold or hot flashes, etc.) may develop panic attacks in response to infusions of 0.5 *M* sodium lactate. Several authors claimed that such a finding is absent in normal controls and suggested that lactate induces actual panic attacks only in clinically susceptible individuals.[28] Because anxiety and depression are interdependent syndromes and overlap to varying degrees we were interested in whether they were differentiable according to their neuroendocrine profile following h-CRH. We selected patients with panic disorder and compared their response patterns with pure depressives and patients who have depression in combination with panic disorder.[29] We further investigated whether response patterns among patients with panic disorder were different from individuals who responded differently following lactate infusion. We observed that ACTH release following h-CRH is blunted

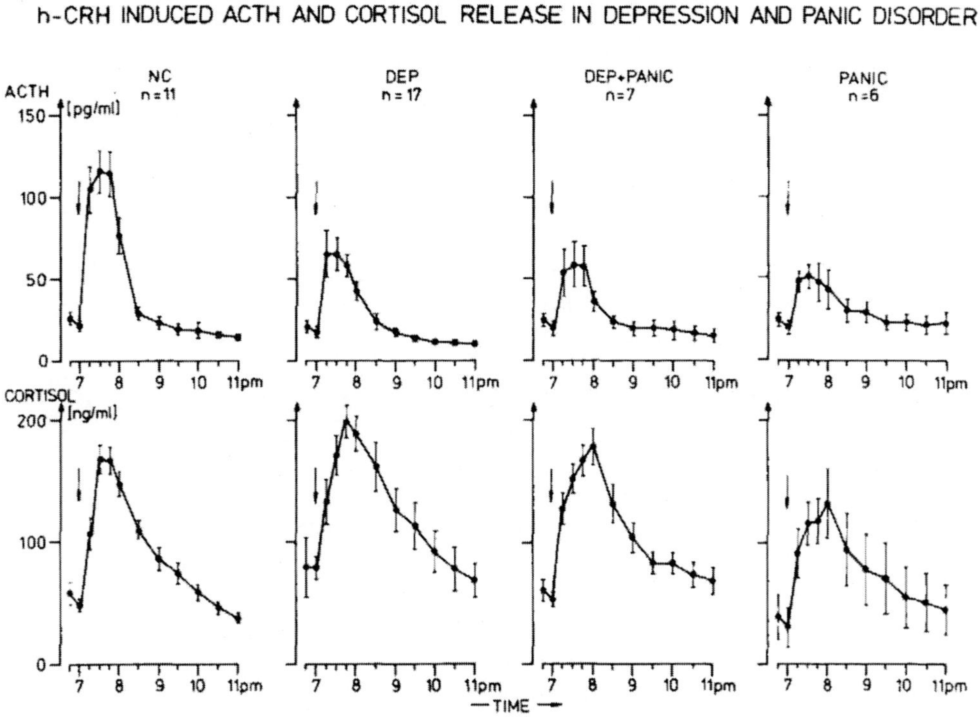

FIGURE 5. ACTH (above) and cortisol (below) secretion (mean ± SEM) at baseline and following (arrows) 100 μg h-CRH administered as an intravenous bolus at 7:00 P.M. to 11 controls (NC), 17 patients with depression (DEP), 7 patients with both major depression and panic attacks (DEP + PANIC) and 6 patients with pure panic disorder (PANIC). (From Holsboer, F., von Bardeleben, U., Buller, R., Heuser, I., and Steiger, A., *Horm. Metab. Res.*, in press. With permission.)

through all three clinical conditions, pure depression, combinations of depression and panic disorder, and pure panic disorder (see Figure 5). In contrast to pure depression and depression associated with panic disorder we observed normal baseline cortisol in pure panic disorder. Therefore, the blunted ACTH response cannot be entirely explained by enhanced cortisol levels at baseline. Though the maximal plasma cortisol levels following h-CRH were markedly lower than in the comparison groups their net cortisol release expressed as area under time-course curves was indistinguishable from that in depressives.

From the present data base, which is limited by the small sample size, we tentatively conclude that patients with panic disorder as a group are biologically heterogeneous at least regarding neuroendocrine regulation of the LHPA-axis. In patients with depression all neuroendocrine data converge to support that LHPA-overactivity is a centrally driven phenomenon which is closely associated with the depressive state. In contrast, the endocrine pattern in panic disorder seems to be less strictly organized and may be characterized by a less stable endocrine responsiveness. This topic can only be studied meaningfully when long-term protocols are applied which avoid carefully any kind of exogenous stressors. This is particularly important in patients with panic disorder, who are inclined to show exaggerated responses in all their behaviors regardless whether verbal psychopathology or endocrine behavior is under study.

In a recent investigation by Roy-Byrne et al.[30] who applied the ovine analog of CRH to eight patients with panic disorder also a blunted ACTH response was observed. In contrast to our study these authors reported a clear-cut elevation of cortisol at baseline. One confounder in the study by Roy-Byrne et al. might be that the indwelling venous catheter was inserted immediately prior to h-CRH administration, while in all studies from our group the catheterization was performed at least 6 h prior to the challenge test. Moreover, we always employ the through-the-wall technique, which leaves all manipulations through the long catheter unrecognized by the patients. The unpleasant experience of the metabolic changes induced by a lactate infusion precipitated panic attacks in two out of six patients with panic disorder, in three out of seven patients having a combination of depression and panic disorder and was absent in five patients with pure endogenous depression. We failed to find any correlation between the endocrine response patterns and the lactate-inducible panic attacks or reported aversive symptoms. Considering the remarkable interest in studies of pathophysiology and treatment of panic disorder a full neuroendocrine evaluation of this disorder is needed to elaborate a more detailed knowledge of the underlying pathophysiology and the biobehavioral response to treatment.

III. DRUG-INDUCED EFFECTS UPON h-CRH STIMULATION RESPONSE

A. EFFECT OF DEXAMETHASONE PRETREATMENT ON RESPONSE TO h-CRH — IMPLICATIONS OF PATHOPHYSIOLOGY UNDERLYING DST NONSUPPRESSION

Nonsuppression of ACTH and cortisol after administration of dexamethasone was originally regarded as a peripheral outprint of a centrally mediated hyperactivity of CRH-secreting neurons. In order to test this hypothesis we further explored the pathophysiology underlying DST nonsuppression. As the first step, we investigated the effect of dexamethasone pretreatment upon hormonal response to h-CRH, lysine-vasopressin or a combination of both neuropeptides.[31] Infusion of LVP (1.2 IU/h) given in the afternoon failed to provoke ACTH and cortisol elevation to levels which are common in DST nonsuppressors. Also, a h-CRH bolus (100 μg) administered at 3:00 P.M. induced only a modest release of ACTH and cortisol. If both peptides were administered in combination, a substantial escape of plasma cortisol from dexamethasone supression became apparent. Since ACTH levels increase with cortisol we concluded that LVP synergizes the effect of h-CRH at pituitary corticotrophs to override dexamethasone suppression which mainly takes place at the pituitary level. This allowed us to speculate that central CRH overproduction is not the only mechanism to explain DST resistance and suggests that other factors including synergizing neuropeptides, altered activity of neural efferent pathways and changes of feedback receptor function are corequired. In line with this contention is a recent study by Nemeroff et al.[32] who measured CRH in cerebrospinal fluid and observed elevated levels in major depression which were not related to the DST status. In a similar study, however, Roy et al.[33] reported different results and showed that in their study sample endogenous CRH elevation in cerebrospinal fluid was absent whereas the CRH levels corresponded to post-DST cortisol levels. The difference may be related to difference in patient characteristics (see also Section II.B). To clarify the contention whether or not h-CRH alone is capable to induce DST nonsuppression in normal controls a more physiological administration of CRH during the early morning hours at minute quantities is needed. Such studies are under way in our laboratory.

Application of a combined DEX/CRH-test in patients with depression provided some unexpected results. In contrast to normal controls, patients with depression responded with elevated ACTH and cortisol, regardless of their actual DST status. After repeatedly challenging depressed patients with the DEX/CRH-test we observed a gradual reduction of releasable ACTH and cortisol after h-CRH which was associated with clinical improvement or a switch into mania (see Figure 6).[34] One patient who showed a failure of dexamethasone to inhibit h-CRH induced cortisol elevation at discharge needed a rehospitalization because a relapse occurred several weeks later. Since we recently demonstrated an association between DST nonsuppression of cortisol and low plasma dexamethasone levels[35] we ruled out that

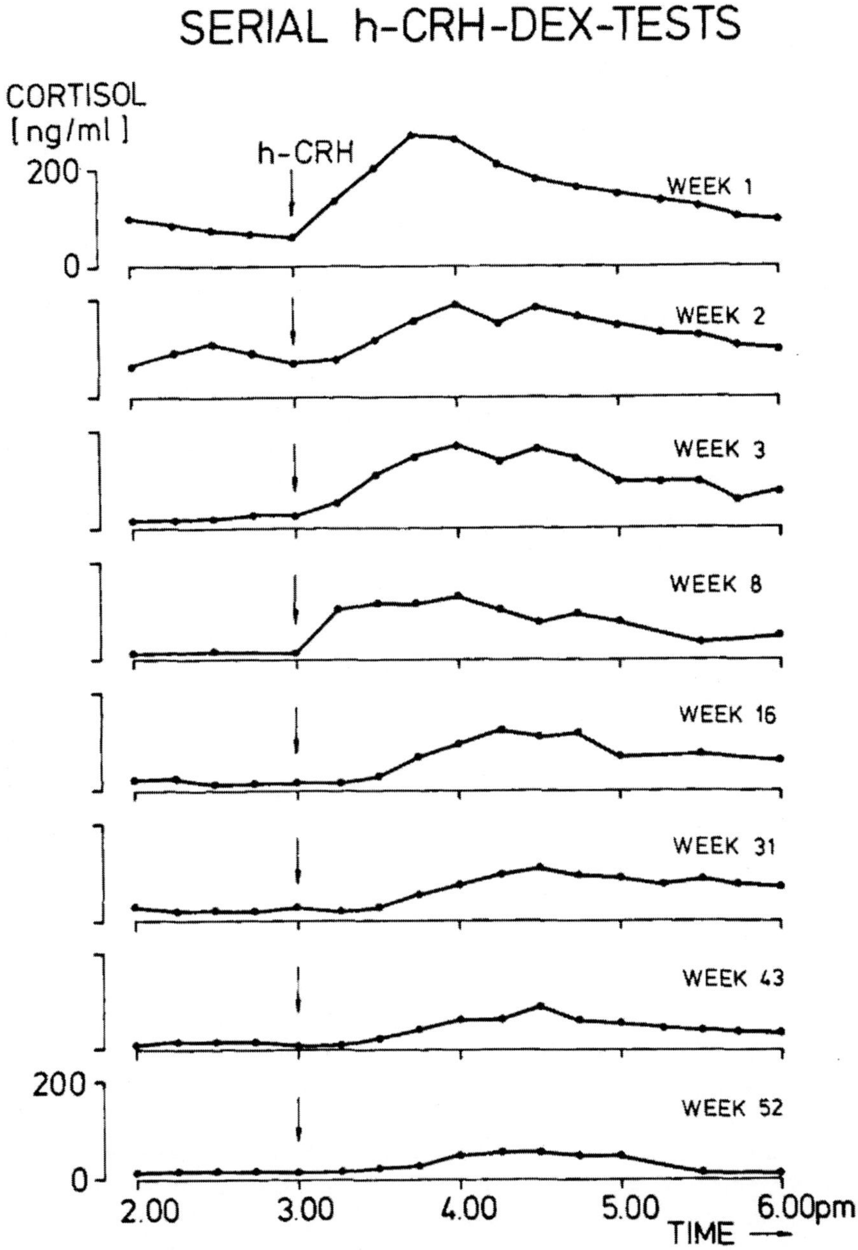

FIGURE 6. Serial application (100 μg h-CRH administered intravenously at 3:00 P.M. after oral ingestion of 1.5 mg dexamethasone at 11:00 P.M. the day before) of the DEX/CRH-test reveals that releasable cortisol in a depressed patient represents a state-dependent phenomenon. (From von Bardeleben, U. et al., unpublished observation.)

individual ACTH and cortisol responses were an artefact of inadequate dexamethasone absorption as we measured comparable amounts of the test drug in the plasma samples at 8:00 A.M. in all patients at all test occasions. While dexamethasone was sufficiently potent to inhibit pituitary adrenal stimulation by h-CRH in normal controls, this was not the case in patients during their depressive episode. Notably, this effect was also seen in those cases where elevation of ACTH and cortisol after h-CRH was associated with nonsuppression of ACTH and cortisol after DST and prior to h-CRH stimulation. In these cases, both classes of steroids, exogenously administered dexamethasone and nonsuppressed endogenous corticosteroids, failed to inhibit stimulatory effects of h-CRH at the pituitary. This is in contrast to the view that elevated corticosteroids in depression may successfully restrain the releasable amount of ACTH following CRH.

One hypothetical mechanism involved in the current observations implies differential glucocorticoid receptor changes and different pharmacodynamic effects of dexamethasone and endogenous corticosteroids. It has been repeatedly demonstrated by the group led by Sapolsky et al.[36] that exogenous administration of glucocorticoids reduces glucocorticoid receptor number in the hippocampus which is an important element for the negative feedback loop of the LHPA-system. This group also showed that repeated stress given to rats reduces the number of glucocorticoid receptors at various brain structures particularly in the hippocampus. If glucocorticoid hypersecretion during depression continues, a stepwise reduction in hippocampal corticosteroid receptor number could develop which results in reduced sensitivity of this feedback element to any circulating corticosteroid. Administration of dexamethasone results primarily in occupation of pituitary glucocorticoid receptors, thus reducing ACTH and subsequently corticosteroid release. At the level of the hippocampus this withdrawal of endogenous corticosteroids is not compensated by dexamethasone since in comparison to pituitary corticotrophic binding its retention by limbic neurons particularly in the hippocampus is different from that of naturally occurring glucocorticoids. The pharmacologic effect of dexamethasone administration upon hippocampal sites can thus be described as a "chemical adrenalectomy" to which hippocampal neurons respond with enhanced activation of the LHPA system. Following dexamethasone, the amount of ACTH which is finally released at baseline or after stimulation with h-CRH depends upon the equilibrium of enforcing (via hippocampus) and suppressing (via pituitary) inputs upon the entire system. In depression, persistent elevation of corticosteroids may increase vulnerability of glucocorticosteroid receptor-containing cells at hippocampal and other central sites. This effect weakens the negative feedback system making the corticotrophic cells more susceptible to stimuli such as h-CRH. After dexamethasone administration the reduced glucocorticosteroid receptor population is no longer capable to inhibit LHPA activation which then can override the suppressive effect of dexamethasone at corticotrophic cells.

In these cases, endogenous corticosteroids in combination with exogenous dexamethasone are probably not sufficiently potent to serve as a break for additional pituitary activation after exogenous h-CRH administration. When central regulation returns to normalcy there is a stepwise increase of receptor sensitivity resulting in an adrenocortical response to h-CRH which is indistinguishable from that in normal controls. In cases where sustained hypercortisolism has produced neuronal loss in central feedback elements, altered pituitary adrenal activity would persist. This mechanism may also account for increased pituitary adrenocortical activity associated with growing age.[36] Further studies are needed to elaborate to which degree alteration of glucocorticoid receptor function induced by ligand hyperexposure is involved in development of depressive symptomatology or in increased susceptibility to relapse into an affective episode. The link between alcoholism and depression, or iatrogenic depression following long-term corticosteroid medication, or major depression as a clinical feature in patients with Cushing's syndrome are just a few selected examples out of the bulk of evidence which speak in favor of such an association.

Our current data base on the combined DEX/CRH test holds promise to uncover important features of DST refractoriness and further supports a central pathophysiology underlying altered LHPA function in a subset of depressives.

B. EFFECT OF METYRAPONE PRETREATMENT UPON RESPONSE TO h-CRH

The most straightforward explanation for blunted ACTH secretion after h-CRH administration in depression implies a negative feedback effect of elevated plasma cortisol sup-

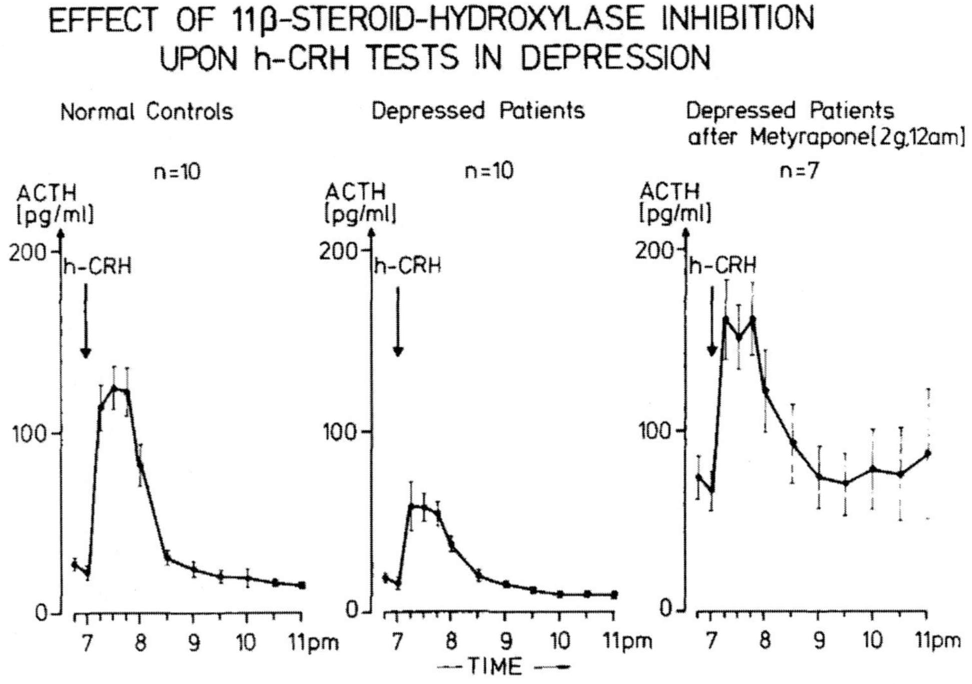

FIGURE 7. ACTH secretion (mean ± SEM) after 100 μg h-CRH administered as a bolus to normal controls, patients with major endogenous depression with or without pretreatment with the 11β-steroid-hydroxylase-inhibiting compound metyrapone. (From von Bardeleben, U., Stalla, G. K., and Holsboer, F., *Biol. Psychiatry*, 24, 787, 1988.

pressing releasable ACTH. However, this conclusion is tentatively, and does not rule out the possibility of other important changes, for example at the CRH receptor level. One possibility to prove that increased levels of circulating cortisol determine the ACTH response to h-CRH in man is to deprive the system from its major feedback inhibiting signal, primarily C-11 hydroxylated corticosteroids. We studied ACTH release following h-CRH in ten unmedicated depressed inpatients (aged 40 to 68), seven depressed inpatients (aged 26 to 67), pretreated with metyrapone to lower cortisol levels by inhibition of the 11 β-steroid hydroxylase and ten normal controls (aged 23 to 62).[38] All patients fulfilled the requirements of a major depressive episode according to DSM-III. As illustrated in Figure 7 we found that in normal controls the ACTH output expressed as the AUCs were: 6.8 ± 0.74 pg/ml/min × 10^3, in unmedicated depressed patients: 2.6 ± 0.6 pg/ml/min × 10^3, in patients pretreated with 2 g metyrapone orally administered at 12:00 A.M. 6.3 ± 1.4 pg/ml/min × 10^3. Thus, ACTH release in unmedicated depressed patients was significantly blunted when compared to normal controls ($p < 0.001$; Whitney U-Test). In contrary, in depressed patients pretreated with metyrapone ACTH blunting was completely avoided. To confirm effectiveness of metyrapone-induced blockade of 11β-steroid-hydroxylase, we determined plasma concentrations of 11-deoxycortisol and found them to be clearly increased after metyrapone (210.2 ± 12.6 ng/ml) when compared to nonmedicated depressives (3.6 ± 0.3 ng/ml) or normal controls (4.0 ± 0.4 ng/ml). A possible desensitization of pituitary CRH receptors due to a longer lasting central CRH hypersecretion is less likely to be the prime cause of blunted ACTH response after exogenous h-CRH in depression because the net ACTH release normalizes in these patients immediately after pretreatment with metyrapone. This finding supports that CRH peptide receptor desensitization is not the determinant mechanism involved in blunted ACTH following h-CRH. *In vitro* studies demonstrated a maximal ACTH response

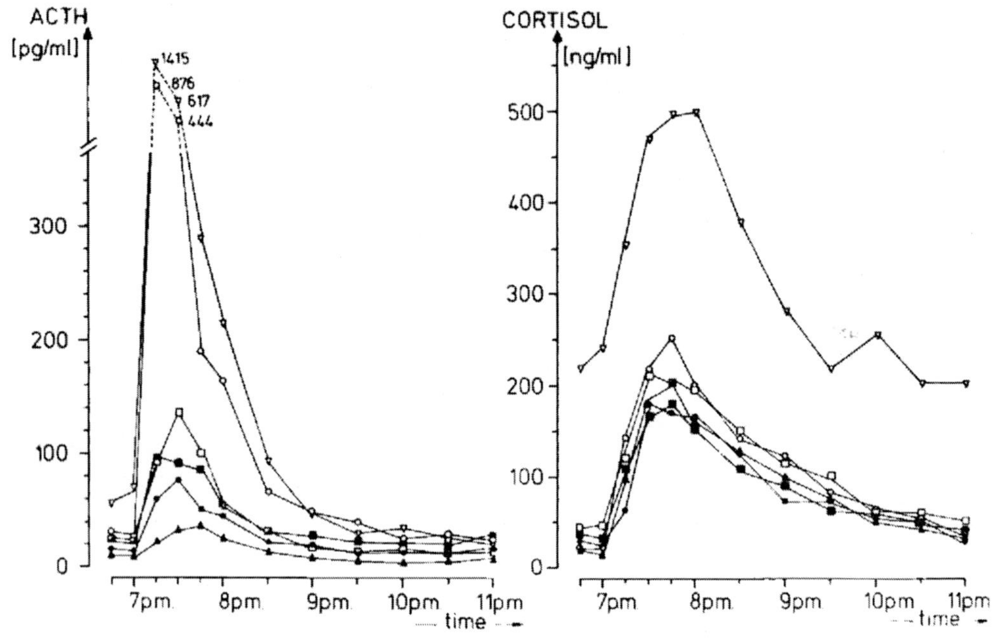

FIGURE 8. ACTH and cortisol (mean ± SEM) secretion after an intravenous bolus of 100 μg h-CRH administered to sic patients with affective disorder, in remission under long-term treatment with carbamazepine. Two patients (open circles, open triangles) show highly abnormal response patterns. (From von Bardeleben, U., Wiedemann, K., Stalla, G. K., Müller, O. A., and Holsboer, F., *Biol. Psychiatry*, 24, 331, 1988. With permission.)

to CRH even if only 50% of the receptors were occupied, indicating that CRH receptors can be classified as reserve receptors. In addition, Young and Akil[39] observed that after chronic stress there develops an increased sensitivity, while after acute stress the releasable amount of ACTH and β-endorphin are blunted. Also, Bruhn et al.[40] reported that pituitary cultures exposed to CRH over a longer time period had exaggerated contents of POMC derived peptides.

C. EFFECT OF CARBAMAZEPINE PRETREATMENT UPON RESPONSE TO h-CRH

Carbamazepine which dampens paroxysmal neuronal activity not only in epilepsy but also in those selected systems which are involved in the etiology of episodic affective disorder has been successfully introduced as an acute and prophylactic treatment of manic depressive illness. Of particular interest is the reported impact of carbamazepine upon LHPA physiology as it induces DST nonsuppression and enhances urinary free cortisol secretion.[41]

To further explore the pathophysiology underlying these phenomena and to possibly analyze drug-induced changes we conducted h-CRH tests in six patients who were in stable remission after a major depressive episode and were under long-term carbamazepine treatment.[42] As illustrated in Figure 8, we observed in one case a 10-fold and in another case a 16-fold increased ACTH response to h-CRH when compared with the other patients. In one case this extremely exaggerated ACTH release was also associated with high cortisol values. Since endocrine diseases such as various forms of Cushing's syndromes were clinically ruled out, the surprising finding was most likely to be attributed to carbamazepine treatment.

Serby et al.[43] recently proposed that somatostatin may be involved in regulating the LHPA axis, because CNS somatostatin was significantly negatively correlated to postdexamethasone plasma cortisol levels in demented man. This intriguing hypothesis might explain our present data, since it has been demonstrated that carbamazepine may reduce CSF somatostatin in a subset of patients during carbamazepine treatment.[44] Somatostatin interacts with CRH and ACTH release as it inhibits central CRH production and decreases ACTH release from mouse pituitary cells challenged with CRH. Since we have not measured CSF somatostatin we can only speculate that profound lowering of this peptide is involved in aberrant hormonal responses to h-CRH in two of our patients. Our previous investigations applying serially o-CRH challenges in patients with major depression during illness and after remission revealed no significant difference of the net ACTH and cortisol outputs between the two tests.[6] An alternative explanation would be that in a susceptible subgroup of patients carbamazepine might hypersensitize pituitary CRH receptors resulting in extremely exaggerated ACTH response following exogenous h-CRH. Further studies which include *in vitro* experiments using pituitary cell cultures are warranted to clarify the significance of these findings and to evaluate the various factors which generate the exaggerated ACTH release. In the light of the ubiquitous extrahypothalamic distribution of somatostatin and pituitary adrenocortical hormones and taking into account their neuroendocrine behavioral effects, we surmise that interference of carbamazepine with these systems is part of its mechanism of action.

IV. EFFECT OF PULSATILE TREATMENT WITH CRH UPON SLEEP-EEG

The present data derived from animal behavior studies allow one to suggest that CRH secretory hyperactivity in the CNS is not only a key factor in mediating stress response but is also involved in etiopathology of depression. Since CRH was found to be elevated in the CSF in severely depressed patients,[32] the conclusions derived from animal behavioral studies may well have clinical implications. For example, CRH induced loss of appetite, sleep disturbance, altered autonomous function, and psychomotor disturbances; anxiety and loss of sexual drive are also yardsticks to diagnose major depression. Of course, observation of depression-like animal behavior (e.g., behavioral despair, etc.) after pharmacological dosages of CRH given intracerebroventricularly does not allow one to conclude that elevated CRH in the CNS of depressives is directly involved in depressive symptom expression. However, methodological drawbacks are also imposed by attempts to quantitate depressive behavior by means of verbal psychopathology. Because sleep and its disturbance can be quantitated by analysis of sleep-EEG structure we conducted sleep-EEG studies in normal controls who were challenged before and during sleep with CRH pulses.[45]

Eleven healthy male control volunteers (age range 20 to 37 years) were infused with 50 μg or placebo four times every hour beginning at 10:00 P.M. All subjects fell asleep after 11:00 P.M. No immediate effects of repetitive h-CRH stimulations during the first part of the night were observed when conventional guidelines for sleep staging were used. However, in the second part of the night, high frequency contents increased while at the same time slow wave sleep (SWS, stage 3 and 4) became significantly reduced. Also, the percentage of REM-sleep decreased following h-CRH whereas REM-density remained unchanged. Importantly, these effects of h-CRH were not associated with significantly increased awakenings, stage shifts, movements or differences in sleep period times.

We do not know for certain to which extent h-CRH itself or elevation of ACTH or corticosteroids account for the observed findings. Because neuropeptide passage through the blood-brain-interface is not fully understood, one expects that the effects of CRH are at least in part mediated by cortisol. However, we cannot rule out direct neuropeptidergic pathways because the median eminence and the circumventricular organs are leaky and even minute peptide quantities may exert CNS effects. In the same direction is a recent study of Ehlers et al.[46] where CRH was injected intracerebroventricularly to rats and resulted in decreased SWS concomitant with significant decreases in spectral power in lower frequencies and increases in higher frequencies. Moreover, after infusion of cortisol Born et al.[47] recorded a significant increase of the total amount of SWS in human controls, indicating that our finding of lower SWS after h-CRH is rather induced by a central action of the neuropeptide and not secondary to pituitary and adrenocortical stimulation. In parallel with sleep-EEG recordings we also collected blood through the long catheter and observed that the nighttime growth hormone release is significantly blunted during h-CRH infusion. A similar finding was reported from spontaneously hypercortisolemic depressives.[48] Also central administration of CRH into the third ventricle of rats induces a blunted growth hormone release. Taking these findings together, we suggest that central hyperactivity of CRH secreting neurons is also involved in the nonendocrine behavioral complex of human depression.

V. CONCLUSION

The current selected examples from on-going investigations employing o- and h-CRH in patients and controls under various experimental conditions strongly support physiological relevance of CRH for the endocrine and nonendocrine behavioral complex of depression. In essence, we conclude the following:

1. The hypothalamus releases excessive CRH which results in enhanced synthesis of ACTH and other POMC-gene-related peptides.
2. Persistent overexposure of the target glands to ACTH and probably other N-terminal POMC products results in functional hyperplasia as evidenced by hyperresponsiveness.
3. The blunted ACTH response to h-CRH in hypercortisolemic depressives is determined by elevated baseline cortisol release because after elimination of cortisol by metyrapone pretreatment the net ACTH output comes back to normal.
4. Blunted aldosterone but normal glucocorticoid response to h-CRH in depression suggest the presence of non-ACTH factors involved in differential fasciculata and glomerulosa cell response to perturbation tests.
5. Persistent excess of CRH, ACTH and corticosteroid release results in central glucocorticoid receptor desensitization and subsequently leads to functional weakening of central glucocorticoid sensitive feedback elements, mainly in the hippocampus. We submit that such functional changes of central glucocorticoid receptors are involved in the DST nonsuppression phenomenon.
6. Some sleep-EEG characteristics of depression such as decreased SWS can be induced after peripheral administration of CRH. There is good evidence that this finding is not secondary to elevated corticosteroid levels.

These studies have provided new insights into pathophysiology of depression and may pave the way to new pharmacotherapeutical strategies such as CRH antagonists, which would challenge the hypothesis of a direct involvement of CRH in the overall clinical phenotype of affective disorder.

ACKNOWLEDGMENT

The current studies from the authors' laboratory were supported by a grant from the Deutsche Forschungsgemeinschaft (Ho 940/1-2). The authors are also indebted to their collegues, Drs. G. K. Stalla and O. A. Müller from the Department of Internal Medicine, University of Munich, for their collaboration in hormonal analysis.

REFERENCES

1. **Axelrod, J. and Reisine, T. D.,** Stress hormones: their interaction and regulation, *Science,* 224, 452, 1984.
2. **Spiess, J., Rivier, J., Rivier, C., and Wale, V.,** Primary structure of corticotropin-releasing factor from ovine hypothalamus, *Proc. Natl. Acad. Sci. U.S.A.,* 78, 6517, 1981.
3. **Shibahara, S., Morimoto, Y., Furutani, Y., Notake, M., Takahashi, H., Shizimus, S., Horikawa, S., and Numa, S.,** Isolation and sequence analysis of the human corticotropin-releasing factor precursor gene, *EMBO J.,* 2, 775, 1983.
4. **Holsboer, F., Müller, O. A., Doerr, H. G., Sippell, W. G., Stalla, G. K., Gerken, A., Steiger, A., Boll, E., and Benkert, O.,** ACTH and multisteroid responses to corticotropin-releasing factor in depressive illness: relationship to multisteroid responses after ACTH stimulation and dexamethasone suppression, *Psychoneuroendocrinology,* 9, 147, 1984.
5. **Holsboer, F., Gerken, A., Steiger, A., Benkert, O., Müller, O. A., and Stalla, G. K.,** Corticotropin-releasing factor induced pituitary-adrenal responses in depression, *Lancet,* 1, 55, 1984.
6. **Holsboer, F., Gerken, A., Stalla, G. K., and Müller, O. A.,** ACTH, cortisol, and corticosterone output after ovine corticotropin-releasing factor challenge during depression and after recovery, *Biol. Psychiatry,* 20, 276, 1985.
7. **Holsboer, F., von Bardeleben, U., Gerken, A., Stalla, G. K., and Müller, O. A.,** Blunted corticotropin and normal cortisol response to human corticotropin-releasing factor (h-CRF) in depression, *N. Engl. J. Med.,* 311, 1127, 1984.
8. **Holsboer, F., Gerken, A., von Bardeleben, U., Grimm, W., Beyer, H., Müller, O. A., and Stalla, G. K.,** Human corticotropin-releasing hormone in depression — correlation with thyrotropin-secretion following thyrotropin-releasing hormone, *Biol. Psychiatry,* 21, 601, 1986.
9. **Schürmeyer, T. H., Avgerinos, P. C., Gold, P. W., Gallucci, W. T., Tomai, T. P., Cutler, G. B., Jr., Loriaux, D. L., and Chrousos, G. P.,** Human corticotropin-releasing factor in man: pharmacokinetic properties and dose-response of plasma adrenocorticotropin and cortisol secretion, *J. Clin. Endocrinol. Metab.,* 59, 1103, 1984.
10. **Lowry, P. J., Estivariz, F. E., Gillies, G. E., Kruseman, A. C. N., and Linton, E. A.,** CRF: its regulation of ACTH and pro-opiomelanocortin peptide release and its extra hypothalamic occurrence, *Acta Endocrinol.,* Suppl. 276, 56, 1986.
11. **Samsoondar, J. and Kudlow, J. E.,** Partial purification of an adrenal growth factor produced by normal bovine anterior pituitary cells in culture, *Endocrinology,* 120, 929, 1987.
12. **Amsterdam, J. D., Winokur, A., Abelman, E., Lucki, I., and Rickels, K.,** Cosyntropin (ACTH 1-24) stimulation test in depressed patients and healthy subjects, *Am. J. Psychiatry,* 140, 907, 1983.
13. **Amsterdam, J. D., Maislin, G., Droba, M., and Winokur, A.,** The ACTH stimulation test before and after clinical recovery from depression, *Psychiatr. Res.,* 20, 325, 1987.
14. **Gerken, A. and Holsboer, F.,** Cortisol and corticosterone response after syn-corticotropin in relationship to dexamethasone suppressibility of cortisol, *Psychoneuroendocrinology,* 11, 185, 1986.
15. **Jaeckle, R. S., Kathol, R. G., Lopez, J. F., Meller, W. H., and Krummel, S. J.,** Enhanced adrenal sensitivity to exogenous cosyntropin (ACTH alpha 1-24) stimulation in major depression. Relationship to dexamethasone suppression test results, *Arch. Gen. Psychiatry,* 44, 233, 1987.
16. **Gold, P. W., Loriaux, D. L., Roy, A., Kling, M. A., Calabrese, J. R., Kellner, C. H., Nieman, L. K., Post, R. M., Pickar, D., Gallucci, W., Augerinos, P., Paul, S., Oldfield, E. H., Cutler, G. B., Jr., and Chrousos, G. P.,** Responses to corticotropin-releasing hormone in the hypercortisolism of depression and Cushing's disease, *N. Engl. J. Med.,* 314, 1329, 1986.
17. **Calabrese, J. R., Kling, M. A., Chrousos, G. P., Khan, I., Tomai, T. P., and Gold, P. W.,** Responses to ovine and human CRH in depression, in Syllabus and Proceedings in Summary Form: The 139th Annual Meeting of the American Psychiatric Assoc., 75, Abstr., 1986.
18. **Lesch, K. P., Laux, G., Schulte, H. M., Pfüller, H., and Beckmann, H.,** Corticotropin and cortisol response to human corticotropin-releasing hormone as a probe for hypothalamic-pituitary-adrenal system integrity in major depressive disorder, *Psychiatry Res.,* 24, 25, 1988.
19. **Hullin, R. P., Jerram, T. C., Lee, M. R., Levell, M. J., and Tyrer, S. P.,** Renin and aldosterone relationships in manic depressive psychosis, *Br. J. Psychiatry,* 131, 575, 1987.
20. **Holsboer, F., Doerr, H. G., and Sippell, W. G.,** Blunted aldosterone response for dexamethasone in female patients with endogenous depression, *Psychoneuroendocrinology,* 7, 155, 1982.
21. **Holsboer, F., Gerken, A., Stalla, G. K., and Müller, O. A.,** Blunted aldosterone and ACTH release after human corticotropin-releasing hormone in depression, *Am. J. Psychiatry,* 144, 229, 1987.
22. **de Kloet, E. R. and Reul, J. M. H. M.,** Feedback action and tonic influence of corticosteroids on brain function: a concept arising from the heterogeneity of brain receptor systems, *Psychoneuroendocrinology,* 12, 83, 1987.

23. **Merinkangas, K. R., Leckman, J. F., Prusoff, B. A., Pausl, D. L., and Weissman, M. M.,** Familiar transmission of depression and alcoholism, *Arch. Gen. Psychiatry,* 42, 367, 1985.
24. **Heuser, I., von Bardeleben, U., Boll, E., and Holsboer, F.,** Response of ACTH and cortisol to human corticotropin-releasing hormone after short-term abstention from alcohol abuse, *Biol. Psychiatry,* 24, 316, 1988.
25. **Hawley, R. J., Major, L. F., Schulman, E. A., and Linnoila, M.,** Cerebrospinal fluid 3-methoxy-4-hydroxyphenylglycol and norepinenephrine levels in alcohol withdrawal, *Arch. Gen. Psychiatry,* 42, 1056, 1985.
26. **Rivier, C., Bruhn, T., and Vale, W.,** Effect of ethanol on the hypothalamic-pituitary-adrenal axis in the rat: role of corticotropin-releasing factor (CRF), *J. Pharmacol. Exp. Ther.,* 229, 127, 1984.
27. **von Bardeleben, U., Stalla, G. K., and Holsboer, F.,** Human CRH stimulation response during acute withdrawal and after medium-term abstention from alcohol abuse, *Psychoneuroendocrinology,* in press.
28. **Liebowitz, M. R., Gorman, J. M., Fyer, A., Dillon, D., Levitt, M., and Klein, D. F.,** Possible mechanisms for lactate's induction of panic, *Am. J. Psychiatry,* 143, 495, 1986.
29. **Holsboer, F., von Bardeleben, U., Buller, R., Heuser, I., and Steiger, A.,** Stimulation response to corticotropin-releasing hormone (CRH) in patients with depression, alcoholism and panic disorder, *Horm. Metab. Res.,* 16, 80, 1987.
30. **Roy-Byrne, P. P., Uhde, T. W., Post, R. M., Gallucci, W., Chrousos, G. P., and Gold, P. W.,** The corticotropin-releasing hormone stimulation test in patients with panic disorder, *Am. J. Psychiatry,* 143, 896, 1986.
31. **von Bardeleben, U., Holsboer, F., Stalla, G. K., and Müller, O. A.,** Combined administration of human corticotropin-releasing factor and lysine vasopressin induces cortisol escape from dexamethasone suppression in healthy subjects, *Life Sci.,* 37, 1613, 1985.
32. **Nemeroff, C. B., Widerlöv, E., Bissette, G., Walleus, H., Karlsson, I., Eklund, K., Kilts, C. D., Loosen, P. T., and Vale, W.,** Elevated concentrations of CSF corticotropin-releasing factor-like immunoreactivity in depressed patients, *Science,* 226, 1342, 1984.
33. **Roy, A., Pickar, D., Paul, S., Doran, A., Chrousos, G. P., and Gold, P. W.,** CSF corticotropin-releasing hormone in depressed patients and normal control subjects, *Am. J. Psychiatry,* 144, 641, 1987.
34. **Holsboer, F., von Bardeleben, U., Wiedemann, K., Müller, O. A., and Stalla, G. K.,** Serial assessment of corticotropin-releasing hormone response after dexamethasone in depression — implications for pathophysiology of DST nonsuppression, *Biol. Psychiatry,* 22, 228, 1987.
35. **Holsboer, F., Haack, D., Gerken, A., and Vecsei, P.,** Plasma dexamethasone concentrations and differential glucocorticoid suppression response in depressives and controls, *Biol. Psychiatry,* 19, 281, 1984.
36. **Sapolsky, R. M., Krey, L. C., and McEwen, B. S.,** The neuroendocrinology of stress and aging: the glucocorticoid cascade hypothesis, *Endocr. Rev.,* 7, 284, 1986.
37. **Sapolsky, R. M. and Pulsinelli, W. A.,** Glucocorticoids potentiate ischemic injury to neurons: therapeutic implications, *Science,* 229, 1397, 1985.
38. **von Bardeleben, U., Stalla, G. K., and Holsboer, F.,** Blunting of ACTH response to h-CRH in depressed patients is avoided by metyrapone pretreatment, *Biol. Psychiatry,* 24, 782, 1988.
39. **Young, E. A. and Akil, H.,** Corticotropin-releasing factor stimulation of adrenocorticotropin and β-endorphin release: effects of acute and chronic stress, *Endocrinology,* 117, 23, 1985.
40. **Bruhn, T. O., Sutton, R. E., Rivier, C. L., and Vale, W.,** Corticotropin-releasing factor regulates proopiomelanocortin messenger ribonucleic acid levels in vivo, *Neuroendocrinology,* 39, 170, 1984.
41. **Rubinow, D. R., Post, R. M., Gold, P. W., and Uhde, T. W.,** Effect of carbamazepine on mean urinary free cortisol excretion in patients with major affective illness, *Psychopharmacology,* 88, 115, 1986.
42. **von Bardeleben, U., Wiedemann, K., Stalla, G. K., Müller, O. A., and Holsboer, F.,** Exaggerated corticotrophic cell response to human corticotropin-releasing hormone in two patients during long-term carbamazepine treatment, *Biol. Psychiatry,* 24, 331, 1988.
43. **Serby, M., Richardson, S. B., Rypma, B., Twente, S., and Rotrosen, J. P.,** Somatostatin regulation of the CRH-ACTH cortisol axis, *Biol. Psychiatry,* 21, 971, 1986.
44. **Rubinow, D. R., Post, R. M., Gold, P. W., Ballenger, J. C., and Reichlin, S.,** Effects of carbamazepine on cerebrospinal fluid somatostatin, *Psychopharmacology,* 85, 210, 1985.
45. **Holsboer, F., von Bardeleben, U., and Steiger, A.,** Effects of intravenous corticotropin-releasing hormone upon sleep-related growth hormone surge and sleep-EEG in man, *Neuroendocrinology,* 48, 62, 1988.
46. **Ehlers, C. L., Reed, T. K., and Henriksen, S. J.,** Effects of corticotropin-releasing factor and growth hormone-releasing factor on sleep and activity in rats, *Neuroendocrinology,* 42, 467, 1986.
47. **Born, J., Kern, W., Bieber, K., Fehm-Wolfsdorf, G., Schiebe, M., and Fehm, H. L.,** Night-time plasma cortisol secretion is associated with specific sleep stages, *Biol. Psychiatry,* 21, 1415, 1986.
48. **Jarrett, D. B., Coble, P., Kupfer, D. J., and Greenhouse, J. B.,** Sleep-related hormone secretion in depressed patients, *Acta Psychiatry Belgium,* 85, 603, 1985.
49. **Rivier, C. and Vale, W.,** Corticotropin-releasing factor (CRF) acts centrally to inhibit growth hormone secretion in the rat, *Endocrinology,* 114, 2409, 1984.

Chapter 23

CORTICOTROPIN-RELEASING FACTOR: CEREBROSPINAL FLUID STUDIES

Garth Bissette and Charles B. Nemeroff

TABLE OF CONTENTS

I. INTRODUCTION

The study of neurochemical alterations in living patients with psychiatric or neurologic disease requires access to a physiologically available tissue that is representative of the CNS environment under scrutiny. In this case, the relevant environment is the extracellular fluid within the synaptic cleft. The presently available tissues for study of this environment are brain biopsy samples, plasma, platelets, or cerebrospinal fluid. Brain biopsy offers the most direct route for assessment of neurochemical status. Unfortunately it is invasive, nonrenewable and, in many cases, the area of interest is not accessable without disruption of overlying structures. Moreover, the anomalous venous supply to the brain renders this technique liable to unexpected morbidity. Plasma has the advantage of being relatively plentiful and renewable, but suffers the serious disadvantages of being on the wrong side of the blood-brain barrier, being remote from the site of interest and, due to its large volume, diluting the substance(s) of interest. In addition, many plasma constituents interfere nonspecifically with assay techniques requiring extraction steps with reduced recovery as well as containing contributions from other tissues that may not be involved in the primary pathological process under investigation. Platelets, like brain tissue, are embryologically derived from neural crest. They have some characteristics of neurons, such as receptors, but lack a nucleus. They possess most of the disadvantages of plasma and are not representative of extracellular fluid. Cerebrospinal fluid at present, offers the most direct access to the internal mileau of the central nervous system. While a lumbar or intracisternal tap is technically an invasive procedure, lumbar taps are now performed routinely with few complications and can be performed more than once in an individual. Cerebrospinal fluid is in direct communication with the extracellular fluid of the brain and moderate amounts of CSF are quickly replaced. However, at present, very few studies have shown a correlation between brain tissue concentrations of a neurotransmitter or its metabolite with their respective concentrations in CSF. Stanley et al.[1] have shown a direct correlation between brain and CSF concentrations of the acidic serotonin metabolite, 5-hydroxyindole acetic acid (5-HIAA). The contribution of spinal cord sources to the component being measured must also be considered when CSF is obtained from lumbar taps. Rostral-caudal gradients have recently been observed for several neurotransmitters in postmortem studies in which both cisternal and lumbar CSF were sampled and in studies in which 30 ml of CSF were obtained in patients or normal control subjects. While these confounding factors must be considered, differences have been reported between several diagnostic groups and normal controls for specific neurochemical substances contained in CSF. One of these substances is corticotropin-releasing factor (CRF). This substance was first isolated from sheep hypothalami and as described elsewhere in this volume has been detected in human brain, plasma, and CSF. Human CRF immunoreactivity in CSF has been characterized by Suda et al.[2] and is indistinguishable from synthetic CRF or hypothalamic CRF by chromatography or radioimmunoassay dilution curves. Oldfield et al.[3] has demonstrated that ^{125}I-CRF is cleared from CSF six times as rapidly as a control substance known to be cleared by bulk flow. Both Garrick et al.[4] and Kalin et al.[5] have reported a circadian rhythm for CRF in monkey CSF that is 180° out of phase with CSF cortisol. Thus CRF peaks at the beginning of the dark cycle while cortisol values have declined to the lowest point in the 24-h cycle. While there does not seem to be any sex difference in CSF concentrations of CRF, there may be an effect of age. In a collaboration with Jan and Thomas Hedner of the University of Gothenburg in Sweden, we found that measurements of CSF from children aged 6 months to 2 years have revealed an exponential decline in CRF concentrations with age (unpublished). Before the age of 2 years, mean CSF concentrations of CRF were twice as high as the amount found in children after the age of 2 years. In the remainder of this chapter we describe the current literature on CSF CRF in neuropsychiatric disorders.

II. CSF STUDIES IN DEPRESSION

Considerable data has accrued indicating that patients who fulfill criteria for major depression, especially those with melancholia or psychotic features, exhibit hyperactivity of the hypothalamic-pituitary-adrenal (HPA) axis. The elevation of plasma cortisol concentrations in depressed patients,[6] the inability of the HPA axis to respond appropriately to exogenous corticosteroids, as measured by the dexamethasone suppression test,[7] and the blunted ACTH response to exogenously administered CRF[8] have all implicated the hypothalamic releasing factor, CRF, as being integral to the pathophysiological response of this axis in depression. There is now considerable evidence that support the hypothesis that CRF is hypersecreted in depression. Some of this evidence has been provided by reports of increased CSF concentrations of this neuropeptide in drug-free depressed patients compared to controls.

In a series of studies, our laboratory has documented increased CSF concentrations of CRF in several geographically diverse populations of patients with DSM-III diagnoses of major depression. In our first study,[9] we compared the CSF concentration of CRF in three groups of patients with those of normal control subjects from Sweden. The mean CSF concentration of CRF was significantly elevated in the depressed group (n = 23) compared to controls (n = 10); there were no differences between the controls and either DSM-III diagnosed schizophrenic patients (n = 11) or patients with dementia (Alzheimer's and/or multi-infarct, n = 29). Eleven of the depressed patients had higher CSF CRF concentrations than the highest normal control value (Figure 1). There was no correlation of age or sex with CRF concentrations in CSF and all patients were drug free for at least 2 weeks prior to lumbar CSF sampling. There was also no correlation between the suppression of cortisol secretion by dexamethasone in the depressed patients or the normal controls and their respective CSF concentration of CRF. However, a negative correlation between CRF and the noradrenergic metabolite MHPG has been established in healthy volunteer control patients (n = 10) while drug-free depressed patients (n = 22) had positive correlations between CSF concentrations of CRF and either HVA or 5-HIAA.[10]

Our findings of higher CSF concentrations of CRF in depression were confirmed and extended in a recent study in Hungarian patients.[11] Patients fulfilling DSM-III criteria for major depression (53 females and 1 male), schizophrenia (n = 23) or bipolar disorder (n = 4, manic) were compared to a large population of neurologic controls (65 males, 73 females) who did not fulfill criteria for any major mental illness. All antidepressant, neuroleptic, and anticonvulsant drugs were withdrawn for at least 2 weeks before CSF sampling and a standard 1 mg dexamethasone suppression test (DST) was administered on the evening following the lumbar puncture to the schizophrenic and depressed patients. The depressed patients had significantly higher mean CSF concentrations of CRF compared to the controls or schizophrenic patients; 24 depressed patients had higher concentrations than the highest normal controls (Figure 2). Schizophrenic patients also had higher mean CSF concentrations of CRF because of three individuals with high CSF CRF concentrations, when compared to controls. Again there was no correlation of DST results and CRF CSF concentrations for either group.

In collaboration with Fink at Stony Brook (unpublished), we measured CSF CRF concentrations in depressed patients before and after a regimen of electroconvulsive shock therapy (ECT). The CRF concentration of the pretreatment CSF sample was higher in most of the patients compared to the concentration in the CSF sample drawn after the sixth ECT. Furthermore, in collaboration with Arato et al.,[12] higher mean CSF concentration of CRF were reported in 22 drug-free, depressed female patients compared to 12 normal controls. Eleven of these depressed patients had attempted suicide; their CRF-CSF concentrations were not different from the nonsuicidal depressed patients. Again, in this population there

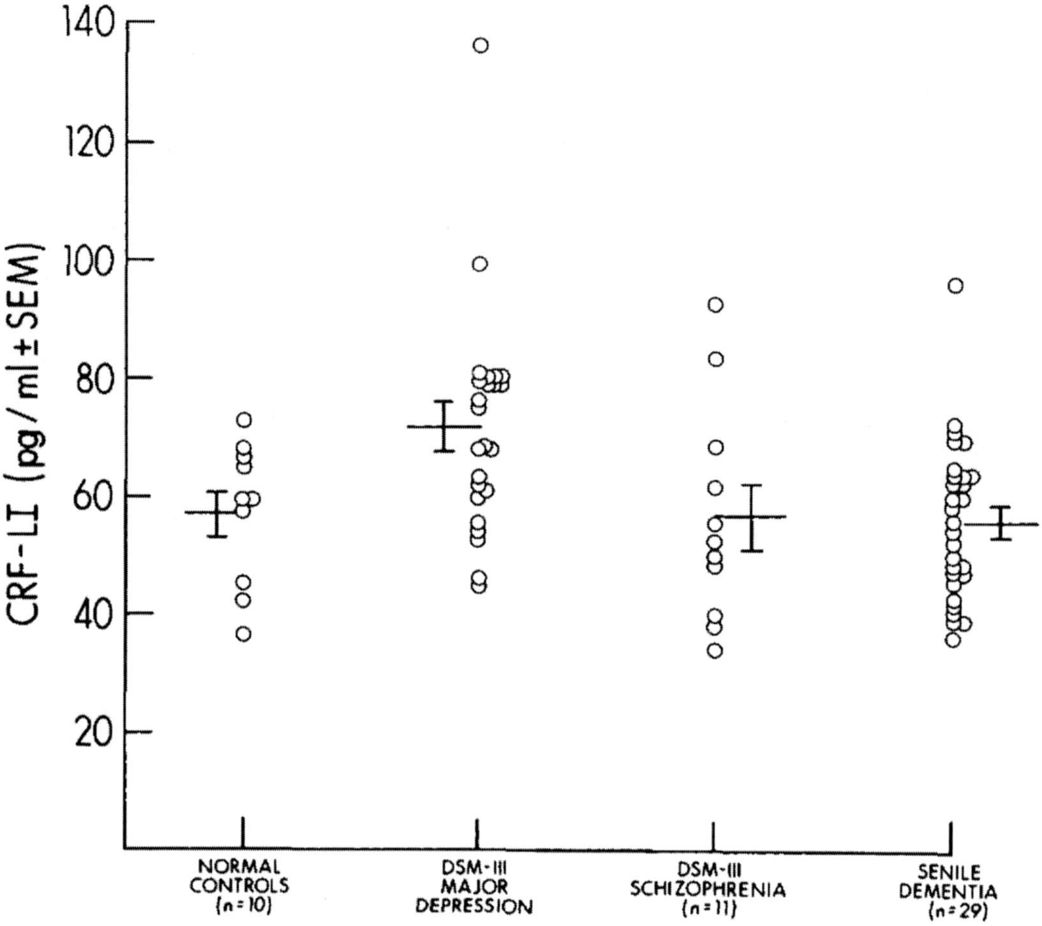

FIGURE 1. The concentration of CRF-like immunoreactivity in the cerebrospinal fluid of normal, healthy volunteers (5 females, 5 males), or patients with DSM-III diagnoses of major depression (13 males, 10 females), schizophrenia (8 females, 3 males) or dementia (20 females, 9 males, multi-infarct or primary degenerative dementia) from Sweden. Mean CSF concentration of CRF-like immunoreactivity in the major depression patients were significantly elevated ($p < 0.05$, ANOVA and Student-Newman-Keuls test; $p < 0.025$, Mann-Whitney U test) compared to the other patient groups or the controls. Details in Reference 9.

was no significant correlation between CSF concentrations of CRF and DST suppression. In a more recent study conducted in collaboration with Arato and Banki we measured the concentration of CRF in cisternal CSF from 19 suicide victims and 19 matched controls (sudden cardiac arrest); the suicide group had markedly elevated mean CSF-CRF concentrations (207 ± 42 pg/ml) compared to the controls (81 ± 10 pg/ml). France et al.[13] have measured CSF concentrations of CRF in chronic pain patients with and without a DSM-III diagnosis of major depression as well as in patients with major depression without pain. The CSF concentration of CRF was significantly higher in the depressed patients with major depression alone (n = 13) compared to patients with depression and chronic back pain (n = 6), patients with chronic back pain only (n = 9) or normal controls (n = 21) without depression or pain. Kurlan et al.[14] measured CSF concentrations of CRF in patients with Huntington's disease (n = 54) and controls receiving spinal anesthesia (n = 21). Higher mean concentrations of CRF were found in the Huntington's disease patients compared to controls. A significant positive correlation between the amount of CRF and the severity of depression was observed in the Huntington's disease patients (n = 24) who fulfilled DSM-III criteria for major depression, though mean group concentration of CRF were not different

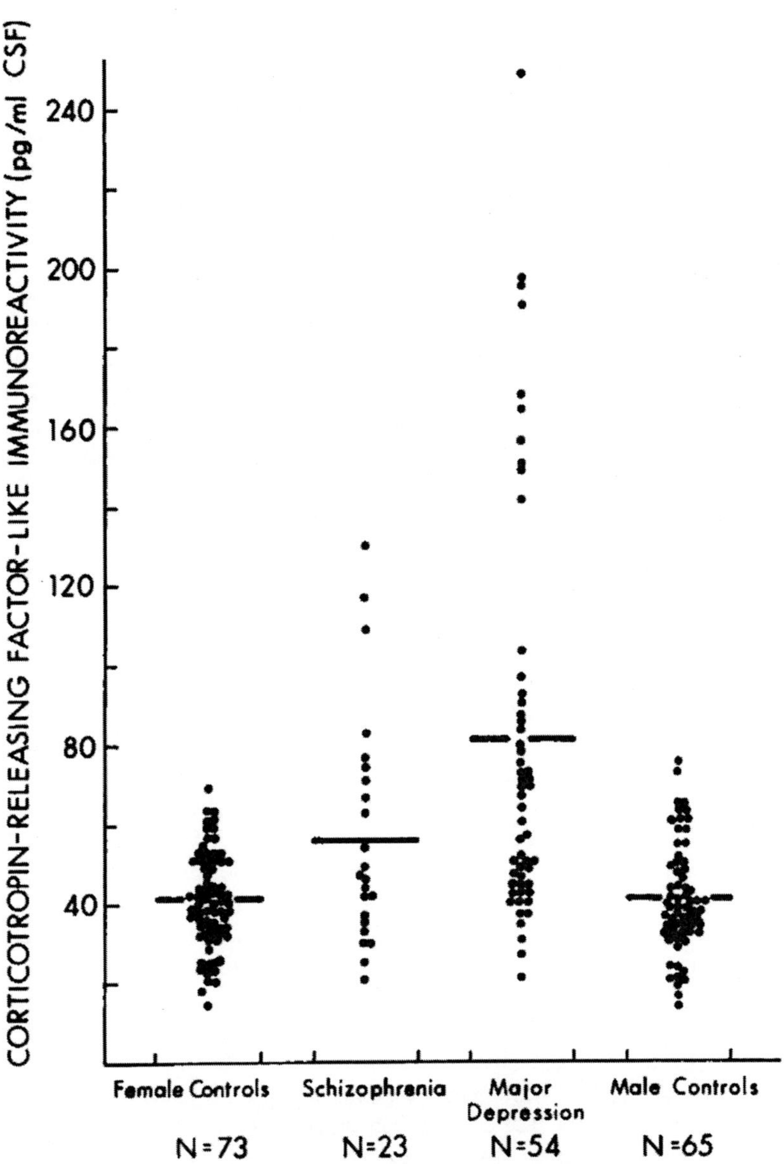

FIGURE 2. The concentration of CRF-like immunoreactivity in cerebrospinal fluid from female neurologic controls, female patients with DSM-III diagnoses of schizophrenia or major depression (one male patient) and male neurologic controls from Hungary. The mean cerebrospinal fluid concentration of CRF-like immunoreactivity was significantly elevated in the depressed patients ($p < 0.001$, ANOVA and Student-Newman-Keuls test) and schizophrenic patients ($p < 0.05$, ANOVA and Student-Newman-Keuls test) compared to either female or male controls. The depressed patients also had significantly higher mean CSF concentration of CRF-like immunoreactivity than the schizophrenic group ($p < 0.01$, ANOVA and Student-Newman-Keuls test). Details in Reference 11.

between the Huntington's disease patients with and without depression. In contrast to our findings in depression, Roy et al.[15] have reported that there was no mean difference in the CSF CRF concentrations among drug-free patients ($n = 27$) with DSM-III diagnoses of major depression compared to 18 psychiatrically normal controls. However, when one depressed patient with an extremely high CSF concentration of CRF was excluded, the patients exhibiting escape from dexamethasone suppression in this population had significantly higher CSF concentrations of CRF than those who exhibited normal DST responses.

III. CSF CRF CONCENTRATIONS IN OTHER NEUROPSYCHIATRIC AND ENDOCRINE DISORDERS

Kling et al.[16] have reported that patients with Cushing's disease (n = 8) had decreased CSF concentrations of CRF compared to patients with a DSM-III diagnosis of major depression (n = 19) or controls (n = 23). The depressed patients in this sample did not differ from the controls while the controls had a significant positive correlation between the amounts of CRF and ACTH in CSF, but this correlation was not seen in the Cushing's disease patients or the depressed patients. Kaye et al.[17] have reported significantly increased CSF concentrations of CRF in underweight women with a diagnosis of anorexia nervosa (n = 20) compared to normal controls (n = 19). CRF concentrations in the anorexic group tended to normalize after weight gain. The underweight women had significantly higher depression scores as rated by three different rating scales. These depression scores were positively correlated with CSF CRF concentrations after weight gain in these patients, but did not do so prior to weight gain.

Klimek et al.[18] have reported that CSF concentrations of CRF are reduced in patients with amyotrophic lateral sclerosis (ALS, n = 5) when compared to controls (n = 10) with discopathy. The highest CSF concentration of CRF in the ALS patients was lower than the lowest control value.

There have been several reports of decreased CRF concentrations in the cerebral cortex and the caudate nucleus of patients with Alzheimer's disease (see Chapter 24). Although we did not observe reduced concentrations of CRF in CSF in our original study[9] or in several other populations of Alzheimer's disease patients (Bissette and Nemeroff, unpublished observations), there are now two reports of decreased concentrations of CRF in CSF of Alzheimer's disease patients. Mouradian et al.[19] reported decreased CRF concentration in CSF of patients with a presumptive clinical diagnosis of Alzheimer's disease (n = 16) compared to neurologically normal controls (n = 9). Seven of the Alzheimer's patients had lower CRF concentrations in CSF than the lowest control value. An increasing gradient of CRF was found in the later (27 to 29 ml) compared to earlier samples (3 to 5 ml) in the controls confirming our observation of a rostral-caudal gradient described above. May et al.[20] have also reported reduced CSF CRF concentrations in Alzheimer's disease patients when compared to controls. In this study, 33 patients with a clinical diagnosis of Alzheimer's disease were compared to 13 healthy, age-matched controls. Both CRF and ACTH concentrations were decreased in CSF of the Alzheimer's patients compared to controls; there were no differences in plasma or urinary free cortisol concentrations.

In summary, CRF is present in readily measurable quantities in human CSF. Although the precise central nervous system source of CRF in CSF remains obscure, alterations in CRF concentrations of CSF have been observed in several pathologic states. The increased CRF concentrations in CSF of some depressed patients may reflect a hypersecretion of this releasing factor which is consistent with other findings, e.g., down-regulation of cerebrocortical CRF receptors.[21] In addition, in laboratory animals, CRF administered centrally, produces many of the signs and symptoms of major depression.[22] The decreases in CSF CRF concentrations reported in patients with chronic neurodegenerative diseases such as Alzheimer's disease and amyotrophic lateral sclerosis may be due to degeneration of CRF neurons in these disorders. Future, well-controlled studies are needed to determine the specificity of these findings.

REFERENCES

1. **Stanley, M., Traskman-Bendz, L., and Dorovini-Zis, K.,** Correlations between aminergic metabolites simultaneously obtained from human CSF and brain, *Life Sci.*, 37, 1279, 1985.
2. **Suda, T., Tozawa, F., Mouri, T., Demura, H., and Shizume, K.,** Presence of immunoreactive corticotropin-releasing factor in human cerebrospinal fluid, *J. Clin. Endocrinol. Metab.*, 57, 225, 1983.
3. **Oldfield, E. H., Schulte, H. M., Chrousos, G. P., Rock, J. P., Kornblith, P. L., O'Neill, D. L., Poplack, D. G., Gold, P. W., Cutler, G. B., and Loriaux, L.,** Active clearance of corticotropin-releasing factor from the cerebrospinal fluid, *Neuroendocrinology*, 40, 84, 1985.
4. **Garrick, N. A., Hill, J. I., Szele, F. G., Tomai, T. P., Gold, P. W., and Murphy, D. L.,** Corticotropin-releasing factor: a marked circadian rhythm in primate cerebrospinal fluid peaks in the evening and is inversely related to the cortisol circadian rhythm, *Endocrinology*, 121, 1329, 1987.
5. **Kalin, N. H., Shelton, S. E., Barksdale, C. M., and Brownfield, M. S.,** A diurnal rhythm in cerebrospinal fluid corticotropin-releasing hormone different from the rhythm of pituitary-adrenal activity, *Brain Res.*, 426, 385, 1987.
6. **Sachar, E. J., Hellman, L., Roffwarg, H. P., Halpern, F., Fukushima, D., and Gallagher, T.,** Disrupted 24-hour patterns of cortisol secretion in psychiatric depression, *Arch. Gen. Psychiatry*, 28, 19, 1973.
7. **Carroll, B. J., Feinberg, M., Greden, J. F., Tarika, J., Albala, A. A., Haskett, R. F., James, N. M., Kronfol, Z., Lohr, N., Steiner, M., Vigne, J. P., and Young, E.,** A specific laboratory test for the diagnosis of melancholia, *Arch. Gen. Psychiatry*, 38, 15, 1981.
8. **Holsboer, F., Gerken, A., Staller, G. K., and Müller, O. A.,** Blunted aldosterone and ACTH release after human CRH administration in depressed patients, *Am. J. Psychiatry*, 144, 229, 1987.
9. **Nemeroff, C. B., Widerlov, E., Bissette, G., Walleus, H., Karlsson, I., Eklund, K., Kilts, C. D., Loosen, P. T., and Vale, W.,** Elevated concentrations of CSF corticotropin-releasing factor-like immunoreactivity in depressed patients, *Science*, 226, 1342, 1984.
10. **Widerlov, E., Bissette, G., and Nemeroff, C. B.,** Monoamine metabolites, corticotropin-releasing factor and somatostatin as CSF markers in depressed patients, *J. Affective Dis.*, 14, 99, 1988.
11. **Banki, C. M., Bissette, G., Arato, M., O'Connor, L., and Nemeroff, C. B.,** Cerebrospinal fluid corticotropin-releasing factor-like immunoreactivity in depression and schizophrenia, *Am. J. Psychiatry*, 144, 873, 1987.
12. **Arato, M., Banki, C. M., Nemeroff, C. B., and Bissette, G.,** Hypothalamic-pituitary-adrenal axis and suicide, *Ann. N.Y. Acad. Sci.*, 487, 263, 1986.
13. **France, R. D., Urban, B., Krishnan, R. R., Bissette, G., Banki, C. M., Nemeroff, C. B., and Spielman, F. J.,** CSF corticotropin-releasing factor-like immunoreactivity in chronic pain patients with and without major depression, *Biol. Psychiatry*, 23, 86, 1988.
14. **Kurlan, R., Shoulson, I., Caine, E., Rubin, A., Nemeroff, C. B., Bissette, G., Zaczek, R., Coyle, J., and Spielman, F. J.,** Cerebrospinal fluid correlates of depression in Huntington's disease, *Soc. Neurosci. Abstr.*, 13, 1473.
15. **Roy, A., Pickar, D., Paul, S., Doran, A., Chrousos, G. P., and Gold, P. W.,** CSF corticotropin-releasing hormone in depressed patients and normal control subjects, *Am. J. Psychiatry*, 144, 641, 1987.
16. **Kling, M. A., Chrousos, G. P., Roy, A., Doran, A. R., Calabrese, J. R., and Gold, P. W.,** CSF, CRH and ACTH in depression and Cushing's disease, *Proc. 139th Annual Meeting of the American Psychiatric Association, Abstr.* 273, 75, 1986.
17. **Kaye, W. H., Gwirtsman, H. E., George, D. T., Ebert, M. H., Jimerson, D. C., Tomai, T. P., Chrousos, G. P., and Gold, P. W.,** Elevated cerebrospinal fluid levels of immunoreactive corticotropin-releasing hormone in anorexia nervosa: relation to state of nutrition, adrenal function and intensity of depression, *J. Clin. Endocrinol. Metab.*, 64, 203, 1987.
18. **Klimek, A., Cieslak, D., Kuberska, J. S., and Stepien, H.,** Reduced lumbar cerebrospinal fluid corticotropin releasing factor (CRF) levels in amyotrophic lateral sclerosis, *Acta Neurol. Scand.*, 74, 72, 1986.
19. **Mouradian, M. M., Farah, J. M., Mohr, E., Fabbrini, G., O'Donohue, T., and Chase, T. N.,** Spinal fluid CRF reduction in Alzheimer's disease, *Neuropeptides*, 8, 393, 1986.
20. **May, C., Rapoport, S. I., Tomai, T. P., Chrousos, G. P., and Gold, P. W.,** Cerebrospinal fluid concentrations corticotropin releasing hormone (CRH) and corticotropin (ACTH) are reduced in patients with Alzheimer's disease, *Neurology*, 37, 535, 1987.
21. **Nemeroff, C. B., Owens, M. J., Bissette, G., Andorn, A., and Stanley, M.,** Reduced corticotropin-releasing factor (CRF) binding sites in the frontal cortex of suicides, *Arch. Gen. Psychiatry*, 45, 577, 1988.
22. **Owens, M. J. and Nemeroff, C. B.,** The neurobiology of corticotropin-releasing factor: implications for affective disorders, in *Hypothalamic-Pituitary-Adrenal Axis Physiology and Pathophysiology*, Schatzberg, A. F. and Nemeroff, C. B., Eds., Raven Press, New York, 1–36.

Chapter 24

ABNORMALITIES IN CORTICOTROPIN-RELEASING FACTOR (CRF) IN NEURODEGENERATIVE DISEASES

Errol B. De Souza, Garth Bissette, Peter J. Whitehouse, Richard E. Powers, Donald L. Price, Wylie W. Vale, and Charles B. Nemeroff

TABLE OF CONTENTS

I. INTRODUCTION

The pathologies occurring in various neurodegenerative disorders are associated with dysfunction and death of neurons in a variety of cell populations, including cholinergic, monoaminergic, and peptidergic systems. Corticotropin-releasing factor (CRF) has been demonstrated to have a broad distribution in the central nervous system (CNS) and to produce a wide spectrum of autonomic, electrophysiologic, and behavioral effects characteristic of a neurotransmitter in the CNS (see various chapters). Recent clinical data suggest that CRF may be of relevance in endocrine, psychiatric, and neurologic illnesses. In this chapter, we summarize some of the data from our recent studies examining changes in pre- and post-synaptic markers for CRF in various brain regions in a variety of neurodegenerative disorders including Alzheimer's disease (AD), Parkinson's Disease (PD), progressive supranuclear palsy (PSP), and Huntington's Disease (HD).

II. CRF IN ALZHEIMER'S DISEASE

Alzheimer's disease is a progressive degenerative disease of the nervous system characterized neuropathologically by the presence of senile plaques and neurofibrillary tangles in amygdala, hippocampus and neocortex.[1,2] Dysfunction and death of basal cholinergic neurons projecting to forebrain targets, are associated with marked decreases in cholinergic markers, including the activity of choline acetyltransferase (ChAT).[3] Although cortical levels of somatostatin and somatostatin receptors have been reported to be reduced in AD, no consistent changes have been reported in other neuropeptide systems.[4,5] We have used radioimmunoassay and immunohistochemical techniques to examine changes in CRF content and morphology, respectively, and receptors binding techniques to determine the consequences of the presynaptic changes in the neuropeptide.

A. RADIOIMMUNOASSAY STUDIES

The concentration of corticotropin-releasing factor-like immunoreactivity (CRF-IR) in the human brain was measured by radioimmunoassay in postmortem tissue from control patients and in those with histologically confirmed AD.[4-6] In the controls, CRF-IR was found in high concentrations in the hypothalamus and frontal cerebral cortex (Brodmann's Area 10), in moderate concentrations in amygdala, substantia innominata, temporal, and parietal cerebral cortex (Brodmann's Areas 38 and 7), and the caudate nucleus, and in low concentrations in posterior hippocampus and nucleus accumbens (Figure 1). A marked reduction in the concentration of CRF-IR was observed in the frontal and temporal cerebral cortex (approximately 50%) as well as in the caudate nucleus (approximately 70%) in the AD group (Figure 1). Additional brain areas affected in AD include the occipital cerebral cortex, with an 80% decrease in CRF content observed in this brain region (Figure 2).

B. RECEPTOR STUDIES

In order to examine the consequences of the decreases in CRF-IR observed in AD, we measured changes in CRF-IR and CRF receptors in four regions of the cerebral cortex. As shown in Figure 2, in individuals with AD the concentrations of CRF-IR were decreased and CRF receptors were reciprocally increased in the temporal, frontal, and occipital cerebral cortex. As mentioned above the change in CRF activity was greatest in the occipital cortex, with a 70 to 80% decrease in the content and doubling in receptor binding; the reduction in CRF-IR and the increase in CRF receptors in the temporal and frontal cortex were all greater than 50% of the corresponding control value (Figure 2). Neither CRF-IR nor CRF receptors were changed significantly in the cingulate cortex. Scatchard analysis data of detailed competition curves carried out in random samples of temporal, frontal and occipital cortex from

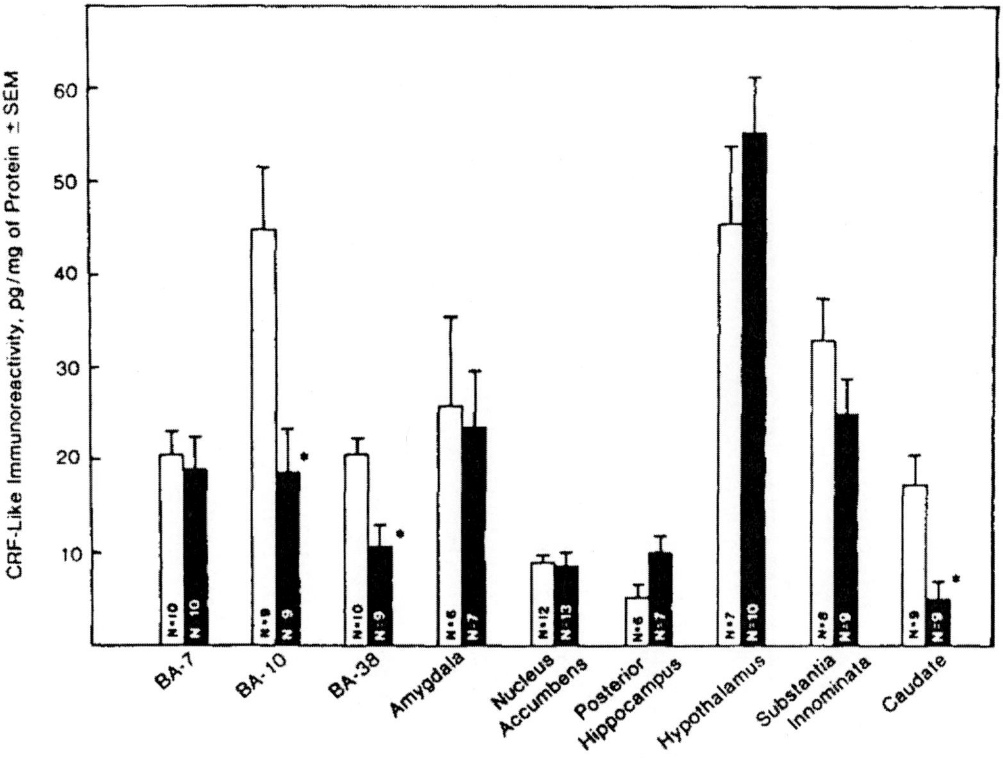

FIGURE 1. Regional brain concentration of corticotropin-releasing factor-like immunoreactivity (CRF-IR) in patients with senile dementia of Alzheimer's type and in controls. Graph represents concentrations of CRF-IR in brain regions from patients dying of senile dementia of Alzheimer's type (solid bars) and controls (open bars). Number of sample from each brain region is shown inside respective bars. Concentration of CRF-IR is shown as mean ± SEM and is reported as picograms per milligrams of protein. Statistical significance was sought by Student's t-test and is represented by asterisk, which indicates $p < 0.01$. (From Bissette, G., Reynolds, G. P., Kilts, C. D., Widerlov, E., and Nemeroff, C. B., *JAMA*, 254, 3067, 1985. With permission.)

both control and AD patients indicated that the increase in CRF receptor binding reflected an up-regulation in receptor number (B_{max}), as the affinity of the ligand for the receptor (K_D) was comparable in control and in AD groups in all the cortical areas examined. A highly significant negative correlation ($r = -0.69$; $p < 0.001$) existed between the change in CRF receptors and CRF-IR in each cortical area (Figure 3). By contrast, in both studies, neither the age, sex of the patient nor the postmortem interval correlated significantly with either CRF-IR nor CRF receptors in any of the cortical regions examined. In addition, no effect of any psychoactive drug medication was noted on CRF receptors in AD, whereas a statistically nonsignificant trend towards a small increase in CRF-IR was noted in those patients treated with neuroleptics.

C. IMMUNOHISTOCHEMICAL STUDIES

In aged nonhuman primates and in humans (both aged controls and individuals with AD), neurites in plaques contain a variety of neurotransmitters.[7] Documentation of CRF-IR neurites in plaques in AD would provide morphological evidence for the involvement of CRF systems in this type of dementia. In the brains of controls and individuals with AD, antisera to CRF were used to immunostain neurons and their processes in order to further define the role of this neuropeptide in AD. In controls, normal-appearing CRF-IR axons were predominantly straight, thin and unbranched (Figure 4A). CRF-immunostained axons were present through the amygdalae of control individuals and were most conspicuous in

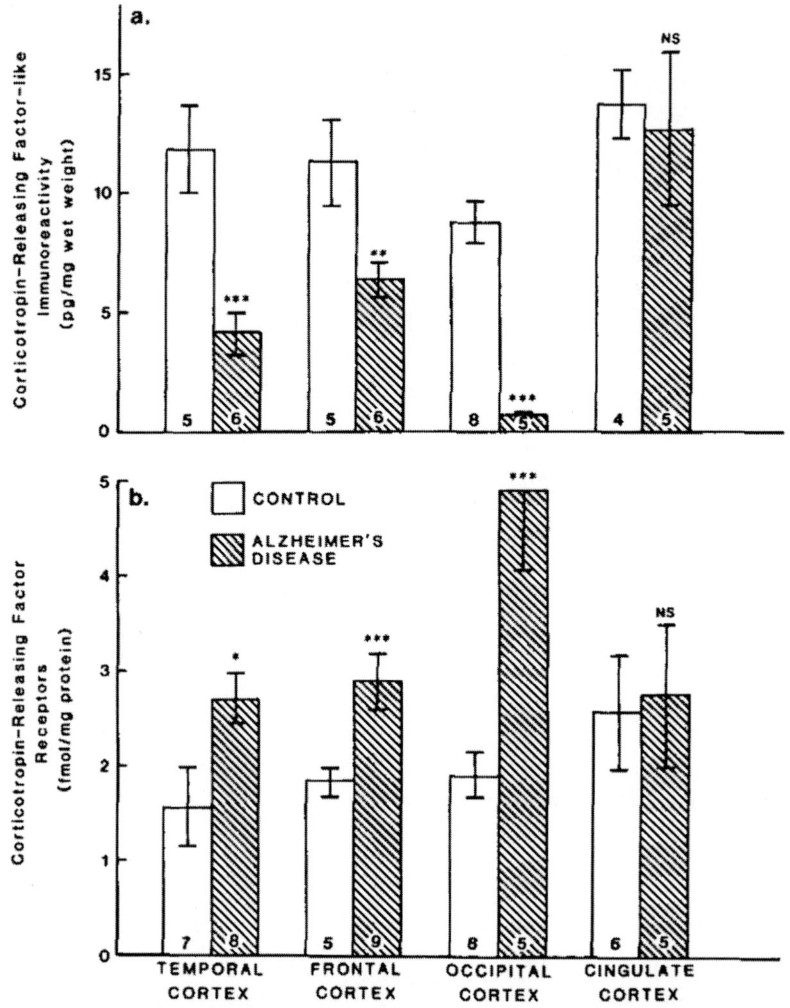

FIGURE 2. CRF-like immunoreactivity (A) and CRF receptor binding (B) in discrete regions of the cerebral cortex of Alzheimer's patients and controls. All values are means ± SEM. The number of subjects in each group is given at the bottom of each histogram. Significant differences from control group at $p < 0.05$, $p < 0.025$, and $p < 0.005$ are denoted by *, **, and ***, respectively. NS, not significant. (From De Souza, E. B., Whitehouse, P. J., Kuhar, M. J., Price, D. L., and Vale, W. W., *Nature*, 319, 593, 1986. With permission.)

the lateral and basolateral nuclei (Figure 5A). The plexus of immunostained axons was most dense in the paraventricular nuclei of the hypothalamus (PVN) (Figure 6) and the bed nucleus of the stria terminalis with fewer fibers present in the amygdala. A sparce population of CRF-IR axons was seen in the rostral hippocampus and entorhinal cortex. In AD, some normal-appearing CRF-IR fibers were identified in all regions, but there were fewer immunoreactive fibers in amygdalae from cases of AD than in age controls (Figure 5B).

In AD, tortuous CRF-immunostained axons, termed fiber abnormalities, were seen individually in the neuropil and in conjunction with amyloid deposits associated with senile plaques (Figure 4B). Amygdalae from individuals with AD contained large numbers of silver-stained senile plaques; however only one or two senile plaques per section contained CRF-IR neurites. Although numerous senile plaques were identified with silver stains in the entorhinal cortex from the AD group, abnormal CRF-stained neurites were not seen in this region.

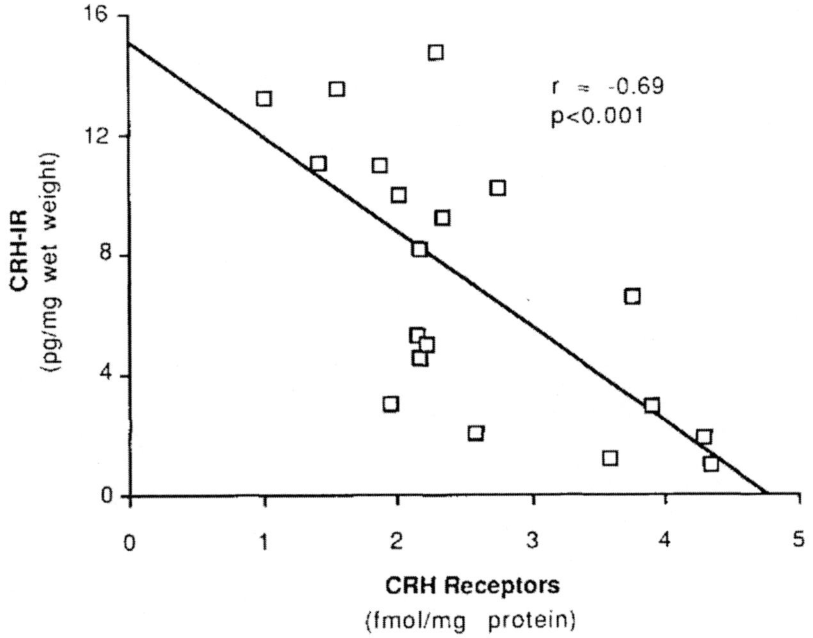

FIGURE 3. Correlation between CRF-like immunoreactivity (CRF-IR) and CRF receptors in cerebral cortex. The values for CRF-IR and CRF receptors were taken from data presented in Figure 2 examining changes in pre- and postsynaptic markers for CRF in Alzheimer's disease. The cerebral cortical regions sampled include frontal, temporal, and occipital cortex. There was a significant negative correlation between CRF-IR and CRF receptors ($r = -0.69$; $p < 0.001$). (From De Souza, E. B., Whitehouse, P. J., Price, D. L., and Vale, W. W., *Ann. N.Y. Acad. Sci.*, 512, 237, 1988. With permission.)

In the central and basolateral nuclei of the amygdala, sparse populations of small CRF-IR, bipolar cells were visualized. No immunostained cells were present in the rostral hippocampus or entorhinal cortex. Lightly immunostained neurons were present in the bed nucleus of the stria terminalis and the PVN of sections from the control group (Figure 6A). Neuronal cell bodies and axons of the PVN in individual with AD showed the most intense immunoreactivity (Figure 6B); senile plaques, neurofibrillary tangles, and cell loss in the PVN were not apparent in these cases.

D. RELEVANCE OF CRF CHANGES IN ALZHEIMER'S DISEASE

The regional distribution of CRF in humans appears to be similar to that occurring in rats, but the distribution of CRF neurons innervating the cerebral cortex is not fully understood in primate brain. In rodents, CRF neurons are intrinsic to the neocortex, with perikarya located in laminae II and III and terminal fields concentrated in laminae I and IV,[8,9] areas rich in CRF receptors.[10,11] The reduction in CRF-IR in the cerebral cortex in AD could result from a selective degeneration of the intrinsic cortical CRF neurons and/or disease of CRF neurons projecting to the cortex from other brain areas. For example, a subpopulation of brain stem, cholinergic neurons containing CRF project to various forebrain targets including the cortex,[12] and pathology of this system could reduce ChAT and CRF-IR. Decreased CRF content in AD could result from cell dysfunction, i.e., decreased synthesis, altered processing, increased degradation, increased release and/or cell death. Increased CRF release appears unlikely, as it might be expected to down-regulate CRF receptors rather than to produce the observed up-regulation. At present, we cannot identify the specific mechanisms leading to reduced CRF levels.

Additional evidence for a role for CRF in AD is provided by the observations demon-

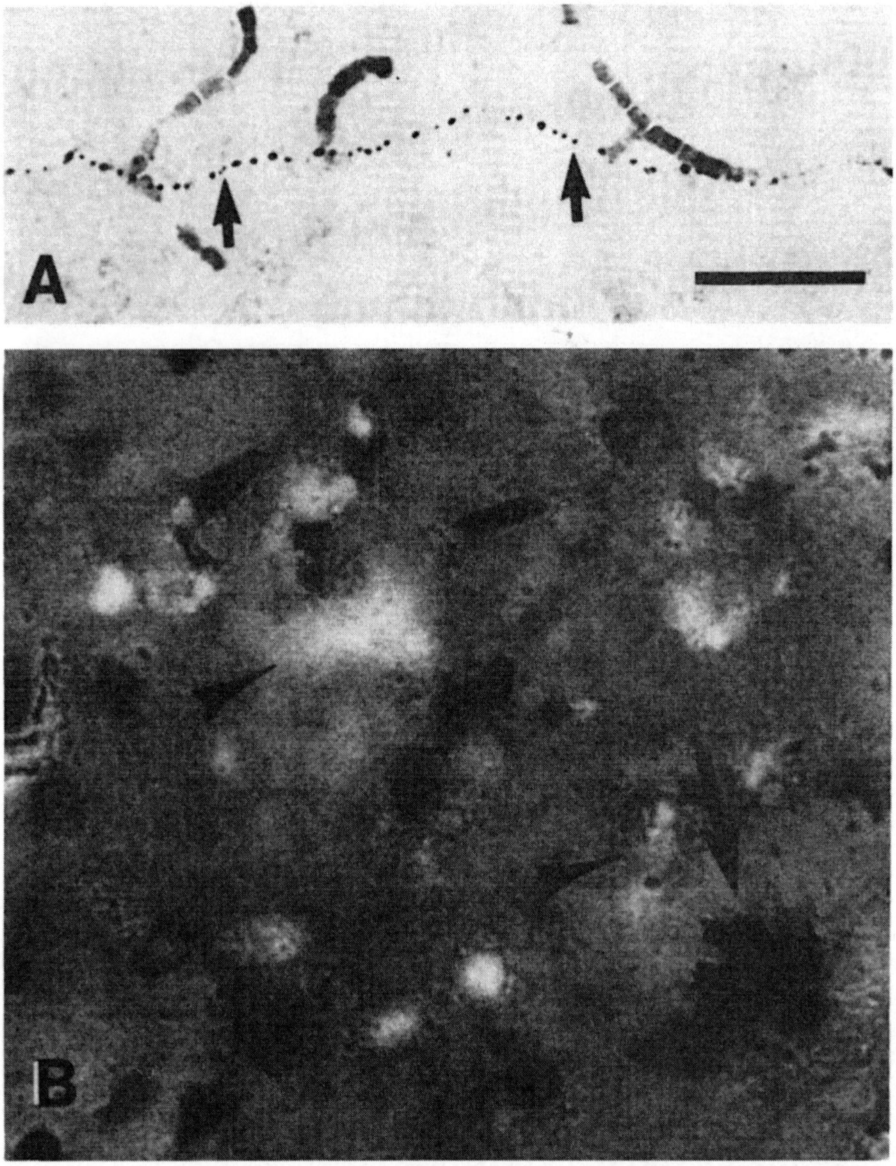

FIGURE 4. Photomicrographs of normal CRF-immunoreactive fibers (A) and a swollen process in a senile plaque (B). Bar = 20 μm. (A) Uniformly thin, beaded, unbranched CRF-immunoreactive fibers (arrows) were present in the lateral nucleus of the amygdala of an aged control. (B) This swollen CRF-immunoreactive neurite (arrow) was present in the amygdala of an individual with AD and was associated with thioflavin-T-stained amyloid deposits (arrowheads). This photograph was taken with combined epifluorescence and low-level transmitted light. (From Powers, R. E., Walker, L. C., De Souza, E. B., Vale, W. W., Struble, R. G., Whitehouse, P. J., and Price, D. L., *Synapse*, 1, 405, 1987. With permission.)

strating decreases in CRF-IR in other brain areas including the caudate and amygdala (see above), decreased concentrations of CRF-IR in the cerebrospinal fluid obtained from patients with AD,[13] and the documentation of the presence of CRF-IR fiber abnormalities in senile plaques in amygdala from individuals with AD. However, the expression of CRF antigen in neurons is not globally reduced in AD. CRF immunostaining in the PVN appeared more intense in the cases of AD in which the hypothalamus was sampled. This observation may

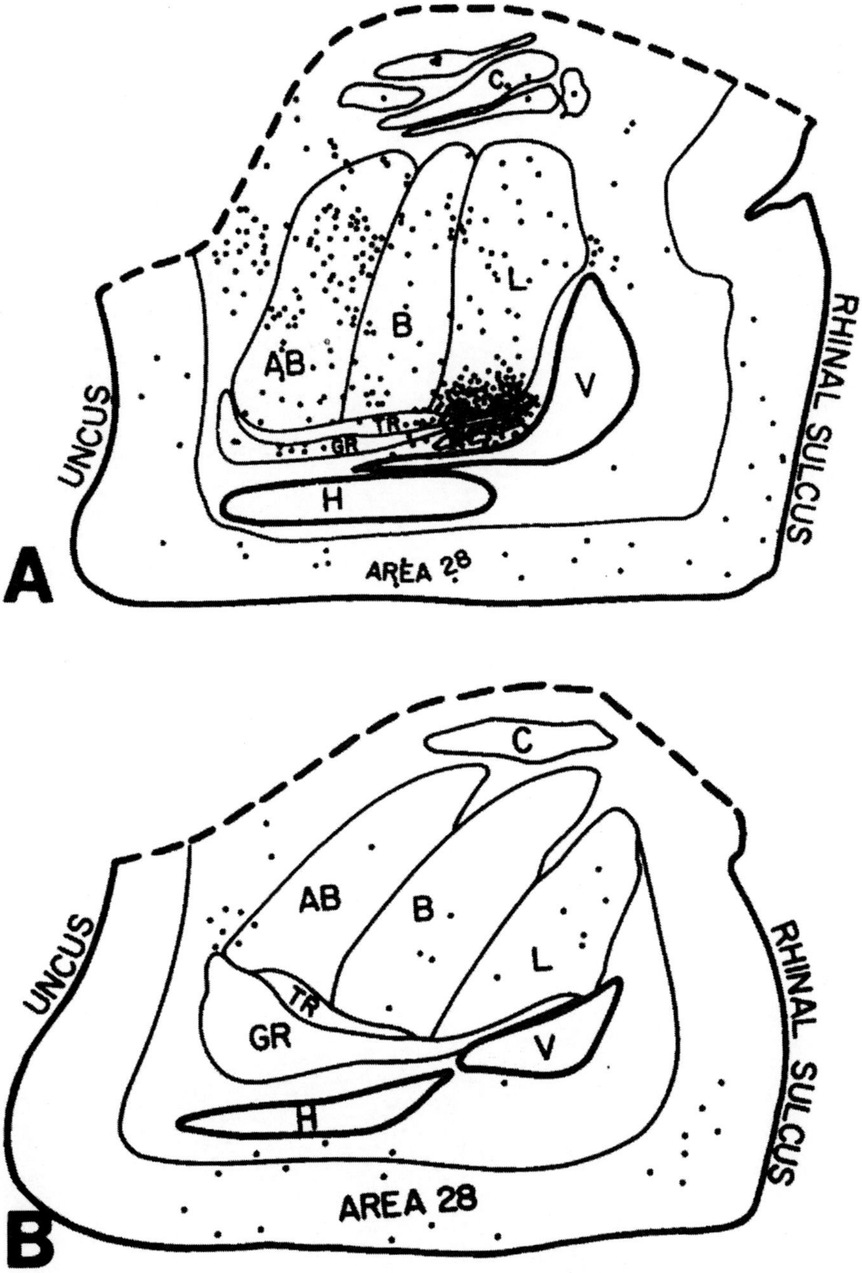

FIGURE 5. Schematic diagrams showing the distribution and frequency of CRF-immunostained fibers in the amygdala of an aged control (A) and an individual with AD (B). Each dot represents an immunostained fiber segment. Abbreviations: AB, accessory basal nucleus; B, basal nucleus; C, central nucleus; GR, granular nucleus; H, rostral hippocampus; L, lateral nucleus; TR, transitional nucleus; V, lateral ventricle, inferior horn. (A) In an aged control, CRF fibers were distributed throughout the amygdala and were less frequent in the adjacent entorhinal cortex and rostral hippocampus. (B) In the amygdala of an individual with AD, there were fewer CRF-immunostained fibers than an aged control (Figure 5A). (From Powers, R. E., Walker, L. C., De Souza, E. B., Vale, W. W., Struble, R. G., Whitehouse, P. J., and Price, D. L., *Synapse*, 1, 405, 1987. With permission.)

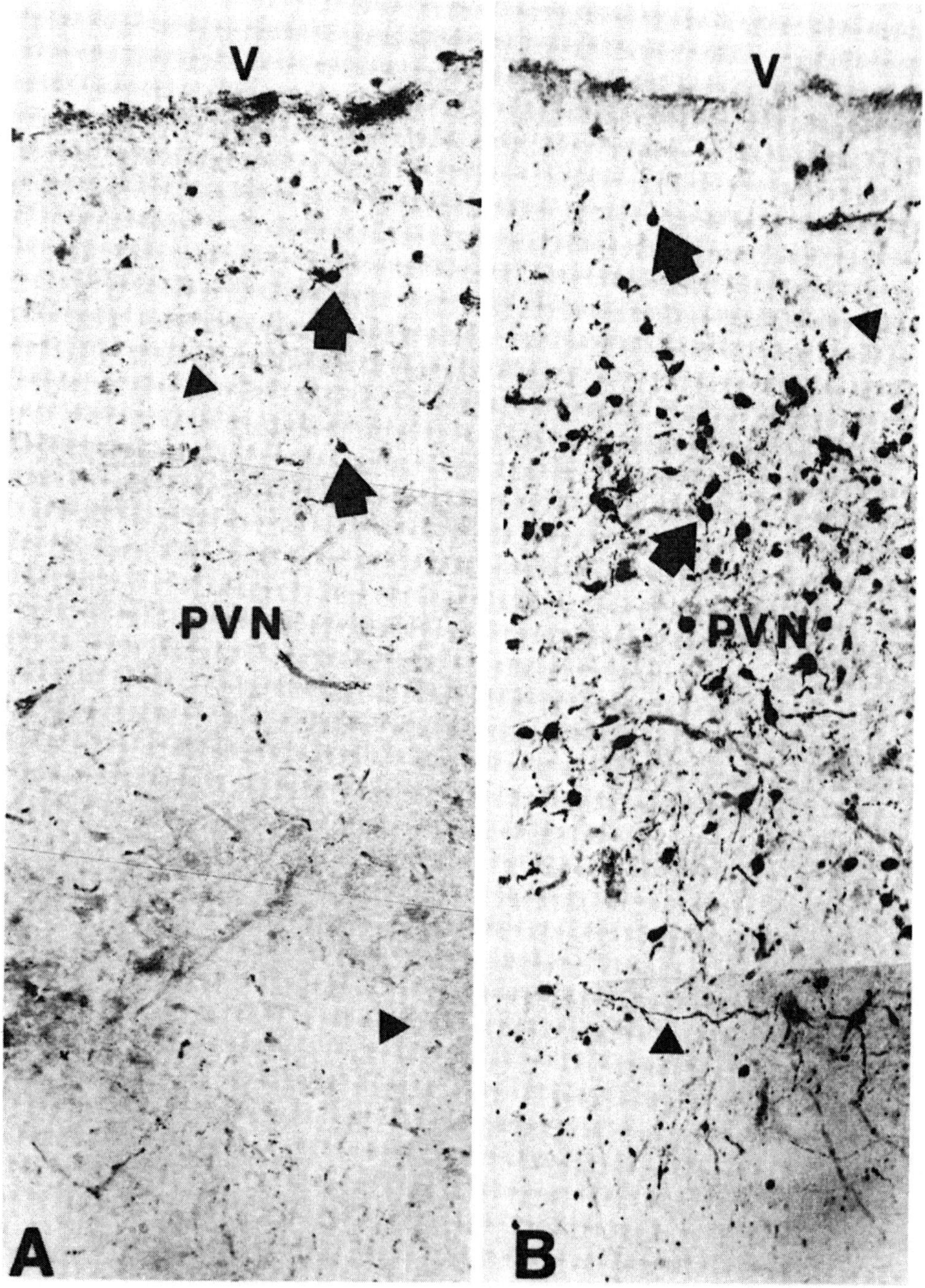

FIGURE 6. Photomicrographs demonstrate patterns of corticotropin-releasing factor (CRF) immunostaining in the paraventricular nucleus (PVN) of an aged control and an individual with Alzheimer's disease (AD) at equivalent levels in the hypothalamus. V, third ventricle, ×400. (A) Lightly stained cell bodies (arrows) and a plexus of thin, beaded CRF-immunoreactive fibers (arrowheads) in the PVN of an individual with AD. Silver stains did not show neurofibrillary tangles or neuritic plaques. (From Powers, R. E., Walker, L. C., De Souza, E. B., Vale, W. W., Struble, R. G., and Price, D. L., *Synapse*, 1, 405, 1987. With permission.)

explain why radioimmunoassay studies show no decrease in CRF levels in hypothalamus in AD (see Figure 1). Increased immunostaining of paraventricular neurons in AD, if truly representative of increased content of CRF, could be related to increased amounts of CRF mRNA in these cells, increased translation of available mRNA, or decreased processing/transport of the peptide. The implications of changes in CRF in the PVN as it relates to function of the hypothalamic-pituitary-adrenocortical axis remains to be demonstrated.

Alterations in CRF receptors in the cortex in AD contrast with previous observations of changes in the cholinergic and somatostatin systems in which reductions in presynaptic cholinergic markers and somatostatin are accompanied by inconsistent changes in cholinergic receptors and decreased somatostatin receptors (see Reference 5). The up-regulation of cortical CRF receptors in AD under conditions in which endogenous peptide is reduced is, to date, a unique finding in the study of neuropeptides in neurodegenerative diseases and suggest that CRF-receptor cells may be preserved in the cerebral cortex in AD.

III. CRF IN PARKINSON'S DISEASE

Parkinson's disease (PD) shares certain clinical and pathological features with AD. A substantial number of PD patients (15 to 40%) eventually develop dementia (see Reference 4). Although several types of cortical pathology occur in demented PD patients, including Lewy body inclusions (Lewy Body Dementia) and so-called "simple" atrophy, some demented patients exhibit the characteristic neuropathological features of AD, i.e., senile plaques and neurofibillary tangles (see Reference 4).

In cases of PD with dementia that also showed pathological features of AD, CRF content was decreased and showed a pattern similar to those cases exhibiting the pathology of AD alone (Figure 7). Specimens from patients with PD who did not have the histopathology characteristic of AD demonstrated statistically significant reductions of CRF content, although the reductions were less marked than in cases of combined AD/PD (Figure 7). No clear-cut relationship between the degree of intellectual impairment or psychiatric symptoms and CRF levels were noted in this group. The reductions in CRF content in PD may result from the same mechanisms as those in AD. Pathological alterations in brains of patients with AD and PD can be quite similar (i.e., senile plaques and neurofibrillary tangles) but only PD patients have extensive Lewy bodies[14-16] and loss of neurons in substantia nigra. The neurochemical pathology in the cortex in PD has not been as extensively studied as in AD. In PD, reductions in cortical ChAT activity and decreased number of neurons in the basal forebrain have been documented, particularly in association with dementia; the changes can occur in the absence of conspicious AD-type pathology.[17,18] In addition, reductions in somatostatin have been noted in PD.[19] Normal CRF levels have been reported in hypothalamus in PD,[20] suggesting that the loss in CRF in cortex is not generalized.

IV. CRF IN PROGRESSIVE SUPRANUCLEAR PALSY

Progressive supranuclear palsy (PSP) is a neurodegenerative disorder that shares certain clinical and pathological features with AD. In PSP, neurofibrillary tangles occur, but the ultrastructure of filaments of neurofibrillary tangles and the distribution of the pathology differ from those occurring in AD.[4] The small number of PSP samples limited the statistical analysis, but CRF-IR was approximately 50% of control values (Figure 7). No relationship between clinical signs and CRF-IR was discernible. The reduction in CRF-IR in the cerebral cortex in the present study could be due to disease of intrinsic CRF neurons or could result from loss of cortical afferents derived from brain stem neurons. In PSP, loss of neurons occurs in the pedunculopontine nucleus pars compacta.[21] Dysfunction in these neurons is probably associated with decriments in ChAT in globus pallidus and subthalamic nucleus.[22,23]

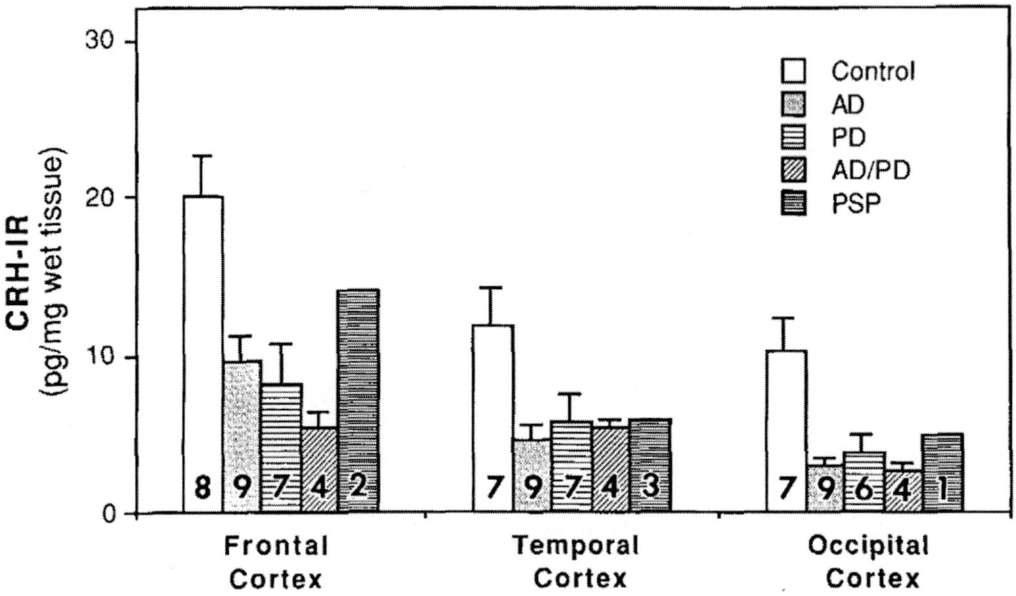

FIGURE 7. CRF-like immunoreactivity (CRF-IR) in discrete regions of cerebral cortex of patients with Alzheimer's disease (AD), Parkinson's disease (PD), Alzheimer's and Parkinson's disease (AD/PD), Progressive Supranuclear Palsy (PSP), and controls. All values are mean ± SEM. The number of subjects in each group is given at the bottom of each histogram. Note consistent reductions across all cortical regions in all diseases; the reductions in CRF-IR in AD, PD, and AD/PD were significantly different from their respective controls. The greatest loss of CRF-IR occurs in AD and AD/PD, with less marked changes occurring in PD and PSP. (Adapted from Whitehouse, P. J., Vale, W. W., Zweig, R. M., Singer, H. S., Mayeux, R., Price, D. L., and De Souza, E. B., *Neurology*, 37, 905, 1987.)

In addition, since a subpopulation of brain stem cholinergic neurons containing CRF project to the cortex,[12] disease of these cells may also contribute to the decreases in cortical CRF-IR in PSP.

V. CRF IN HUNTINGTON'S DISEASE

Huntington's disease (HD) is an autosomal dominant disorder of the CNS characterized clinically by major motor dysfunction, dementia, and emotional disturbance.[24] HD is characterized neuropathologically by profound neuronal cell loss in the caudate and putamen (striatum) with less prominent cell loss occurring in other regions of the basal ganglia and in cerebral cortex.[25,26] Dysfunction and death of cells in the striatum in HD are associated with marked decreases in a variety of classic neurotransmitters including gamma-aminobutyric acid and acetylcholine, and a variety of putative peptide neurotransmitters including substance P, cholecystokinin, methionine-enkephalin, and prodynorphin-derived peptides (see Reference 27). In contrast, the concentrations of somatostatin, thyrotropin-releasing hormone, neurotensin, and neuropeptide Y are increased in the striatum in HD, while the concentrations of vasoactive intestinal polypeptide and vasopressin are relatively unaffected by the disease (see Reference 27). To further define the role of CRF in neurodegenerative diseases, the concentrations of CRF-IR were measured in regions of the basal ganglia and cerebral cortex from postmortem control and HD brains.[27]

The concentrations of CRF-IR in various regions of the cerebral cortex and basal ganglia of controls and patients with HD are shown in Figure 8. In HD, there was a trend towards a small, statistically nonsignificant, reduction (10 to 20% decrease) in the concentrations of CRF-IR in the parietal, frontal and cingulate cerebral cortex; the concentration of CRF-IR was relatively unaffected in the temporal cortex and slightly increased in the occipital cortex.

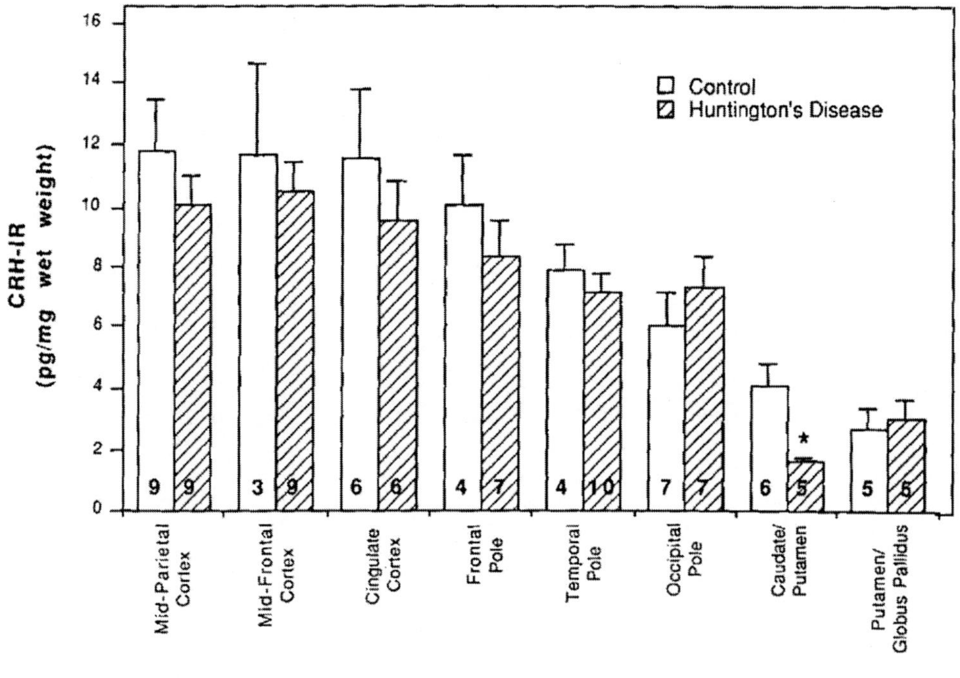

FIGURE 8. Corticotropin-releasing factor (CRF)-like immunoreactivity in discrete regions of brain of Huntington's disease patients and controls. All values are mean ± SEM. The number of subjects in each group is given at the bottom of each histogram. Data were analyzed for differences using a Student's t-test. *Significant difference from control group at $p < 0.05$. (From De Souza, E. B., Whitehouse, P. J., Folstein, S. E., Price, D. L., and Vale, W. W., *Brain Res.*, 437, 355, 1987. With permission.)

In contrast, there was a significant reduction in the concentration of CRF-IR (less than 40% of control) in the caudate/putamen of patients with HD (Figure 8), whereas, the concentration of somatostatin-like immunoreactivity (SLI) (pg/mg wet weight; mean ± SEM) was significantly greater ($p < 0.05$) in HD patients (58 ± 10) when compared to controls (32 ± 4). There was a significant negative correlation between the concentrations of CRF-IR and SLI in the caudate/putamen ($r = -0.63$, $p < 0.05$). No significant change in the concentration of CRF-IR was evident in HD in the globus pallidus.

The major cell type within the striatum is the medium-sized spiny neuron which comprises approximately 90% of the caudate and putamen cells; these cell represent the major projection neurons of the striatum, and it is these neurons that degenerate in HD.[25,28] In contrast, the medium-sized aspiny neurons with deeply indented nuclei comprise primarily the interneurons within the striatum and represent only a small percent of the total striatal cells; these neurons are relatively preserved in HD.[29] The data of the present study demonstrating decreases in CRF-IR in the caudate/putamen in HD suggest that CRF is present in the spiny striatal neurons that degenerate in the disease. In addition, the concommitant increase in SLI in the caudate/putamen in HD confirmed previous radioimmunoassay and immunocytochemical data,[29] demonstrating survival of basal ganglia somatostatin interneurons in HD. While immunocytochemical studies in rodents have identified CRF-containing fibers in the caudate/putamen originating from the amygdala, the relative distribution within the striatum appears to be sparse.[8,9,30] CRF neurons probably have a similar distribution in the human basal ganglia. Decreased CRF content in the striatum in HD could result from cell dysfunction, that is, decreased synthesis, altered processing, increased degradation, increased release, and/or cell death. At present, we cannot identify the specific mechanisms leading to reduced striatal CRF levels in HD.

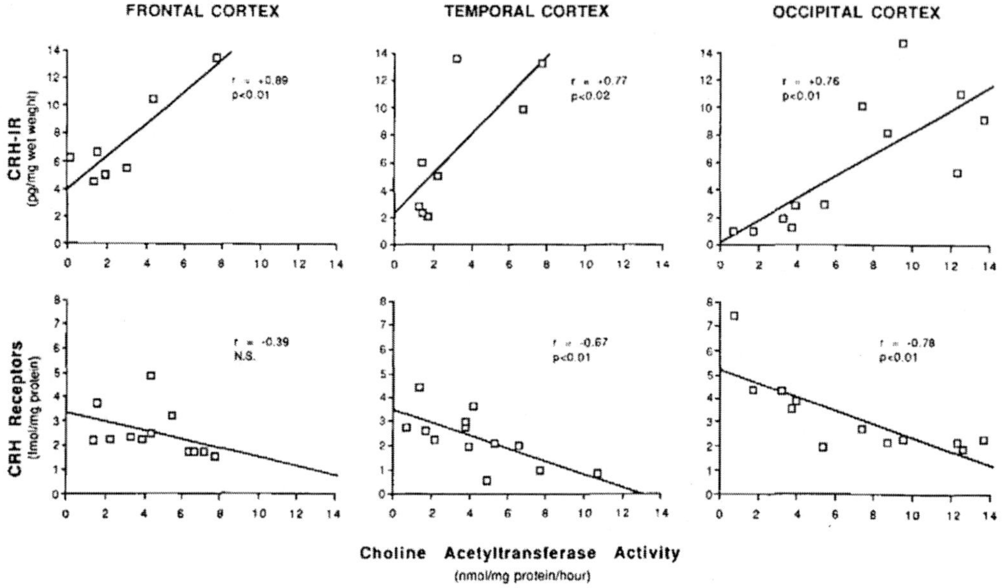

FIGURE 9. Correlations between choline acetyltransferase (ChAT) and CRF-like immunoreactivity (CRF-IR) (upper panels) and ChAT activity and CRF receptors (lower panels) in the frontal, temporal, and occipital cortex. (From De Souza, E. B., Whitehouse, P. J., Price, D. L., and Vale, W. W., *Ann. N.Y. Acad. Sci.*, 512, 237, 1988. With permission.)

In contrast to the major reductions in CRF-IR seen in the various cerebral cortical regions in neurodegenerative disorders such as AD, PD, and PSP (see above), no significant changes in CRF-IR were observed in HD in any region of the cerebral cortex examined in the present study. Thus, HD represents the first dementia in which cerebral cortical levels of CRF-IR are relatively unaffected. While neurochemical changes have been reported in cerebral cortical regions in HD, the most severe pathological changes in the disease are confined to the striatum.[25,26] In contrast, AD, PD, and PSP are degenerative disorders that share certain clinical and morphological features including several types of cerebral cortical pathology. The localization of changes in CRF in various neurodegenerative disorders described above, only to brain regions affected by the respective disease, suggests an important role for CRF in the pathophysiology of these various disorders.

VI. INTERACTIONS OF CRF AND ACETYLCHOLINE

At present, the cerebral cortical cholinergic deficiency appears to be one of the most severe and consistent neurochemical deficit associated with AD.[3] Dysfunction and death of basal forebrain cholinergic neurons projecting to forebrain targets[31] are associated with marked decreases in cholinergic markers, including the activity of ChAT.[2,3] In PD, reductions in cortical ChAT activity and decreased number of neurons in the basal forebrain have been documented.[17,18] In PSP, reductions in ChAT activity occur in the cortex to a variable degree[22,23] in association with mild, neuronal loss in the cholinergic basal forebrain.[32,33] We examined possible interactions between cholinergic deficits in AD, PD, and PSP and alterations in CRF markers by examining the correlation between ChAT activity with CRF-IR and with CRF receptors in each cortical area. As shown in Figure 9, in AD, there were significant positive relationships between ChAT activity and CRF-IR levels in the frontal (r $= +0.89$; $p < 0.01$), temporal (r $= +0.77$; $p < 0.02$) and occipital (r $= +0.76$; $p < 0.01$) cortex and significant negative correlations were detected between ChAT activity and CRF receptors in the temporal ($r = -0.67$; $p < 0.01$) and occipital ($r = -0.78$; $p < 0.01$) cortex. A similar negative correlation existed between ChAT activity and CRF receptors in the frontal cortex ($r = -0.39$; n.s.); however, this was not statistically significant. In PD, CRF content and ChAT activity were decreased to a similar degree with a highly significant negative correlation present between CRF content and ChAT activity ($r = +0.75$; $p < 0.03$). ChAT activity also appeared to be mildly reduced in PSP.

Recent anatomical and behavioral studies in rodents have indicated interactions between CRF and acetylcholine in brain.[12] Specifically, colocalization of CRF and acetylcholinesterase were detected by retrograde tracing and immunocytochemical staining in some brain stem nuclei projecting to the frontal cortex.[12] In behavioral studies, CRF injected into the rat frontal cortex did not initiate, but significantly inhibited the effects of carbachol, a muscarinic cholinergic receptor agonist, on a stereotyped motor "boxing" behavior.[12] The demonstration of significant correlations between changes in CRF and ChAT activity in the human cortex in a variety of neurodegenerative disorders (see above), complemented by similar preliminary studies in rodents, suggest that CRF and cholinergic systems may interact.

To investigate the functional consequences of cholinergic deficits on the regulation of brain CRF neuronal activity, we designed a study in which rats were treated chronically with atropine, a muscarinic cholinergic receptor antagonist, and CRF receptors were measured in various regions of brain.[34] No significant changes in the concentration of CRF receptors were observed in olfactory bulb, cerebellum, striatum, and hippocampus.[34] In contrast, there was a significant and selective increase in CRF receptors in the frontoparietal cortex of atropine-treated rats. These results demonstrating a selective up-regulation of CRF receptors in the cerebral cortex of atropine-treated rats suggest that the reported reciprocal changes in pre- and postsynaptic markers of CRF in the cerebral cortex of patients with AD and possibly PD and PSP may be, in part, a consequence of deficits in cholinergic projections to the cerebral cortex in these diseases. Additional studies are necessary to determine the functional significance of the interaction between CRF and cholinergic systems.

VII. SUMMARY AND CONCLUSIONS

CRF-IR was significantly reduced in the cerebral cortex of individuals with AD, PD, and PSP but not in HD. In HD, CRF-IR was significantly reduced in the caudate-putamen. The decreases in CRF-IR in AD were accompanied by reciprocal increases in CRF receptors in affected cortical areas. In AD, we identified abnormal CRF-immunoreactive axons in the neuropil and in associations with amyloid-containing senile plaques. The number of CRF-IR fibers was also decreased in individuals with AD. In contrast, CRF immunoreactivity was markedly increased in some neurons located within the PVN of the hypothalamus in AD. The changes in pre- and postsynaptic markers for CRF were significantly correlated with decreases in ChAT activity. The demonstration of an up-regulation of CRF receptors following a decrease in CRF-IR indicates a physiological relevance of the receptor site and is consistent with the concept that CRF acts as a neurotransmitter in regulating normal cortical functions and that disease of this peptidergic system may be important in certain clinical manifestations of dementia. While the clinical consequences of the changes in CRF in these various disorders are unclear, future therapies directed at increasing CRF levels in brain may prove useful for treatment.

ACKNOWLEDGMENTS

We thank Terrie Pierce and Priscilla Heeter for manuscript preparation. The studies at Johns Hopkins University School of Medicine were supported by PHS Grants MH 25951, AG 03359, AG 05146, NS 15080, NS 15721, NS 16375, NS 20471, and NS 07179, grants from the McKnight and Sloan Foundations, Robert L. and Clara G. Patterson Trust and Claster Family Funds. Studies at Duke University Medical Center were supported by grants from the NIMH (MH 40524) and The Alzheimer's Diseases Research Center for the National Institute on Aging. Research of the Clayton Foundation Laboratories for Peptide Biology was supported in part by PHS Grants AN 26741 and HD 13527. Research was conducted in part by the Clayton Foundation for Research, California Division. W. W. V. is a senior Clayton Foundation investigator.

REFERENCES

1. **Terry, R. D. and Katzman, R.,** Senile dementia of the Alzheimer's type, *Ann. Neurol.,* 14, 497, 1983.
2. **Price, D. L., Whitehouse, P. J., and Struble, R. G.,** Alzheimer's disease, *Annu. Rev. Med.,* 36, 349, 1985.
3. **Coyle, J. T., Price, D. L., and De Long, M. R.,** Alzheimer's disease: A disorder of cortical cholinergic innervation, *Science,* 219, 1184, 1983.
4. **Whitehouse, P. J., Vale, W. W., Zweig, R. M., Singer, H. S., Mayeux, R., Price, D. L., and De Souza, E. B.** Reduction in corticotropin-releasing factor-like immunoreactivity in cerebral cortex in Alzheimer's disease, Parkinson's Disease, and Progressive Supranuclear Palsy, *Neurology,* 37, 905, 1987.
5. **De Souza, E. B., Whitehouse, P. J., Kuhar, M. J., Price, D. L., and Vale, W. W.,** Reciprocal changes in corticotropin-releasing factor (CRF)-like immunoreactivity and CRF receptors in cerebral cortex of Alzheimer's disease, *Nature,* 319, 593, 1986.
6. **Bissette, G., Reynolds, G. P., Kilts, C. D., Widerlov, E., and Nemeroff, C. B.,** Corticotropin-releasing factor-like immunoreactivity in senile dementia of the Alzheimer type, *JAMA,* 254, 3067, 1985.
7. **Powers, R. E., Walker, L. C., De Souza, E. G., Vale, W. W., Struble, R. G., Whitehouse, P. J., and Price, D. L.,** Immunohistochemical study of neurons containing corticotropin-releasing factor in Alzheimer's Disease, *Synapse,* 1, 405, 1987.
8. **Olschowka, J. A., O'Donohue, T. L., Mueller, G. P., and Jacobowitz, D. M.,** The distribution of corticotropin-releasing factor-like immunoreactive neurons in rat brain, *Peptides,* 3, 995, 1982.
9. **Swanson, L. W., Sawchenko, P. E., Rivier, J., and Vale, W. W.,** Organization of ovine corticotropin-releasing factor and immunoreactive cells and fibers in the rat brain: an immunohistochemical study, *Neuroendocrinology,* 36, 165, 1983.
10. **De Souza, E. B., Perrin, M. H., Insel, T. R., Rivier, J., Vale, W. W., and Kuhar, M. J.,** Corticotropin-releasing factor receptors in rat forebrain: autoradiographic identification, *Science,* 224, 1449, 1984.
11. **De Souza, E. B., Insel, T. R., Perrin, M. H., Rivier, J., Vale, W. W., and Kuhar, M. J.,** Corticotropin-releasing factor receptors are widely distributed within the rat central nervous system: an autoradiographic study, *J. Neurosci.,* 5, 3189, 1985.
12. **Crawley, J. N., Olschowka, J. A., Diz, D. I., and Jacobowitz, D. M.,** Behavioral investigation of the co-existence of substance P, corticotropin-releasing factor, and acetylcholinesterase in lateral dorsal tegmental neurons projecting to the medial frontal cortex of the rat, *Peptides,* 6, 891, 1985.
13. **Mouradian, M. M., Farah, J. M., O'Donohue, T. L., and Chase, T. N.,** Decreased immunoreactive corticotropin-releasing factor (CRF) in CSF of Alzheimer's disease, *Soc. Neurosci. Abstr.,* 12, 98, 1986.
14. **Alvord, E. C., Jr., Forno, L. S., Kusske, J. A., Kauffman, R. J., Rhodes, J. S., and Goetowski, C. R.,** The pathology of parkinsonism: a comparison of degenerations in cerebral cortex and brain stem, in *Second Canadian-American Conference on Parkinson's Disease, Advances in Neurology,* Vol. 5, McDowell, F.H. and Barbeau, A., Eds., Raven Press, New York, 1974, 175.
15. **Jellinger, K. and Grisold, W.,** Cerebral atrophy in Parkinson syndrome, *Exp. Brain Res. Suppl.,* 5, 26, 1982.
16. **Yoshimura, M.,** Cortical changes in parkinsonian brain: a contribution to the delineation of "diffuse Lewy body disease", *J. Neurol.,* 229, 17, 1983.
17. **Perry, E. K., Curtis, M., Dick, D. J., Candy, J. M., Atack, J. R., Bloxham, C. A., Blessed, G., Fairbairn, A., Tomlinson, B. E., Perry, R. H.,** Cholinergic correlates of cognitive impairment in Parkinson's disease: comparisons with Alzheimer's disease, *J. Neurol. Neurosurg. Psychiatry,* 48, 413, 1985.
18. **Ruberg, M., Ploska, A., Javoy-Agid, F., and Agid, Y.,** Muscarinic binding and choline acetyltransferase activity in parkinsonian subjects with reference to dementia, *Brain Res.,* 232, 129, 1982.
19. **Epelbaum, J., Ruberg, M., Moyse, E., Javoy-Agid, F., Dubois, B., and Agid, Y.,** Somatostatin and dementia in Parkinson's disease, *Brain Res.,* 278, 376, 1983.

20. **Conte-Devolx, B., Grino, M., Neioullon, A., Javoy-Agid, F., Castanas, E., Guillaume, V., Tonon, M. C., Vaudry, H., and Oliver, C.,** Corticoliberin, somatocrinin and amine contents in normal and parkinsonian human hypothalamus, *Neurosci. Lett.*, 56, 217, 1985.
21. **Zweig, R. M., Whitehouse, P. J., Casanova, M. F., Walker, L. C., Jankel, W. R., and Price, D. L.,** Loss of putative cholinergic neurons of the pedunculopontine nucleus in progressive supranuclear palsy, *Ann. Neurol.*, 18, 144, 1985.
22. **Kish, S. H., Chang, L. J., Mirchandani, L., Shannak, K., and Hornykiewicz, O.,** Progressive supranuclear palsy: relationship between extrapyramidal disturbance, dementia and brain neurotransmitter markers, *Ann. Neurol.*, 18, 530, 1985.
23. **Ruberg, M., Javoy-Agid, F., Hirsch, E., Scatton, B., L'Heureux, R., Hauz, J. J., Duychaerts, C., Gray, F., Morel-Maroger, A., Rascol, A., and Agid, Y.,** Dopaminergic and cholinergic lesions in progressive supranuclear palsy, *Ann. Neurol.*, 18, 523, 1985.
24. **Martin, J. B.,** Huntington's disease: new approaches to an old problem, *Neurology (Cleveland)*, 34, 1059, 1984.
25. **Bruyn, G.,** Neuropathological changes in Huntington's chorea, in *Huntington's Chorea 1872—1972*, Barbeau, A., Chase, T. N., and Paulson, G. W., Eds., Raven Press, New York, 1973, 399.
26. **Bruyn, G. W., Bots, G. T. A. M., and Dom, R.,** Huntington's chorea: current neuropathological status, in *Huntington's Disease, Advances in Neurology, Vol. 23*, Chase, T. N., Wexler, N. X., and Barbeau, A., Eds., Raven Press, New York, 1979, 83.
27. **De Souza, E. B., Whitehouse, P. J., Folstein, S. E., Price, D. L., and Vale, W. W.,** Corticotropin-releasing hormone (CRH) is decreased in the basal ganglia in Huntington's disease, *Brain Res.*, 437, 355, 1987.
28. **Graveland, G. A., Williams, R. S., and DiFiglia, M.,** Evidence for degenerative and regenerative changes in neostriatal spiny neurones in Huntington's disease, *Science*, 227, 770, 1985.
29. **Marshal, P. and Landis, D. M. D.,** Somatostatin in the neostriatum and substantia nigra in normal human brain and in Huntington's disease, *Neurosci. Abstr.*, 8, 507, 1982.
30. **Merchenthaler, I., Vigh, S., Petrusz, P., and Schally, A. V.,** Immunocytochemical localization of corticotropin-releasing factor (CRF) in the rat brain, *Am. J. Anat.*, 165, 385, 1982.
31. **Whitehouse, P. J., Price, D. L., Struble, R. G., Clark, A. W., Coyle, J. T., and De Long, M. R.,** Alzheimer's disease and senile dementia: loss of neurons in the basal forebrain, *Science*, 215, 1237, 1982.
32. **Tagliavini, F., Pilleri, G., Gemignani, F., and Lechi, A.,** Neuronal loss in the basal nucleus of Meynert in progressive supranuclear palsy, *Acta Neuropathol.*, 61, 157, 1983.
33. **Rogers, J. D., Brogan, D., and Mirra, S. S.,** The nucleus basalis of Meynert in neurological disease: a quantitative morphological study, *Ann. Neurol.*, 17, 163, 1985.
34. **De Souza, E. B. and Battaglia, G.,** Increased corticotropin-releasing factor receptors in rat cerebral cortex following chronic atropine treatment, *Brain Res.*, 397, 401, 1986.
35. **De Souza, E. B., Whitehouse, P. J., Price, D. L., and Vale, W. W.,** Abnormalities in corticotropin-releasing hormone (CRH) in Alzheimer's disease and other human disorders, *Ann. N.Y. Acad. Sci.*, 512, 237, 1988.

INDEX

A

B

C

D

E

H

I

J

K

L

M

T